S0-BXD-944

Ecological Studies, Vol. 192

Analysis and Synthesis

Edited by

Ecological Studies

Volumes published since 2001 are listed at the end of this book.

Richard B. Aronson

Editor

Geological Approaches
to Coral Reef Ecology

 Springer

Richard B. Aronson
Dauphin Island Sea Lab
101 Bienville Boulevard
Dauphin Island, Alabama 36528
USA
and
Department of Marine Sciences
University of South Alabama
Mobile, Alabama 36688
USA

Cover: A coral reef near Discovery Bay on the north coast of Jamaica, photographed at 6 m depth in July 2003. The branching species is staghorn coral, *Acropora cervicornis*, and the massive corals belong to the *Montastraea annularis* species complex. *A. cervicornis* populations in Jamaica were killed by disease and hurricanes in the early 1980s and began to recover circa 2002. The inset photographs at the top show extraordinarily well-preserved fossil *A. cervicornis* (left) and *Montastraea* (right) from the early- to middle-Holocene reef complex in the Enriquillo Valley, Dominican Republic. The fossil corals are part of an 11-m vertical section composed almost entirely of *A. cervicornis*, which has been radiocarbon-dated to 9400 to 7400 years before present. Geological approaches broaden our understanding of the threats facing modern coral reefs, increasing our predictive power in a rapidly changing world. Photos by William F. Precht (Jamaica) and H. Allen Curran (Dominican Republic); montage design by Ryan M. Moody.

Library of Congress Control Number: 2006924285

ISSN: 0070-8356
ISBN-10: 0-387-33538-2   e-ISBN-10: 0-387-33537-4   Printed on acid-free paper.
ISBN-13: 978-0-387-33538-4   e-ISBN-13: 978-0-387-33537-7

10 9 8 7 6 5 4 3 2 1

springer.com

For my wife
Lisa Young
with love

Preface

The worsening condition of the world's coral reefs and their deteriorating prospects for recovery have erased any remaining distinction between pure and applied research on these most diverse and vital of marine ecosystems. Geophysics, paleontology, geochemistry, and physical and chemical oceanography provide insights into the workings of coral reefs that complement what we can learn from physiology, molecular biology, and community ecology. The larger spatio-temporal view afforded by the geosciences—reef geology construed in its loosest sense—will be critical to crafting realistic environmental policy. As Greenstein emphasizes in Chapter 2 and Pandolfi and Jackson underscore in Chapter 8, geological study of coral reefs as they were prior to the human onslaught provides the only reliable baseline for evaluating current conditions and predicting future trajectories. The chapters in this book cover a medley of pertinent topics through which reef geology serves and can continue to serve the goals of conservation and management.

Many authors have pointed out that the entire Caribbean region is about the size of the Great Barrier Reef. Nevertheless, much of our understanding of reef ecology and paleoecology comes from studies carried out in the Caribbean. In this volume the Caribbean region receives greater emphasis than the Indo-Pacific, not because of a conceptual imbalance but because conditions are so much worse in the Caribbean at present. Unfortunately Indo-Pacific reefs are currently entering a period of accelerated decline as well (Bellwood et al. 2004; Buddemeier, Kleypas, and Aronson 2004), so several chapters consider processes on reefs in that far larger swath of tropical and subtropical ocean.

Part I, *Coral Reefs in Context*, consists entirely of Wood's Chapter 1. Wood introduces the book's main themes by placing the current state of coral reefs in the perspective of deep time. A key point is that Caribbean reefs are particularly vulnerable to strong perturbations, which are now occurring with increasing frequency on ecological time scales. Caribbean reef communities entered the present-day environmental crisis predisposed to a low level of resilience, because the coral fauna is still recovering from Neogene episodes of elevated taxic turnover. Being in the recovery phase limits the ability of coral assemblages to bounce back from the multitude of interacting stresses and disturbances, both natural and anthropogenic, that figure prominently throughout the book. Some of these perturbations have no analogs in the geological record, a point Kleypas (Chapter 12) and Wood emphasize particularly with respect to the geochemical implications of rapidly rising concentrations of carbon dioxide in the atmosphere. The complexity of interactions leads Wood to advocate an adaptive, Bayesian approach to detecting and predicting pattern and process on coral reefs in an uncertain and highly variable world.

In Part II, *Detecting Critical Events*, five contributions explore the question of prior occurrence. What observational and statistical tools can we bring to bear to determine whether current conditions or recent events are (or are not) unprecedented? Some ecologically critical events, such as the Caribbean-wide mass mortality in the early 1980s of the herbivorous sea urchin, *Diadema antillarum*, are apparently undetectable as layers of skeletal remains in reef sediments, due to rapid and extensive bioturbation. The same can be said for detecting past outbreaks of the corallivorous crown-of-thorns starfish, *Acanthaster planci*, in the Indo-Pacific, although there may be some subtle ways around the taphonomic problems in that case (Greenstein, Pandolfi, and Moran 1995).

Fortunately, other epistemological barriers are not so impassable as one might think. Greenstein grapples with the taphonomy of corals in Chapter 2. Understanding how individual coral skeletons degrade and quantitatively assessing the degree of taphonomic information loss in death assemblages of reef corals will make possible standardized comparisons among living, subfossil, and fossil assemblages. Aronson and Ellner (Chapter 3), like Greenstein, take on the challenge of putting bounds on information loss. Aronson and Ellner use probability theory to interpret event preservation in sets of hierarchically sampled reef cores. How many cores must we sample to show that coral mortality did or did not occur on a large spatial scale at a particular time? What proportion of those cores must show evidence of a putative event, before we can safely conclude that something did or did not happen? As Greenstein points out, conservation policy cannot go forward without a realistic picture of the pristine condition of reefs prior to human interference. Both chapters conclude that the dramatic ecological changes witnessed on Caribbean reefs in the last three decades—specifically the catastrophic decline of acroporid corals throughout the region and the ensuing collateral effect of a phase shift to dominance by seaweeds—were unprecedented on a multimillennial time scale.

DeVantier and Done (Chapter 4) follow with a novel approach to detecting past *Acanthaster* outbreaks on the Great Barrier Reef. They use feeding scars left by the starfish on living coral heads as indicators of previous, sublethal *Acanthaster* attacks. By combining censuses of the positions of feeding scars on large, old colonies; the growth rates of the corals; and sclerochronological data from cores of the colonies, DeVantier and Done show that outbreaks have become more frequent and spatially extensive since the 1960s. As with the decline of acroporids in the Caribbean, which was due primarily to the regional outbreak of white-band disease, the increase in *Acanthaster* outbreaks is coincident with a period of increased anthropogenic perturbation. Even if we cannot establish direct causal connections between specific human activities and outbreaks of infectious disease or *Acanthaster*, we must at least act to reduce synergistic interactions between these sources of coral mortality and negative human impacts, which we know are intensifying year by year.

The next two chapters, by Deslarzes and Lugo-Fernandez (Chapter 5) and Halley and Hudson (Chapter 6), are about two of those well-established human impacts: nutrient loading and climate change. Deslarzes and Lugo-Fernandez synthesize oceanographic information and data on fluorescent banding in coral heads to demonstrate that river runoff affects the near-pristine reefs of the Flower Garden Banks in the northern Gulf of Mexico, 190 km off the coast of Texas. The fact that terrigenous input poses a threat to reefs located so far offshore is a reminder that environmental forcing operates on scales that far exceed the sizes of individual reefs or reef systems. Halley and Hudson follow with some more sobering news: high-temperature bleaching prevents corals from laying down skeletal growth bands, and missing bands in coral skeletons from Florida are clustered in the years after 1986. Their results corroborate ecological observations that the frequency and severity of bleaching events have increased recently in response to global increases in sea temperature.

Part III, *Patterns of Reef Development and Their Implications*, begins with a contribution by Macintyre (Chapter 7) on the responses of western Atlantic reefs to rising sea level following the last glaciation. During the past 18,000 years, reef growth has stalled during intervals of accelerated sea-level rise and when terrigenous materials have washed onto flooded shelves. Nevertheless, coral taxa persisted and reassembled to form new reef communities when and where conditions improved, so reef development has continued. Macintyre's historical lessons are important, because both reef drowning and reef poisoning are worrying possibilities in an era when sea-level rise and shelf-flooding are expected to accelerate once again. Precht and Miller (Chapter 9) flesh out Macintyre's points in their exhaustive treatment of the Quaternary history of the Florida reef tract. The structure of reef-building coral assemblages has changed radically as conditions on the Florida shelf have switched from being less favorable to framework-constructing acroporid corals, to being more favorable, and back again. In a similar vein, Pandolfi and Jackson (Chapter 8) report on a hierarchical study of the structure of Pleistocene reef-coral assemblages along a geographic gradient in the southern Caribbean. They invoke differences in environmental variability at different spatial scales to account for the scale-dependent variability they observe in the coral faunas.

All the chapters in this book are directed toward predicting what reefs will look like in the next decades to centuries, but the three reviews that comprise Part IV,

Coral Reefs and Global Change, wrestle directly and explicitly with the complexities of prediction. Riegl (Chapter 10) examines the extent to which extreme climatic events, particularly tropical cyclones and El Niño–Southern Oscillation (ENSO) events, can be expected to damage corals and coral reefs in a warming world. Wellington and Glynn (Chapter 11) review the effects of ENSO on coral reefs of the eastern Pacific, where a severe El Niño in the early 1980s first made the scientific community aware of the regional-scale devastation that such events are capable of causing. The radically different response of eastern Pacific coral assemblages to a second severe El Niño in the late 1990s is telling us a great deal about their scope for adjusting to climate change. Finally, Kleypas (Chapter 12) discusses what we know and what we do not know about the geochemical effects on coral reefs of increasing carbon dioxide concentrations in the atmosphere. Riegl points out that reefs are fairly resilient to physical damage alone; however, he, Wellington and Glynn, and Kleypas all worry that the increasing likelihood of interactions among higher frequencies of severe cyclone damage and severe ENSO events, warming background temperatures, deleterious changes in ocean chemistry, and more frequent outbreaks of infectious diseases could well be disastrous for corals and the reefs they build.

The conclusions and recommendations from these disparate geological studies, tentative though some of them may be, have withstood the rigors of scientific review. Other, equally interesting topics can best be described as works in progress. Here I touch on three of these emerging ideas, my guess being that they will be worth watching in the literature.

One geologically driven prediction about coral reefs is that they, and marine ecosystems in general, are racing along a degradational trajectory to dominance by microbial mats (Hallock 2001; Jackson et al. 2001; Bellwood et al. 2004). This notion of a bacterial end-state comes from the observation that Phanerozoic reef biotas have repeatedly given way to stromatolites and other sorts of microbialites in the wake of mass extinction events, representing a reversion to quasi-Proterozoic community structure (Copper 1988, 1994; Wood 1999; Calner 2005; see also Xie et al. 2005). Whether or not one accepts the scenario that benthic reef assemblages are in danger of devolving to bacterial slime—and there are suggestions that this is already happening (Pandolfi et al. 2005)—it is a geological hypothesis about progressive biotic homogenization, which is testable on a variety of spatial and temporal scales (Aronson et al. 2005).

A second controversial idea is the African dust hypothesis. Increased transport of dust from North Africa (and Asia) on a decadal scale may have been at least partially responsible for reef degradation in the Caribbean by introducing pathogens, micronutrients, and contaminants beginning in the late 1970s (Shinn et al. 2000; Garrison et al. 2003). The pathogen responsible for episodic outbreaks of aspergillosis, a fungal infection of gorgonian sea fans, has been isolated from samples of African dust, and trace iron supplied by the dust could promote the pathogenicity of other disease agents (for background on coral diseases, see Richardson and Aronson 2002; Weil, Smith, and Gil-Agudelo 2006). Causal connections between African dust and disease outbreaks have yet to be firmly established, but at this point the hypothesis cannot be dismissed out of hand. Shinn et al. (2000) also suggested that African dust could have been responsible for the disease outbreaks that caused regional mass mortalities of *Diadema* in 1983 to 1984 and acroporid corals from the late 1970s through the early 1990s. These two diseases have had enormous impacts on the ecology of Caribbean reef communities over the past few decades (Lessios 1988; Aronson and Precht 2001), but considering that the pathogens have not been fully characterized in either case, their connection to African dust must for the moment remain speculative.

The third not-so-preposterous idea is coral creep: as the climate warms, some cold-sensitive coral species will expand their ranges from tropical to subtropical latitudes. Precht and Aronson (2004) suggested that rapid warming at subtropical latitudes could explain recent discoveries of dense thickets of staghorn coral, *Acropora cervicornis*, and colonies of elkhorn coral, *A. palmata*, off the eastern coast of the Florida Peninsula. Until a few years ago the northern limit of these species was near Miami, more than 50 km to the south. Although acroporid corals are important components of reef frameworks in the Caribbean, they have not occurred off the east coast of Florida in any appreciable abundance since the warm conditions of the early to middle Holocene, 10,000 to 6000 years ago. Precht and Aronson (2004) also reported that in July 2003 the first known colony of *A. palmata* was discovered on the Flower Garden Banks in the northern Gulf of Mexico.

There is strong evidence that both recent and past warming trends have triggered range expansions of Indo-Pacific corals. When the idea was applied to living acroporids in the western Atlantic, however, it met with strong opposition from some colleagues who doubted that climate change could be responsible. Others said just the opposite: range expansions of coral species are not surprising, and they are exactly what one would expect in a warming world (e.g., Hughes et al. 2003). Adding to the debate over coral creep, Zimmer et al. (2006) discovered a second colony of *A. palmata* on the Flower Garden Banks in June 2005.

Clearly, many causal connections that currently drive reef dynamics cannot be addressed through local management actions such as establishing marine protected areas (MPAs). MPAs, which foster populations of herbivorous fish, could potentially reduce algal cover and as a consequence might be able to accelerate the recovery of coral populations (e.g., Hughes et al. 2003; Bellwood et al. 2004); however, MPAs will only produce these desirable effects if they are employed in conjunction with environmental policies that directly confront climate change and related, global-scale sources of coral mortality.

Many of the critical processes are geological in nature, and the scale of their effects is far greater than the largest conceivable MPA. Geological approaches to coral reef ecology are vital to mitigating and ultimately reversing reef degradation. We cannot save coral reefs unless we are geologists as well as chemists, biologists, ecologists, managers, policymakers, and human beings.

Richard B. Aronson
Dauphin Island, Alabama
February 2006

References

Aronson, R.B., and W.F. Precht. 2001. White-band disease and the changing face of Caribbean coral reefs. *Hydrobiologia* 460:25–38.

Aronson, R.B., I.G. Macintyre, S.A. Lewis, and N.L. Hilbun. 2005. Emergent zonation and geographic convergence of coral reefs. *Ecology* 86:2586–2600.

Bellwood, D.R., T.P. Hughes, C. Folke, and M. Nyström. 2004. Confronting the coral reef crisis. *Nature* 429:827–833.

Buddemeier, R.W., J.A. Kleypas, and R.B. Aronson. 2004. *Coral Reefs and Global Climate Change: Potential Contributions of Climate Change to Stresses on Coral Reef Ecosystems.* Arlington, Virginia: Pew Center on Global Climate Change.

Calner, M. 2005. A Late Silurian extinction event and anachronistic period. *Geology* 33:305–308.

Copper, P. 1988. Ecological succession in Phanerozoic reef ecosystems: Is it real? *Palaios* 3:136–152.

Copper, P. 1994. Ancient reef ecosystem expansion and collapse. *Coral Reefs* 13:3–11.

Garrison, V.H., E.A. Shinn, W.T. Foreman, D.W. Griffin, C.W. Holmes, C.A. Kellogg, M.S. Majewski, L.L. Richardson, K.B. Ritchie, and G.W. Smith. 2003. African and Asian dust: From desert soils to coral reefs. *BioScience* 53:469–480.

Greenstein, B.J., J.M. Pandolfi, and P.J. Moran. 1995. Taphonomy of crown-of-thorns starfish: Implications for recognizing ancient population outbreaks. *Coral Reefs* 14:91–497.

Hallock, P. 2001. Coral reefs, carbonate sediments, nutrients, and global change. In *The History and Sedimentology of Ancient Reef Systems*, ed. G.D. Stanley, Jr., 387–427. New York: Kluwer Academic/Plenum.

Hughes, T.P., A.H. Baird, D.R. Bellwood, S.R. Connolly, C. Folke, R. Grosberg, O. Hoegh-Guldberg, J.B.C. Jackson, J. Kleypas, J.M. Lough, P. Marshall, M. Nyström, S.R. Palumbi, J.M. Pandolfi, B. Rosen, and J. Roughgarden. 2003. Climate change, human impacts, and the resilience of coral reefs. *Science* 301:929–933.

Jackson, J.B.C., M.X. Kirby, W.H. Berger, K.A. Bjorndal, L.W. Botsford, B.J. Bourque, R.H. Bradbury, R. Cooke, J. Erlandson, J.A. Estes, T.P. Hughes, S. Kidwell, C.B. Lange, H.S. Lenihan, J.M. Pandolfi, C.H. Peterson, R.S. Steneck, M.J. Tegner, and R.R. Warner. 2001. Historical overfishing and the recent collapse of coastal ecosystems. *Science* 293:629–638.

Lessios, H.A. 1988. Mass mortality of *Diadema antillarum* in the Caribbean: What have we learned? *Ann. Rev. Ecol. Syst.* 19:371–393.

Pandolfi, J.M., J.B.C. Jackson, N. Baron, R.H. Bradbury, H.M. Guzman, T.P. Hughes, C.V. Kappel, F. Micheli, J.C. Ogden, H.P. Possingham, and E. Sala. 2005. Are U.S. coral reefs on the slippery slope to slime? *Science* 307:1725–1726.

Precht, W.F., and R.B. Aronson. 2004. Climate flickers and range shifts of reef corals. *Front. Ecol. Environ.* 2:307–314.

Richardson, L.L., and R.B. Aronson. 2002. Infectious diseases of reef corals. *Proc. Ninth Int. Coral Reef Symp., Bali* 2:1225–1230.

Shinn, E.A., G.W. Smith, J.M. Prospero, P. Betzer, M.L. Hayes, V. Garrison, and R.T. Barber. 2000. African dust and the demise of Caribbean coral reefs. *Geophys. Res. Lett.* 27:3029–3032.

Weil, E., G. Smith, and D. L. Gil-Agudelo. 2006. Status and progress in coral reef disease research. *Dis. Aquat. Org.* 69:1–7.

Wood, R. 1999. *Reef Evolution.* Oxford: Oxford University Press.

Xie, S., R.D. Pancost, H. Yin, H. Wang, and R.P. Evershed. 2005. Two episodes of microbial change coupled with Permo/Triassic faunal mass extinction. *Nature* 434:494–497.

Zimmer, B., W. Precht, E. Hickerson, and J. Sinclair. 2006. Discovery of *Acropora palmata* at the Flower Garden Banks National Marine Sanctuary, northwestern Gulf of Mexico. *Coral Reefs*. 25:192.

Acknowledgments

The chapters in this volume were peer reviewed by Robert Buddemeier, Allen Curran, Mark Eakin, Evan Edinger, Peter Edmunds, Ron Etter, Dennis Hubbard, Janice Lough, Peter Mumby, Robbie Smith, Sally Walker, the authors of other chapters, and several individuals who wish to remain anonymous. Each contribution was improved by the hard work of the reviewers, and I thank them all for participating.

My sincerest appreciation goes to our editors at Springer. In 1998, Robert Badger encouraged me to go forward with this project, just as he has encouraged me on so many other occasions over the many decades we have been friends. Although he is a chemist, Bob immediately saw the academic and practical value of a volume that combines geological and ecological approaches to understand what is happening to coral reefs. When Bob retired from Springer, Janet Slobodien took over the job of editing the book. Janet has shown remarkably good humor through innumerable delays.

Special thanks are due Bill Precht and Ian Macintyre for all they have taught me about reef geology through twenty years of friendship and collaborative research. Dennis Hubbard has also shown me a great deal of kindness and a great deal of geology over the years. Robert Ginsburg and Eugene Shinn deserve mention as well, because their influence on reef geology is manifest throughout this book. I am particularly indebted to Peter Glynn, who continues to show by example that uncompromising integrity is what distinguishes a great scientist from the rest. Peter has set an extraordinarily high standard of conduct to which we all should aspire.

Finally, words cannot express how grateful I am to my wife and true companion, Lisa Young, for supporting me in so many ways. My sons Ben and Max do not know it yet, but their sweet little faces keep me walking the walk.

Contents

Contributors

Richard B. Aronson

Dauphin Island Sea Lab
101 Bienville Boulevard
Dauphin Island, Alabama 36528
USA
and
Department of Marine Sciences
University of South Alabama
Mobile, Alabama 36688
USA

Kenneth J. P. Deslarzes

Geo-Marine, Inc.
2201 K Avenue
Plano, Texas 75074
USA

Lyndon M. DeVantier

Australian Institute of Marine Science
PMB No. 3
Townsville Mail Centre
Queensland 4810
Australia

Terence J. Done

Australian Institute of Marine Science
PMB No. 3
Townsville Mail Centre
Queensland 4810
Australia

Stephen P. Ellner

Department of Ecology and Evolutionary
    Biology and
Center for Applied Mathematics
Cornell University
E145 Corson Hall
Ithaca, New York 14853
USA

Peter W. Glynn

Division of Marine Biology and Fisheries
Rosenstiel School of Marine and
    Atmospheric Science
4600 Rickenbacker Causeway
Miami, Florida 33149
USA

Benjamin J. Greenstein

Department of Geology
Cornell College
600 First Street SW
Mount Vernon, Iowa 52314
USA

Robert B. Halley

United States Geological Survey
Florida Integrated Science Center
St. Petersburg, Florida 33701
USA

J. Harold Hudson

National Oceanic and Atmospheric
    Administration
Florida Keys National Marine Sanctuary
P.O. Box 1083
Key Largo, Florida 33037
USA

Jeremy B. C. Jackson

Marine Biology Research Division
Scripps Institution of Oceanography
University of California, San Diego
La Jolla, California 92093
USA
and
Smithsonian Tropical Research Institute
P.O. Box 2072
Balboa
Republic of Panama

Joan A. Kleypas

Institute for the Study of Society and
    Environment
National Center for Atmospheric Research
P.O. Box 3000
Boulder, Colorado 80302
USA

Alexis Lugo-Fernández

U.S. Department of the Interior
Minerals Management Service
Environmental Sciences Section
1201 Elmwood Park Boulevard
New Orleans, Louisiana 70123
USA

Ian G. Macintyre

Department of Paleobiology, MRC 121
National Museum of Natural History
Smithsonian Institution
P.O. Box 37012
Washington, D.C. 20013
USA

Steven L. Miller

The University of North Carolina at
    Wilmington
Center for Marine Science
515 Caribbean Drive
Key Largo, Florida 33037
USA

John M. Pandolfi

The Centre for Marine Studies and
Department of Earth Sciences
The University of Queensland
St. Lucia
Queensland 4072
Australia

William F. Precht

Ecological Sciences Program
PBS & J
2001 NW 107th Avenue
Miami, Florida 33172
USA

Bernhard Riegl

National Coral Reef Institute
Oceanographic Center
Nova Southeastern University
8000 North Ocean Drive
Dania, Florida 33004
USA

Gerard M. Wellington

Department of Biology and Biochemistry
University of Houston
Houston, Texas 77204
USA

Rachel Wood

School of GeoSciences
Grant Institute
University of Edinburgh
King's Buildings
West Mains Road
Edinburgh EH9 3JW
United Kingdom
and
Schlumberger Cambridge Research
High Cross
Madingley Road
Cambridge CB3 0EL
United Kingdom

Part I. Coral Reefs in Context

1. The Changing Fate of Coral Reefs: Lessons from the Deep Past

Rachel Wood

1.1 Introduction

The more clearly we can focus our attention on the wonders and realities of the universe about us, the less taste we shall have for destruction.

—Rachel Carson (1954)

When Rachel Carson spoke these words, she did not know that her chilling and prescient analysis of the decline of terrestrial communities could equally be applied to life in the sea. We now know that reefs worldwide are undergoing dramatic and far-reaching change, and that many of these changes are historically recent phenomena (see, e.g., Hughes 1994; Jackson et al. 2001; Pandolfi et al. 2003). Most notable is the increase of soft-bodied algal cover and biomass, and the decline of acroporid corals in the Caribbean. The phase shift from coral to algal dominance has also led to a dramatic reduction in coral biodiversity over whole regions and a marked decline in rates of calcification (see Hughes 1994; Kleypas et al. 1999; Gardner et al. 2003; Kleypas, Chapter 12). In addition, the demise of *Acropora cervicornis* and *Acropora palmata* (in some cases replaced by other corals) over the last few decades has removed the zonation patterns considered characteristic of Caribbean reefs and now known to have been present since the Pleistocene (Jackson 1991, 1992; Aronson and Ellner, Chapter 3; Pandolfi and Jackson, Chapter 8).

That these changes are taking place is unequivocal; but the causes and controls
are far less clear. The relative importance of factors such as reduced herbivory and
the loss of higher predators, nutrient overloading, or the rise of pathogens and outbreak populations is still unresolved (see, e.g., Littler and Littler 1984; Aronson
and Precht 2000; Jackson et al. 2001; Wapnick et al. 2004; DeVantier and Done,
Chapter 4), but all proposed causes have been variously attributed to anthropogenic
impacts, either directly as a result of changing land use, overfishing, and pollution,
or indirectly due to a multitude of effects caused by global warming (Glynn 1993;
Richardson 1998; Kleypas et al. 1999; Jackson et al. 2001). Of particular concern
is the frequency of bleaching events in response to episodes of rapid temperature
rise that has led to significant and sustained global degradation of reefs (see Brown
1997; Wellington and Glynn, Chapter 11). In addition, the predicted rise in CO_2
concentration will increase oceanic acidification and so reduce aragonite saturation,
leading to a notable decline in shallow marine carbonate production (Kleypas,
Chapter 12). The effects of these environmental changes will be global, impacting
all reefs even those currently protected from direct anthropogenic stress. With the
likelihood of accelerated degradation as a result of synergies between these multiple
factors, coral reefs clearly face an uncertain future.

Most researchers agree that many of these changes have only occurred over the
last 10 to 30 years (Aronson et al. 2002), although this recent accelerated decline
appears to be the dramatic finale of centuries of degradation due to increasing
anthropogenic disturbance, promoted particularly by the onset of colonial occupation and development (Pandolfi et al. 2003). Nevertheless, there still remain some
uncertainties as to whether recent changes may yet be part of longer-term cycles
unrelated to anthropogenic disturbance (Bak and Nieuwland 1995). Abundant data
gathered from Pleistocene reefs worldwide, however, are persuasive in demonstrating that the community structure of reefs has, as far as the fidelity of the
record allows, been remarkably stable for the past 125 kyr (see review in Pandolfi
and Jackson, Chapter 8).

Much of our current knowledge of coral reef decline is still based upon limited
local observation, anecdote, or simple ad hoc correlation, with as yet little understanding of the underlying and unifying causal processes. We are increasingly aware
of the issue of potential synergies, as well as nonlinear responses, threshold effects,
and the importance of the historical sequence of environmental events in relationship
to biological response. Many of our data sets are also small, so that sampling error
can be high. It also remains uncertain as to whether differences between sites reflect
only locally important factors rather than universal responses. Many data sets are
also not comparable or indeed compatible. For example, species abundance patterns
over large areas are often studied in modern reefs (e.g., Hughes et al. 1999; Murdoch
and Aronson 1999), whereas although the global record of Pleistocene reefs is extensive and reliable (Greenstein, Chapter 2), the temporal resolution is comparatively
coarse and relatively few localities have been studied (Pandolfi and Jackson, Chapter
8). Indeed, many ecological phenomena operate at multiple spatial and temporal
scales, so a major prerequisite to understanding the large-scale dynamics of coral
reefs must be to analyze the nature of this scale dependence and variability, and to

integrate small-scale observations into a unified regional or global theoretical framework that allows for the prediction of long-term change.

What is clear is that the ecological changes now documented in living coral reefs are unprecedented in the geological history of reefs. This volume presents a series of contributions that explore changes in living coral reef community structure: together they highlight issues of scale and process, so bridging the methodological chasm between those who study the ancient and modern. But they all seek to answer the same central question: How can processes currently observable on human time scales together with the fossil record be extrapolated to predict and manage the future health of coral reefs?

In this chapter, I address how processes that operate over different scales may be expressed and recognized in the geological record so that we may learn from the deep past. First, the sources of data available from the fossil record are considered, together with their fidelities and biases. Each data set covers a particular range of temporal and spatial scales, with fidelity often decreasing inversely with scale. With these caveats in mind, each source of data is then analyzed to reveal how different processes (with their variable rates and scales of effects) have operated in the past, and how these might be used to predict the effects of current and future global change.

1.2 Probing the Past: The Nature of the Record

Reefs have been present on Earth for over 3.5 billion years, and the record presents a highly complex pattern of origination, expansion, collapse, and eventual extinction of a series of different ecosystems. The fossil record therefore offers a rich and potentially highly informative data set for understanding reef ecosystem dynamics in the absence of anthropogenic influence.

Many studies in this volume have searched the fossil and geological records in order to understand current ecological change. There are, however, many problems in extrapolating ecological processes to their manifestation in these records. The poorly known environmental and ecological demands of extinct biota as well as the difficulty of determining the relative importance of multifarious controls on the growth of any individual fossil reef, makes detailed understanding of patterns highly problematic: inference of cause and effect requires correlation between independent measures of environmental conditions and biological change. In addition, the coarse stratigraphic resolution of the data means that the results of experiments that operate over ecological time scales are difficult to apply to the fossil record. This results in reduced variability becoming apparent over broader temporal and spatial scales. Such issues impose an apparent uniformity on community structure that was in fact far more dynamic and labile.

Any application of our understanding of the controls on the growth of ancient reefs to aid mitigation of ongoing living coral reef destruction continues, therefore, to be hampered by the dual straightjackets of uniformitarianism and incompatible temporal scales of inquiry.

Pleistocene coral communities have been widely heralded as offering a record of pre-anthropogenic reef community ecology (e.g., Macintyre 1988; Jackson 1992; Greenstein et al. 1998). While there is considerable ecological information preserved in Pleistocene reefs, numerous taphonomic processes operating on modern reefs conspire to change, degrade, or remove the evidence of events from future fossil communities that appear vital to understanding the functioning of present-day reefs (see Greenstein and Moffat 1996). Knowledge of which processes can be justifiably explored by analysis of the fossil record—and those that cannot—is therefore vital before any conclusions can be drawn.

Fossil data sets range from outcrop data of diversity and abundance (Pandolfi and Jackson, Chapter 8), to cores through living, relict, and fossil reefs (Greenstein, Chapter 2; Aronson and Ellner, Chapter 3; Macintyre, Chapter 7; Precht and Miller, Chapter 9), to analyses of patterns of skeletal growth, geochemistry, and bioerosion (DeVantier and Done, Chapter 4; Deslarzes and Lugo-Fernandez, Chapter 5; Halley and Hudson, Chapter 6). In the following section I consider the fidelity, biases, and scales of resolution of data derived from the geological and fossil record.

1.2.1 Fidelity, Scale, and Process

One insurmountable problem is that the fossil record is virtually mute on many key ecological players and processes. Consider these three examples: First, causes of mortality for either individuals or communities are often difficult or impossible to identify in fossil skeletal material; second, soft-bodied fauna and fleshy and fila-mentous algae leave virtually no fossil record (but see Wapnick et al. (2004) on potential ways to infer past periods of macroalgal dominance); and third, the record of reef fish and higher predators is highly patchy and incomplete. These factors mean that the operation of many processes known to be of profound ecological importance on modern reefs can only be inferred indirectly to have been present in the past. Methods for such inference are explored in several chapters in this volume.

Each source of data encompasses a different spatial and temporal resolution. These are summarized in Table 1.1, together with the type of information that can be derived from each data source, and the potential processes that each allows to be explored. They are compared to two other sources of ecological data: historical/cultural information, and the direct ecological observation of living reefs.

Outcrop data have extensive spatial resolution over many kilometers, allowing for regional and potentially global correlation of coeval stratigraphic units over long periods of geological time. Fossil data collected from outcrops allow analysis of species and generic diversity of preserved biota, and associated sedimentolog-ical data can be used to analyze all aspects of the physicochemical environment, such as relative rates of carbonate production (e.g., Opdyke and Wilkinson 1993). Origination and extinction analysis can be used to reveal patterns of community development and turnover over time, and regional paleoenvironmental data can be used to explore any correlation of these patterns with regional or global environmental events (e.g., Budd 2000).

Table 1.1. Sources of geological and ecological data, with their spatial and temporal resolutions, the type of information that can be derived from each source, and the potential processes that each allows to be explored

Data source	Outcrop	Core	Skeletal growth patterns	Historical/cultural data	Living ecological observation
Spatial resolution (m)	10^{-3}–10^3; potentially global through correlation	10^{-4}	10^{-4}	10–10^3; potentially global through correlation	10^{-2}–10^5
Temporal resolution (years)	10^4; 10^9 through compiled data sets	10^3; 10^7 through compiled data sets	10^{-3}; 10^5 through compiled data sets	10–10^4	1–10^2 (human lifetime)
Data type	Skeletal biota diversity through time and space; sedimentological data	Diversity and abundance of skeletal biota through time; sedimentological data	Skeletal growth rates; geochemical signatures	Diversity of some biota in time and space; regional to global patterns	Diversity and abundance of total biota
Potential process(es) explored	Dispersal and recruitment; environmental setting and perturbation; regional patterns; turnover events; carbonate production	Taphonomic processes; event detection	Causes of mortality; event detection; climatic trends	Changes in historical diversity; possible causes	All ecological dynamics (e.g., predation; competition; recruitment); spatial patterns

Temporal resolution is, however, coarse, rarely allowing resolution of individual beds or assemblages to less than 10^4 to 10^5 years. Compiled composite data sets can extend over considerable periods of geological time, but this coarseness increases further back in geological time.

A considerable proportion of modern reefs are preserved in the geological record as rubble, sediment, and voids as a result of physical and biological destruction (Hubbard et al. 1990). In addition, storms often remove reef sediment from its origin, redistributing and reincorporating material within the reef interior (Hubbard 1992). Reef organisms also vary greatly in their rates of relative skeletal material production as well as their durability in the face of a multitude of destructive forces. As a result, patterns of fidelity and time-averaging are highly complex (Kidwell 1998), and there may be no general rules that can be applied consistently to all ancient reefs. Analyses show that the resolution provided by the fossil record will vary in different environments and within each habitat, and facies must be evaluated individually. Specific questions will also require specific techniques. For example, DeVantier and Done (Chapter 4) offer an approach in evaluating the frequency of feeding scars of starfish on living coral heads, potentially permitting the detection of outbreaks in the geological record.

Greenstein (Chapter 2) reviews the sources of taphonomic bias in coral reefs using outcrop and core data and explores whether short-term but nevertheless highly critical ecological events are likely to be preserved in the fossil record. Relative abundance data are available in fossil reefs and can be used to determine ecological patterns over broad temporal and spatial scales (Pandolfi and Jackson, Chapter 8), but other potential sources of data may be highly biased. Greenstein discusses that any analysis must compensate for the facts that coral growth forms are differentially susceptible to degradation, with massive corals being subject to the greatest loss, and that there is also an inverse relationship between wave energy and taphonomic alteration. Greenstein and Moffat (1996) have also demonstrated that *A. cervicornis* growing in Pleistocene high-energy facies were significantly less degraded than these species from modern death assemblages. Notwithstanding recent ecological changes, branching growth forms can be overrepresented in death assemblages, due mainly to far higher rates of growth and fragmentation.

On a small scale, reef communities are clearly dynamic and to a large extent unpredictable (see Aronson 1994), but on larger scales (over tens of kilometers and centuries to millennia) patterns that show considerable consistency become apparent (e.g., Pandolfi 1996, 2002). Using outcrop data, variation at the smallest scales has been shown to be higher than even biogeographic differences (Pandolfi and Jackson, Chapter 8). This suggests that "order" in reef-coral communities is lowest at smaller scales, highest at intermediate scales, and intermediate at the broadest spatial scales within the same biogeographic province. Similar trends in predictability are apparent over varying temporal scales (e.g., Tanner et al. 1994; Pandolfi 1996; Aronson and Precht 1997; Connell 1997).

Core data can provide some temporal resolution, often down to 10^3 to 10^4 years for individual horizons which, even given the time-averaging of the samples, allows analysis of changing species abundance together with changes in sedimentological setting or climate. An increasing number of studies seek to identify the signatures of historic abundance from reef core material. For example, Pandolfi (1996) has applied univariate and multivariate methods to increase the confidence of analysis of data from Pleistocene coral communities, and Aronson and Ellner (Chapter 3) present a statistical approach for calculating the probability that contemporaneous layers in a sample of cores are coincidental. This enables limits to be set on the detectability of large-scale events in the fossil record and can be used to estimate the size of sample required to detect these events with some degree of confidence and certainty. Although the record is virtually mute as to the causes of death, there are also some data to suggest that particle analysis of ancient reef sediments may yield evidence for episodes of widespread coral mortality (Greenstein, Chapter 2; Precht and Miller, Chapter 9).

Analysis of skeletal growth patterns in conjunction with geochemical methods has been used to detect or infer past environments and environmental perturbations down to an annual resolution, and these data can be complied into composite data sets that now extend over millennia. Deslarzes and Lugo-Fernandez (Chapter 5) have used such data to explore terrigenous—oceanic and exposed—protected environmental gradients within and between reef complexes. Glynn (2000) and Wellington and Glynn (Chapter 11) outline a variety of potential indicators of past mass bleaching events that might be applied to fossil material. These include isotopic and trace metal markers in coral cores indicative of ENSO events, alterations in skeletal banding, protuberant growths on massive corals, and accelerated bioerosion in reef sediments. There is also evidence that some bleached corals fail to secrete a growth band (Halley and Hudson, Chapter 6). All of these perturbations may, however, be caused by factors other than bleaching, which would limit their utility. To date, no historical or fossil record of mass bleaching events at regional scales has been identified prior to 1982 (Glynn 1993).

1.2.2 Compatibility of Data

One major issue is how to combine disparate data from different reefs within a single frame of reference. One possible way forward here is offered by the work of Dawson (2002), Ninio and Meekan (2002), Pandolfi (2002), Deslarzes and Lugo-Fernandez (Chapter 5), and Pandolfi and Jackson (Chapter 8). These workers have shown that different coral reef communities can be distinguished within the Great Barrier Reef (GBR) and the Caribbean that correspond to terrigenous—oceanic and exposed—protected environmental gradients. These criteria are also useful for understanding the occurrence of reef biota within marginal or nonreef settings at high latitudes (Harriott and Banks 2002). Clearly, the Caribbean region has been subject to more terrigenous influences than the GBR. These gradients can be determined from paleogeographic data, offering a universal framework for the study of such influential controls of reef biota as they move from reef to marginal to nonreef status on a given temporal scale.

1.3 Lessons from the Phanerozoic Record

What does the deep past tell us about the response of reefs to global change? Here I consider two sets of data: first, the response of reefs and carbonate systems to past mass extinction events, particularly the end-Permian event which is proposed to be caused by an episode of rapid global warming, and second, how analysis of the record of Cenozoic corals and reef communities reveals the major controlling processes that were responsible for the assembly of the modern coral reef fauna over evolutionary time scales.

1.3.1 Past Episodes of Rapid Global Warming: The End-Permian Extinction Event

Reefs have been a focus for much work on mass extinctions and their subsequent recoveries. Many authors have proposed that reef ecosystems are particularly susceptible to mass extinctions and require longer periods to recover than other communities, taking some 2 to 10 Myr to reappear in the geological record (Copper 1994). This contention has not been accepted by all, however, as no such gaps are apparent in some successions where carbonate platforms persist across the extinction boundary (Wood 2000) and, in one study, selective extinction was noted in all communities from carbonate settings, not just those associated with reefs (Smith and Jeffrey 1997).

It has also been suggested that carbonate production declines substantially in the aftermath of mass extinctions due to loss of carbonate-producing biota (Bosscher and Schlager 1993). An alternative explanation is that carbonate production itself is often suppressed due to the extreme environmental perturbations that cause mass extinctions (Raup and Jablonski 1993; Smith and Jeffrey 1997; Wood 2000). In this view, the apparent lag in reef recovery may reflect a delay in reestablishing an appropriate carbonate platform environment rather than any inherent ecological lag necessary for the reassembly of a reef ecosystem; that is, reef formation may have a strong physicochemical basis rather than being driven by the diversity of available reef-builders. This in turn suggests that the reformation of reef communities can be local phenomena, and that apparent stratigraphic gaps are therefore not due to ecologically imposed delays in recovery (Wood 1999). Such a hypothesis is corroborated by studies that have shown that all low-latitude biota, not just those found in carbonate settings, appear to be preferentially susceptible to some extinction events, and that within carbonate settings, extinction occurs across all habitats, not just those associated with reefs (Smith and Jeffrey 1997).

The causes of mass extinction events are multifarious, but many are associated with rapid sea-level changes, global cooling, anoxia, or some other massive environmental perturbation, so few parallels can be drawn between these events and current scenarios of global warming. By contrast, the end-Permian extinction (251 Ma), which eliminated up to 80% to 95% of all marine species (Erwin 1993), is now thought to have been due to a geological rapid and catastrophic episode of global warming caused by an as yet unresolved mechanism (Erwin et al. 2002). This extinction event selectively removed tropical biota, a cold-adapted, high-latitude flora (which was replaced by warm, temperate floras), and diverse tetrapods, which were replaced by low-diversity, pandemic forms (Erwin 1993). Reefs, which were abundant globally in the latest Permian (e.g., the extensive reef complex at Laolongdong, in Sichuan, China), disappeared abruptly in the late Changxingian (Wood 1999).

The end-Permian extinction also provides perhaps the classic example of a delay before the onset of biotic recovery (Erwin 1993), as signs of global recovery do not appear until the end of the Early Triassic, perhaps 5 Myr after the extinction event. The reasons for this delay into the Mesozoic are unclear, and have been variously attributed to the persistence of hostile environmental conditions, ecological disturbance, and preservation failure (Erwin 1993).

Paleoecological studies reveal that most of the Early Triassic is characterized by low-diversity, highly cosmopolitan assemblages of opportunistic forms, and reefs did not reappear until the late Scythian to Anisian in either island refugia, or mid- and deep-ramp settings (see summary in Wood 1999). Many Lazarus taxa of a Permian cast have been proposed in Triassic recovery biotas, but studies have failed to confirm the presence of any such taxa or true "holdovers" in Triassic reefs. It has been suggested that a disaster biota of stromatolites spread into many "normal" marine environments during the early Triassic for the first time since the Ordovician (Schubert and Bottjer 1993), but Pratt (1995) has reinterpreted these forms as occupying deep-water settings. In fact, reefs appeared at the same time as other communities; brachiopod, bryozoan, and many bivalve radiations did not start until the Ladinian or later (Erwin 1993). Scleractinian corals appeared at the base of the Anisian with a high standing diversity. This suggests that this order of corals had some history of diversification prior to the onset of skeletonization, and that their calcification was triggered by some environmental cue that developed in the early Anisian (see summary in Wood 1999).

Empirical and modeling studies emphasize that not all survivors of mass extinctions are eurytopic, generalized, or opportunistic taxa (Harries et al. 1996). Indeed such models offer limited utility in predicting the range of processes that drive survival and recovery processes, as they exceed that which can be reliably determined from the fossil record. Although some studies have shown that widespread species are more resistant to extinction than endemic species (Jablonski 1989), the construction of explicitly phylogenetic frameworks that correctly identify clade survival of echinoderms across the K/T boundary removes this apparent preference (Smith and Jeffrey 1997). Such detailed assessments are largely absent from studies of pre-Cenozoic reef biota recoveries after mass extinctions.

1.3.2 Assembly of the Modern Reef Ecosystem: Key Patterns and Processes

The modern coral reef ecosystem is geologically very young. Scleractinian corals appeared in the mid-Triassic and had almost certainly acquired photosymbionts by the Late Triassic (Stanley and Swart 1995). After the K/T extinction, modern reef-coral genera had appeared in the Caribbean by the Eocene (50 Ma) with total generic diversity increasing due to dispersal from the Mediterranean fauna until the Miocene (22 Ma) (Budd 2000). New genera that appeared after this time prevailed in marginal deepwater or seagrass settings, with most appearing during the 5-Myr interval preceding the closure of the Isthmus of Panama. Teleostean reef fish of a modern cast also appeared about 50 Ma (Bellwood 1997), but the oldest record of parrotfish remains are from Miocene sediments dated at 14 Ma (Bellwood and Schultz 1991).

Three long intervals (up to ~5 Myr) of turnover have been recognized in Caribbean faunas for the Cenozoic (middle to late Eocene, late Oligocene to early Miocene, and Pliocene to Pleistocene), each with a distinctive coral composition. All three intervals correspond to marked environmental perturbations, particularly global cooling, sometimes accompanied by regionally increased upwelling and turbidity (Budd 2000). The compression of climatic belts and the rise of the Isthmus of Panama created the two distinct Caribbean and Indo-Pacific reef provinces, with the final closure occurring approximately 3.5 Ma.

Two extinctions served to further enhance differences between the two provinces. First, as a probable result of climatic cooling and possibly habitat loss, a major episode of turnover of coral taxa ensued between 4 and 1 Ma in the Caribbean (Budd et al. 1994b). Extinction peaked during the Plio-Pleistocene as climate deteriorated in response to the onset of Northern Hemisphere glaciation. Extinction of genera in the families Pocilloporidae and Agariciidae was marked in the Caribbean, while many of these genera continued to persist in the Indo-Pacific (Budd et al. 1994b). Second, an extinction of reef fish broadly coincident with that of corals appears to have taken place in the Atlantic (Bellwood 1997). This explains the absence of large excavating scarids, herbivorous siganids, and planktivorous caesionids in Caribbean reef communities.

Although acroporid corals appeared in the Eocene, pocilloporids appear to have dominated Caribbean reefs from 6 to 5 Ma, but following a 1-Myr transition period of mixed acroporid–pocilloporid assemblages, acroporids had become the dominant corals in reef communities by the early Pleistocene, approximately 1.6 Ma. Today, over 150 species of acroporids have been described, but only three are dominant on Caribbean reefs: *Acropora palmata, A. cervicornis,* and *A. prolifera* (a hybrid of *A. palmata* and *A. cervicornis*). Acroporids may not, however, have achieved levels of extreme abundance until the late Pleistocene, some 0.5 Ma (Budd and Kievman 1994). With this rise to dominance of branching *Acropora* during the Pleistocene glaciations, suggested to be due to their high growth rates and reproduction dominated by fragmentation (Jackson 1994), and a corresponding decline in massive corals, coral reef communities with a completely modern aspect had formed by about 0.5 Ma. Except for the extinction of *Pocillopora* in the Caribbean at about 60 ka (Budd et al. 1994b), the patterns of community membership and dominance of coral species appear to have been highly predictable for at least the past 125 kyr (Pandolfi and Jackson 1997).

The fossil record of Caribbean corals shows that dispersal and recruitment, and the size and structure of metapopulations, were fundamental to the development of the fauna. First, the Caribbean region was populated mainly by dispersal from the Mediterranean fauna until this ceased in the Miocene (22 Ma) due to changing circulation patterns (Budd 2000). Second, while the compositions of Caribbean reef communities were not limited to a fixed set of species, the total number of species in any assemblage at one location appears to have been limited (between 40 and 60 species), perhaps due to restrictions of dispersal and recruitment (Budd 2000). Third, regional generic diversity and its recovery, following losses in response to environmental perturbations, also appear to have been strongly correlated with the size of the dispersal pool and spacing between populations (Rosen 1984).

Analysis of the Late Oligocene–Early Miocene turnover event shows that cold-tolerant, eurytopic reef-coral species that brood larvae preferentially survived (Edinger and Risk 1994); in the Plio-Pleistocene turnover event, however, coral species with large colony sizes and longer generation times were more likely to survive (Johnson et al. 1995). It took 5 to 10 Myr to recover full species richness after the first two Cenozoic turnover events in the Caribbean; faunas have yet to recover from the Plio-Pleistocene event.

This historical summary suggests that living Caribbean reefs have a very low resilience to further anthropogenic disturbance. The modern coral fauna is already in a state of diversity recovery, and has been honed by successive and selective adaptations to a series of cooling events that favor the dominance of cold-tolerant species with long generation times. Such a fauna may not be well prepared for a future scenario of rapid global warming (but see Precht and Aronson 2004).

1.4 Predicting the Response to Environmental Change

How many of the processes that operated in the absence of anthropogenic change in the deep past can be usefully applied today? The following section concentrates on patterns and inferred processes which can be derived with some certainty from the geological record, and which are known to be important agents of change and destruction in modern reefs.

1.4.1 Loss of Herbivores and Higher Predators

Many researchers have summarized the case for the importance of herbivores and large marine vertebrates to the healthy functioning of coral reefs. Jackson et al. (2001) present multiple historical data over a range of scales and biogeographic realms to show how overfishing of key marine vertebrates has been the major cause of the profound ecological changes seen on coral reefs (and other coastal ecosystems). These authors argue that overfishing may also be a necessary pre-condition for additional sources of degradation—such as eutrophication, and out-breaks of disease or gregarious species—to occur. This proposal is not, however, universally accepted (see Aronson and Precht 2000; Precht and Miller, Chapter 9), but what is clear is that the superimposition of multiple factors leads to feedbacks that cause increased vulnerability due to complex synergies, and these are far from understood.

The importance of herbivores and other predators to the healthy ecology of modern coral reefs is corroborated by analysis of the fossil record. A dramatic escalation of new predators with innovative and destructive feeding methods occurred from the mid-Jurassic to Miocene (see summary in Wood 1999), with key evolutionary events being the appearance of deep-grazing limpets in the Jurassic, sea urchins with camerodont lanterns in the Cretaceous, and grazing reef fish in the Eocene–Miocene. The appearance of these herbivores was coincident with profound changes in reef ecology, including the rise of well-defended, highly tolerant coralline algae (Steneck 1983), a notable increase in branching corals since the Late Cretaceous (Jackson and McKinney 1991), and the loss of many functional groups that proved to be intolerant to excavatory attack (see summary in Wood 1999). This suggests a cause–effect system where adaptation to predatory attack has been intimately bound to the origin and assembly of the modern reef ecosystem (Vermeij 1987; Wood 1999).

In such scenarios, biodiversity in the form of ecological redundancy becomes important. That is, if a key predator can be replaced by another that performs the same ecological function, then the potentially devastating effects of its removal can be mitigated. However, there are now data to suggest that for some keystone species no functional replacement may be possible. Bellwood et al. (2003) have shown that despite the high fish diversity present on Indo-Pacific reefs (~3000 species), the removal of just one species, the giant humphead parrotfish *Bolbometopon muricatum*, will have far-reaching and multifaceted effects. This result demonstrates the weaknesses of the supposed link between biodiversity and ecological resilience, as there appears to be no direct functional replacement for parrotfish in coral reef ecosystems. The giant humphead parrotfish is also a notable bioeroder (a single male can remove up to 5 tonnes of reef per year), and so plays a major role in maintaining the steady-state calcification rates present on many modern reefs (Bellwood 1996). The loss of this single species may therefore result not only in the promotion of fast-growing, grazing-resistant corals, but also in a shift to a state of net carbonate accumulation.

1.4.2 Changing Storm Patterns and Land Use

The behavior of hurricanes and storms has been reviewed by Riegl (Chapter 10). The frequency of Atlantic hurricanes appears to follow 15- to 20-year cycles, and since the mid-1990s a period of more vigorous hurricane activity has begun. Riegl suggests that the frequency of such storms is not predicted to increase under conditions of global warming, but peak intensities and their relative moisture content may increase, which will notably increase their powers of destruction. Tropical cyclone basins may also shift, so exposing more (or less) reef areas to their effects. This is likely to increase damage until acclimatization can take place.

Population increase in watersheds will certainly lead to an increase in clastic sediment input and attendant nitrification of waters over the next decades. Excess nutrients destabilize coral photosymbiosis (Falkowski et al. 1993), reduce reproductive capacity (Koop et al. 2001), and promote the growth of fleshy algae and bioeroders. These detrimental effects will be reflected in community dynamics and ultimately reef-building potential on a local to regional scale, and can be predicted with considerable accuracy (e.g., Hallock 2001).

1.4.3 Rise in Sea Level

Sea level is expected to rise by about 0.5 m during this century (Houghton et al. 2001), some two orders of magnitude less than the 120-m rise since the last glacial maximum. Reefs are probably not directly threatened by sea-level rise in terms of drowning due to decreasing light-dependent calcification rates, but there may, however, be many indirect effects of sea-level rise that could have an impact on some reefs. For example, nutrients and sediments released from newly flooded coastlines could lead to degradation of water quality.

Macintyre (Chapter 7) argues that the geological record of reefs paints a picture of extraordinary robustness in the fate of catastrophic sea-level change. Reefs are known to have survived very rapid rates of sea-level rise during the Pleistocene, such as during periods of ice-sheet collapse (Blanchon and Shaw 1995), where corals rapidly colonized new areas where there was sufficient substrate available. Such regrowth will not be possible in the absence of available accommodation space, however, and this is likely to be the fate of coral reefs that occupy low-level islands, protected embayments, and areas such as the Florida Keys (see Precht and Miller, Chapter 9).

Buddemeier and Fautin (2002) raise the possibility of creating a new model to study the relative influence of sea-level rise on continental versus isolated islands or platforms. During periods of sea-level lowstand, suitable habitat (and substrate) for coral reefs often decreases on continental shelves, but it increases in island settings due to the exposure of broad submerged banks (Kleypas 1997). This allows for the construction of predictive models and the testing of hypotheses of changing habitat availability in different settings using past and projected sea-level oscillations.

1.4.4 Rises in CO_2 and Global Temperature

According to the IPCC's Special Report on Emission Scenarios (Nakićenović and Swart 2000), atmospheric CO_2 concentrations are predicted this century to reach between about 555 and 825 ppmV; other greenhouse gases (CH_4, N_2O, H_2O) will increase as well (Houghton et al. 2001). Such a rise would represent a doubling of the preindustrial concentration by the middle of this century. The range of predicted temperature increase is 1.4 to 5.8 °C for the period 1990 to 2100 (Houghton et al. 2001), with most coupled models indicating greater warming at higher latitudes than at the tropics (Kleypas, Chapter 12).

Kleypas (Chapter 12) suggests that the current rapid rate of increase in atmospheric CO_2 concentration is potentially catastrophic for regulation of Earth's climate and carbonate system, as the time scales of natural feedbacks required to return these systems to equilibrium are far greater than the time scale of fossil fuel burning. There is also the possibility that emergent pathogens that thrive in warmer oceans will increase. This may lead to a synergistic effect as such pathogens may preferentially attack already vulnerable or weakened biota.

It is now generally accepted that atmospheric pCO_2 was much higher during the Cretaceous and has declined throughout the Tertiary. The Vostok ice core indicates that atmospheric CO_2 concentration remained between 180 and 300 ppmV through nearly a half-million years and over several major glaciations (Petit et al. 1999). Levels of atmospheric CO_2 levels close to those predicted for the middle of this century (probably at least 500 ppmV) occurred during the Paleocene and Eocene (65–35 Ma: Pagani et al. 1999; De La Rocha and DePaolo 2000; Pearson and Palmer 2000; Pearson et al. 2001). Kleypas (Chapter 12) suggests that reefs that formed during the Paleocene and Eocene may therefore provide important clues in terms of certain physical reef characteristics such as calcification rates and distribution patterns, but they are probably less useful as analogs for ecological response because most of the dominant modern coral reef species, notably acroporids, had not appeared by that time.

The massive coral reef bleachings of the last two decades are probably unprecedented within this century and for several preceding centuries (Aronson et al. 2000). They are closely associated with abnormally warm sea surface temperature, and the clear inference is that global warming is their cause, exacerbated by other factors such as subaerial exposure, increased penetration of UV light, and decreased water circulation. Temperature and the nonanthropogenic exacerbating factors are all associated with climatic features of ENSO, with the three most notable periods of coral bleaching occurring during or shortly after ENSO events (1982/1983, 1987, 1997/1998; Glynn 1993, 2000; Wellington and Glynn, Chapter 11). Estimates of future elevated sea surface temperatures have predicted bleaching events with notable accuracy (Carriquiry et al. 2001); such events are now expected to become nearly annual on reefs worldwide over the next few decades.

There is evidence that reef-building corals and other symbiotic organisms can adapt to increasing temperatures through a range of mechanisms, including short-term acclimation, medium-term acclimatization, and even natural selection (Coles 2001). Some living corals such as those in the Red Sea and Persian Gulf currently thrive under extreme temperature regimes. Many tolerate far greater temperature ranges than conspecifics in the Indo-Pacific (Coles and Fadlallah 1991), suggesting adaptation over long periods of time, including natural selection.

Buddemeier and Fautin (1993) proposed that bleaching is an adaptive mechanism by which corals adapt to a changing environment by expelling existing zooxanthellae and then acquiring another suite better suited to the new temperature–light regime. This has been partially confirmed by experimentation (Baker 2001; Kinzie et al. 2001). Adaptive bleaching is, however, more difficult to demonstrate in the field, mainly because of the difficulty of removing the effects of other environmental factors.

The temperature range to which a coral at a particular location will acclimatize may be derived from the average annual maximum temperature experienced over some past interval of time (Ware 1997). Kleypas (Chapter 12) proposes that corals with relatively short acclimation periods (<25 years) should experience fewer and milder bleaching episodes than those that require longer acclimation periods. This means that the acclimation period could be derived theoretically for different coral species and then used to determine the probability that bleaching will occur in the future (see Ware 1997). It is doubtful, however, that many shallow-water corals will be capable of such rapid acclimation; in such a scenario, only low-diversity communities of "disaster" species may persist.

1.4.5 Changes in Seawater Chemistry

Compared to the uncertainty of how surface temperature will respond to increasing atmospheric CO_2, the response of ocean chemistry to future CO_2 increases is more predictable over the next century (Kleypas, Chapter 12), although this is complicated by biological responses and feedback mechanisms. This is because surface ocean seawater chemistry responds rapidly (within about a year) to changes in gas concentrations in the atmosphere.

Atmospheric pCO_2 estimates cannot, however, be used directly to infer ocean carbonate chemistry during the Tertiary because other key controls, such as alkalinity, may also have been different. Pearson and Palmer (2000) used boron isotopes of deep-sea foraminiferans to determine the paleo-pH of the ocean for periods extending back 49 Myr, but were able to use their data to infer atmospheric concentrations only because they assumed that ocean alkalinity changed little during this period.

Atmospheric pCO_2 prior to the Miocene probably remained higher than today but the magnesium-to-calcium ratio was probably lower (Wilson and Opdyke 1996), so that the ocean chemistry of the near future cannot be adequately compared to any past Cenozoic time period (B.N. Opdyke, quoted in Kleypas, Chapter 12). Ocean chemistry of the near future will be unique and extraordinary, mainly because the rapidity of the increase in atmospheric CO_2 will drive the system out of equilibrium.

Although details of how corals might adapt to changing sea chemistry are not clear, experimental work both on natural reefs and under artificial conditions shows that calcification rates (both biologically and inorganically mediated) decrease and dissolution rates increase as calcium carbonate saturation state declines (Gattuso et al. 1996; Suzuki and Kawahata 1999; Kayanne et al. 2003). Other studies, however, illustrate that temperature, latitude, and calcification rate are strongly correlated (e.g., Lough and Barnes 1997, 2000; Bessat and Buigues 2001). Biogenic calcification rates are already 10% to 20% lower than they were under preindustrial conditions, and aragonite saturation is estimated to decline by 30% when atmospheric CO_2 concentrations reach double preindustrial levels (Kleypas et al. 2001). This has also been estimated to mark the point when the dissolution rate will equal the calcification rate on some reefs (Halley and Yates 2000), so reducing reef-building potential globally. It is not clear how this effect could be counter-acted by the loss of parrotfish bioeroders from reefs, which will induce a coun-terwise shift to a state of net carbonate accumulation on reef crests and fore reefs (Bellwood et al. 2003).

The dominant form of precipitated crystalline $CaCO_3$ has oscillated during the geological past, with both inorganic and organic production of aragonite and high-magnesium calcite dominating carbonate formation during cool periods (icehouse conditions), and low-magnesium calcite during warm periods (green-house conditions) (Sandberg 1983; Stanley and Hardie 1998). Such mineralogi-cal shifts are interpreted as markers for major changes in seawater chemistry. Two hypotheses have been suggested. Sandberg (1983) proposed that carbonate min-eralogy has oscillated in concert with changes in atmospheric CO_2 and/or sea level change, with aragonite precipitation occurring under low pCO_2 conditions (icehouse conditions), and low-magnesium calcite forming preferentially when pCO_2 was high (greenhouse conditions). Alternatively, Stanley and Hardie (1998) propose that it is shifts in the Mg:Ca ratio that have exerted a powerful influence on the predominance of calcite versus aragonite, in both inorganic carbonate and reef-building calcifiers, due to the inhibiting effect of high Mg^{2+} concentrations on the secretion of calcite.

Scleractinian corals were dominant reef-builders in the Jurassic, but they did not build reefs during the greenhouse period of the Cretaceous. During the Cretaceous, coral diversity remained high but with lower abundance on carbon-ate platforms compared to the Jurassic, and with a distribution shifted to outer platform settings and higher latitudes (~35°N–45°N; Rosen and Turnsek 1989). There are many hypotheses offered to explain these observations, including that the high temperatures, restricted circulation, and unstable sediment conditions of Cretaceous platforms favored the growth of calcite-producing rudist bivalves over aragonite corals (see summary in Wood 1999).

It is likely, however, that greenhouse conditions at high pCO_2 and low Ca:Mg ratios will be detrimental to calcification rate across all mineralogies. Under pre-dicted scenarios of sea chemistry changes, a marked global decrease in carbonate production is probable, at least until the dissolution of calcium carbonate returns ocean alkalinity to equilibrium with atmospheric conditions, which is predicted to take many millennia to restore (Archer et al. 1997).

1.5 What Data Should We Be Collecting?

A key question in understanding the response of reefs to environmental change is what are the rules of assembly of reef communities? Are reef communities flexible or static, and on what geographic scales do different processes operate? How, and at what rate, do these processes control the way reef communities respond to change, and do these changes occur in unison or independently? Additionally, how do longer-term influences, for example the physiological tolerances of individual taxa and patterns of larval recruitment, control the spatial variation within coral communities?

If coral reef communities are flexible, then corals will show a wide range of susceptibilities to environmental change. Over the longer term, this will influence the ability of different coral species to migrate from unfavorable to favorable regions. In this section, I consider those research topics and environmental settings that might yield the greatest insight into predicting the response of coral reefs to future global change.

1.5.1 The Patchiness of Reefs

During the Plio-Pleistocene, some 4 to 1.5 Ma, a major turnover of corals and molluscs in the Caribbean occurred. Despite these dramatic changes, species richness changed little, and reef communities remained common throughout the Caribbean (Budd et al. 1994a, 1998). In one case study from Curaçao, members of the post-turnover faunas were already well established in the species pool before members of the pre-turnover fauna disappeared (Budd et al. 1998). Such a patchy distribution pattern for most reef coral species means that most species may occur at relatively few localities, even though they may be widely distributed geographically across a region (Budd et al. 1998). While the compositions of Caribbean reef communities were not limited to a fixed set of species, the total number of species at any assemblage in one location appears to have been limited (40–60 species) (Budd 2000). On this scale, observed high variability between different localities suggests that ecological interactions are not tightly bound, and that observed faunal change may have proceeded by varying dispersal and colonization rates at different spatial and temporal scales. Such "patchiness" may explain how reef faunas appear more or less "stable" at time scales less than 1 to 2 Myr (Jackson 1992; Pandolfi 1996).

1.5.2 Reefs Growing in Extreme Conditions

Many analyses suggest that real insight into rates and styles of adaptation and acclimatization of corals to new environmental regimes may be gained from the study of reefs that already occupy extreme environments at the limits of tolerance.

Reefs that grow at higher latitudes are subject to numerous nonoptimal conditions. For example, reefs growing in Florida have undergone many large-scale disturbances that have reorganized community structure throughout the Quaternary (Precht and Miller, Chapter 9). Reefs that live at such extremes not only show how reef communities change in the face of environmental degradation, but also the resilience of such communities to rapid environmental change.

Budd (2000) shows that rates of change during Cenozoic turnover events in the Caribbean were more rapid in exposed settings than protected reef environments, and that turnover may have proceeded more rapidly at northern locations, presumably those communities that were growing at their minimum temperature tolerance. Corals that occupy semienclosed seas, from small tidal pools to whole regions such as the Red Sea, already tolerate higher than normal sea-surface temperatures and often salinities. Such settings may provide useful analogs for scenarios that involve the colonization of successively higher latitudes by reefs or reef biotas as global warming persists. Analysis of the characteristics that allow for their higher temperature tolerance may aid determination of the adaptational time scales of different species of corals to future temperature increases.

1.5.3 The Importance of Nonreef Habitats: Potential Refugia?

Veron and Kelley (1988) have calculated that rates of origination and extinction of corals in the Indo-Pacific over the last 2 Myr have been similar, and so conclude that the dramatic sea-level and climatic fluctuations that occurred during this time had little effect upon the diversity of reef corals. But while coral diversity remained high, reefs were not always present throughout this interval; the record is predominantly of reefs formed during highstands that are preserved nearshore or exposed on land. A similar pattern is found in the Cretaceous record, suggesting that coral distribution and diversity alone are not reliable indicators of climatic stability in tropical shelf environments, and that nonreef habitats may provide potential refugia for otherwise reef-building taxa through periods when reef-building is not favored. Indeed, Budd (2000) in her analysis of Cenozoic Caribbean coral evolution found that the new genera that appeared after 22 Ma prevailed in marginal deepwater or seagrass settings. Similarly, non-reef-building coral communities that already exist under high pCO_2 conditions, such as high-latitude reefs and those in the Galápagos, may provide clues as to the importance of carbonate saturation state to reef-building.

Although reef fish evolved in shallow tropical seas, throughout their 50-Myr history they have always maintained strong links to nonreef habitats. The geographic range of all extant coral reef fish families extends beyond coral reef habitats and, in at least two major clades, the acathurids and scarids, occupation of associated habitats such as seagrasses appears to be the primitive condition that was only followed later by a move into coral reefs (Bellwood 1994). This analysis also suggests that, like corals, reef fish may be able to find future refuge in nonreef habitats.

1.6 Summary and Conclusions

Most of the key ecological and physicochemical phenomena that we now believe to control the functioning and distribution of modern reef communities can only be observed directly on human time scales. The geological record of reefs shows that reef communities may take 2 to 10 Myr to re-form after mass extinctions, particularly where carbonate environments are disrupted or lost. This suggests that reef formation may have a strong physicochemical basis that is a condition of local environment factors rather than being driven by the diversity of available reef-builders.

Analysis of the fossil record suggests that today's Caribbean reefs have a very low resilience to further anthropogenic disturbance. The modern coral fauna is already in a state of diversity recovery, which in the past has taken up to 5 to 10 Myr to be fully restored, and its composition has been successively selected by a series of cooling events that favor the dominance of cold-tolerant species and those with long generation times. Such a fauna will not be well-prepared for future global warming.

The fossil record also shows that dispersal and recruitment, and the size and structure of metapopulation and spacing between populations (patchiness) were fundamental to the development of the Caribbean coral reef fauna. Rates of change have been consistently more rapid in exposed settings than in protected reef environments, and turnover may have proceeded more rapidly at northern limits of reef-building; in addition, new recruits to the coral fauna seem to have preferentially appeared in marginal settings (Budd 2000).

The case, based on both recent and fossil data, for maintaining the diversity of reef herbivores and higher predators in order to promote coral growth is persuasive. The loss of potentially reef frame-building corals and herbivorous fish, even if some species are able to persist in patches or in nonreef refugia, will certainly lead to substantial changes in ecological function, which might be irrecoverable on human time scales.

All predictions show that the current rapid rate of increase in atmospheric CO_2 concentration is potentially catastrophic for regulation of Earth's climate and carbonate system, and will create a suite of conditions for which there is no known analog in Earth's history. When atmospheric CO_2 concentrations reach double preindustrial levels, a marked global decrease in carbonate production is likely, probably across all carbonate mineralogies. Rates of dissolution may also equal calcification rates. Without doubt, reef-building potential will be significantly reduced globally.

The record of the deep past, however, offers no true ecological analogues for these predicted changes. Even though reefs that formed under similar CO_2 conditions during the Paleocene and Eocene may offer clues as to calcification rates, these communities lacked many of the most important modern coral reef species, most notably acroporids.

Current understanding from both paleoecological and ecological sources suggests that reefs are flexible and patchy, and that individual coral species respond to environmental change in differing ways. Concentration of our efforts upon the ecological tolerances and adaptations of coral species from reefs already growing in extreme conditions and nonreef environments may yield valuable data that will allow prediction of how new reef communities may form in these potential refugia in the future.

The multitude of controls, their potential and complex interactions, as well as the considerable degree of uncertainty, call for a synthetic approach to prediction of the response of reefs to environmental change. Here, use of Bayesian approaches may help promote understanding. The essence of the Bayesian approach is to provide a mathematical rule that explains how existing beliefs should be modified in the light of new evidence, thus allowing scientists to combine new data with their existing knowledge or expertise. Bayes' rule is commonly viewed in terms of updating a belief about a hypothesis. As a result, the Bayesian approach can represent the subjective probability held by an individual expert, thus illuminating some of the processes that affect the evolution of expert opinion by incorporating the importance of prior information, allowing for the limits of quantification, and capturing imprecise or ranked judgements of probabilities. Most importantly, expert systems based on Bayesian networks incorporate statistical techniques that tolerate both subjectivity and small data sets.

The use of such techniques may, therefore, enable us to overcome many of the outlined issues of complexity and multiscaled processes that are required in order to formulate a more unified theoretical framework that allows for the prediction of long-term change of coral reef ecosystems.

Acknowledgements. I am grateful for the support of Schlumberger Cambridge Research, and to Ms. Frances Collin, Trustee of the Estate of Rachel Carson, for permission to quote from the writings of Rachel Carson. Earth Sciences Contribution No. 8033.

References

Archer, D., H. Kheshgi, and E. Maier-Reimer. 1997. Multiple timescales for neutralization of fossil fuel CO_2. *Geophys. Res. Lett.* 24:405–408.

Aronson, R.B. 1994. Scale–independent biological interactions in the marine environment. *Oceanogr. Mar. Biol. Ann. Rev.* 32:435–460.

Aronson, R.B., I.G. Macintyre, W.F. Precht, T.J.T. Murdoch, and C.M. Wapnick. 2002. The expanding scale of species turnover events on coral reefs in Belize. *Ecol. Monogr.* 72:233–249.

Aronson, R.B., and W.F. Precht. 1997. Stasis, biological disturbance, and community structure of a Holocene coral reef. *Paleobiology* 23:326–346.

Aronson, R.B., and W.F. Precht. 2000. Herbivory and algal dynamics on the coral reef at Discovery Bay, Jamaica. *Limnol. Oceanogr.* 45:251–255.

Aronson, R.B., W.F. Precht, I.G. Macintyre, and T.J.T. Murdoch. 2000. Coral bleach-out in Belize. *Nature* 405:36–38.

Bak, R.P.M., and G. Nieuwland. 1995. Long-term change in coral communities along depth gradients over leeward reefs in the Netherlands Antilles. *Bull. Mar. Sci.* 56:609–619.

Baker, A.C. 2001. Reef corals bleach to survive change. *Nature* 411:765–766.

Bellwood, D.R. 1994. A phylogenetic study of the parrotfishes family Scaridae (Pisces: Labroidei) with a revision of genera. *Rec. Aust. Mus. Suppl.* 20:1–86.

Bellwood, D.R. 1996. Coral reef crunchers. *Nature Australia* Autumn: 49–55.

Bellwood, D.R. 1997. Reef fish biogeography; habitat associations, fossils and phylogenies. *Proc. Eighth Int. Coral Reef Symp., Panama* 2:1295–1300.

Bellwood, D.R., A.S. Hoey, and J.H. Choat. 2003. Limited functional redundancy in high diversity systems: Resilience and ecosystem function on coral reefs. *Ecol. Lett.* 6:281–285.

Bellwood, D.R., and O. Schultz. 1991. A review of the fossil record of the parrotfishes (family Scaridae) with a description of a new *Calatomus* species from the Middle Miocene (Badenian) of Austria. *Naturhist. Mus. Wein* 92: 55–71.

Bessat, F., and A.D. Buigues. 2001. Two centuries of variation in coral growth in a massive *Porites* colony from Moorea (French Polynesia): A response of ocean-atmosphere variability from south central Pacific. *Palaeogeogr. Palaeoclimatol. Palaeoecol.* 175:381–392.

Blanchon, P., and J. Shaw. 1995. Reef drowning during the last deglaciation: Evidence for catastrophic sea-level rise and ice-sheet collapse. *Geology* 23:4–8.

Bosscher, H., and W. Schlager. 1993. Accumulation rates of carbonate platforms. *J. Geol.* 10:345–355.

Brown, B.E. 1997. Coral bleaching: Causes and consequences. *Coral Reefs* 16:S129–S138.

Budd, A.F. 2000. Diversity and extinction in the Cenozoic history of Caribbean reefs. *Coral Reefs* 19:25–35.

Budd, A.F., K.G. Johnson, and T.A. Stemann. 1994a. Plio-Pleistocene extinctions and the origin of the modern Caribbean reef-coral fauna. In *Proceedings of the Colloquium on Global Aspects of Coral Reefs: Heath, Hazards and History,* ed. R.N. Ginsberg, Rosenstiel School of Marine and Atmospheric Science, University of Miami.

Budd, A.F., and C.M. Kievman. 1994. Coral assemblages and reef environments in the Bahamas Drilling Project cores. In *Final draft report of the Bahamas Drilling Project,* 3. Coral Gables, Florida. Rosenstiel School of Marine and Atmospheric Science, University of Miami.

Budd, A.F., R.A. Petersen, and D.F. McNeill. 1998. Stepwise faunal change during evolutionary turnover: A case study from the Neogene of Curaçao, Netherlands Antilles. *Palaios* 13:170–188.

Budd, A.F., T.A. Stemann, and K.G. Johnson. 1994b. Stratigraphic distribution of genera and species of Neogene to Recent Caribbean reef corals. *J. Paleontol.* 68:951–977.

Buddemeier, R.W., and D.G. Fautin. 1993. Coral bleaching as an adaptive mechanism — a testable hypothesis. *BioScience* 43:320–326.

Buddemeier, R.W., and D.G. Fautin. 2002. Large-scale dynamics: The state of the science, the state of the reef, the research issues. *Coral Reefs* 21:1–8.

Carriquiry, J. D., A.L. Culpul-Magana, F. Rodriguez-Zaragoza, and P. Medina-Rosas. 2001. Coral bleaching and mortality in the Mexican Pacific during the 1997-1998 El Nino and prediction from a remote sensing approach. *Bull. Mar. Sci.* 69:237–249.

Carson, R. 1954. The real world around us. In *Lost Woods: The Discovered Writing of Rachel Carson,* ed. L. Lear, 1998. Beacon Press. Reproduced with permission from the Estate of Rachel Carson.

Coles, S.L. 2001. Coral bleaching: What do we know and what can we do? In *Proceedings of the Workshop on mitigating coral bleaching impact through MPA design,* May 29-31, 2001, Honolulu, HI, 25–35.

Coles, S.L., and Y.H. Fadallah, 1991. Reef coral survival and mortality at low temperatures in the Arabian Gulf: New species-specific lower temperature limits. *Coral Reefs* 9:231–237.

Connell, J.H. 1997. Disturbance and recovery of coral assemblages. *Coral Reefs* 16:101–113.

Copper, P. 1994. Ancient reef ecosystem expansion and collapse. *Coral Reefs* 13:3–11.

Dawson, J.P. 2002. Biogeography of azooxanthellae corals in the Caribbean and surrounding areas. *Coral Reefs* 21:27–40.

De La Rocha, C.L., and D.J. DePaolo. 2000. Isotopic evidence for variations in the marine calcium cycle over the Cenozoic. *Science* 289:1176–1178.

Edinger, E.N., and M.J. Risk. 1994. Oligocene-Miocene extinction and geographic restriction of Caribbean reef corals: Roles of temperature, turbidity, and nutrients. *Palaios* 9:576–598.

Erwin, D.H. 1993. *The Great Paleozoic Crisis*. New York: Columbia University Press.

Erwin, D.H., S.A. Bowring, and Y.G. Jin. 2002. The end-Permian mass extinctions. In *Catastrophic Events and Mass Extinctions: Impacts and Beyond*, eds. C. Koeberl and K.G. MacLeod, 363–383. Geological Society of America Special Paper 356.

Falkowski, P.G., Z. Dubinsky, L. Muscatine, and L. McCloskey. 1993. Population control in symbiotic corals. *BioScience* 43:606–611.

Gardner, T.A., I.M. Cote, J.A. Gill, A. Grant, and A.R. Watkinson. 2003. Long-term region-wide declines in Caribbean corals. *Science* 301:958–960.

Gattuso, J.-P., M. Pichon, B. Delesalle, C. Canon, and M. Frankignoulle. 1996. Carbon fluxes in coral reefs. I. Lagrangian measurement of community metabolism and resulting air-sea CO_2 disequilibrium. *Mar. Ecol. Prog. Ser.* 145:109–121.

Glynn, P.W. 1993. Coral reef bleaching: ecological perspectives. *Coral Reefs* 12:1–7.

Glynn, P.W. 2000. El Niño-Southern Oscillation mass mortalities of reef corals: a model of high temperature marine extinctions? In *Carbonate Platform Systems: Components and Interactions*, eds. E. Insalaco, P.W. Skelton, and T.J. Palmer, 117–133. Geological Society of London, Special Publication 178.

Greenstein, B.J., H.A. Curran, and J.M. Pandolfi. 1998. Shifting ecological baselines and the demise of *Acropora cervicornis* in the Western Atlantic and Caribbean Province: A Pleistocene perspective. *Coral Reefs* 17:249–261.

Greenstein, B.J., and H.A. Moffat. 1996. Comparative taphonomy of modern and Pleistocene corals, San Salvador, Bahamas. *Palaios* 11:57–63.

Halley, R.B., and K.K. Yates. 2000. Will reef sediments buffer corals from increased global CO_2? *Ninth Int. Coral Reef Symp.*, *Bali*, Abstract.

Hallock, P. 2001. Coral reefs, carbonate sediment, nutrients, and global change. In *Ancient Reef Ecosystems. Their Evolution, Paleoecology and Importance in Earth History*, ed. G.D. Stanley, 388–427. New York: Kluwer Academic/Plenum Publishers.

Harries, P.J., E.G. Kauffman, and T.A. Hansen. 1996. In *Biotic Recovery from Mass Extinction Events*, ed. M.B. Hart, 41-60. Geology Society of London, Special Publication 102.

Harriott, V.J., and Banks, S.A. 2002. Latitudinal variation in coral communities in eastern Australia: A qualitative biophysical model of factors regulating coral reefs. *Coral Reefs* 21:83–94.

Houghton, J.T., Y. Ding, D.J. Griggs, M. Noguer, P.J. van der Linden, and D. Xiaosu (eds.). 2001. *IPCC Third Assessment Report: Climate Change 2001: The Scientific Basis*. Cambridge University Press, UK. [Also see: *Summary for Policymakers and Technical Summary.*]

Hubbard, D.K. 1992. Hurricane-induced sediment transport in open-shelf tropical systems — An example from St. Croix, US Virgin Islands. *J. Sediment. Petrol.* 62:325–338.

Hubbard, D.K., A.I. Miller, and D. Scaturo. 1990. Production and cycling of calcium carbonate in a shelf-edge reef system (St Croix, US Virgin Islands): Applications to the nature of reef systems in the fossil record. *J. Sedimentol.* 60:335–360.

Hughes, T.P. 1994. Catastrophes, phase shifts and large-scale degradation of a Caribbean coral reef. *Science* 265:1547–1551.

Hughes, T.P., A.H. Baird, E.A. Dinsdale, N.A. Moltschaniwskyj, M.S. Pratchett, J.E. Tanner, and B.L. Willis. 1999. Patterns of recruitment and abundance of corals along the Great Barrier Reef. *Nature* 397:59–63.

Jablonski, D. 1989. The biology of mass extinctions: A paleontological view. *Philos. Trans. R. Soc. London Ser. B* 325:357–368.

Jackson, J.B.C. 1991. Adaptation and diversity of reef corals. *BioScience* 41:475–482.

Jackson, J.B.C. 1992. Pleistocene perspectives of coral reef community structure. *Am. Zool.* 32:719–731.

Jackson, J.B.C. 1994. Community unity. *Science* 264:1412–1413.

Jackson, J.B.C., M.X. Kirby, W.H. Berger, K.A. Bjorndal, L.W. Botsford, B.J. Bourque, R.H. Bradbury, R. Cooke, J. Erlandson, J.A. Estes, T.P. Hughes, S. Kidwell, C.B. Lange, H.S. Lenihan, J.M. Pandolfi, C.H. Peterson, R.S. Steneck, M.J. Tegner, and R.R. Warner. 2001. Historical overfishing and the recent collapse of coastal ecosystems. *Science* 293:629–638.

Jackson, J.B.C., and F.K. McKinney. 1991. Ecological processes and progressive macroevolution of marine clonal benthos. In *Causes of Evolution*, eds. R.M. Ross and W.D. Allmon, 173-209. Chicago: University of Chicago Press.

Johnson, K.G., A.F. Budd, and T.A. Stemann. 1995. Extinction selectivity and ecology of Neogene Caribbean reef corals. *Paleobiology* 21:52–73.

Kayanne, H., S. Kudo, H. Hata, H. Yamano, K. Nozaki, K. Kato, A. Negishi, H. Saito, F. Akimoto, and H. Kimoto. 2003. Integrated monitoring system for coral reef water pCO_2, carbonate system and physical parameters. *Proc. Ninth Int. Coral Reef Symp., Bali.*

Kidwell, S.M. 1998. Time-averaging in the marine fossil record: overview of strategies and uncertainties. *Géobios* 30:977–95.

Kinzie, R.A., III, M. Takayama, S.R. Santos, and M.A. Coffroth. 2001. The adaptive bleaching hypothesis: experimental tests of critical assumptions. *Biol. Bull.* 200:51–58.

Kleypas, J.A. 1997. Modeled estimates of global reef habitat and carbonate production since the last glacial maximum. *Paleoceanography* 12:533–545.

Kleypas, J.A., R.W. Buddemeier, D. Archer, J.-P. Gattuso, D. Langdon, and B.N. Opdyke. 1999. Geochemical consequences of increased atmospheric CO_2 on coral reefs. *Science* 284:118–120.

Kleypas, J.A., R.W. Buddemeier, and J.-P. Gattuso. 2001. The future of coral reefs in an age of global change. *Int. J. Earth Sci.* 90:426–437.

Koop, K., D. Booth, A. Broadbent, J. Brodie, et al. 2001. ENCORE: the effect of nutrient enrichment on coral reefs: Synthesis of results and conclusions. *Mar. Pollut. Bull.* 42:91–120.

Littler, M.M., and D.S. Littler. 1984. Models of tropical reef biogenesis. *Prog. Phycol. Res.* 3:117–129.

Lough, J.M., and D.J. Barnes. 1997. Several centuries of variation in skeletal extension, density and calcification in massive *Porites* colonies from the Great Barrier Reef: A proxy for seawater temperature and a background of variability against which to identify unnatural change. *J. Exp. Mar. Biol. Ecol.* 211:29–67.

Lough, J.M., and D.J. Barnes. 2000. Environmental controls on growth of the massive coral *Porites*. *J. Exp. Mar. Biol. Ecol.* 245:225–243.

Macintyre, I.G. 1988. Modern coral reefs of Western Atlantic: New geological perspective. *AAPG Bull.* 72:1360–1369.

Murdoch, T.J.T., and R.B. Aronson. 1999. Scale-dependent spatial variability of coral assemblages along the Florida Reef Tract. *Coral Reefs* 18:341–351.

Nakićenović, N., and R. Swart (eds.). 2000. *Emission Scenarios 2000. Special Report of the IPCC.* Cambridge: Cambridge University Press.

Ninio, R., and M.G. Meekan. 2002. Spatial patterns in benthic communities and the dynamics of a mosaic ecosystem on the Great Barrier Reef, Australia. *Coral Reefs* 21:95–103.

Opdyke, B.N., and B.H. Wilkinson. 1993. Oceanic carbonate saturation and cratonic carbonate accumulation. *Am. J. Sci.* 293:217–234.

Pagani, M., M.A. Arthur, and K.H. Freeman, 1999. Miocene evolution of atmospheric carbon dioxide. *Paleoceanography* 14:273–292.

Pandolfi, J.M. 1996. Limited membership in Pleistocene reef coral assemblages from the Huon Peninsula, Papua New Guinea: constancy during global change. *Paleobiology* 22:152–176.

Pandolfi, J.M. 2002. Coral community dynamics at multiple scales. *Coral Reefs* 21:13–23.

Pandolfi, J.M., R.H. Bradbury, E. Sala, T.P. Hughes, K.A. Bjorndal, R.G. Cooke, D. Macardle, L. McClenahan, M.J.H. Newman, G. Paredes, R.R. Warner, and J.B.C. Jackson. 2003. Global trajectories of the long-term decline of coral reef ecosystems. *Science* 301:955–958.

Pandolfi, J.M., and J.B.C. Jackson. 1997. The maintenance of diversity on coral reefs: examples from the fossil record. *Proc. Eighth Int. Coral Reef Symp., Panama* 1:397–404.

Pearson, P.N., P.W. Ditchfield, J. Singano, K.G. Harcourt-Brown, C.J. Nicholas, R.K. Olsson, N.J. Shackleton, and M.A. Hall. 2001. Warm tropical sea surface temperatures in the Late Cretaceous and Eocene epochs. *Nature* 413:481–487.

Pearson, P.N., and M.R. Palmer. 2000. Atmospheric carbon dioxide concentrations over the past 60 million years. *Nature* 406:695–699.

Petit, J.R., J. Jouzel, D. Raynaud, N.I. Barkov, J.-M. Barnola, I. Basile, M. Bender, J. Chappellaz, M. Davis, G. Delaygue, M. Delmotte, V.M. Kotlyakov, M. Legrand, V.Y. Lipenkov, C. Lorius, L. Pepin, C. Ritz, E. Saltzman, and M. Stievenard. 1999. Climate and atmospheric history of the past 420,000 years from the Vostok ice core, Antarctica. *Nature* 399:429–436.

Pratt, B.R. 1995. The origin, biota and evolution of deep-water mud-mounds. In *Carbonate Mud Mounds: Their Origin and Evolution,* eds. C.L.V. Monty, D.W.J. Bosence, and P.H. Bridges, 49–123. International Association of Sedimentologists Special Publication 23. Oxford: Blackwell Science.

Precht, W.F., and R.B. Aronson. 2004. Climate flickers and range shifts of reef corals. *Front. Ecol. Environ.* 2:307–314.

Raup, D.M., and J. Jablonski. 1993. Geography of end-Cretaceous marine bivalve extinction. *Science* 260:971–973.

Richardson, L.L. 1998. Coral diseases: What is really known? *Trends Ecol. Evol.* 13:438–443.

Rosen, B.R. 1984. Reef coral biogeography and climate through the Late Cainozoic: Just islands in the sun or a critical pattern of islands? In *Fossils and Climate*, ed. P. Brenchley, 201–262. New York: John Wiley.

Rosen, B.R., and D. Turnsek. 1989. Extinction patterns and biogeography of scleractinian corals across the Cretaceous/Tertiary boundary. *Mem. Assoc. Aust. Paleontol.* 8:355–370.

Sandberg, P.A. 1983. An oscillating trend in Phanerozoic non-skeletal carbonate mineralogy. *Nature* 305:19–22.

Schubert, J.K., and D.J. Bottjer. 1993. Early Triassic stromatolites as post-mass extinction disaster forms. *Geology* 20:883–886.

Smith, A.B., and C.H. Jeffrey. 1997. Selectivity of extinction among sea urchins at the end of the Cretaceous. *Nature* 392:69–71.

Stanley, G.D., and P.W. Swart. 1995. Evolution of the coral-zooxanthellae symbiosis during the Triassic: A geochemical approach. *Paleobiology* 21:179–199.

Stanley, S.M., and L.A. Hardie. 1998. Secular oscillations in the carbonate mineralogy of reef-building and sediment-producing organisms driven by tectonically forced shifts in seawater chemistry. *Palaeogeogr. Palaeoclimatol. Palaeoecol.* 144:3–19.

Steneck, R.S. 1983. Escalating herbivory and resulting adaptive trends in calcareous algal crusts. *Paleobiology* 9:44–61.

Suzuki, A., and H. Kawahata. 1999. Partial pressure of carbon dioxide in coral reef lagoon waters: comparative study of atolls and barrier reefs in the Indo-Pacific Oceans. *J. Oceanogr.* 55:731–745.

Tanner, J.E., T.P. Hughes, and J.H. Connell. 1994. Species co-existence, keystone species, and succession: a sensitivity analysis. *Ecology* 75:2204–2219.

Vermeij, G.J. 1987. *Evolution and Escalation: An Ecological History of Life.* Princeton: Princeton University Press.

Veron, J.E.N., and R. Kelley. 1988. Species stability in reef corals of Papua New Guinea and the Indo Pacific. *Assoc. Aust. Palaeontol.* 6:1–19.

Wapnick, C.M., W.F. Precht, and R.B. Aronson. 2004. Millennial-scale dynamics of staghorn coral in Discovery Bay, Jamaica. *Ecol. Lett.* 7:354–361.

Ware, J.R. 1997. The effect of global warming on coral reefs: Acclimate or die. *Proc. Eighth Int. Coral Reef Symp.* 1:527–532.

Wilson, P.A., and B.N. Opdyke. 1996. Equatorial sea surface temperatures for the Maastrichtian revealed through remarkable preservation of metastable carbonate. *Geology* 24:555–558.

Wood, R. 1999. *Reef Evolution.* Oxford: Oxford University Press.

Wood, R.A. 2000. Novel paleoecology of a post-extinction reef: The Famennian (Late Devonian) of the Canning Basin, Northwestern Australia. *Geology* 28:987–990.

Part II. Detecting Critical Events

2. Taphonomy: Detecting Critical Events in Fossil Reef-Coral Assemblages

Benjamin J. Greenstein

2.1 Introduction

The rapid and dramatic changes affecting the world's coral reef systems have alternately alarmed and frustrated the marine scientific community. This is particularly true in the Caribbean and tropical western Atlantic province, where live corals have been replaced by fleshy algae in many areas (see Carpenter 1985; de Ruyter van Stevenick and Bak 1986; Liddell and Ohlhorst 1986; Hughes, Reed, and Boyle 1987; Wilkinson 1993). While the rapidity of the observed changes is a prominent source of alarm (Curran et al. 1994; Precht and Aronson 1995; Wilkinson 2000), the lack of long-term ecological data sets confounds any attempt to place these changes into a temporal perspective. Researchers generally acknowledge that patterns demonstrated to have occurred over 10, 20, and even 30 years (Hughes 1994) may simply represent part of longer-term cycles that operate over geologic time scales (Bak and Nieuwland 1995), or at least over the generation time and longevity of reef-building corals, many of which easily exceed the length of a typical scientific research career. This fact has made it difficult for workers to determine whether the observed transitions are natural components of long-term ecological cycles or an unprecedented phenomenon resulting from primarily anthropogenic disturbances (Grigg and Dollar 1990; Brown 1997).

An additional source of frustration is the phenomenon known as "shifting baseline syndrome" (Pauly 1995; Sheppard 1995). Jackson (1997) argued eloquently that coastal Caribbean ecosystems were severely degraded by humans long before ecologists began to study them and, thus, no standard, "pristine" coral reef exists to which current floral and faunal transitions can be compared. Establishing such a standard is a particularly urgent priority since it may still be possible to save Caribbean coral reefs (because ". . . as far as we can tell almost all the reef species are still there almost everywhere" [Jackson 1997, p. S30]).

The fact that careful study of Pleistocene fossil reef-coral assemblages may be applied to these issues has been recently recognized. Exposed Pleistocene coral reefs commonly exhibit spectacular preservation (Greenstein and Moffat 1996; Greenstein and Curran 1997), and consequently tantalize paleoecologists seeking to exploit the wealth of information archived in Pleistocene strata. For example, preserved fossil coral reefs provide documentation of the response of coral communities to environmental perturbations over geologic time intervals. Their apparent stability during Pleistocene climatic and sea-level fluctuations and over geologic time scales (see Jackson 1992; Jackson, Budd, and Pandolfi 1996; Pandolfi 1996) is in stark contrast to results obtained from traditional, modern ecological studies performed over limited temporal and spatial scales.

Additionally, Pleistocene reefs were thriving long before any anthropogenic disturbance could have occurred and contain coral species virtually identical to those composing modern coral reefs. With this premise in mind, Greenstein, Curran, and Pandolfi (1998) suggested that examination of Pleistocene reefal deposits could mitigate problems of a shifting ecological baseline. Moreover, such deposits might provide an appropriate database to test whether many of the ecological transitions observed today have a Pleistocene precedent and hence help to resolve the issue of anthropogenic versus nonanthropogenic sources of disturbance on modern reefs.

Given the above, one could surmise that the Pleistocene fossil record is a panacea for better understanding the ecological problems confronting the world's coral reefs. Indeed, in his call for increased application of paleoecological data to modern reef ecology, Jackson (1992) stated, "Our only recourse is the fossil record." In a series of studies, my colleagues and I have sought to demonstrate the ecologic fidelity of Pleistocene fossil reef-coral assemblages to their modern counterparts, and consequently justify their use in approaching a variety of modern ecological problems (Greenstein and Curran 1997; Greenstein and Pandolfi 1997; Pandolfi and Greenstein 1997a,b; Greenstein, Curran, and Pandolfi 1998; Greenstein, Pandolfi, and Curran 1998). Given the results of those studies, the lack of preserved evidence for changes in reef-coral community structure similar to that observed today (see Aronson and Precht 1997; Greenstein, Curran, and Pandolfi 1998) is particularly sobering.

Are short-term ecological events—outbreaks of disease, bleaching, widespread coral mortality, or phase shifts to a new dominant taxon—likely to be manifested in the reef sedimentary record given the spectacular preservation exhibited by many Pleistocene reef-coral assemblages? Are some types of perturbations more likely to be preserved than others? Do short-term transitions in dominant taxa simply represent ecological "noise," ultimately filtered out by processes of time averaging, or do fossil coral reefs provide such remarkable resolution that changes occurring over limited temporal scales are likely preserved? Resolving these questions requires study of the taphonomic processes affecting the inhabitants of coral reef communities and applying the results to the reef sedimentary record. In this chapter, I outline some examples of this research, and conclude that while we have made an exciting and important start, much additional work needs to be done.

2.2 Reef-Coral Taphonomy

Relative to other marine invertebrates that possess potentially preservable hard parts, coral preservation has until very recently received surprisingly little attention from paleontologists (Table 2.1). There are a number of reasons for this, perhaps foremost among them an a priori assumption by many workers that massive, framework-building corals are less susceptible to taphonomic bias than the molluscan-dominated communities inhabiting temperate and tropical systems

Table 2.1. Compilation of studies of reef-coral taphonomy. Studies of coral taphonomy can be divided into four general categories

Type of study	Location	Reference
Taphonomic biasing of reef corals	Florida Keys	Gardiner, Greenstein and Pandolfi (1995) and Greenstein and Pandolfi (2003)
	Bahamas	Greenstein and Moffat (1996)
	Great Barrier Reef	Pandolfi and Greenstein (1997a)
Fidelity of coral life assemblages to death assemblages	Indo-Pacific	Pandolfi and Minchin (1995)
	Florida Keys, shallow reefs	Greenstein and Pandolfi (1997)
	Florida Keys, deep reefs	Pandolfi and Greenstein (1997b)
	Belize	Gamble and Greenstein (2001)
	Bahamas	Bishop and Greenstein (2001)
Comparison of reef coral life, death and fossil assemblages	Bahamas, Florida Keys	Greenstein and Curran (1997), Greenstein, Curran, and Pandolfi (1998), Greenstein, Pandolfi, and Curran (1998), and Greenstein, Harris, and Curran (1998)
	Indo-Pacific	Edinger, Pandolfi, and Kelley (2001)
Application to modern reef ecology	Belize	Aronson and Precht (1997)
	Jamaica	Precht and Aronson (1997)
	Bahamas	Greenstein, Harris, and Curran (1998)
	Florida Keys/Bahamas	Greenstein and Curran (1997), Greenstein, Curran, and Pandolfi (1998), and Greenstein, Pandolfi, and Curran (1998)
	Barbados	Perry (2001)
	Curaçao, Bahamas	Meyer et al. (2003)

(see Ager 1963; Schäfer 1972). Additionally, Scoffin (1992) pointed out that reefs are biological mosaics comprising a diverse assemblage of skeletal architectures and mineralogies, autecologies, and trophic groups. Although each organism follows a unique taphonomic pathway, the preserved assemblage is the product of the relationships between many different taphonomic pathways, thus making taphonomic studies of coral reefs particularly complex. Nevertheless, taphonomic research on coral reefs to date has yielded a variety of important results which fall into four general categories: (1) sources of taphonomic bias affecting corals; (2) fidelity of coral death assemblages to life assemblages; (3) comparisons of the taxonomic composition of coral life, death, and fossil assemblages; and (4) application of paleontological data to understand better modern ecological perturbations affecting coral reefs (Table 2.1).

2.2.1 Taphonomic Bias

Corals are no less subject to the physical, chemical, and biological agents of destruction than other reef inhabitants. Pandolfi and Greenstein (1997a) investigated the influences of growth form and environment on the degree of degradation suffered by corals in death assemblages on the Great Barrier Reef, Australia. Eleven biological variables reflecting the intensity of activity of boring and encrusting organisms, and four physical variables reflecting combined physical and chemical effects were measured on each of three coral growth forms: massive, branching, and free living. The corals were obtained from shallow (2–3 m) and deep (6–7 m) substrata present on leeward and windward fringing reefs surrounding Orpheus Island on the Great Barrier Reef. The sampling regimen was designed to test for three different effects: reef environment, water depth, and coral growth form. Coral taxonomy was not considered in the study since the authors were interested in taxonomy-independent factors. Kruskal–Wallis one-way analyses of variance demonstrated that massive colony growth forms consistently suffered higher amounts of degradation by both biological and physicochemical processes than their free-living and branching counterparts (Table 2.2). Moreover, coral death assemblages occurring in deep water and in leeward fringing reef environments were more degraded than corals collected from shallow water or windward fringing reefs.

Although these results seem counterintuitive, they are apparently produced by the residence time of dead coral material in the taphonomically active zone (TAZ) of Davies, Powell, and Stanton (1989). The majority of physical and biological destruction that occurs to skeletal material postmortem does so within the TAZ, which extends from the sediment–water interface down to several centimeters below the substrate. For corals, both intrinsic (coral colony growth form) and extrinsic (environmental energy) factors influence residence time in the TAZ. For reasons having to do with skeletal density (Highsmith 1981), extent of colony coverage by living tissue (Highsmith, Lueptow, and Schönberg 1983) and growth rate relative to other colony growth forms (Pandolfi and Greenstein 1997a), massive coral colonies are more prone to accumulate the variety of variables listed in Table 2.2 while still maintaining their skeletal integrity. Thus, in any reef environment, they survive in the TAZ longer than their branching and free-living counterparts.

Table 2.2. Results of Kruskal–Wallis nonparametric one-way analysis of variance of taphonomic variables on the reef at Orpheus Island, Great Barrier Reef. Growth form, depth, and site preferences for individual variables are indicated where $P < 0.05$. Coral growth form codes as follows: M = massive, F = free living, B = branching. ns = not significant. Modified from Pandolfi and Greenstein (1997b)

Variable	Growth form preference	Depth preference	Site preference
Borers			
Worms	Massive	ns	ns
Bivalves	Massive	ns	ns
Sponges	Free-living	Deep	ns
Encrusters			
Worm tubes	ns	Deep	Leeward
Coralline algae	ns	ns	ns
Foraminiferans	ns	Shallow	Windward
Bryozoans	ns	ns	Leeward
Bivalves	ns	Deep	Leeward
Sponges	Branching	Deep	ns
Biological interactions	Massive	Deep	ns
Diversity	Massive	Deep	ns
Fragmentation	–	ns	Leeward
Preservation class	Massive	ns	ns
Dissolution	Massive	ns	Leeward
Abrasion	ns	ns	Leeward

Similarly, low-energy environments (in this study, either leeward reefs *or* deeper sites) allow any colony growth form to survive in the TAZ longer than high-energy environments. Dead coral skeletons in high-energy environments are destroyed, transported, or buried before extensive taphonomic alteration can occur. Hence, when collected from the substrate in these environments, coral skeletons are in relatively pristine condition.

The results from the Indo-Pacific were corroborated and extended by Gardiner, Greenstein, and Pandolfi (1995) who conducted a similar study on patch-reef and reef-tract environments of the Florida Keys. Using the same methods of data capture and analysis, Gardiner, Greenstein, and Pandolfi (1995) documented that coral death assemblages in low-energy patch-reef environments consistently displayed higher levels of degradation than death assemblages accumulating in higher-energy reef-tract environments (Table 2.3). Additionally, by identifying the coral species present in the death assemblages, they were able to explore further the relative importance of extrinsic versus intrinsic factors in the degradation of potential fossil material. Analysis of massive (*Favia fragum*), branching (*Acropora cervicornis*), and encrusting (*Millepora alcicornis*) growth forms common to both environments revealed insignificant differences in taphonomic alteration between the two environments (Table 2.4). Since no environmental effect on taphonomic alteration was observed for the selected taxa, Gardiner, Greenstein, and Pandolfi (1995) concluded that intrinsic factors (which corals were present) were more important than extrinsic factors (environmental energy) in determining the extent of degradation. Specifically, the presence of more diverse and abundant

Table 2.3. Results of Kruskal–Wallis nonparametric one-way analysis of variance of taphonomic variables for corals obtained from patch-reef and reef-tract environments of the Florida Keys. Reef preferences for individual variables are indicated where $P < 0.05$. ns = not significant. Fragmentation was not measured. Modified from Gardiner, Greenstein, and Pandolfi (1995)

Variable	Kruskal–Wallis results
Borers	
Worms	Patch reef > reef tract
Bivalves	Patch reef > reef tract
Sponges	Patch reef > reef tract
Encrusters	
Worm tubes	ns
Coralline algae	Patch reef > reef tract
Foraminiferans	Patch reef > reef tract
Bryozoans	Patch reef > reef tract
Bivalves	Patch reef > reef tract
Sponges	Patch reef > reef tract
Biological interactions	Patch reef > reef tract
Diversity	Patch reef > reef tract
Preservation class	Patch reef > reef tract
Dissolution	ns
Abrasion	Patch reef > reef tract

Table 2.4. Results of Kruskal–Wallis nonparametric one-way analysis of variance of taphonomic variables for corals with three distinct colony growth forms. Reef preferences for individual variables are indicated where $P < 0.05$. ns = not significant. Note that when colony growth form is held constant between reefs, few significant differences between reefs exist. Modified from Gardiner, Greenstein, and Pandolfi (1995)

Variable	F. fragum (massive)	M. alcicornis (encrusting)	A. cervicornis (branching)
Borers			
Worms	ns	ns	Patch reef
Bivalves	Patch reef	ns	ns
Sponges	Reef tract	ns	ns
Encrusters			
Worm tubes	ns	Patch reef	ns
Coralline algae	ns	ns	ns
Foraminiferans	ns	ns	ns
Bryozoans	ns	ns	ns
Bivalves	ns	ns	ns
Sponges	ns	ns	ns
Biological interactions	ns	ns	ns
Diversity	ns	ns	Patch reef
Fragmentation	ns	ns	ns
Preservation class	ns	ns	ns
Dissolution	Patch reef	ns	Patch reef
Abrasion	ns	ns	Patch reef

massive coral colony growth forms in patch-reef environments was responsible for the differences observed.

Greenstein and Pandolfi (2003) extended the taphonomic analysis in the Florida Keys to include deep-reef (20 m and 30 m) environments (Table 2.5). Similar to the results they obtained from the Indo-Pacific, physical and biological taphonomic attributes measured from coral specimens with massive, branching, and platy growth forms showed great variability with respect to reef environment. Physicochemical degradation (smoothing and dissolution) was greatest in reef-crest and patch-reef environments. With the exception of encrusting foraminifera, coverage by epi- and endobionts was higher in deep-reef environments (20 m and 30 m). They suggested that variability in dissolution and smoothing was likely the result of the different energy regimes present in the reef habitats they examined. Variability in biological attributes was suggested to be the result of a combination of increased residence time of coral skeletons on substrates in deep-reef environments, higher overall coral skeletal densities of corals inhabiting deep-reef environments, and

Table 2.5. Summary of ANOVA of average taphonomic scores among habitats and coral colony growth forms in shallow and deep-reef environments of the Florida Keys. Where differences are significant ($P < 0.05$), the results for sites nested within habitats and three distinct colony growth forms are listed. Results are given for pairwise comparisons using LSD. RC = reef crest; PR = patch reef; ns = not significant. Modified from Greenstein and Pandolfi (2003)

Variable	Habitat preference	Growth form preference
Borers		
Worms	ns	
Bivalves	30 m = 20 m > RC; 20 m > PR = PR $F_{(3,4)} = 7.12; P = 0.0441$	ns
Sponges	ns	ns
Encrusters		
Worm tubes	20 m = 30 m > RC = PR $F_{(3,4)} = 62.94; P = 0.0008$	ns
Coralline algae	20 m > RC = PR = 30 m $F_{(3,4)} = 7.21; P = 0.0431$	ns
Foraminiferans	RC = PR > 20 m; RC > 30 m $F_{(3,4)} = 6.56; P = 0.0503$	ns
Bryozoans	30 m > 20 m = RC = PR $F_{(3,4)} = 6.43; P = 0.0521$	ns
Bivalves	20 m = 30 m > RC = PR $F_{(3,4)} = 30.84; P = 0.0032$	ns
Sponges	20 m = 30 m > RC = PR $F_{(3,4)} = 108.82; P = 0.0003$	ns
Biological interactions	ns	ns
Diversity	ns	ns
Preservation class	RC = PR > 20 m = 30 m $F_{(3,4)} = 108.82; P = 0.0343$	ns
Dissolution	PR > RC > 20 m = 30 m $F_{(3,4)} = 50.58; P = 0.0012$	ns
Abrasion	RC = PR > 20 m = 30 m $F_{(3,4)} = 27.20; P = 0.0040$	ns

increased nutrient availability in the deep reefs sampled. Clear gradients in the degree of taphonomic alteration of reef corals with reef habitat were also indicated. In contrast to shallow-water reefs on the Great Barrier Reef (Pandolfi and Greenstein 1997a), taphonomic alteration of corals in the Florida Keys was equitable across growth forms.

Results of these studies suggest that detecting short-term changes in reef community structure, mass mortality events, and other ecological perturbations in fossil reef assemblages at least in part depends on the type of reef facies that is studied. Since there may be an inverse relationship between wave energy and taphonomic alteration, high-energy reef facies could produce well-preserved fragile coral colony growth forms should burial occur (the "either preserved well or not at all" phenomenon). Greenstein and Moffat (1996) demonstrated that specimens of *Acropora palmata* and *A. cervicornis* preserved in Pleistocene high-energy reef facies were significantly less degraded than the same species collected from modern death assemblages accumulating in patch reefs offshore of San Salvador Island, Bahamas. They reasoned that rapid burial of both live and dead coral colonies produced the fossil assemblage. The foregoing studies emphasize that short-term changes in reef community structure may be preserved only under sedimentation regimes that favor rapid entombment of both living and dead reef corals such as occurs during rapid deposition during reef accretionary events on "keep-up" reefs during rapid sea-level change (e.g., Chappell and Polach 1976).

2.2.2 Fidelity of Death Assemblages

The ability of the reef sedimentary record to preserve the array of ecological perturbations witnessed today also depends on the degree to which the fossil assemblages reflect once-living coral assemblages. Attempts to ascertain this initially began with comparisons of live and dead reef-coral assemblages by Pandolfi and Minchin (1995) on fringing reefs in Madang Lagoon, Papua New Guinea. Live–dead comparisons of coral assemblages were subsequently performed by Greenstein and Pandolfi (1997) and Pandolfi and Greenstein (1997b) on shallow and deep reefs of the Florida Keys, respectively. More recently, Gamble and Greenstein (2001) and Bishop and Greenstein (2001) evaluated whether major storm events were preserved in coral death assemblages in Belize (Hurricane Mitch) and the Bahamas (Hurricane Floyd), respectively. Implicit in all of these studies is the assumption that reef-coral death assemblages represent the first step in the transition from biosphere to lithosphere. They therefore provide a conservative estimate of the differences in coral community composition between life and fossil assemblages that result from taphonomic processes. However, comparison of life, death, and fossil (Holocene) assemblages in Madang Lagoon by Edinger, Pandolfi, and Kelley (2001) revealed that the death assemblage was not ". . . a progressive precursor to the fossil assemblage" (p. 682). Careful comparisons of reef-coral life, death, and fossil assemblages suggest this may also be true for Pleistocene reef assemblages in the Caribbean region (see next section).

Although the data are limited, patterns of fidelity exhibited by coral death assemblages are complex. Unlike molluscan shelly assemblages, which show fairly consistent patterns of fidelity both within and between habitats (see Kidwell and Bosence 1991; Kidwell and Flessa 1995; Kidwell 2001), fidelity patterns for corals are highly variable between the Indo-Pacific and tropical western Atlantic provinces as well as between shallow- and deep-reef environments of the Florida Keys (Table 2.6). Interprovincial differences are likely related to differences in live coral diversity, especially of branching species of *Acropora*, which are

Table 2.6. Comparison of three fidelity measures tabulated for shallow-reef environments of the Florida Keys (Greenstein and Pandolfi 1997), deep-reef environments of the Florida Keys (Pandolfi and Greenstein 1997a) with values obtained by Pandolfi and Minchin (1995) and compiled for nonreef marine environments by Kidwell (2001). All values listed are means (standard errors). Values for the Florida Keys are means (standard errors) of replicate sites within four environments (reef tract, patch reef, 20 m depth, and 30 m depth), and means (standard errors) pooled for each environment. Values from Madang Lagoon comprise three sites with varying wave energy regimes. For reef environments, n = number of transects used to calculate index. For nonreef environments, n = number of studies from which data were obtained

	% live species found dead	% dead species found live	% dead individuals found live
Coral reef environments (corals)			
Shallow Florida Keys			
(Greenstein and Pandolfi 1997)			
Overall reef tract (n = 16)	72 (3.7)	50 (2.5)	65 (4.0)
Grecian Dry Rocks (n = 8)	80 (4.1)	51 (3.5)	57 (4.4)
Little Carysfort Reef (n = 8)	65 (5.1)	50 (3.8)	73 (5.6)
Overall patch reef (n = 16)	61 (4.4)	64 (3.3)	74 (4.7)
Horseshoe Reef (n = 8)	60 (7.7)	57 (4.5)	68 (4.3)
Cannon Patch Reef (n = 8)	62 (5.0)	72 (3.2)	79 (8.3)
Shallow reef mean (n = 32)	67 (3.0)	57 (2.4)	69 (3.2)
Deep Florida Keys			
(Pandolfi and Greenstein 1997a)			
Overall 20 m (n = 13)	66 (3.6)	54 (2.9)	64 (6.4)
North 20 m (n = 7)	63 (4.7)	49 (3.5)	59 (11.3)
South 20 m (n = 6)	71 (5.4)	60 (3.3)	70 (4.3)
Overall 30 m (n = 14)	87 (2.2)	54 (2.0)	49 (6.1)
North 30 m (n = 7)	84 (3.6)	55 (2.8)	48 (8.1)
South 30 m (n = 7)	89 (2.6)	53 (2.9)	50 (9.8)
Deep reef mean (n = 27)	77 (2.8)	54 (1.7)	56 (4.6)
Papua New Guinea			
(Pandolfi and Minchin 1995)			
Madang Lagoon (n = 30)	54 (4.0)	90 (1.3)	94 (6.1)
Nonreef environments			
(mollusks; Kidwell 2001)			
Marshes	91 (17); n = 22	61 (29); n = 22	93 (11); n = 22
Intertidal	86 (17); n = 14	42 (23); n = 14	76 (26); n = 12
Coastal embayments	95 (8); n = 25	38 (27); n = 25	88 (12); n = 17
Open marine	85 (13); n = 19	45 (24); n = 19	72 (20); n = 17
Nonreef mean	89 (5); n = 80	46 (10); n = 80	82 (10); n = 68

difficult or impossible to distinguish as coral rubble. The source of the differences observed between shallow and deep environments of the Florida Keys is less clear.

In Papua New Guinea, Pandolfi and Minchin (1995) found that high-energy reef environments showed a greater loss in fidelity of coral composition between life and death assemblages than low-energy reef environments. Thus, although high-energy environments might produce the best-preserved corals (see above), they also potentially preserve a more biased assemblage. Moreover, in high-, intermediate-, and low-energy reef environments, Pandolfi and Minchin (1995) found life assemblages to be more diverse than death assemblages, a result exactly opposite to those obtained from analyses of molluscan assemblages and one that suggests the conservative manner in which fossil data should be used for investigating changes in reef community structure. Pandolfi and Minchin (1995) suggested that both of these observations were, in part, the result of the high diversity of delicate colony growth forms in the Indo-Pacific. Although present in the life assemblage, such growth forms are either absent from or unidentifiable in the death assemblage.

Shallow-water reef-tract and patch-reef death assemblages of the Florida Keys are similar to their Indo-Pacific counterparts in that they do not capture the species richness of the life assemblage as effectively as molluscan death assemblages (compare the percentage of live species contributing to dead material, the first column in Table 2.6). However, shallow reefs of the Keys are different from the reefs studied in the Indo-Pacific in three important ways:

1. Diversity of life and death assemblages—No significant difference in diversity between life and death assemblages exists (Fig. 2.1). Note also that a greater percentage of live species is found in the death assemblage in the Florida Keys than in Madang Lagoon (first column in Table 2.6). Given the higher diversity of delicate coral colony growth forms in the Indo-Pacific than the western Tropical Atlantic, this observation is likely a result of the failure of their skeletons to survive even limited amounts of time on reef substrates.

2. Percentage of dead-only species—Reefs of the Florida Keys contain a much higher percentage of species found dead-only than in Madang Lagoon. Examine the second column in Table 2.6; the mean value for the percentage of dead species found live is 57% for shallow Floridian reefs. In Madang Lagoon, the mean value for the percentage of dead species also found alive is 90%. Hence the death assemblage in the Florida Keys contains many species unrecorded in the life assemblage. In a taphonomically edifying study, Kidwell (2001) showed that large numbers of dead-only species in molluscan death assemblages were primarily the result of undersampling of the living fauna. However, the sampling protocol was the same for the coral life and death assemblages in the Indo-Pacific and the Florida Keys, suggesting the importance of taphonomic bias (enrichment of the death assemblage by exotic or relict species) in producing this result.

3. Contribution to the death assemblage—Fewer individuals present in the coral death assemblages are derived from coral species documented alive in the same habitat. As the third column in Table 2.6 indicates, virtually all (94%) of the dead individuals obtained from Madang Lagoon belong to species that

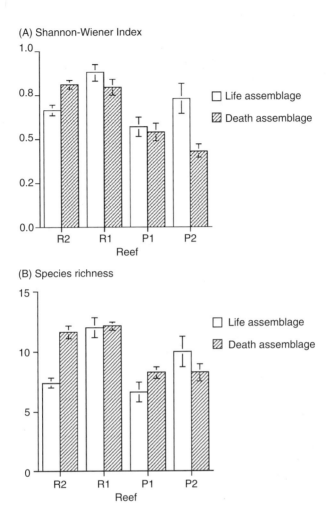

Figure 2.1. Average diversity of life and death assemblages present on shallow-water reefs of the Florida Keys using **(A)** Shannon-Wiener index and **(B)** species richness. R1 and R2 are replicate reef tract sites (Little Carysfort Reef and Grecian Dry Rocks, respectively). P1 and P2 are replicate patch-reef sites (Horseshoe Reef and Cannon Patch Reef, respectively). $n = 8$ at each site. Error bars are standard errors. No clear diversity difference is evident. The death assemblage is more diverse than the life assemblage only at Grecian Dry Rocks. The Shannon-Wiener index alone suggests that the life assemblage is more diverse than the death assemblage at Cannon Patch Reef. From Greenstein and Pandolfi (1997), reproduced with permission.

were found alive. The percentage is much lower for reef-tract and patch-reef environments of the Florida Keys. This result may be due to the recent widespread mortality of corals common to the Florida reef tract (Porter and Meier 1992), which are now present only in the death assemblage (note that this reasoning can also be invoked to explain item 2 above).

Using the same methods of data capture and analysis employed in Madang Lagoon and the Florida Keys Reef Tract, Pandolfi and Greenstein (1997b) demonstrated a third pattern of diversity and fidelity for deep-water (20 m and 30 m) coral assemblages in the Florida Keys. Deep-water death assemblages are more diverse than their living counterparts (Fig. 2.2). Additionally, the death assemblage at 30 m depth captures the highest percentage of species richness of any of the reefs studied.

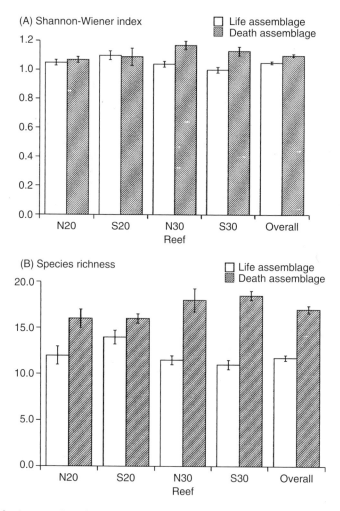

Figure 2.2. Average diversity of life and death assemblages present at each site and over all sites and depths in deep-water reefs of the Florida Keys using **(A)** Shannon-Wiener index and **(B)** species richness. $n = 8$ at each site; error bars are standard errors. Overall, death assemblages are more diverse than life assemblages for both measures of diversity. In 20 m, however, Shannon-Wiener indices are similar between life and death assemblages. Modified from Pandolfi and Greenstein (1997b) with permission from the American Society of Limnology and Oceanography.

The first column in Table 2.6 shows that the coral death assemblage at 30 m retains species richness about as effectively as molluscan death assemblages in nonreef environments (87% of living coral species were also found dead compared to 89% of molluscan species). A final departure for deep-reef fidelity values is that fewer individuals in the death assemblages belong to species documented alive in the same habitat. This is especially true for the death assemblage at 30 m, where 49% of the individuals are derived from species found live (compare the percentages listed in the third column of Table 2.6).

Three alternative, though not mutually exclusive, reasons for this result were suggested by Pandolfi and Greenstein (1997b). First, due to limited light availability, coral growth rates are slower in deep-reef environments. Lower sedimentation rates may result, allowing dead corals to accumulate on the sea floor prior to entombment, and produce an overall high degree of time averaging. Second, the reefs at 20 and 30 m depth occurred at the top and base, respectively, of a slope. Many of the dead corals composing the death assemblage were likely exotic taxa, derived from the reef in shallower water.

Finally, the study was conducted on deep reefs that had witnessed the same drastic reduction of *A. cervicornis* that has occurred throughout the Caribbean region. The two shallow-water studies discussed (Greenstein and Pandolfi 1997; Pandolfi and Greenstein 1997b) were conducted on reefs within a marine protected area and that were much less degraded (although they were more degraded than the reefs sampled in Madang Lagoon). Since the shallow-water reefs showed no greater diversity in death than in life assemblages, the dramatic differences between Caribbean reefs three decades ago (e.g., Goreau 1959; Goreau and Wells 1967) and reefs in the Florida Keys today suggest that ecological reef degradation also played an important role in producing the results obtained from the deep-reef environments. The fact that the deep-water reefs (especially at 30 m) had the lowest percentage of dead individuals also found live (Table 2.6) supports this interpretation. A low value for this fidelity metric indicates that the coral species found live are represented by few dead individuals—exactly what might occur following a recent drastic reduction of a once dominant coral taxon.

A final, consistent characteristic of coral death assemblages in all environments has been observed in the Florida Keys: when compared to their relative abundance in life assemblages, species with branching growth forms are over-represented in death assemblages, while massive-colony growth forms are under-represented (Greenstein and Pandolfi 1997; Pandolfi and Greenstein 1997b). Three potential reasons for this observation are outlined below:

1. Growth rates and time averaging—Branching coral species typically grow at a greater rate than their massive counterparts and have a much shorter generation time and life span. In addition, they may have a greater susceptibility to mortality during storms and a greater ability to quickly regenerate after them (Woodley et al. 1981; Knowlton, Lang, and Keller 1990; Massell and Done 1993). Thus, their flow to the death assemblage may be occurring at a faster rate than for massive corals. In contrast, the life assemblage contains only a small

portion of the number of colonies and time represented in the death assemblage; consequently branching corals are overrepresented in the death assemblage.

2. Influence of postmortem residence time on coral colony degradation—Because they possess more robust skeletons than their branching counterparts, massive coral colonies might be able to survive for longer intervals of time in the taphonomically active zone. Therefore, they accumulate a variety of physical, chemical, and biological agents of degradation while still exhibiting their colony form. But once the corallites of a massive colony are obscured, it becomes very difficult to distinguish it conclusively from other coral species that possess a massive colony growth form.

   Branching coral colonies, however, are more rapidly reduced to essentially unrecognizable grains of carbonate sand; when present in the death assemblage, they are found in less degraded condition because the skeleton does not survive long enough to accumulate extensive features of degradation. Thus in the death assemblages in the Florida Keys, branching coral colonies are more often identifiable to species level than massive coral colonies.

   This identification bias is also related to the species diversity of branching corals. Greenstein and Pandolfi (1997) pointed out that degraded branching coral rubble obtained from shallow-water reefs in Florida could usually be identified to the level of species based on overall colony morphology because there are relatively few potential species from which to choose. This situation does not exist in the Indo-Pacific, which supports a much more diverse coral assemblage. Consequently, degraded branching-coral rubble is virtually impossible to identify to the level of species, and thus the diversity and fidelity values obtained by Pandolfi and Minchin (1995) for coral life and death assemblages in Madang Lagoon differ substantially from those obtained for corals in the Florida Keys.

3. Sampling protocol—Methods of obtaining coral specimens from the death assemblages may have accentuated this phenomenon since they did not allow for identification of in situ massive colony growth forms whose surfaces were degraded. Moreover, samples of dead coral rubble removed from the reef were necessarily biased toward those taxa and growth forms not cemented directly on the reef substrate (e.g., many branching corals).

   The passage of Hurricanes Mitch in Belize and Floyd in the Bahamas provided an opportunity to test whether either storm was recognizable in coral death assemblages accumulating adjacent to reefs on the Belize Barrier Reef and at several locations around San Salvador Island, Bahamas. Less than one year after Hurricane Mitch passed south of the area, Gamble and Greenstein (2001) surveyed live and dead coral assemblages from 13 reefs occurring in high- and low-energy environments in the northern and central Belize reef tract. They compared their results with the fidelity values obtained by Greenstein and Pandolfi (1997) for shallow reefs of the Florida Keys. Gamble and Greenstein (2001) obtained fidelity values similar to those from the Florida Keys (Table 2.7). Fewer dead individuals were derived from species found alive on the reefs in Belize (53% for all reef environments,

Table 2.7. Fidelity measures obtained from three reef environments of the Belize Barrier Reef. Values are means (standard errors) pooled from several replicate sites within each environment, n = number of transects pooled for each environment. After Gamble and Greenstein (2001)

	% live species found dead	% dead species found live	% dead individuals found live
Fore reef (6 replicates; $n = 18$)	71 (1.8)	63 (1.4)	53 (0.9)
Reef ridge (3 replicates; $n = 9$)	73 (4.0)	62 (2.7)	61 (2.0)
Patch reef (4 replicates; $n = 12$)	75 (3.0)	61 (2.0)	53 (0.6)

compared to 65% and 74% for the reef tract and patch reefs in Florida; compare the third column in Table 2.6 with the third column in Table 2.7). Since Hurricane Mitch produced limited observable damage to the reefs studied, the authors attributed their results to the widespread coral mortality in response to the severe El Niño event of 1998, rather than a near miss by the hurricane later that year.

In a serendipitous study, Bishop and Greenstein (2001) were able to compare fidelity values obtained from patch-reef and reef-tract environments adjacent to San Salvador Island, Bahamas, within one year before, and after, the island suffered a direct hit from Hurricane Floyd (Table 2.8). They found that all of the fidelity metrics increased to some degree, and suggested each reef had witnessed a pulse of once-living coral material into the death assemblage. This was particularly true for reefs offshore to the south (French Bay) and north (Gaulin's Reef) that received the brunt of the storm impact (Table 2.8: compare the before/after values for percent dead individuals from species found alive at French Bay and the two sites at Gaulin's Reef). While the authors concluded that their comparative study yielded evidence for the hurricane, they were skeptical that the signature would be preservable unless *both* life and death assemblages were preserved, and could be distinguished, in the fossil record.

Finally, detailed examination of the orientation of corals preserved in Pleistocene reef assemblages exposed in the northern (Bahamas) and southern (Netherlands Antilles) Caribbean regions showed that the percentage of coral colonies in growth position increased in areas which today experience lower frequency of hurricanes

Table 2.8. Fidelity measures obtained before/after Hurricane Floyd battered San Salvador Island during September 1999. Values are means obtained from transects examined in June 1998/June 2000, n = number of transects. After Bishop and Greenstein (2001)

	Snapshot Reef (patch reef; $n = 6$)	French Bay (patch reef; $n = 6$)	Gaulin's I (leeward reef tract; $n = 6$)	Gaulin's II (windward reef tract; $n = 6$)
% live species found dead	83/100	92/82.4	90/100	100/100
% dead species found live	62/63	58/74	59/68	44/65
% dead individuals found live	88/86	69/94	79/95	75/9

(Meyer et al. 2003). Hence coral colony orientation may be a more suitable metric for recognizing past storm events preserved in the reef fossil record.

The implications of these various studies for our recognition of critical ecological events in the reef sedimentary record are threefold. First, although the controls on degradation of coral skeletons are highly variable, processes of preservation can differ substantially in similar reef environments; thus, the resolution provided by preserved fossil reef-coral assemblages may vary with different reef facies. Second, for reefs with relatively low coral diversity (e.g., in the Florida Keys) the absence of coral species from the death assemblages may be ecologically meaningful. Third, the likelihood that subtle changes in coral community composition will be preserved must be evaluated on a per-habitat and per-facies basis. In order to undertake such an evaluation, taphonomic studies of corals must be conducted in modern environments that are as similar as possible to the facies interpreted for the fossil assemblage under study.

Finally, the fidelity studies discussed above suggest that signatures of major storm events, coral mortality events, or degradation of reefs in general may be recognized in some reef-coral death assemblages. These studies, conducted in modern environments, do not test whether death assemblages ultimately become part of the fossil record, and hence cannot alone be used to justify the application of paleoecological data to the ecological problems confronting modern reefs. A third avenue of reef taphonomic research, discussed below, has provided illuminating results in this regard.

2.2.3 Comparison of Coral Life, Death, and Fossil Assemblages

Given that the rationale for taphonomic research on corals is a better understanding of the fossil record, any study of taphonomic processes affecting modern corals or any examination of the fidelity of coral death assemblages is justified by the following assumption: modern reef-coral death assemblages provide a reasonable proxy for an eventual fossil assemblage. A third avenue of reef taphonomic research has been to test this assumption by systematically censusing live, dead, and fossil reef-coral assemblages, and comparing the diversity and taxonomic composition between them. Results suggest that simply assuming a coral death assemblage will become a fossil assemblage is unwarranted.

Most comparative studies between modern coral life and death assemblages and their ancient counterparts involve fossil assemblages that were flourishing coral reefs during various intervals of Pleistocene time but all fall somewhere within the interval 119 to 140 ka, substage 5e of the marine oxygen isotope scale (see Harrison and Coniglio 1985). An additional study by Edinger, Pandolfi, and Kelley (2001) compared modern reef-coral life and death assemblages with Holocene fossil assemblages in Papua New Guinea. These studies began with the work of Greenstein and Curran (1997) who compared coral life and death assemblages of shallow-reef environments of the Florida Keys to those preserved in Pleistocene strata exposed on Great Inagua Island, Bahamas. Concerned that comparisons in such disparate geographic regions required carefully qualified conclusions, Greenstein, Pandolfi, and Curran (1998) censused the corals preserved in the Key Largo Limestone exposed on Key Largo and Windley Key, Florida, and compared their results to those obtained from modern reef-tract and patch-reef environments of the Florida Keys. Subsequently, Greenstein, Harris, and Curran (1998) conducted a comparative analysis using a modern patch reef adjacent to San Salvador Island, Bahamas, and a Pleistocene fossil reef exposed nearby on the island itself.

Greenstein and Curran (1997) demonstrated that the diversity and composition of the assemblages of corals currently living in patch-reef and reef-tract environments of the Florida Keys were more accurately represented by analogous facies preserved in Pleistocene strata exposed on Great Inagua Island, Bahamas, than by the contemporary death assemblage accumulating around them in Florida (Fig. 2.3). They concluded that, since the fossil and *life* assemblages were the most similar, the assumption that coral death assemblages represent reasonable "protofossil assemblages" (the raison d'être for most, if not all, studies of death-assemblage fidelity) is not entirely warranted. Further, the authors suggested that, given the congruence between the living and fossil reef-coral assemblages, the Pleistocene fossil record of reef-building corals might be useful for addressing a variety of ecological problems ranging from the response of reef-corals to environmental disturbances to more general questions, such as the degree to which coral communities are biologically integrated systems.

By conducting essentially the same study entirely in the Florida Keys region, Greenstein, Pandolfi, and Curran (1998) removed their concern for the geographic incongruence of the Greenstein and Curran (1997) work. However, they obtained results very similar to those discussed above: the taxonomic composition of corals in the life assemblages of the patch-reef and that of fossil assemblages exposed in the patch-reef facies of the Key Largo Limestone were essentially indistinguishable from one another (Fig. 2.4). In addition to supporting the patch-reef interpretation for the facies preserved in the Pleistocene strata, the authors reemphasized the utility of the Pleistocene fossil record of reef-forming coral assemblages for applied paleoecological studies.

Results from San Salvador Island, Bahamas (Greenstein, Harris, and Curran 1998), contrasted starkly with the two earlier comparative surveys. Censuses of coral life and death assemblages contained by a small midshelf patch reef adjacent to the island were compared to Pleistocene reef-tract and patch-reef facies exposed nearby in the Cockburn Town Quarry. In this study, the modern coral *death* assemblage was more similar to the fossil assemblage, while the living and fossil coral assemblages were more variable in terms of the corals that had high relative abundances (Fig. 2.5). The authors acknowledged that the habitats represented by the modern and ancient reef-coral assemblages were not an exact match (fossil reef-tract and patch-reef facies were compared to the modern patch reef), but suggested that the Pleistocene facies were sufficiently similar for meaningful comparisons to be attempted. They concluded that the recent history of the modern reef was responsible for the results. Unlike the modern reefs surveyed in the Florida Keys, the Bahamian reef had, within the last 10 years, undergone a

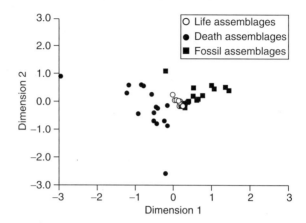

Figure 2.3. Two-dimensional global nonmetric multidimensional scaling ordination of coral life and death assemblages from the Florida Keys reef tract and patch reef and fossil assemblages from Great Inagua Island, Bahamas. Points closest to one another represent samples (transects) that are more similar in taxonomic composition than points farther away from one another. Note that samples from fossil assemblages are more similar to samples obtained from life assemblages. The minimum stress value for the two-dimensional analysis was 0.20. After Greenstein and Curran (1997).

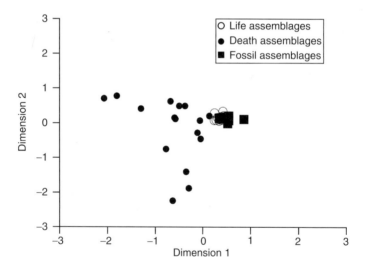

Figure 2.4. Two-dimensional global nonmetric multidimensional scaling ordination of coral life and death assemblages from modern patch reefs of the Florida Keys and fossil assemblages preserved in the Key Largo Limestone (Florida). Points closest to one another represent samples (transects) that are more similar in taxonomic composition than points farther away from one another. Note that samples from the living patch reef are virtually superimposed from those obtained from the fossil assemblages. The minimum stress value for the two-dimensional analysis was 0.17. After Greenstein, Curran, and Pandolfi (1998), reproduced with permission from Springer-Verlag.

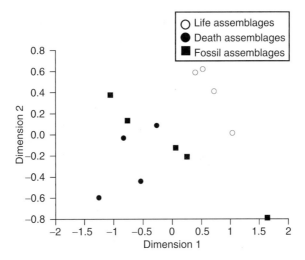

Figure 2.5. Two-dimensional global nonmetric multidimensional scaling ordination of coral life, death, and fossil assemblages from San Salvador, Bahamas. Points closest to one another represent samples (transects) that are more similar in taxonomic composition than points farther away from one another. Note that each assemblage occupies a distinct portion of ordination space, and that coral death and fossil assemblages are more similar to one another. The minimum stress value for the three-dimensional analysis was 0.09. After Greenstein, Harris, and Curran (1998).

profound transition in that the dominant coral, *A. cervicornis*, had been entirely replaced by *Porites porites*. They further demonstrated that the former *A. cervicornis*-dominated community was present in the death assemblage and was responsible for the greater similarity of the death assemblage to the fossil assemblage exposed on the island (Fig. 2.6). Finally, the authors offered the pessimistic hypothesis that, since the observed decimation and replacement of the *A. cervicornis*-dominated community apparently had no historical precedent preserved in Pleistocene age strata, the recent demise of *A. cervicornis* throughout the wider Caribbean was a singular event on a scale of a hundred thousand years or more.

Considering the results of the studies discussed above, it seems that the assumption of a stepwise transition from live to dead to fossil coral reef assemblages is largely false; reef-coral death assemblages generally are not reasonable proxies for fossil assemblages. Rather, "healthy" living reefs (i.e., those that conform to those described by Goreau (1959), Ginsburg (1964), and Multer (1977)) *and* the death assemblages associated with them apparently are preserved in Pleistocene strata. The results obtained from San Salvador by Greenstein, Harris, and Curran (1998) are particularly revealing: recall that Greenstein and Moffat (1996) demonstrated that fossil *A. cervicornis* obtained from the Cockburn Town fossil reef were generally better preserved than those obtained from the same patch reef discussed above. They concluded that, based on taphonomic data, live *and* dead corals had been entombed to form the fossil reef at Cockburn

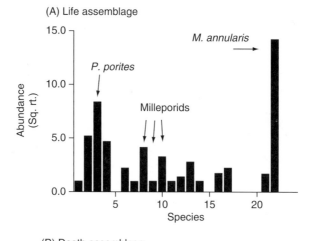

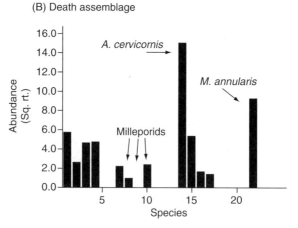

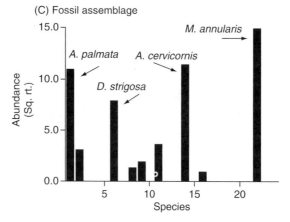

Figure 2.6. Histogram of the frequency distribution of common coral taxa in life (**A**), death (**B**), and fossil (**C**) assemblages preserved on San Salvador. Abundance data are square-root transformed. Note the relatively high abundance of *P. porites*, milleporids,

Town. However, when they compared the species composition of life and death assemblages on a reef that witnessed a change in the dominant coral taxa to the assemblage exposed at Cockburn Town, Greenstein, Harris, and Curran (1998) demonstrated that only the death assemblage, dominated by a coral that once flourished in the region (*A. cervicornis*), was closely comparable to the fossil reef.

Edinger, Pandolfi, and Kelley (2001) compared coral species diversity, taxonomic composition, and relative abundance of growth forms between life, death, and Holocene fossil assemblages in Madang Lagoon, Papua New Guinea. They also calculated fidelity indices from comparisons of the living and fossil reef. The fossil assemblages were clearly amalgamations of the coral life *and* death assemblages, and Edinger, Pandolfi, and Kelley (2001) suggested that such "composite" assemblages might be a general phenomenon for fossil reefs in general. Additionally, Edinger, Pandolfi, and Kelley (2001) demonstrated that the fossil assemblages preserved reef community structure integrated over ecological time. Similar to the "healthy" reefs of the Florida Keys discussed above, the modern reefs in Madang Lagoon have not witnessed large-scale changes in reef community composition, hence the reef-coral fossil assemblages could potentially be used to provide comparative taxonomic data for reefs that have suffered degradation from human disturbances.

Based on the discussion above, the reader should not conclude that studies of reef-coral death assemblages are unimportant; the previous sections demonstrate clearly the wealth of taphonomic information obtainable from their examination. Moreover, the comparative studies discussed above, which included death

←——————————————————————————————

and *Montastraea* "*annularis*" in the life assemblage. The death assemblage comprises *A. cervicornis*, *A. palmata*, and lower abundances of *M.* "*annularis*." The fossil assemblage also comprises high abundances of these taxa. Data codes (*x*-axis) are as follows:

1. *Acropora palmata*	13. *Siderastrea siderea*
2. *Porites astreoides*	14. *Acropora cervicornis*
3. *Porites porites*	15. *Porites furcata*
4. *Agaricia agaricites*	16. *Mycetophyllia lamarckiana*
5. *Millepora* sp.	17. *Montastraea cavernosa*
6. *Diploria strigosa*	18. *Mycetophyllia danaana*
7. *Favia fragum*	19. *Colpophyllia natans*
8. *Millepora squarrosa*	20. *Dichocoenia stokesii*
9. *Millepora complanata*	21. *Diploria labyrinthiformis*
10. *Millepora alcicornis*	22. *Montastraea* "*annularis*"
11. *Diploria clivosa*	23. *Meandrina meandrites*
12. *Siderastrea radians*	24. *Solenastrea bournoni*
	25. *Solenastrea hyades*

Note that the code for *Montastraea* "*annularis*" includes colony growth forms now recognized as sibling species. From Greenstein, Harris, and Curran (1998), reproduced with permission from Springer-Verlag.

assemblage data, have demonstrated the feasibility of applying paleoecological studies of the reef sedimentary record (at least Pleistocene age strata) to the variety of ecological perturbations presently affecting modern reefs. Such applications are discussed in the following section.

2.2.4 Application of Paleontological Data

Shifting Ecological Baseline

Jackson (1997) suggested that Caribbean reef-coral assemblages from which the classic coral zonation patterns had been described (Goreau 1959) had begun to degrade long before ecologists began studying them; studies of reef ecology were therefore confounded by the "shifting baseline syndrome." In a synthesis of some of the above work, Greenstein, Curran, and Pandolfi (1998) extended the application of comparative studies of reef-coral life, death, and fossil assemblages to test Jackson's (1997) assertion. Greenstein, Curran, and Pandolfi (1998) reasoned that, since Pleistocene reefs were thriving long before any anthropogenic disturbance could have occurred, they might provide an appropriate ecological baseline to which modern reef communities could be compared. They emphasized the results of earlier work (Greenstein, Pandolfi, and Curran 1998) in which they demonstrated that Pleistocene reef assemblages were essentially indistinguishable from modern reefs whose taxonomic composition conformed to early pre-1980 descriptions of the majority of Caribbean patch reefs. They concluded that, at least for Floridian reefs, Jackson's (1997) concern might not be warranted. More importantly, the study demonstrated that concern over the shifting baseline syndrome might be mitigated where a sufficient fossil record occurs.

Pleistocene Precedent for Changes in Reef-Coral Community Structure

Because corals have been replaced by fleshy algae in most cases where transitions have been observed in Caribbean regions (Carpenter 1985; de Ruyter van Stevenick and Bak 1986; Liddell and Ohlhorst 1986; Hughes, Reed, and Boyle 1987; Wilkinson 1993), the preservation potential of these events is extremely low (Kauffman and Fagerstrom 1993). However, in at least three areas of the Caribbean, the dominant corals have died and been replaced by other species of coral rather than fleshy macroalgae. Replacement by corals has been observed in Belize (Aronson and Precht 1997) and the Bahamas (Curran et al. 1994), where dense thickets of *A. cervicornis* have been replaced by *Agaricia tenuifolia* and *P. porites*, respectively. In a coastal lagoon in Panama, *P. furcata* was replaced by *A. tenuifolia* (Aronson et al. 2004). Because corals are more likely to survive the vagaries of the preservation process than are fleshy algae, the fossil record should yield some insight as to whether a historical precedent exists for either event. With this premise in mind, Aronson and Precht (1997) examined cores taken through the Holocene sediments accumulating on the Belize reef and concluded that no recognizable signals (abrupt changes in coral taxa, or evidence of an essentially monospecific death assemblage) of similar transitions were present, suggesting that the present drastic reduction of *A. cervicornis* has no precursor in the recent geologic past at least within the last 3800 years (see also Aronson et al. 2002; Wapnick, Precht, and Aronson 2004; Aronson and Ellner, Chapter 3). Recall that Greenstein, Curran, and Pandolfi (1998) reached essentially the same conclusion based on the results of their earlier (Greenstein, Harris, and Curran 1998) study. These studies suggest that, although the *A. cervicornis*-dominated coral association persisted during Pleistocene climatic fluctuations throughout the Caribbean and tropical western Atlantic regions, it appears vulnerable to the array of perturbations currently being inflicted on it. Similar inferences apply to the *P. furcata* populations in Panama.

This conclusion was corroborated by Perry (2001) using *A. palmata*-dominated coral communities preserved in Pleistocene reef facies exposed in marine terraces on Barbados. Using taphonomic data (colonization sequences by a variety of epibionts), Perry was able to distinguish between *A. palmata* horizons that represented storm deposition and those that had accumulated through reef accretion. He further demonstrated that, over multiple events, storm horizons were followed by the same reef succession, ultimately culminating in *A. palmata*. Perry contrasted the situation he observed in Pleistocene strata with that observed today, where disease and anthropogenic stresses have been documented to delay coral recovery following natural disturbances. Subsequent major community shifts have been observed (the reefs in Jamaica provide an example, see Woodley et al. 1981; Perry 1996).

Clearly, more work needs to be done to investigate this issue further, including analyses of aspects of fossil reef sequences that encompass more data than simply the corals that are preserved in them. For example, Precht and Aronson (1997) monitored the composition of sediments accumulating on LTS Reef (Discovery Bay, Jamaica) over a 20-year interval. During that time (1978–1997) several perturbations produced profound changes in the coral community. The most significant of these was a decline of live coral cover (from 75% to 5%) and increase of fleshy algae (<5% to >65%). The sedimentary response was a change in the relative abundance of the two primary constituents: *Halimeda* and coral fragments. Prior to 1981, reef sediments were principally composed of *Halimeda* (>50%) followed by coral fragments (<35%). Subsequent to 1981, sediments have been composed primarily of coral fragments as a consequence of widespread coral mortality and ensuing bioerosion. In a similar applied study, Lidz and Hallock (2001) demonstrated an increase in coral grains relative to algal and molluscan grains over a 37-year interval in surficial sediments obtained from numerous shelf localities adjacent to the Florida Keys. They suggested that the relative proportion of these sediment constituents provided a proxy for coral reef vitality. For corals at least, constituent particle analyses of ancient reef sediments might yield evidence for ancient episodes of widespread coral mortality.

Taphonomic analyses of fossil coral material might also suggest past widespread coral mortality. Rothfus and Greenstein (2001) analyzed the taphonomic condition of several coral taxa collected at centimeter-scale intervals from a Pleistocene fossil coral reef on San Salvador Island, Bahamas. Given the spectacular preservation of corals in the sequence (Greenstein and Moffat 1996), their working hypothesis was that an ancient widespread coral mortality event (and therefore coral death assemblage) might be represented as a horizon of highly degraded fossil coral material. However, they did not find a layer in the fossil assemblage in which coral preservation was notably worse than in surrounding layers, or a particular coral taxon that exhibited consistently poor preservation. They concluded that there was no evidence of widespread coral mortality preserved in the fossil reef and emphasized the importance of taphonomic studies to mitigate information loss due to time averaging in Pleistocene strata.

Perturbations Involving Additional Reef Inhabitants

Although evidence of coral mortality might be preserved in the reef sedimentary record as an increase in coral fragments (Precht and Aronson 1997; Lidz and Hallock 2001) evidence for fluctuations in populations of other reef inhabitants has proved elusive. For example, the mass mortality of *Diadema antillarum* throughout the Caribbean and tropical western Atlantic during 1983 to 1984 continues to affect coral reef communities adversely in many areas (Smith and Ogden 1994). However, when constituent particle analyses of reef sediments obtained after the mortality event occurred in Bonaire, Netherlands Antilles, and Andros Island, Bahamas, were compared to published reports of the composition of reef sediments examined prior to the mortality event (Greenstein 1989; Greenstein and Meyer 1990), no evidence of a layer rich in echinoderm material was obtained. These results underscore the rapidity with which echinoid material was reworked and diluted by additional sediment derived from the reef, considering that innumerable echinoids, each possessing a skeleton of thousands of elements, were observed to decay and disarticulate on the reef substrate over a relatively short interval of time (Greenstein 1989).

Similarly, the ability of the reef sedimentary record to preserve evidence of population explosions of the crown-of-thorns starfish, *Acanthaster planci*, on the Great Barrier Reef has been the subject of intensive debate for two decades (Frankel 1977, 1978; Moran 1986; Moran, Reichelt, and Bradbury 1986; Walbran et al. 1989a, b; theme issues of *Coral Reefs* in 1990 and 1992; Greenstein, Pandolfi, and Moran 1995). At issue is not whether this animal has been part of the Great Barrier Reef system for many thousands of years (Wilkinson and Macintyre 1992) but rather whether the sedimentary record has the resolution to provide conclusive evidence that the damaging population explosions observed in the last 40 years have historical precedent. Greenstein, Pandolfi, and Moran (1995) synthesized much of the current controversy, and applied field and laboratory experiments of crown-of-thorns taphonomy to the problem. They concluded that while it was possible for the reef sedimentary record to preserve past outbreaks, such preservation had yet to be demonstrated. In retrospect, evidence for both echinoid and asteroid mass mortality might be detectable in the form of increased coral fragments in the sediments for reasons similar to those suggested by Precht and Aronson (1997). However, in all cases, the source of coral mortality that produced an influx of their skeletal fragments would not be detectable. This drawback would also be true should one detect highly degraded fossil coral material, a hardground within a reef sequence or other evidence of interruption of reef growth. A promising approach to detecting past *Acanthaster* outbreaks is to examine the frequency of feeding scars on living coral heads (DeVantier and Done, Chapter 4). This methodology could also be applied to grazing traces of echinoids or herbivorous fish.

Summary and Future Research Priorities

Over the last ten years, vigorous study of a variety of facets of coral reef taphonomy has demonstrated the importance of this type of interdisciplinary program to research problems in basic and applied paleobiology. Additional research along several separate but interrelated avenues should be attempted.

Field study of coral taphonomy. To date, our knowledge of the relationships between skeletal density, coral growth form, and reef environment in determining the way in which corals might be preserved as fossils is limited simply to knowing that differences exist. Tests for the effects of these variables need to be conducted. Field studies should involve emplacing dead coral species exhibiting a wide variety of growth forms and skeletal densities at various depths in sediments at selected reef localities. Since differences in live coral diversity between the Caribbean and Indo-Pacific provinces were invoked to explain differences in fidelity and diversity in coral death assemblages (Greenstein and Pandolfi 1997; Pandolfi and Greenstein 1997b), burial experiments should involve reciprocal transplants of dead coral material emplaced in selected sites in both provinces. Indo-Pacific species exhibiting a variety of colony growth forms should be placed in both Indo-Pacific and Caribbean reef environments. Caribbean species exhibiting the same growth forms should also be placed in both Caribbean and Indo-Pacific environments.

Fidelity studies in widely differing reef environments. Given the disparity between the results of the few studies of fidelity that have thus far been reported, there is clearly a need to continue this research agenda in many different reef environments. Studies of fidelity need not be undertaken solely to investigate live versus dead coral assemblages; however, they could be coupled with established reef-survey programs. For example, the current global effort to assess the condition of reef corals (Ginsburg et al. 1996) provides an opportunity to "tack on" studies of fidelity in a wide variety of reef environments.

Continued application of paleontological data. This facet of reef taphonomic research is clearly the most directly applicable to the variety of ecological issues facing reefs today. Study of the fossil record can (1) provide a temporal context for changes observed on reefs over the last 30 years; (2) test for a shifting ecological baseline; (3) determine what constitutes a "pristine" coral reef (and thus inform conservation policy); and (4) detect ancient examples of transitions in coral taxa described for modern reefs. To achieve these objectives, study of fossil reef assemblages must continue, and include horizons that exhibit a wide range of coral preservation (e.g., Pandolfi and Jackson, Chapter 8). This avenue of research is particularly important. Recall that studies of reef coral fidelity indicated that short-term perturbations (storms, coral mortality, reef degradation) might, in some reef environments, be interpreted from comparing live and dead coral assemblages. In what really amounted to a pilot study, Rothfus and Greenstein (2001) used taphonomic condition of corals preserved in the Pleistocene of San Salvador Island, Bahamas, to distinguish between colonies that had likely been alive at the time of burial from those that were already dead. They then calculated species richness for each defined assemblage and demonstrated that diversity differences between the fossilized life and death assemblages were similar to those published for modern shallow reef-coral life and death assemblages in the Florida Keys and Bahamas. If such information can be obtained reliably from fossil assemblages, it can potentially be combined with study of modern life and death assemblages to ascertain whether short-term transitions in reef-coral communities, such as those witnessed today, have a precedent in the geologic past.

References

Ager, D.V. 1963. *Principles of Paleoecology*. New York: McGraw-Hill.

Aronson, R.B., I.G. Macintyre, W.F. Precht, T.J.T. Murdoch, and C.M. Wapnick. 2002. The expanding scale of species turnover events on coral reefs in Belize. *Ecol. Monogr.* 72:233–249.

Aronson, R.B., I.G. Macintyre, C.M. Wapnick, and M.W. O'Neill. 2004. Phase shifts, alternative states, and the unprecedented convergence of two reef systems. *Ecology* 85:1876–1891.

Aronson, R.B., and W.F. Precht. 1997. Stasis, biological disturbance, and community structure of a Holocene coral reef. *Paleobiology* 23:326–346.

Bak, R.P.M., and G. Nieuwland. 1995. Long-term change in coral communities along depth gradients over leeward reefs in the Netherlands Antilles. *Bull. Mar. Sci.* 56:609–619.

Bishop, D., and B.J. Greenstein. 2001. The effects of Hurricane Floyd on the fidelity of coral life and death assemblages in San Salvador, Bahamas: Does a hurricane leave a signature in the fossil record? *Geol. Soc. Am. Abstr. Prog.* 33(4):A51.

Brown, B.E. 1997. Disturbances to reefs in recent times. In *Life and Death of Coral Reefs*, ed. C. Birkeland, 354–379. New York: Chapman and Hall.

Carpenter, R.C. 1985. Sea urchin mass mortalities: Effects on reef algal abundance, species composition, and metabolism and other coral reef herbivores. *Proc. Fifth Int. Coral Reef Congr., Tahiti* 4:53–60.

Chappell, J., and H.A. Polach. 1976. Holocene sea-level change and coral-reef growth at Huon Peninsula, Papua New Guinea. *Geol. Soc. Am. Bull.* 87:235–240.

Curran, H.A., D.P. Smith, L.C. Meigs, A.E. Pufall, and M.L. Greer. 1994. The health and short-term change of two coral patch reefs, Fernandez Bay, San Salvador Island, Bahamas. In *Proceedings Colloquium on Global Aspects of Coral Reefs: Health, Hazards, and History, 1993*, ed. R.N. Ginsburg, 147–153. Miami: Rosenstiel School of Marine and Atmospheric Science, University of Miami.

Davies, D.J., E.N. Powell, and R.J. Stanton, Jr. 1989. Relative rates of shell dissolution and net sediment accumulation—a commentary: Can shell beds form by the gradual accumulation of biogenic debris on the sea floor? *Lethaia* 22:207–212.

de Ruyter van Stevenick, E.D., and R.P.M. Bak. 1986. Changes in abundance of coral reef bottom components related to mass mortality of the sea urchin *Diadema antillarum*. *Mar. Ecol. Prog. Ser.* 34:87–94.

Edinger, E.N., J. M. Pandolfi, and R.A. Kelley. 2001. Community structure of Quaternary coral reefs compared with Recent life and death assemblages. *Paleobiology* 27:669–694.

Frankel, E. 1977. Previous *Acanthaster* aggregations in the Great Barrier Reef. *Proc. Third Int. Coral Reef Symp., Miami* 1:201–208.

Frankel, E. 1978. Evidence from the Great Barrier Reef of ancient *Acanthaster* aggregations. *Atoll Res. Bull.* 220:75–93.

Gamble, V.C., and J. J. Greenstein. 2001. Fidelity of coral life and death assemblages on Belizean reefs in the wake of Hurricane Mitch. In *Proceedings of the Tenth Symposium on the Geology of the Bahamas and other Carbonate Regions*, eds. B.J. Greenstein and C.K. Carney, 142–151. San Salvador: Gerace Research Center.

Gardiner, E.S., B.J. Greenstein, and J. M. Pandolfi. 1995. Taphonomic analysis of Florida reef corals: The effect of habitat on preservation. *Geol. Soc. Am. Abstr. Prog.* 27:46.

Ginsburg, R.N. 1964. South Florida Carbonate Sediments. *Geological Society of America Annual Meeting, Guidebook for Field Trip #1.* Boulder: Geological Society of America.

Ginsburg, R.N., R.P.M. Bak, W.E. Kiene, E. Gischler, and V. Kosmynin. 1996. Rapid assessment of reef condition using coral vitality. *Reef Encounter* 19:12–14.

Goreau, T.F. 1959. The ecology of Jamaican coral reefs. I. Species composition and zonation. *Ecology* 40:67–90.

Goreau, T.F., and J. W. Wells. 1967. The shallow water Scleractinia of Jamaica: Revised list of species and their vertical distribution range. *Bull. Mar. Sci.* 17:442–453.

Greenstein, B.J. 1989. Mass mortality of the West-Indian echinoid *Diadema antillarum* (Echinodermata: Echinoidea): A natural experiment in taphonomy. *Palaios* 4:487–492.

Greenstein, B.J., and H.A. Curran, 1997. How much ecological information is preserved in fossil reefs and how reliable is it? *Proc. Eighth Int. Coral Reef Symp., Panama* 417–422.

Greenstein, B.J., H.A. Curran, and J. M. Pandolfi. 1998. Shifting ecological baselines and the demise of *Acropora cervicornis* in the western North Atlantic and Caribbean Province: A Pleistocene perspective. *Coral Reefs* 17:249–261.

Greenstein, B.J., L.A. Harris, and H.A. Curran. 1998. Comparison of recent coral life and death assemblages to Pleistocene reef communities: Implications for rapid faunal replacement observed on modern reefs. *Carbonates Evaporites* 13:23–31.

Greenstein, B.J., and D.L. Meyer. 1990. Mass mortality of *Diadema antillarum* adjacent to Andros Island, Bahamas. In *Proceedings of the Fourth Symposium on the Geology of the Bahamas*, eds. J. Mylroie and D. Gerace, 159–168. San Salvador: Bahamian Field Station.

Greenstein, B.J., and H.A. Moffat. 1996. Comparative taphonomy of Holocene and Pleistocene corals, San Salvador, Bahamas. *Palaios* 11:57–63.

Greenstein, B.J., and J. M. Pandolfi. 1997. Preservation of community structure in modern reef coral life and death assemblages of the Florida Keys: Implications for the Quaternary record of coral reefs. *Bull. Mar. Sci.* 19:39–59.

Greenstein, B.J., and J. M. Pandolfi. 2003. Taphonomic alteration of reef corals: Effects of reef environment and coral growth form. II: The Florida Keys. *Palaios* 18:495–509.

Greenstein, B.J., J. M. Pandolfi, and H.A. Curran. 1998. The completeness of the Pleistocene fossil record: Implications for stratigraphic adequacy. In *The Adequacy of the Fossil Record*, eds. S.K. Donovan and C.R.C. Paul, 75–110. London: John Wiley & Sons.

Greenstein, B.J., J. M. Pandolfi, and P.J. Moran. 1995. Taphonomy of crown-of-thorns starfish: Implications for recognizing ancient population outbreaks. *Coral Reefs* 14:91–97.

Grigg, R.W., and S.J. Dollar. 1990. Natural and anthropogenic disturbances on coral reefs. In *Ecosystems of the World 25: Coral Reefs*, ed. Z. Dubinshky, 439–452. Amsterdam: Elsevier.

Harrison, R.S., and M. Coniglio. 1985. Origin of the Pleistocene Key Largo Limestone, Florida Keys. *Bull. Can. Soc. Petrol. Geol.* 33:350–358.

Highsmith, R.C. 1981. Coral bioerosion: Damage relative to skeletal density. *Am. Nat.* 117:193–198.

Highsmith, R.C., R.L. Lueptow, and S.C. Schönberg. 1983. Growth and bioerosion of three massive corals on the Belize barrier reef. *Mar. Ecol. Prog. Ser.* 13:261–271.

Hughes, T.P. 1994. Catastrophes, phase shifts, and large-scale degradation of a Caribbean coral reef. *Science* 265:1547–1551.

Hughes, T.P., D.C. Reed, and M.J. Boyle. 1987. Herbivory on coral reefs: Community structure following mass mortalities of sea urchins. *J. Exp. Mar. Biol. Ecol.* 113:39–59.

Jackson, J.B.C. 1992. Pleistocene perspectives on coral reef community structure. *Am. Zool.* 32:719–731.

Jackson, J.B.C. 1997. Reefs since Columbus. *Coral Reefs* 16:S23–S32.

Jackson, J.B.C., A.F. Budd, and J.M. Pandolfi. 1996. The shifting balance of natural communities? In *Evolutionary Paleobiology*, eds. D. Erwin, D. Jablonski, and J. Lipps, 89–122. Chicago: University of Chicago Press.

Kauffman, E.G., and J.A. Fagerstrom. 1993. The Phanerozoic evolution of reef diversity. In *Species Diversity in Ecological Communities: Historical and Geographical Perspectives*, eds. R.E. Ricklefs and D. Schluter, 315–329. Chicago: University of Chicago Press.

Kidwell, S.M. 2001. Ecological fidelity of molluscan death assemblages. In *Organism–Sediment Interactions*, eds. J. Y. Aller, S.A. Woodin, and R.C. Aller, 199–222. Belle W. Baruch Library in Marine Science, No. 21. Columbia: University of South Carolina Press.

Kidwell, S.M., and D.W.J. Bosence. 1991. Taphonomy and time averaging of marine shelly faunas. In *Taphonomy: Releasing the Data Locked in the Fossil Record*, eds. P.A. Allison and D.E.G. Briggs, 115–209. New York: Plenum Press.

Kidwell, S.M., and K. Flessa. 1995. The quality of the fossil record: Populations, species, and communities. *Ann. Rev. Ecol. Syst.* 26:269–299.

Knowlton, N., J.C. Lang, and B.D. Keller. 1990. Case study of natural population collapse: Post-hurricane predation on Jamaican staghorn corals. *Smithson. Contrib. Mar. Sci.* 31:1–25.

Liddell, W.D., and S.L. Ohlhorst. 1986. Changes in benthic community composition following the mass mortality of *Diadema* at Jamaica. *J. Exp. Mar. Biol. Ecol.* 95:271–278.

Lidz, B.H., and P. Hallock. 2001. Sedimentary petrology of a declining reef ecosystem, Florida reef tract (U.S.A.). *J. Coast. Res.* 16:675–697.

Massell, S.R., and T.J. Done. 1993. Effects of cyclone waves on massive coral assemblages on the Great Barrier Reef: Meteorology, hydrodynamics and demography. *Coral Reefs* 12:153–166.

Meyer, D.M., J.M. Bries, B.J. Greenstein, and A.O. Debrot. 2003. Preservation of in situ reef framework in regions of low hurricane frequency: Pleistocene of Curaçao and Bonaire, southern Caribbean. *Lethaia* 36:273–286.

Moran, P.J., 1986. The *Acanthaster* phenomenon. *Oceanogr. Mar. Biol. Ann. Rev.* 24:379–480.

Moran, P.J., R.E. Reichelt, and R.H. Bradbury. 1986. An assessment of the geological evidence for previous *Acanthaster* outbreaks. *Coral Reefs* 4:235–238.

Multer, H.G. 1977. *Field Guide to some Carbonate Rock Environments: Florida Keys and Western Bahamas.* Dubuque, Iowa: Kendall Hunt Publishing Co.

Pandolfi, J.M. 1996. Limited membership in Pleistocene reef coral assemblages from the Huon Peninsula, Papua New Guinea. Constancy during global change. *Paleobiology* 22:152–176.

Pandolfi, J.M., and B.J. Greenstein. 1997a. Taphonomic alteration of reef corals: Effects of reef environment and coral growth form. I: The Great Barrier Reef. *Palaios* 12:27–42.

Pandolfi, J.M., and B.J. Greenstein. 1997b. Preservation of community structure in death assemblages of deep water Caribbean reef corals. *Limnol. Oceanogr.* 42:1505–1516.

Pandolfi, J.M., and P.R. Minchin. 1995. A comparison of taxonomic composition and diversity between reef coral life and death assemblages in Madang Lagoon, Papua New Guinea. *Palaeoceanogr. Palaeoclimatol. Palaeoecol.* 119:321–341.

Pauly, D. 1995. Anecdotes and the shifting baseline syndrome of fisheries. *Trends Ecol. Evol.* 10:430.

Perry, C.T. 1996. The rapid response of reef sediments to changes in community composition: Implications for time averaging and sediment accumulation. *J. Sediment. Res.* 66:459–467.

Perry, C.T. 2001. Storm-induced coral rubble deposition: Pleistocene records of natural reef disturbance and community response. *Coral Reefs* 20:171–183.

Porter, J.W., and O.W. Meier. 1992. Quantification of loss and change in Floridian reef coral populations. *Am. Zool.* 32:625–640.

Precht, W.F., and R.B. Aronson. 1995. Disease induced community change on a coral reef; first time in 3,400 years. In *Linked Earth Systems: Congress Program and Abstracts; Volume 1,* eds. A.C. Hine and R.B. Halley, 101. Tulsa: Society of Economic Paleontologists and Mineralogists.

Precht, W.B., and R.B. Aronson. 1997. Compositional changes in reef sediments related to changes in coral reef community structure. *AAPG Bull.* 81:1561.

Rothfus, T.A., and B.J. Greenstein. 2001. Taphonomic evidence for Late Pleistocene transitions in coral reef community composition, San Salvador, Bahamas. In *Proceedings of the Tenth Symposium on the Geology of the Bahamas and other Carbonate Regions,* eds. B.J. Greenstein and C.K. Carney, 152–162. San Salvador: Gerace Research Center.

Schäfer, W. 1972. *Ecology and Paleoecology of Marine Environments.* Chicago: University of Chicago Press.

Scoffin, T.P. 1992. Taphonomy of coral reefs: A review. *Coral Reefs* 11:57–77.

Sheppard, C. 1995. The shifting baseline syndrome. *Mar. Pollut. Bull.* 30:766–767.

Smith, S.R., and J.C. Ogden (eds). 1994. Status and recent history of coral reefs at the CARICOMP network of Caribbean marine laboratories. In *Proceedings of the Colloquium on Global Aspects of Coral Reefs: Health, Hazards, and History,* 1993, ed. R.N. Ginsburg, 43–49. Miami: Rosenstiel School of Marine and Atmospheric Science, University of Miami.

Walbran, P.W., R.A. Henderson, J.W. Faithful, H.A. Polach, R.J. Sparks, G. Wallace, and D.C. Lowe. 1989a. Crown-of-thorns starfish outbreaks on the Great Barrier Reef: A geological perspective based upon the sediment record. *Coral Reefs* 8:67–78.

Walbran, P.W., R.A. Henderson, A.J.T. Jull, and J.M. Head. 1989b. Evidence from sediments of long-term *Acanthaster planci* predation on corals of the Great Barrier Reef. *Science* 245:847–850.

Wapnick, C.M., W.F. Precht, and R.G. Aronson. 2004. Millennial-scale dynamics of staghorn coral in Discovery Bay, Jamaica. *Ecol. Lett.* 7:354–361.

Wilkinson, C.R. 1993. Coral reefs of the world are facing widespread devastation: Can we prevent this through sustainable management practices? *Proc. Seventh Int. Coral Reef Symp., Guam* 1:11–21.

Wilkinson, C.R. 2000. Executive summary. In *Status of the Coral Reefs of the World: 2000*, ed. C.R. Wilkinson, 7–20. Townsville: Australian Institute of Marine Science.

Wilkinson, C.R., and I.G. Macintyre. 1992. Preface to *Acanthaster* theme issue. *Coral Reefs* 11:51–52.

Woodley, J.D., E.A. Chornesky, P.A. Clifford, J.B.C. Jackson, L.S. Hoffman, N. Knowlton, J.C. Lang, M.P. Pearson, J.W. Porter, M.L. Rooney, K.W. Rylaarsdam, B.J. Tunnicliffe, C.M. Wahle, J.L. Wulff, A.S.G. Curtis, M.D. Dallmeyer, B.P. Jupp, A.R. Koehl, J. Neigel, and E.M. Sides. 1981. Hurricane Allen's impact on Jamaican coral reefs. *Science* 214:749–755.

3.   Biotic Turnover Events on Coral Reefs: A Probabilistic Approach

Richard B. Aronson and Stephen P. Ellner

3.1 Introduction

Coral reefs of the western Atlantic and Caribbean region have changed dramatically since the 1970s. Stands of branching and massive corals have given way to fields of coral rubble covered with *Lobophora*, *Dictyota*, and other seaweeds. Fleshy and filamentous macroalgae have replaced scleractinian corals, algal turfs, and coralline algae as the primary occupants of open reef surfaces at most localities (Aronson and Precht 2001a; Gardner et al. 2003). Attempts to explain this transition have focused on reduced herbivory, increased nutrient input, or combinations of top-down and bottom-up control (Littler and Littler 1985; Hughes 1994; Steneck 1994; McCook 1996; Lapointe 1997; Aronson and Precht 2000; Lirman 2001; Smith, Smith, and Hunter 2001; and many others). Regardless of the trophic mechanism or mechanisms involved, however, coral mortality has played a central role in the shift to macroalgal dominance, because the death of corals (and other sessile benthos) provides space for macroalgae to settle (Adey et al. 1977; Done 1992; Ostrander et al. 2000; McCook, Jompa, and Diaz-Pulido 2001; Williams and Polunin 2001; Williams, Polunin, and Hendrick 2001).

Two species of acroporid corals, *Acropora palmata* (elkhorn coral) and *Ac. cervicornis* (staghorn coral), were ecological dominants and important framework builders in the shallow (0–5 m) and intermediate (5–25 m) depths, respectively, of wave-exposed fore reefs in Florida, the Bahamas, and the Caribbean (Goreau and Goreau 1973; Graus and Macintyre 1989). *Ac. cervicornis* ranged into

61

shallower habitats on more protected fore reefs, and it was also found in back-reef and lagoonal habitats (Geister 1977; Rützler and Macintyre 1982; Hubbard 1988). Both species of *Acropora* have experienced high levels of mortality over the last several decades, and this mortality has eliminated the zonation patterns that were considered typical of Caribbean reefs (Jackson 1991; Hughes 1994). A number of authors have advanced the idea that high coral cover is the normal condition of Caribbean reefs and that the current, macroalga-dominated situation reflects a novel combination of natural and anthropogenic disturbances and stresses (Jackson 1992; Greenstein, Curran, and Pandolfi 1998; Nyström, Folke, and Moberg 2000; see Abram et al. 2003 for an example from Indonesia).

Testing the hypothesis of a recent and unprecedented change in dominance of Caribbean reefs has proven difficult. Fleshy and filamentous macroalgae generally do not preserve in fossil reef deposits, so their absence from the record cannot by itself be construed as having ecological meaning. On the other hand, thick Holocene accumulations of well-preserved *Acropora* skeletons, which are only lightly abraded, encrusted, and bored, support the notion that the current situation is unusual. These deposits, which date to 9000 ybp and younger, suggest rapid and generally continuous upward growth, deposition, and burial of coral skeletons with little time for algal colonization at the sediment–water interface (Macintyre, Burke, and Stuckenrath 1977; Westphall 1986; Stemann and Johnson 1995; Wapnick, Precht, and Aronson 2004; Hubbard et al. 2005).

The most significant sources of mortality for *Acropora* populations over the past 30 years have been white-band disease (WBD) and hurricane damage. WBD refers to at least two syndromes that appear to be specific to the genus *Acropora* (Santavy and Peters 1997; Richardson 1998). Hurricanes have repeatedly caused catastrophic damage at a number of localities, but this damage is limited to discrete areas around the storm tracks and is patchy on spatial scales ranging from kilometers along a coastline to thousands of kilometers across the Caribbean region (Hubbard et al. 1991; Rogers 1993). Some areas, such as Panama, are rarely or never affected by hurricanes on a centennial time scale, whereas others, such as Jamaica, are affected many times each century (Neumann et al. 1987; Lugo, Rogers, and Nixon 2000). The effects of WBD, in contrast, have been regional in scope (Rogers 1985). Epizootics of WBD have destroyed *Acropora* populations on reefs all over the western Atlantic, to the extent that the disease has been the most important cause of *Acropora* mortality in recent times (Aronson and Precht 2001b). Since *Acropora* spp. were dominants on many reefs, their decline substantially reduced coral cover. White pox, which is another emergent disease (Patterson et al. 2002); predation by corallivorous snails (Miller 2001; Baums, Miller, and Szmant 2003); and other factors have further reduced the abundance of *Acropora* on Caribbean reefs.

Aronson and Precht (1997) observed that reefs in the central shelf lagoon of the Belizean barrier reef system responded differently to the mortality of *Acropora* than did most Caribbean reefs. The mass mortality of *Ac. cervicornis* from WBD, which began after 1986 in the central lagoon, was followed by a dramatic and opportunistic increase in the cover of another coral species: the lettuce coral *Agaricia tenuifolia*. *Ag. tenuifolia* came to dominate these reefs because grazing by an abundant echinoid, *Echinometra viridis*, controlled macroalgal cover and biomass.

The transition from *Ac. cervicornis* to *Ag. tenuifolia* after 1986 was recorded in the Holocene reef sediments of the central lagoon, including the well-studied reef at Channel Cay (16°38′N, 88°10′W; Fig. 3.1), as an uppermost layer of skeletal plates of *Agaricia*. Cores extracted from Channel Cay were composed almost entirely of *Ac. cervicornis* in good taphonomic condition; the cores showed no evidence of a transition to *Ag. tenuifolia* in at least the last 3000 years, until the recent

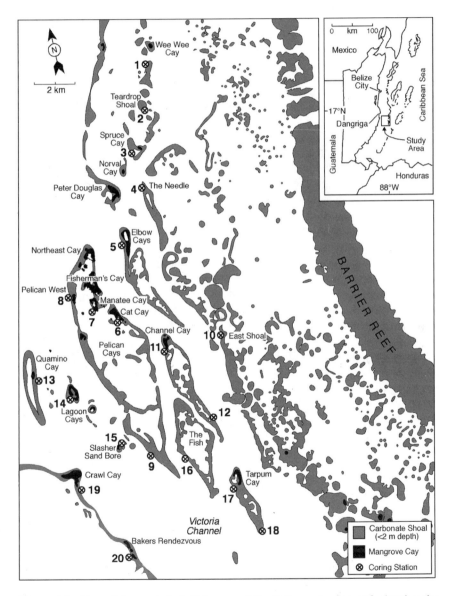

Figure 3.1. Map of the central shelf lagoon of the Belizean barrier reef, showing the rhomboid shoals and the 20 coring stations used in the analysis.

episode. This result suggested that the recent turnover event was a novel occurrence in the late Holocene history of Channel Cay (Aronson and Precht 1997).

The generality of Aronson and Precht's (1997) conclusion was limited by the small spatial scale of their study. Furthermore, their inferences were based on negative evidence. The absence of layers in the cores indicating an *Acropora*-to-*Agaricia* turnover could have meant that those types of events did not occur in the past. Alternatively, it is possible that such events had occurred, but that the signature layers had been lost from the fossil record through transport and other taphonomic processes.

Aronson et al. (2002a) increased the spatial scale by extracting cores over a 375-km^2 area of the central lagoon. The expanded study also diminished the problem of negative evidence, because Aronson et al. (2002a) detected discrete layers indicating turnover events, against the background of *Ac. cervicornis*, well below the surface layers. Radiocarbon dating placed these anomalous event layers in the hundreds to thousands of years before present.

The 95% confidence intervals of the dates of event layers in some of the cores fell within the 95% confidence intervals of event layers in other cores. These layers thus raised a question of statistical interpretation: How many contemporaneous or nearly contemporaneous layers are required from a set of cores to conclude that they represent large-scale species turnover, as opposed to representing the coincidence of multiple, local turnover events? In other words, what are the upper bounds of taphonomic information loss (Kidwell 2001)?

In this paper we develop probabilistic models to evaluate whether approximately contemporaneous event layers in a sample of cores are connected and reflect a large-scale turnover event, or whether they are coincidental. First, a binomial sampling model is derived and parameterized to test the hypothesis that the contemporaneous layers reflect a large-scale event that also occurred at the other core locations but was not detected due to taphonomic information loss. This model sets bounds on the detectability of large-scale events in the fossil record and can be used to determine the sample sizes necessary to detect those events with statistical rigor. Second, a Poisson point-process model is derived to test the hypothesis that the contemporaneous layers are coincidental. This model allows us to compute the number of contemporaneous layers (within the resolution of the radiocarbon dating) that would be expected to occur by chance in the absence of any large-scale trends in species abundance.

3.2 Geology and Ecology of the Rhomboid Shoals

The central sector of the shelf lagoon of the Belizean barrier reef system is characterized by numerous atoll-like "rhomboid shoals" (Fig. 3.1). The geometries of these diamond-shaped reef complexes are due to their underlying, antecedent topographies, which are probably the result of reef growth around the edges of fault blocks (Purdy 1974a,b; James and Ginsburg 1979; Lara 1993; Precht 1997; Esker, Eberli, and McNeill 1998). The narrow, steep-sided reefs reach sea level and surround sediment-filled basins (Macintyre and Aronson 1997; Macintyre, Precht, and Aronson 2000). The Holocene deposits, which are up to 20 m thick, accreted over the past 8000 to 9000 years (Westphall and Ginsburg 1984; Westphall 1986).

As a result of low-energy conditions in the central lagoon, there is little or no submarine cementation on the rhomboid shoals (James and Macintyre 1985; Purser and Schroeder 1986; Macintyre and Marshall 1988; Macintyre and Aronson 2006). The corals are stabilized by the interlocking of their skeletons (Shinn et al. 1979; Westphall 1986), and space is normally opened when coral colonies grow to the point of oversteepening and roll downslope. Debris fans at the bases of the rhomboid shoals (22–30 m water depth) suggest occasional storm disturbance, but Hurricanes Greta (September 1978), Mitch (October 1998), and Keith (October 2000) had only minor effects on the living communities and the Holocene sediments of these reefs (Westphall 1986; Aronson et al. 2000; R.B. Aronson and W.F. Precht unpublished data). The effects of Hurricane Iris (October 2001) are discussed in Section 3.5.

Before the mid-1980s, the living reef communities of the rhomboid shoals were dominated by *Acropora cervicornis* from 2- to 15-m depth (Shinn et al. 1979; Westphall 1986). *Agaricia tenuifolia* and other agariciids were minor components of the community. During the 1980s, WBD essentially extirpated the *Ac. cervicornis* populations on the rhomboid shoals (Aronson and Precht 1997). The dead *Ac. cervicornis* colonies collapsed from the weakening effects of bioerosion, and most macroalgae that colonized the skeletons were consumed by the *E. viridis*. Agariciids readily recruited to the grazed *Ac. cervicornis* rubble and the cover of *Agaricia* spp., especially *Ag. tenuifolia*, increased dramatically. Abnormally high water temperatures in the summer and fall of 1998 caused almost all coral colonies in the central lagoon to expel their zooxanthellae and bleach. By January 2000, the *Ag. tenuifolia* and other coral populations had died catastrophically, dropping to near-zero cover (Aronson et al. 2000). A few small agariciid colonies survived in deeper water (≥15 m), but there were no signs of increasing coral cover as late as April 2004. Skeletal rubble of *Ag. tenuifolia* came to dominate the surface sediments in the wake of the mass coral mortality.

3.3 The Cores

Push-cores, extracted by hand from 5 to 9 m water depth in 1995 to 2000, were used to reconstruct the history of species turnover on the rhomboid shoals during the late Holocene (Aronson et al. 2002a). For each core, divers forced a 4- to 5-m long segment of 7.6-cm (3-in) diameter aluminum tubing into the reef. The tube was capped, sealed, and pulled from the reef. It was then transported to the Smithsonian Institution's field station at nearby Carrie Bow Cay for extrusion and analysis. The coring methodology is described in detail by Dardeau et al. (2000).

The recent shift in dominance from *Acropora cervicornis* to *Agaricia tenuifolia* left a signature in the fossil record. The cores contained an uppermost layer of

Ag. tenuifolia plates overlying a thick accumulation of *Ac. cervicornis*. *Ac. cervicornis* rubble in a thin layer just beneath the *Ag. tenuifolia* was abraded, encrusted, and bored by clionid sponges and other invertebrates, indicating relatively long exposure at the sediment–water interface (Greenstein and Moffat 1996; Greenstein, Chapter 2). This layer was deposited following the mortality of *Ac. cervicornis* in the WBD epizootic, and the abrasion was caused by the grazing activity of *E. viridis*. Many of the pieces of *Ac. cervicornis* were encrusted with juvenile *Ag. tenuifolia* or fragments of larger *Ag. tenuifolia* colonies, corresponding to our observation that *Ag. tenuifolia* recruited to the *Ac. cervicornis* rubble after 1986. Neither this layer of taphonomically degraded *Ac. cervicornis* rubble nor the overlying layer of *Ag. tenuifolia* plates was observed at the tops of cores extracted from Channel Cay by other investigators prior to 1986 (Shinn et al. 1979; Westphall 1986). Below the layer of degraded *Acropora*, the subsurface Holocene consisted primarily of fresh-looking *Ac. cervicornis* skeletons. This skeletal material was not extensively abraded, encrusted, or bored, indicating rapid deposition and burial under further accumulations of *Ac. cervicornis*.

Radiocarbon analysis of corals from the bottoms of the cores yielded variable ages, with a maximum uncorrected ^{14}C date of 3480 ± 70 years (^{14}C years ± 1 standard error, or SE). (We use uncorrected dates here for consistency with Aronson et al. (2002a).) There was no evidence of large-scale bioturbation. *E. viridis* spines were common throughout the cores, suggesting that herbivory was an important process for at least the last three to four millennia.

Eight cores in a sample of 36 contained anomalous layers, which departed from the background composition of dominance by *Ac. cervicornis* in good taphonomic condition. The anomalous layers indicated complete or incomplete species turnover during the previous 3000 years (Fig. 3.2, Table 3.1). Such a layer in a single core reflected a turnover event at the scale of the individual coral colony or coring station (square meters to tens of square meters), or at a larger scale, ranging up to the entire study area and beyond (hundreds of square kilometers or more).

Layers of *Ag. tenuifolia* plates within the *Ac. cervicornis*-dominated cores recorded temporary *Acropora*-to-*Agaricia* transitions at some spatial scale. Layers of eroded *Ac. cervicornis* branches encrusted with *Agaricia* recruits represented turnover events at some scale that did not reach completion before the growth of *Ac. cervicornis* resumed (Aronson et al. 2002a). Core 97-8 from Elbow Cay, for example, contained a layer of *Ag. tenuifolia* plates near the bottom. A coral sample from just beneath this layer, representing the time of transition to *Ag. tenuifolia*, dated to 1500 ± 40 years. Another core, 97-12 from Bakers Rendezvous, contained a layer of degraded *Ac. cervicornis* branches, some of which were encrusted with *Agaricia* recruits. This layer dated to 380 ± 60 years.

In a number of cases the 95% confidence intervals of anomalous layers in different cores overlapped (95% confidence intervals calculated as $\pm 1.96 \times$ SE; Table 3.1). We compared the measured dates of the anomalous layers pairwise to determine which dates were significantly different and which were not. We used

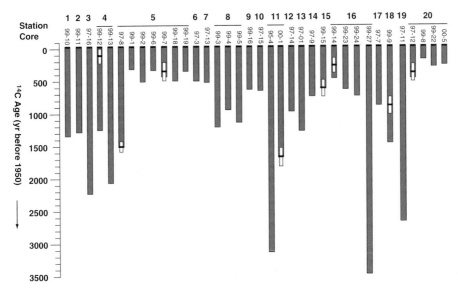

Figure 3.2. Age distribution of the cores. Gray shading represents ecological dominance by *Acropora cervicornis*. Horizontal black bars indicate complete or incomplete *Acropora-to-Agaricia* transitions. White bars represent the 95% confidence intervals about the radiocarbon dates for the pre-1986 events, calculated as ±1.96 standard errors (SE). For clarity of presentation, error bars are omitted from the bottom dates of the cores. Radiometric age determinations were performed by Beta Analytic, Inc., Miami, Florida. Dates are reported as uncorrected radiocarbon years before 1950, using a Libby half-life of 5568 years and a modern standard based on 95% of the activity of the National Bureau of Standards' oxalic acid. Modified from Aronson et al. (2002a).

Table 3.1. Records of anomalous layers in the cores. Uncorrected radiocarbon dates, reported as estimated ages before 1950 ± SE, are for coral samples from the bases of the layers. The 95% confidence intervals are calculated as ±1.96×SE, rounded to the nearest 10 years. Types of layers: Ag, plates of *Agaricia tenuifolia*; Rec, eroded branches of *Acropora cervicornis* encrusted with *Agaricia* recruits. Data are from Fig. 3.2 and Aronson et al. (2002a)

Core	Type of layer	Date of anomalous layer	
		^{14}C age	95% confidence interval
97-8	Ag	1500 ± 40	1420–1580
97-12	Rec	380 ± 60	260–500
99-7	Ag	340 ± 70	200–480
99-9	Rec	890 ± 60	770–1110
99-12	Rec	110 ± 60	0–230
99-14	Ag	270 ± 60	150–390
99-15	Ag	610 ± 60	490–730
00-1	Ag	1670 ± 70	1530–1820

the standard error of each date to compute the standard error of the difference, $\{SE_{diff} = [(SE_1)^2 + (SE_2)^2]^{0.5}\}$. Pairs of anomalous layers were considered significantly different at the 0.05 level if the difference in measured dates exceeded $1.96 \times SE_{diff}$. By this criterion, the anomalous layers found in four of the cores were temporally unique: the 1500-year layer in core 97-8, the 890-year layer in 99-9, the 610-year layer in 99-15, and the 1670-year layer in 00-1. The four remaining layers formed two sets of three overlapping layers: (1) 380 ± 60 years in 97-12, 340 ± 70 years in 99-7, and 270 ± 60 years in 99-14 and (2) 340 ± 70 years in 99-7, 270 ± 60 years in 99-14, and 110 ± 60 years in 99-12.

The possibility that all four layers actually occurred at the same time can be rejected (see Appendix). Does each of these sets of three layers represent coincidence? Alternatively, is one of these sets of three layers the signal of a large-scale turnover event? Either of these sets of three, but not both, could represent a large-scale event. Here we analyze the first case: Did a large-scale turnover event occur at 270 to 380 years? The reasoning for the second case (a putative event at 110 to 340 years) is the same.

3.4 A Hierarchical Sampling Design

Pandolfi (1996), Greenstein and Pandolfi (1997), Pandolfi and Jackson (2001, Chapter 8), Greenstein (Chapter 2), and others have applied the univariate and multivariate statistical approaches used in community ecology to the paleoecology of subaerially exposed Pleistocene and Holocene coral reefs, greatly increasing the confidence of their conclusions. The keys to ecological-style analysis of pattern in the fossil record are (1) observing phenomena at multiple scales and (2) obtaining sample sizes large enough to yield sufficient power. Complex logistics usually constrain reef-coring studies to low sample sizes; therefore, investigators cannot, in general, associate statistical probabilities (P-values) with their reconstructions of paleocommunity dynamics. Rotary drilling and vibracoring, for example, require barges, tripods, and other hardware (e.g., Westphall 1986; Hubbard, Gladfelter, and Bythell 1994; Macintyre, Littler, and Littler 1995). These methods are necessary for coring cemented reefs and tightly packed peats and muds, respectively, but because large sample sizes generally are not attainable the resulting ecological interpretations must in the end remain primarily qualitative.

In contrast, the diver-operated push-coring method used in this study of uncemented reef frameworks is logistically simple enough and sufficiently transportable to satisfy both requirements of ecological-style studies (Dardeau et al. 2000; Aronson et al. 2004). Although penetration is limited to 3 to 4 m, push-coring is adequate for multiscale, statistically rigorous sampling of the last several millennia of uncemented reef deposits. Percussion coring (Halley et al. 1977) and vibracoring (Westphall 1986) have achieved deeper penetrations in the central shelf lagoon of Belize, but again the logistics are more complex.

The statistical design of this study was analogous to a one-way analysis of variance (ANOVA). It involved sampling at two hierarchical levels: a higher, "groups" level and a lower, "error" level. The groups were coring stations, and the error level consisted of multiple cores taken within stations, meters to tens of meters apart. In the study in Belize, 20 viable coring stations were spread over a 375-km² area of the central lagoon, separated by hundreds of meters to kilometers (Fig. 3.1).

Distributing the coring stations over the entire study area provided the spatial coverage necessary to draw general conclusions about the among-station variability of paleocommunity dynamics in the area. Even so, it was important to ensure that the stations could legitimately be treated as independent replicates in the analysis. One approach would have been to locate each station on a separate reef or shoal, but assessing the geomorphological discreteness of reefs can be problematic. A map of the study area in Belize (Fig. 3.1) highlights this ambiguity: Wee Wee, Spruce, Norval, and Peter Douglas Cays are separated at present, but they are clearly components of a large rhomboid shoal. The solution was to recognize that statistical independence in an ecological/paleoecological sense is different from independence in a geomorphological context. Considering the dispersal abilities of the planula larvae of corals (reviewed in Aronson and Precht 2001a) and the reefwide to regional scales of WBD outbreaks, bleaching episodes, and other disturbances and stresses, there is no reason to expect that coring stations established on the same rhomboid shoal, separated by hundreds of meters to kilometers, were ecologically or paleoecologically more similar to each other than they were to stations established on other shoals, which were in many cases situated at closer distances. Separate stations on the same shoal were, therefore, considered replicates at the scale of the study area.

3.5 Testing the Hypothesis of an Areawide Event

Our goal in this section is to test the hypothesis that the set of three overlapping anomalous layers (at 270 to 380 years) constitute the signal of an areawide event, which was not observed elsewhere due to taphonomic information loss. The analysis works from a tentative acceptance of this hypothesis: We assume for the sake of argument that these layers are evidence of events that occurred at the scale of the study area. As a corollary, we assume that the absence of contemporaneous event layers in other cores at the same or other stations was because of information loss from taphonomic processes such as degradation and transport. Under the hypothesis that an areawide event actually affected all the cores, can taphonomic processes account for the large number of cores in which no anomalous layer was observed in that range of ages?

To answer this question, we need an estimate of the within-station (i.e., per-core) failure rate, f, which is the probability of an areawide event layer failing to appear in any one particular core from a station. Aronson et al. (2002a) used the cores themselves to derive an estimate of $f = 0.75$. This value was then used to calculate the cumulative, one-tailed binomial probability of 3 or fewer successes out of 20 stations, where a success was defined as a case in which at least one core from the station displayed the 270- to 380-year layer. From the binomial calculation, Aronson et al. (2002a) rejected the hypothesis of an areawide event at $P = 0.044$.

Strictly speaking, if we tentatively accepted the hypothesis of an areawide event and its corollary, then f should have been calculated as the fraction of cores without an anomalous layer. The resulting high value of f, however, would have raised the type II error rate to the point of rendering the binomial test essentially powerless. Furthermore, sampling a larger number of stations that did not contain the layer would have increased f, potentially raising the P-value. This would have been the opposite of our expectation, because a larger number of stations lacking the layer should have given us greater confidence that the hypothesis of an areawide event is false. Instead, Aronson et al. (2002a) averaged the stations at which one or more cores actually displayed the putative layer, which provided an estimate ($f = 0.75$) that was more realistic than the proportion of cores without the layer. The estimate of f was biased, however, because even if the event causing the layers had not been areawide in scale, some stations affected by the event would have been expected to show no evidence of it (so long as $f > 0$). Those stations also should have been included in the data used to estimate f, but we had no way of knowing which stations those were. Thus, using the cores themselves to estimate f was problematic but at the time there was no alternative. A recent hurricane in Belize provided more accurate estimates of f, which freed us from the tautological, biased estimates derived from the layers in the cores.

Hurricane Iris, a Category 4 storm, struck the central and southern barrier reef in October 2001. The eye of the storm passed 15 to 17 km to the south of Channel Cay and hurricane-force winds extended approximately 30 km from the eye, with the strongest, northeastern quadrant passing over the study area. Two additional storms affected Belize in the fall of 2001: Hurricane Michelle, which tracked to the east of the Belizean barrier reef in October, and a "norther," which blew from the northwest in early November. Both storms produced high winds and storm waves in central and southern Belize.

Sedimentological impacts on the rhomboid shoals varied with exposure to storm waves, and it was clear that the main effects resulted from the direct hit by Hurricane Iris. During Hurricane Iris, storm waves came from the east-northeast over the barrier reef crest and into the central lagoon. Observations by W.F. Precht in November 2001, after the norther had passed, suggested that some reworking, scour, and winnowing of sediment occurred on the windward-facing flank of the Channel Cay shoal (station 12) down to 8 to 9 m depth. On the leeward flank (station 11), sand and some coral heads were brought from the narrow platform at the top of the reef down to 7 to 8 m depth, forming a thin leeward drape at those depths (cf. Logan 1969). The latter effect was strongly attenuated in the lee of the islands that comprise Channel Cay itself (north of station 11). Within the protected confines of Cat Cay (station 6) no effects of the hurricane were detected at all. Despite the impacts of the hurricane, which varied among reef areas depending on exposure, the uppermost layer of *Agaricia tenuifolia* remained largely coherent at the depths from which the cores had been extracted. The sedimentological effects of Hurricane Iris were used to estimate the extent of information loss resulting from a large-scale disturbance event.

The cores in this study were taken at water depths ranging from 4.2 to 11.5 m, with a mean of 7.11 ± 1.79 SD m (Aronson et al. 2002a, Table 1). Point-count data recorded along permanent transects at Channel Cay in November 2001, one month after Hurricane Iris, were used to calculate the percent cover of various living and nonliving benthic components at the sediment–water interface at those depths. Since essentially all the *Acropora cervicornis* had been killed between 1986 and the early 1990s, the percent cover of *Ac. cervicornis* rubble provided an estimate of the proportion of the area over which the topmost layer of *Ag. tenuifolia* plates was lost in the time since *Agaricia* became dominant, plus the proportion of the area over which it never grew over the *Acropora*. In November 2001, the cover of *Ac. cervicornis* rubble on the windward (eastern) side of Channel Cay declined with depth: it was 25.0 to 26.5% at 3 to 6 m depth, 16.4 to 18.4% at 6 to 9 m, 0.0% at 9 to 12 m, and 0.0% at 12 to 15 m. The lower values were from station 12 and the higher values were from an area at the northeastern portion of the cay. At station 11 on the leeward side of Channel Cay, the maximum cover of *Ac. cervicornis* rubble in November 2001 was 7.8% at 3 to 6 m depth, dropping to 2.0 to 4.3% at greater depths down to 15 m. Corals and coral rubble other than *Ag. tenuifolia* and *Ac. cervicornis* accounted for a maximum of 5% cover in these posthurricane surveys. A core extracted from a random position on the exposed side of Channel Cay in November 2001 would thus have been at most 32% likely to miss the uppermost *Agaricia* layer. The cover of *Ac. cervicornis* rubble was either lower or not appreciably different at these stations/depths in July 2002. To summarize, following a decade of *Ag. tenuifolia* growth, termination by the 1998 bleaching event, and the passage of Hurricane Iris the loss of signal was minimal even at the most exposed sites.

The conservatively estimated failure rate of 0.32 is considerably lower than the 0.75 failure estimate of Aronson et al. (2002a), implying an even stronger rejection of the hypothesis of an areawide event. Two Iris-type events would be expected to result in a maximum failure rate of $\{f = 0.32 + 0.32(1 - 0.32) = 0.54\}$. Even this figure is conservative for the exposed habitat from which it is derived. The rate of burial and stabilization of dead *Agaricia* skeletons on the rhomboid shoals and in other lagoonal environments of the western Caribbean is rapid, at approximately 10 cm of vertical accretion per decade (Aronson and Precht 1997; Aronson et al. 2004). Direct hits from intense hurricanes, which occur less frequently than once per decade, cannot be expected to cause significant information loss anywhere but at the shallowest depths at the most exposed sites (Aronson et al. 2004). We adopted $f = 0.54$ as a conservative estimate of the within-station failure rate.

The probability that no cores from a given station would have recorded the putative large-scale event is $q = f^c$, where c is the number of cores that were sampled at the station. Ideally, the number of cores extracted should have been equal at each station, so that c would have been constant across stations. As with large-scale ecological studies (e.g., Murdoch and Aronson 1999), however, a balanced design is often not possible. Even if the same number of cores are extracted at each station in a coring study, it is unlikely that they will all date to an age older than the age of a putative event. This virtually guarantees an unbalanced design for all but the youngest putative events (Aronson et al. 2004). In the Belizean study, 30 cores distributed among the 20 stations dated to $\geq$388 years (270 years + 1.96 × SE, where SE = 60 years, in core 99-14, the youngest of the three roughly contemporaneous layers under consideration). We calculated f^c for each station, where c is the number of cores at that station dating to at least 388 years. The average failure rate for the 20 stations is thus $q = 0.454$, and the average probability that at least one core at the average station will record the event is $p = 1.000 - 0.454 = 0.546$.

Testing the hypothesis of an areawide event requires calculating the probability of the observed number of stations or fewer exhibiting the supposed large-scale event in at least one core. This is the one-tailed, cumulative binomial probability of s or fewer successful preservations out of n stations:

$$P = \sum_{i=1}^{s} \left[\frac{n!}{i!(n-i)!} \right] p^i q^{(n-i)}$$

Because the model assumes $s > 0$, the binomial probabilities are summed over the range $\{i = 1 \text{ to } s\}$ rather than over the usual range $\{i = 0 \text{ to } s\}$. In practice, the probabilities corresponding to $i = 0$ were negligible and had no effect on the cumulative P-values.

The probability of observing an episode of large-scale turnover as an event layer at 3 or fewer stations out of 20, given the failure estimate q, is $P = 0.00032$. The hypothesis of an areawide turnover event, the signal of which was lost at 17 (or more) stations, is therefore rejected with far greater confidence than before. Instead, the analysis supports the null hypothesis that the three nearly contemporaneous layers are the coincidental result of local processes, leading to localized turnover events, at the stations in question.

The one-tailed, cumulative probability of 6 or fewer successes out of 20, with $q = 0.454$ and $p = 0.546$, is $P = 0.023$. The probability of 7 or fewer successes out of 20 is $P = 0.062$. In other words, using the binomial approach with $f = 0.54$ we would not have been able to falsify the hypothesis of an areawide event had 7 or more stations contained at least one core with the layer.

Recent coring work has shown that the information loss caused by Hurricane Iris is far lower than the estimates used above. Cores taken in April 2004 at stations 1, 4, 5, 6, 7, 8, 11, 12, 15, 17, 19, and 20 (Fig. 3.1) were compared to cores extracted from those 12 stations prior to Hurricane Iris. The uppermost *Ag. tenuifolia* layer survived at all coring stations except station 1 (Aronson, Macintyre, and Precht 2005). That station was the only one not in the lee of a continuous stretch of the Belizean barrier reef. Stabilization and burial of the recent *Ag. tenuifolia* layer were disrupted neither by the hurricane, nor by the cessation of production of *Agaricia* plates after the bleaching episode of 1998. The upshot of this new information is that if we restrict ourselves to the leeward stations, f is essentially zero. This means that the cores can be read literally and without recourse to probability analysis, boosting our confidence in the conclusion that contemporaneous layers are coincidental.

The effects of Hurricane Iris are consistent with previous observations. The impacts of hurricanes and tectonic events on reef systems are patchy at scales smaller than the 375-km^2 size of the study area (see Hubbard et al. (1991) on hurricane effects in St. Croix; Phillips and Bustin (1996) on an earthquake affecting a coastal lagoon in Panamá; Esker, Eberli, and McNeill (1998) on tectonics in the Belizean lagoon). Information loss, therefore, will be patchy in these types of situations. On the other hand, a sudden, temporary drop in sea level could have exposed the living communities subaerially at all stations, leading to massive erosion and information loss throughout the study area. This possibility is excluded, however, because sea level has been rising slowly over the past several millennia in Belize and has been nearly stable for the past 3000 to 4000 years (Macintyre, Littler, and Littler 1995; Toscano and Macintyre 2003). Combining these observations with the foregoing analysis, it is unrealistic to suppose that a few contemporaneous layers in cores scattered over a large area of the Belizean lagoon represent the severely attenuated signal of a large-scale event.

3.6 Testing the Null Hypothesis that Overlapping Layers Are Coincidental

Having now firmly rejected the hypothesis that the three roughly contemporaneous layers are evidence of a large-scale event encompassing the study area, we turn to the opposite hypothesis, which is that the observed degree of overlap is the coincidental result of localized events that happened to occur at more or less the same time and were preserved. How many stations with overlapping anomalous layers would be required to falsify the null hypothesis that they represent the coincidence of localized events? The whole-station failure estimates (q-values) were derived either by assuming the areawide hypothesis (Aronson et al. 2002a) or by considering the effects of processes that could account for a high rate of areawide taphonomic loss (this chapter). To avoid the circularity of falsifying the hypothesis of coincidence under the assumption that it is false, we used a different approach: looking at the observed temporal distribution of event layers in the cores and asking whether the dates of those layers overlap to a significantly greater degree than expected under the appropriate null hypothesis. The occurrence of three anomalous layers at similar and possibly identical times might indicate a "spike" in *Agaricia tenuifolia* abundance 270 to 380 years ago. We ask whether or not this cluster of layers provides evidence for temporal variability in community composition, against a statistical null hypothesis of perfectly constant *Agaricia* abundance. Note that in this analysis the alternative hypothesis is simply that the contemporaneous layers are more than a coincidence; they represent an ecological phenomenon at *some* spatial scale, although not necessarily an areawide event.

We conceptualize the data as a set of lines of variable duration running backwards in time from the present, along which occur eight events (the anomalous layers). Each event time is surrounded by a 95% confidence interval based on the standard error associated with its radiocarbon date. The statistical null hypothesis is that events occurred randomly in time and space; formally, the assignment of event times is a Poisson point process with constant intensity across time and cores. If the sampling intensity (the number of cores) had been constant over time, then this statistical null hypothesis would imply that the set of event times has a uniform distribution over the interval between the present and the earliest time sampled (Karlin and Taylor 1975). The null hypothesis could then be tested by comparing the observed event times to a uniform distribution. Because sampling intensity was not constant, however, we have to apply a correction to recover a uniform distribution. The correction is to associate events with the total amount of elapsed sampling time (EST) summed over all cores, starting from the earliest time sampled in any of the cores. This places all events onto a single "sampling time line," the length of which is the total duration of all the cores. Under the null hypothesis, the event ESTs on the sampling time line are a Poisson point process with constant intensity, because (1) each segment of the sampling time line represents a segment of time on one of the cores and (2) the null expectation is that all such segments are independent and have an equal chance of containing an event.

A variety of statistics could be used to test whether event ESTs are consistent with a uniform distribution. The simplest is the Kolmogorov–Smirnov (KS) one-sample test. However, the KS test is a general-purpose test intended to have some power against any departures from the null, and it is therefore not especially powerful for the alternative of interest here (clustering). In addition, the KS test involves the conventional assumption that the data are measured without error. We used Monte Carlo simulation to perform statistical tests that targeted event clusters and took into account the imprecision of event dating. We considered the following test statistics:

1. S = the number of pairs of event dates that could really be identical, based on overlapping 95% confidence intervals for the two dates.
2. $d2$ = the smallest difference between the ith and $(i+1)$th EST, $i = 1,2,\ldots,7$.
3. $d3$ = the smallest difference between the ith and $(i+2)$th EST, $i = 1,2,\ldots,6$.

The third of these, $d3$, specifically concerns the cluster of three events that prompted the analysis, which makes it particularly useful in our case for testing the null hypothesis.

The distribution of the test statistics under the null hypothesis was obtained by Monte Carlo simulation of event ESTs (i.e., by a parametric bootstrap) as follows:

1. Sets of eight simulated ESTs were generated as independent draws from a uniform distribution on $[0,T]$, where T is the total sampling duration of all cores.
2. The randomly generated data sets were completed by adding to the simulated ESTs a simulated observation error. In the real data, the standard error of EST was linearly related to the estimated EST ($r^2 = 0.98$ for linear regression of standard error on estimated EST), with estimated slope = 0.072. The simulated observation errors were Gaussian, with standard deviation equal to 0.072 times the simulated "true EST."

3. For 1000 sets of eight simulated ESTs, we calculated the value of the test statistics S, $d2$, and $d3$.

Point 2 deserves some clarification. There is a positive relationship between the estimated EST for a turnover event and the standard error of the EST for that event. This positive relationship results from the fact that the number of cores available for analysis increases over elapsed time. EST begins at the oldest date on any of the cores, so initially EST and real time are the same. Times closer to the present are sampled by more cores, and a single unit of real time that is sampled by m cores generates m units of EST. A roughly constant uncertainty in the real-time radiocarbon date of a turnover event, therefore, translates into greater uncertainty in the EST of the event as EST increases (i.e., the closer the tally of elapsed time comes to the present).

For all of the Monte Carlo tests, the real data are typical of data generated under the null hypothesis of constant *Agaricia* abundance (Fig. 3.3). The null hypothesis, therefore, cannot be rejected. The near-contemporaneity of the three layers appears to have been coincidental.

Under what circumstances could we have rejected the null hypothesis? Two patterns, which we did not observe in the data, could have led to rejection. The first

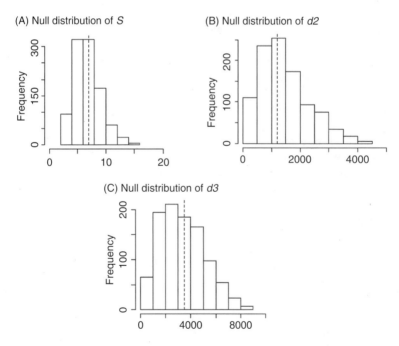

Figure 3.3. Histograms of the distributions of the test statistics S, $d2$, and $d3$ under the null hypothesis of a uniform distribution of anomalous layers, based on 1000 Monte Carlo replicates as described in the text. The dashed lines indicate the values of the test statistics derived from the real data.

is a larger cluster of anomalous layers with similar measured dates. A cluster of six layers with overlapping 95% confidence intervals would have yielded at least $S = 15$ overlapping pairs, which would have been above the 95th percentile of the null distribution of S (Fig. 3.3A). A single cluster of five layers would not have been sufficient for rejection of the null hypothesis (minimum $S = 10$). On the other hand, if the 95% confidence intervals of the oldest and youngest dates in a cluster of five layers each had overlapped the confidence intervals of another layer that lay outside the rest of the cluster, that would have yielded $S = 10 + 2 = 12$, which would have been outside the 95th percentile of the distribution of S. Two separate clusters of four layers also would have yielded $S = 12$ (minimum $S = 6$ for each cluster). Any of these patterns would have suggested an event at a scale larger than the coring station, although not necessarily at a scale as large as the study area.

The second pattern that could have led to rejection of the null hypothesis is if the three possibly contemporaneous layers had been even more similar in their measured dates than observed. Given the standard errors of the measured dates (Table 3.1), however, the null distribution of $d3$ (Fig. 3.3C) shows that a cluster of three dates much tighter than those observed would have to have been a lucky accident. That is, rejection of the null hypothesis due to an extremely tight cluster of three measured dates would almost certainly have been a type I error (spurious rejection of the null hypothesis).

These requirements for rejection of the null hypothesis raise the possibility that our failure to reject was a type II error; it could be argued that our analysis lacked power due to the small number of anomalous layers. Power becomes a concern when the data cast doubt on the null hypothesis, but statistical significance is not achieved due to low sample size or other limitations. That was not the case here: the observed distribution of dates of anomalous layers was in fact consistent with expectations under the null hypothesis.

Finally, uncorrected radiocarbon dates were used in the Monte Carlo simulations. Applying corrections to the dates could have affected the outcome. Correcting the dates for isotopic fractionation would have added 410 years to the uncorrected dates, assuming $\delta^{13}C = 0.0$ for marine carbonate. An increase in the magnitudes of all the dates would have decreased the sizes of the standard errors relative to EST. This could have increased the probability of three nearly contemporaneous events representing something more than a coincidence, meaning that there would have been a greater probability of falsifying the null hypothesis. The most realistic dates to use in the simulations, however, would have been dates calibrated to calendar years before present, not the conventional (isotopically corrected) dates. Calibration of the conventional dates to calendar years gives estimates within several decades of the uncorrected dates, but the errors associated with calendar-year calibrations are asymmetrical. Uncorrected dates, therefore, are more meaningful than conventional dates and statistically more tractable than calendar dates. Using conventional dates rather than uncorrected dates for the binomial tests in Section 3.5 would have had no effect on the results of that analysis.

3.7 Conclusion

The transition from *Acropora cervicornis* to *Agaricia tenuifolia* in the Belizean shelf lagoon after 1986 was clearly mediated by WBD. Other possibilities, including hurricanes, changes in sea level, and growth of the reefs to sea level, do not explain the observed patterns (Aronson and Precht 1997; Aronson, Precht, and Macintyre 1998; Aronson et al. 2002a). A probabilistic analysis of reef cores, which dated to several millennia before present and which were sampled in a spatially hierarchical design, supported the hypothesis that the recent areawide excursion from the *Acropora*-dominated situation was unprecedented in the lagoon. For the first time in more than 3000 years an episode of species turnover occurred on a scale of hundreds of square kilometers. Previous turnover events were localized to tens of square meters or smaller areas.

Can the coral-to-coral turnover in the Belizean lagoon be viewed as a preservable proxy for the regional mortality of *A. cervicornis* and its replacement by macroalgae? The coral-to-macroalgal replacement occurred throughout the Caribbean at approximately the same time, and the cause was WBD in most places (Aronson and Precht 2001a,b). Ecological observations on the fore reef at Discovery Bay, Jamaica, suggest that, analogous to turnover dynamics in Belize, the scale of transitions from coral to macroalgal dominance has increased greatly. In the 1970s, large stands of *Ac. cervicornis* at Discovery Bay were interrupted by small patches of algal growth, which were produced by the territorial activities of pomacentrid damselfish (Kaufman 1977). By the 1990s, however, coral cover was extremely low (<5%) on the fore reef at Discovery Bay, and most of the surfaces were covered by macroalgae (Aronson et al. 1994; Edmunds and Bruno 1996). The loss of coral cover principally involved mortality of *Ac. cervicornis* from hurricanes, WBD, and predation (Knowlton, Lang, and Keller 1990). The subsequent increase in macroalgal cover at Discovery Bay was exacerbated by the regional loss of the herbivorous sea urchin *Diadema antillarum* and local overfishing of herbivorous parrotfish (Labridae: Scarinae) and surgeonfish (Acanthuridae). The scale of turnover clearly increased at Discovery Bay, supporting the conclusion that this reef is currently in an unusual configuration, or alternative community state (Knowlton 1992). Whether or not the coral and macroalgal states are stable (Petraitis and Dudgeon 2004), a coring study has shown that the macroalgal state is unprecedented at Discovery Bay on a centennial to millennial scale (Wapnick, Precht, and Aronson 2004).

The scale of turnover has also increased regionally. *Acropora* populations formerly were quite volatile, fluctuating semi-independently in ecological time (e.g., Shinn et al. 1989). Now they have been drastically reduced throughout Florida, the Bahamas, and the Caribbean, and the short-term, short-distance fluctuations observed earlier are no longer possible.

Whether or not the post-*Acropora* situation in the Belizean lagoon represents or will represent an alternative community state is unknown at present, because the benthic assemblages are still dynamic in the wake of the 1998 bleaching event (Aronson et al. 2002b). The loss of *Acropora* spp. from reefs in Belize and elsewhere in the Caribbean could persist for decades (but see Idjadi et al. 2006). It is unclear how global climate change will interact with disease and other disturbances and stresses to influence recovery (Brown 1997; Harvell et al. 1999; Buddemeier, Kleypas, and Aronson 2004), but warming sea temperatures appear to be promoting the expansion of *Acropora* spp. northward along the east coast of the Florida Peninsula (Precht and Aronson 2004).

Studying the fossil record is our best hope for understanding the dynamics of coral reefs on the appropriate time scales (Jackson 1992; Greenstein, Curran, and Pandolfi 1998; Pandolfi and Jackson 2001). The quantitative approach to analyzing core data developed in this chapter makes it possible (1) to assign probabilities to hypotheses about temporal patterns in the scale of species turnover events and (2) to calculate the sample sizes (numbers of stations and numbers of cores per station) required for statistically powerful tests of those hypotheses. As an example of a testable hypothesis, Hubbard et al. (2005) suggested that the accumulation of *Ac. palmata* was interrupted regionally ~6000 and ~3000 ybp, possibly coinciding with abrupt climatic changes (Noren et al. 2002; Thompson et al. 2002). Large enough sample sizes will not be attainable in all situations. Nevertheless, we can at least gain some idea of the confidence with which we can view qualitative reconstructions from core data.

3.8 Summary

Populations of *Acropora palmata* and *Ac. cervicornis*, two of the most important framework-building species of western Atlantic and Caribbean reefs, have died throughout the region in recent decades, substantially reducing coral cover and providing substratum for algal growth. Although hurricanes have devastated some local populations of *Acropora* spp., white-band disease (WBD) has been the primary source of mortality. For example, *Ac. cervicornis* was the dominant ecological and geological constituent of reefs in the central shelf lagoon of the Belizean barrier reef until it was nearly extirpated by WBD after 1986. Another coral species, *Agaricia tenuifolia*, replaced *Ac. cervicornis* throughout an area of hundreds of square kilometers.

Cores were extracted by hand from the uncemented reefs of the central Belizean lagoon and radiocarbon dated to determine if any episodes of species turnover at a large spatial scale had occurred previously in the late Holocene (the last three millennia). Probabilistic approaches were developed to test hypotheses about the spatial extent of event layers appearing in the cores. The analyses showed that previous turnover events were confined to small areas of square meters to tens of square meters. Contemporaneous or nearly contemporaneous layers indicating turnover in the cores were the product of coincidence. The models are simple and powerful, and they can be used to determine the sample sizes necessary for rigorous tests of hypotheses pertaining to the long-term dynamics of coral reefs.

Acknowledgments. We thank Bill Precht for valuable discussion and for his field observations of the effects of Hurricane Iris. Ian Macintyre, Thad Murdoch, and

Cheryl Wapnick also provided helpful advice. They and Adam Gelber assisted with the field work that formed the basis of this analysis. Ron Etter and John Pandolfi provided valuable comments on the manuscript. Funding was provided by the National Science Foundation (grants OCE-9901969 and EAR-9902192), the Smithsonian Institution's Caribbean Coral Reef Ecosystems (CCRE) Program, the National Geographic Society (grant 7041-01), the University of South Alabama Research Council, and the Dauphin Island Sea Lab (DISL). The Monte Carlo simulations were performed at the U.S. National Center for Ecological Analysis and Synthesis (NCEAS) as part of the 2001 Working Group on Diseases in the Ocean. NCEAS is funded by the National Science Foundation (grant DEB-9421535), the University of California at Santa Barbara, and the State of California. This is DISL Contribution No. 346 and CCRE Contribution No. 652.

Appendix

We consider here the possibility that the four youngest anomalous layers all occurred simultaneously. If the SEs of all the dates had been similar, we could have used a Studentized Range statistic to identify which subsets of layer dates were significantly different. Because the SEs are not constant, however, we used a parametric bootstrap approach to compare the four measured dates in question with the null hypothesis that all four events were simultaneous. Using the SEs of the four measured dates, we simulated the null distribution of the difference between the earliest and latest measured dates. Based on 10,000 replicates, the actual difference (270 years) falls at the 98.7th percentile of the null distribution, so the null hypothesis that the four layers are contemporaneous is rejected at $P < 0.02$.

References

Abram, N.J., M.K. Gagan, M.T. McCulloch, J. Chappell, and W.S. Hantaro. 2003. Coral reef death during the 1997 Indian Ocean Dipole linked to Indonesian wildfires. *Science* 301:952–955.

Adey, W.H., P.J. Adey, R. Burke, and L. Kaufman. 1977. The Holocene reef systems of eastern Martinique, French West Indies. *Atoll Res. Bull.* 218:1–40.

Aronson, R.B., P.J. Edmunds, W.F. Precht, D.W. Swanson, and D.R. Levitan. 1994. Large-scale, long-term monitoring of Caribbean coral reefs: Simple, quick, inexpensive techniques. *Atoll Res. Bull.* 421:1–19.

Aronson, R.B., I.G. Macintyre, and W.F. Precht. 2005. Event preservation in lagoonal reef systems. *Geology* 33:717–720.

Aronson, R.B., I.G. Macintyre, W.F. Precht, T.J.T. Murdoch, and C.M. Wapnick. 2002a. The expanding scale of species turnover events on coral reefs in Belize. *Ecol. Monogr.* 72:233–249.

Aronson, R.B., I.G. Macintyre, C.M. Wapnick, and M.W. O'Neill. 2004. Phase shifts, alternative states, and the unprecedented convergence of two reef systems. *Ecology* 85:1876–1891.

Aronson, R.B., and W.F. Precht. 1997. Stasis, biological disturbance, and community structure of a Holocene coral reef. *Paleobiology* 23:326–346.

Aronson, R.B., and W.F. Precht. 2000. Herbivory and algal dynamics on the coral reef at Discovery Bay, Jamaica. *Limnol. Oceanogr.* 45:251–255.

Aronson, R.B., and W.F. Precht. 2001a. Evolutionary paleoecology of Caribbean coral reefs. In *Evolutionary Paleoecology: The Ecological Context of Macroevolutionary Change*, eds. W.D. Allmon and D.J. Bottjer, 171–233. New York: Columbia University Press.

Aronson, R.B., and W.F. Precht. 2001b. White-band disease and the changing face of Caribbean coral reefs. *Hydrobiologia* 460:25–38.

Aronson, R.B., W.F. Precht, and I.G. Macintyre. 1998. Extrinsic control of species replacement on a Holocene reef in Belize: The role of coral disease. *Coral Reefs* 17:223–230.

Aronson, R.B., W.F. Precht, I.G. Macintyre, and T.J.T. Murdoch. 2000. Coral bleach-out in Belize. *Nature* 405:36.

Aronson, R.B., W.F. Precht, M.A. Toscano, and K.H. Koltes. 2002b. The 1998 bleaching event and its aftermath on a coral reef in Belize. *Mar. Biol.* 141:435–447.

Baums, I.B., M.W. Miller, and A.M. Szmant. 2003. Ecology of a corallivorous gastropod, *Coralliophila abbreviata*, on two scleractinian hosts. I. Population structure of snails and corals. *Mar. Biol.* 142:1083–1091.

Brown, B.E. 1997. Disturbances to reefs in recent times. In *Life and Death of Coral Reefs*, ed. C. Birkeland, 354–379. New York: Chapman and Hall.

Buddemeier, R.W., J.A. Kleypas, and R.B. Aronson. 2004. *Coral Reefs and Global Climate Change: Potential Contributions of Climate Change to Stresses on Coral Reef Ecosystems*. Arlington, Virginia: Pew Center on Global Climate Change.

Dardeau, M.R., R.B. Aronson, W.F. Precht, and I.G. Macintyre. 2000. Use of a hand-operated, open-barrel corer to sample uncemented Holocene corals. In *Diving for Science in the 21st Century: 20th Annual Symposium of the American Academy of Underwater Sciences*, eds. P. Muller and L. French, 6–9. St. Petersburg: University of South Florida.

Done, T.J. 1992. Phase shifts in coral reef communities and their ecological significance. *Hydrobiologia* 247:121–132.

Edmunds, P.J., and J.F. Bruno. 1996. The importance of sampling scale in ecology: Kilometer-wide variation in coral reef communities. *Mar. Ecol. Prog. Ser.* 143:165–171.

Esker, D., G.P. Eberli, and D.F. McNeill. 1998. The structural and sedimentological controls on the reoccupation of Quaternary incised valleys, Belize southern lagoon. *AAPG Bull.* 82:2075–2109.

Fairbanks, R.G. 1989. A 17,000-year glacio-eustatic sea level record: Influence of glacial melting rates on the Younger Dryas event and deep-ocean circulation. *Nature* 342:637–642.

Gardner, T.A., I.M. Côté, J.A. Gill, A. Grant, and A.R. Watkinson. 2003. Long-term region-wide declines in Caribbean corals. *Science* 301:958–960.

Geister, J. 1977. The influence of wave exposure on the ecological zonation of Caribbean coral reefs. *Proc. Third Int. Coral Reef Symp., Miami* 1:23–29.

Goreau, T.F., and N.I. Goreau. 1973. The ecology of Jamaican coral reefs. II. Geomorphology, zonation and sedimentary phases. *Bull. Mar. Sci.* 23:399–464.

Graus, R.R., and I.G. Macintyre. 1989. The zonation of Caribbean coral reefs as controlled by wave and light energy input, bathymetric setting and reef morphology: Computer simulation experiments. *Coral Reefs* 8:9–18.

Greenstein, B.J., H.A. Curran, and J.M. Pandolfi. 1998. Shifting ecological baselines and the demise of *Acropora cervicornis* in the Western Atlantic and Caribbean Province: A Pleistocene perspective. *Coral Reefs* 17:249–261.

Greenstein, B.J., and H.A. Moffat. 1996. Comparative taphonomy of modern and Pleistocene corals, San Salvador, Bahamas. *Palaios* 11:57–63.

Greenstein, B.J., and J.M. Pandolfi. 1997. Preservation of community structure in modern reef coral life and death assemblages of the Florida Keys: Implications for the Quaternary record of coral reefs. *Bull. Mar. Sci.* 19:39–59.

Halley, R.B., E.A. Shinn, J.H. Hudson, and B. Lidz. 1977. Recent and relict topography of Boo Bee Patch Reef, Belize. *Proc. Third Int. Coral Reef Symp., Miami* 2:29–35.

Harvell, C.D., K. Kim, J.M.B. Burkholder, R.R. Colwell, P.R. Epstein, D.J. Grimes, E.E. Hofmann, E.K. Lipp, A.D.M.E. Osterhaus, R.M. Overstreet, J.W. Porter, G.W. Smith, and G.R. Vasta. 1999. Emerging marine diseases—Climate links and anthropogenic factors. *Science* 285:1505–1510.

Hubbard, D.K. 1988. Controls of modern and fossil reef development: Common ground for biological and geological research. *Proc. Sixth Int. Coral Reef Symp., Australia* 1:243–252.

Hubbard, D.K., E.A. Gladfelter, and J.C. Bythell. 1994. Comparison of biological and geological perspectives of coral-reef community structure at Buck Island, U.S. Virgin Islands. In *Proceedings of the Colloquium on Global Aspects of Coral Reefs: Health, Hazards, and History, 1993*, ed. R.N. Ginsburg, 201–07. Miami: Rosenstiel School of Marine and Atmospheric Science, University of Miami.

Hubbard, D.K., K.M. Parsons, J.C. Bythell, and N.D. Walker. 1991. The effects of Hurricane Hugo on the reefs and associated environments of St. Croix, U.S. Virgin Islands. *J. Coast. Res. Spec. Iss.* 8:33–48.

Hubbard, D.K., H. Zankl, I. van Heerden, and I.P. Gill. 2005. Holocene reef development along the northeastern St. Croix Shelf, Buck Island, U.S. Virgin Islands. *J. Sediment. Res.* 75:97–113.

Hughes, T.P. 1994. Catastrophes, phase shifts and large-scale degradation of a Caribbean coral reef. *Science* 265:1547–1551.

Idjadi, J.A., S.C. Lee, J.F. Bruno, W.F. Precht, L. Allen-Requa, and P.J. Edmunds. 2006. Rapid phase-shift reversal on a Jamaican coral reef. *Coral Reefs* 25:209–211.

Jackson, J.B.C. 1991. Adaptation and diversity of reef corals. *BioScience* 41:475–482.

Jackson, J.B.C. 1992. Pleistocene perspectives of coral reef community structure. *Am. Zool.* 32:719–731.

James, N.P., and R.N. Ginsburg. 1979. The geological setting of Belize reefs. In *The Seaward Margin of Belize Barrier and Atoll Reefs*, eds. N.P. James and R.N. Ginsburg, 1–14. International Association of Sedimentologists Special Publication 3. Oxford: Blackwell Scientific Publications.

James, N.P., and I.G. Macintyre. 1985. Carbonate depositional environments: Modern and ancient. Part I: Reefs: Zonation, depositional facies, and diagenesis. *Colo. Sch. Mines Q.* 80(3):1–70.

Karlin, S., and H.M. Taylor. 1975. *A First Course in Stochastic Processes,* Second Edition. New York: Academic Press.

Kaufman, L. 1977. The three spot damselfish: Effects on benthic biota of Caribbean coral reefs. *Proc. Third Int. Coral Reef Symp., Miami* 1:559–564.

Kidwell, S.M. 2001. Preservation of species abundance in marine death assemblages. *Science* 294:1091–1094.

Knowlton, N. 1992. Thresholds and multiple stable states in coral reef community dynamics. *Am. Zool.* 32:674–682.

Knowlton, N., J.C. Lang, and B.D. Keller. 1990. Case study of a natural population collapse: Post-hurricane predation on Jamaican staghorn corals. *Smithson. Contrib. Mar. Sci.* 31:1–25.

Lapointe, B.E. 1997. Nutrient thresholds for bottom-up control of macroalgal blooms on coral reefs in Jamaica and southeast Florida. *Limnol. Oceanogr.* 42:1119–1131.

Lara, M.E. 1993. Divergent wrench faulting in the Belize southern lagoon: Implications for Tertiary Caribbean Plate movements and Quaternary reef distribution. *AAPG Bull.* 77:1041–1063.

Lirman, D. 2001. Competition between macroalgae and corals: Effects of herbivore exclusion and increased algal biomass on coral survivorship and growth. *Coral Reefs* 19:392–399.

Littler, M.M., and D.S. Littler. 1985. Factors controlling relative dominance of primary producers on biotic reefs. *Proc. Fifth Int. Coral Reef Congr., Tahiti* 4:35–39.

Logan, B.W. 1969. Coral reefs and banks, Yucatán Shelf, Mexico (Yucatán Reef Unit). In *Carbonate Sediments and Reefs, Yucatán Shelf, Mexico,* eds. B.W. Logan, J.L. Harding, W.M. Ahr, J.D. Williams, and R.G. Snead, 129–198. AAPG Memoir 11. Tulsa: American Association of Petroleum Geologists.

Lugo, A.E., C.S. Rogers, and S.W. Nixon. 2000. Hurricanes, coral reefs and rainforests: Resistance, ruin and recovery in the Caribbean. *Ambio* 29:106–114.

Macintyre, I.G., and R.B. Aronson. 1997. Field guidebook to the reefs of Belize. *Proc. Eighth Int. Coral Reef Symp., Panama* 1:203–222.

Macintyre, I.G., and R.B. Aronson. 2006. Lithified and unlithified Mg-calcite precipitates in tropical reef environments. *J. Sediment. Res.* 76:81–90.

Macintyre, I.G., R.B. Burke, and R. Stuckenrath. 1977. Thickest recorded Holocene reef section, Isla Pérez core hold, Alacran Reef, Mexico. *Geology* 5:749–754.

Macintyre, I.G., M.M. Littler, and D.S. Littler. 1995. Holocene history of Tobacco Range, Belize, Central America. *Atoll Res. Bull.* 430:1–18.

Macintyre, I.G., and J.F. Marshall. 1988. Submarine lithification in coral reefs: Some facts and misconceptions. *Proc. Sixth Int. Coral Reef Symp., Australia* 1:263–272.

Macintyre, I.G., W.F. Precht, and R.B. Aronson. 2000. Origin of the Pelican Cays ponds, Belize. *Atoll Res. Bull.* 466:1–11.

McCook, L.J. 1996. Effects of herbivores and water quality on *Sargassum* distribution on the central Great Barrier Reef: Cross-shelf transplants. *Mar. Ecol. Prog. Ser.* 139:179–92.

McCook, L.J., J. Jompa, and G. Diaz-Pulido. 2001. Competition between corals and algae on coral reefs: A review of evidence and mechanisms. *Coral Reefs* 19:400–417.

Miller, M.W. 2001. Corallivorous snail removal: Evaluation of impact on *Acropora palmata. Coral Reefs* 19:293–295.

Murdoch, T.J.T., and R.B. Aronson. 1999. Scale-dependent spatial variability of coral assemblages along the Florida Reef Tract. *Coral Reefs* 18:341–351.

Neumann, C.J., B.R. Jarvinen, A.C. Pike, and J.D. Elms. 1987. *Tropical Cyclones of the North Atlantic Ocean, 1871–1986,* Third Revision. Asheville, NC: NOAA National Climatic Data Center.

Noren, A.J., P.R. Bierman, E.J. Steig, A. Lini, and J. Southon. 2002. Millennial-scale storminess variability in the northeastern United States during the Holocene epoch. *Nature* 419:821–824.

Nyström, M., C. Folke, and F. Moberg. 2000. Coral reef disturbance and resilience in a human-dominated environment. *Trends Ecol. Evol.* 15:413–417.

Ostrander, G.K., K.M. Armstrong, E.T. Knobbee, T. Gerace, and E.P. Scully. 2000. Rapid transition in the structure of a coral reef community: The effects of coral bleaching and physical disturbance. *Proc. Natl. Acad. Sci. USA* 97:5297–5302.

Pandolfi, J.M. 1996. Limited membership in Pleistocene reef coral assemblages from the Huon Peninsula, Papua New Guinea: Constancy during global change. *Paleobiology* 22:152–176.

Pandolfi, J., and J.B.C. Jackson. 2001. Community structure of Pleistocene coral reefs of Curaçao, Netherlands Antilles. *Ecol. Monogr.* 71:49–67.

Patterson, K.L., J.W. Porter, K.B. Ritchie, S.W. Polson, E. Mueller, E.C. Peters, D.L. Santavy, and G.W. Smith. 2002. The etiology of white pox, a lethal disease of the Caribbean elkhorn coral, *Acropora palmata. Proc. Natl. Acad. Sci. USA* 99:8725–8730.

Petraitis, P.S., and S.R. Dudgeon. 2004. Detection of alternative stable states in marine communities. *J. Exp. Mar. Biol. Ecol.* 300:343–371.

Phillips, S., and R.M. Bustin. 1996. Sedimentology of the Changuinola peat deposit: Organic and clastic sedimentary response to punctuated coastal subsidence. *Geol. Soc. Am. Bull.* 108:794–814.

Precht, W.F. 1997. Divergent wrench faulting in the Belize southern lagoon: Implications for Tertiary Caribbean Plate movements and Quaternary reef distribution: Discussion. *AAPG Bull.* 81:329–333.

Precht, W.F., and R.B. Aronson. 2004. Climate flickers and range shifts of reef corals. *Front. Ecol. Environ.* 2:307–314.

Purdy, E.G. 1974a. Karst-determined facies patterns in British Honduras: Holocene carbonate sedimentation model. *AAPG Bull.* 58:825–855.

Purdy, E.G. 1974b. Reef configurations: Cause and effect. In *Reefs in Time and Space*, ed. L.F. Laporte, 9–76. SEPM Special Publication 18. Tulsa: Society of Economic Paleontologists and Mineralogists.

Purser, B.H., and J.H. Schroeder. 1986. The diagenesis of reefs: A brief review of our present understanding. In *Reef Diagenesis*, eds. J.H. Schroeder and B.H. Purser, 424–446. Berlin: Springer-Verlag.

Richardson, L.L. 1998. Coral diseases: What is really known? *Trends Ecol. Evol.* 13:438–443.

Rogers, C.S. 1985. Degradation of Caribbean and western Atlantic coral reefs and decline of associated fisheries. *Proc. Fifth Int. Coral Reef Congr., Tahiti* 6:491–496.

Rogers, C.S. 1993. A matter of scale: Damage from Hurricane Hugo (1989) to U.S. Virgin Islands reefs at the colony, community, and whole reef level. *Proc. Seventh Int. Coral Reef Symp., Guam* 1:127–133.

Rützler, K., and I.G. Macintyre. 1982. The habitat distribution and community structure of the barrier reef complex at Carrie Bow Cay, Belize. In *The Atlantic Barrier Reef Ecosystem at Carrie Bow Cay, Belize, I. Structure and Communities*, eds. K. Rützler and I.G. Macintyre, 9–45. Smithsonian Contributions to the Marine Sciences 12. Washington, DC: Smithsonian Institution Press.

Santavy, D.L., and E.C. Peters. 1997. Microbial pests: Coral diseases in the western Atlantic. *Proc. Eighth Int. Coral Reef Symp., Panama* 1:607–612.

Shinn, E.A., R.B. Halley, J.H. Hudson, B. Lidz, and D.M. Robbin. 1979. Three-dimensional aspects of Belize patch reefs (abstract). *AAPG Bull.* 63:528.

Shinn, E.A., B.H. Lidz, R.B. Halley, J.H. Hudson, and J.L. Kindinger. 1989. *Reefs of Florida and the Dry Tortugas.* Field Trip Guidebook T176, 28th International Geological Congress. Washington, DC: American Geophysical Union.

Smith, J.E., C.M. Smith, and C.L. Hunter. 2001. An experimental analysis of the effects of herbivory and nutrient enrichment on benthic community dynamics on a Hawaiian reef. *Coral Reefs* 19: 332–342.

Stemann, T.A., and K.G. Johnson. 1995. Ecologic stability and spatial continuity in a Holocene reef, Lago Enriquillo, Dominican Republic (abstract). *Geol. Soc. Am. Ann. Mtg. Abstr. Prog.* 27:A166.

Steneck, R.S. 1994. Is herbivore loss more damaging to reefs than hurricanes? Case studies from two Caribbean reef systems (1978–1988). In *Proceedings of the Colloquium on Global Aspects of Coral Reefs: Health, Hazards and History, 1993*, ed. R.N. Ginsburg, 220–226. Miami: Rosenstiel School of Marine and Atmospheric Science, University of Miami.

Thompson, L.G., E. Mosley-Thompson, M.E. Davis, K.A. Henderson, H.H. Brecher, V.S. Zagorodnov, T.A. Mashiotta, P.-N. Lin, V.N. Nikhalenko, D.R. Hardy, and J. Beer. 2002. Kilimanjaro ice core records: Evidence of Holocene climate change in tropical Africa. *Science* 298:589–593.

Toscano, M.A., and I.G. Macintyre. 2003. Corrected western Atlantic sea-level curve for the last 11,000 years based on calibrated ^{14}C dates from *Acropora palmata* and intertidal mangrove peat. *Coral Reefs* 22:257–270.

Wapnick, C.M., W.F. Precht, and R.B. Aronson. 2004. Millennial-scale dynamics of staghorn coral in Discovery Bay, Jamaica. *Ecol. Lett.* 7:354–361.

Westphall, M.J. 1986. *Anatomy and History of a Ringed-Reef Complex, Belize, Central America* (M.S. thesis). Coral Gables: University of Miami.

Westphall, M.J., and R.N. Ginsburg. 1984. Substrate control and taphonomy of Recent (Holocene) lagoon reefs from Belize, Central America. In *Advances in Reef Science*, eds. P.W. Glynn, P.K. Swart, and A.M. Szmant-Froelich, 135–136. Miami: Rosenstiel School of Marine and Atmospheric Science, University of Miami.

Williams, I.D., and N.V.C. Polunin. 2001. Large-scale associations between macroalgal cover and grazer biomass on mid-depth reefs in the Caribbean. *Coral Reefs* 19:358–366.

Williams, I.D., N.V.C. Polunin, and V.J. Hendrick. 2001. Limits to grazing by herbivorous fishes and the impact of low coral cover on macroalgal abundance on a coral reef in Belize. *Mar. Ecol. Prog. Ser.* 222:187–196.

4. Inferring Past Outbreaks of the Crown-of-Thorns Seastar from Scar Patterns on Coral Heads

Lyndon M. DeVantier and Terence J. Done

4.1 The Crown-of-Thorns Phenomenon

Since the 1960s, dense aggregations representing population outbreaks of the coral-feeding crown-of-thorns seastar *Acanthaster planci* (L.) have caused extensive mortality of reef-building corals on many reefs throughout the Indo-Pacific region. The outbreaks, previously unknown to science (Chesher 1969), occurred more or less synchronously at widespeard locations across the Indo-Pacific (Moran 1986; Birkeland and Lucas 1990 for reviews). On Australia's Great Barrier Reef (GBR), the first reported outbreak occurred at Green Island Reef in 1962 and spread to many reefs of the central third of the GBR by the mid-1970s (Endean and Stablum 1975; Endean 1976). A second series of outbreaks occurred from 1979 to the late 1980s (Moran et al. 1988). Presently (in 2005) a third series of outbreaks is underway, having begun in the mid-1990s (Sweatman 1997; Sweatman et al. 1998). Since the 1960s, the outbreaks have recurred on many individual reefs after ~15 to 17 years. They have caused repeated losses of up to 90% of living coral cover and major short-term reductions in coral diversity, although there has been considerable variability in the intensity of predation, both within and among reefs (Done 1985, 1987; Moran, Bradbury, and Reichelt 1988).

The crown-of-thorns phenomenon has engendered more controversy and scientific debate than any other issue related to Indo-Pacific coral reefs (reviewed by Sapp 1999). The controversy initially centered on whether the outbreaks were natural events or were caused by human activities (Chesher 1969; Randall 1972;

Vine 1973; Endean 1976; Potts 1981). More recently, debate has focused on whether human activities have altered the frequency and intensity of outbreaks (Walbran et al. 1989a,b; Cameron et al. 1991a,b; Keesing et al. 1992).

Acanthaster planci is one of the most widely distributed of all Indo-Pacific reef species and is thought to have existed for more than a million years (Lucas, Nash, and Nishida 1985). However, there remains no unequivocal scientific evidence of major seastar outbreaks prior to the 1950s, even though considerable effort has been directed at finding such historical evidence (Dana 1970; Birkeland 1981; and see Moran 1986; Birkeland and Lucas 1990; Sapp 1999 for reviews). Various attempts have been made to resolve the controversy since the 1970s. Some studies have been based in ecological theory and the life-history traits of the seastar (Cameron 1977; Moore 1978; Cameron and Endean 1982). Others have focused on its metabolic requirements (Birkeland 1982; Olson 1987; Birkeland and Lucas 1990; Ayukai, Okaji, and Lucas 1997; Brodie et al. 2005) or on changes in predation pressure on the seastar (Endean and Stablum 1975; Endean 1976; McCallum, Endean, and Cameron 1989; Ormond et al. 1990; Sweatman 1996). Another approach has examined the taphonomic evidence of *Acanthaster* skeletal remains in reef sediments (Frankel 1975, 1977; Walbran et al. 1989a,b). Still other studies have examined the effects of seastar predation on coral communities, focusing on the population structures of massive (mound, boulder, and brain) corals (Done 1987, 1988; Done, Osborne, and Navin 1988; Endean, Cameron, and DeVantier 1988; Cameron, Endean, and DeVantier 1991a,b).

4.1.1 Theoretical Perspectives

Initially the debate focused on life-history theory. *A. planci* was considered an "r-selected" opportunistic species (Moore 1978), its large numbers of eggs and potentially high fecundity favoring large natural fluctuations in population sizes. In highlighting various life-history traits, Moore (1978) suggested that fluctuations in larval settlement could play a major role in determining adult population numbers and outbreaks. Alternatively, *A. planci* was considered to be "K-selected," a rare and specialized carnivore, whose populations were historically regulated by predation (Cameron 1977; Cameron and Endean 1982). From this viewpoint, the outbreaks arose following human interference in reef ecology, specifically the overcollecting of gastropod and piscine predators: the giant triton *Charonia tritonis* and lethrinid and balistid fishes (see also Ormond and Campbell 1974; McCallum 1987; Endean and Cameron 1990; Ormond et al. 1990).

From the perspective of disturbance theory, the outbreaks have been characterized as examples of intermediate disturbance of a type that maintains high levels of coral diversity on reefs. High coral diversity and even enhanced rates of coral speciation have been attributed to the effects of putative past outbreaks (Walbran et al. 1989a). Rates of speciation of reef corals, however, have been low during the likely evolutionary history of *A. planci* (Veron and Kelly 1988; Veron 1995). Furthermore, the effects of predation on coral species diversity are not consistent, but rather depend on complex interactions of feeding intensity, coral community composition, and prey preference and defenses (Glynn 1973, 1974, 1982; Colgan 1982, 1987; Done, Osborne, and Navin 1988; Endean and Cameron 1990; DeVantier 1995). The scale of disturbance caused by major seastar outbreaks is orders of magnitude greater than that usually considered in intermediate disturbance patch dynamics (Connell 1978). In short, theoretical approaches have not provided definitive answers to why outbreaks occur.

4.1.2 Metabolic Requirements of *Acanthaster*

During gametogenesis and larval development, *A. planci* exhibits particular temperature and salinity tolerances and food requirements (Lucas 1982; reviewed in Birkeland and Lucas 1990). High larval survivorship and subsequent population outbreaks have been linked with increased food abundance, enhanced by nutrient enrichment following terrestrial runoff (Birkeland 1982). Birkeland demonstrated a positive correlation between flood runoff from high Pacific islands following drought and subsequent seastar outbreaks. He emphasized that these conditions may occur naturally or may be enhanced or exacerbated by human activities—nutrient enrichment and deforestation—on the adjacent land. However, such correlations were less clear-cut in other regions with outbreaks, and larval rearing studies have also proven equivocal (compare, e.g., Olson 1987; Ayukai, Okaji, and Lucas 1997).

The outbreaks have spread annually throughout the central GBR over most of the past 40 years (1963–1979, 1981–1992, 1997–2005 and continuing), largely attributable to larval dispersal from parent populations on upstream "source" reefs to downstream "sink" reefs (Dight, James, and Bode 1988; Moran, Bradbury, and Reichelt 1988; Reichelt, Bradbury, and Moran 1990). This suggests that water quality and food availability are not currently limiting the survival of *A. planci* larvae in the region. The question remains: Were these factors ever limiting? River nutrient levels and chlorophyll levels in the central GBR lagoon are enriched significantly (two- to fourfold) above those present prior to the 20th century (Bell 1991; Bell and Elmetri 1995; Brodie 1995; Pulseford 1996; Brodie et al. 2001; Furnas and Mitchell 2001; Furnas 2003). This enrichment may have provided an environment more suited to survival of seastar larvae than that present prior to the agricultural, urban, and industrial development of the Queensland coast. Potential effects of changing water quality (lowered salinity and increased nutrient concentrations and chlorophyll levels) remain key questions and are important areas of current research, notably mathematical modeling of larval survivorship and outbreak frequency under different water quality conditions (K. Day, Queensland Department of Natural Resources, personal communication 2001, G. De'ath, Australian Institute of Marine Science, personal communication; Brodie et al. 2005).

4.1.3 Release from Predation Pressure

A variety of predators of juvenile, subadult, and adult *A. planci* have been identified, including mollusks, shrimps, worms, crabs, and reef fishes (reviewed by Moran 1986). Mathematical modeling suggests that predation can play a key role in regulating the abundance of *A. planci* (McCallum 1987; Ormond et al. 1990; see also Keesing and Halford 1992). Field studies have also demonstrated the importance of predation in influencing *A. planci* abundance in the Red Sea (Ormond and Campbell 1974) and the Eastern Pacific (Glynn 1982). The 1960s' outbreaks on the GBR followed a documented decline in commercial catches of lethrinid fish, some of which are known to prey on *A. planci* (Ormond et al. 1990). However, subsequent field studies on the GBR have proven equivocal (McCallum, Endean, and Cameron 1989; Sweatman 1996), and the importance of predation in regulating seastar populations on the GBR, particularly prior to increases in fishing pressure in the latter part of the 20th century, remains controversial.

4.1.4 Coral Population Dynamics

Several studies addressed the impact of the 1980s' outbreaks on population structures of massive corals, and assessed the sustainability of the coral populations if outbreaks continued. Many species of massive coral exhibit life-history characteristics of slow growth and low recruitment. Colonies of these species may be extremely long-lived—on the order of several centuries—persisting through disturbances that kill most other corals in their vicinity (Cameron and Endean 1985; Potts et al. 1985; DeVantier 1995). On the GBR massive corals are generally less preferred as prey by *A. planci* than corals of other growth forms, particularly branching and tabular colonies in the genus *Acropora* (Keesing and Lucas 1992; De'ath and Moran 1998).

On some reefs subjected to low to moderate levels of seastar predation during the 1980s' outbreaks, massive coral populations were little affected (DeVantier 1995). On other reefs, massive corals were heavily preyed upon (Done 1987, 1988; Endean, Cameron, and DeVantier 1988; Cameron, Endean, and DeVantier 1991a,b), especially where more preferred species were locally uncommon or had already been extirpated by predation. Such intense predation caused severe injury or death of entire ancient coral heads. Centuries-old massive corals became overgrown by fast-growing, more ephemeral corals and other taxa, leading to a fundamental change in coral community structure (Done and DeVantier 1990). Such observations shaped the view that assemblages of long-lived, massive corals could not have developed under past seastar outbreak regimes of similar frequency and intensity to those since the 1960s.

On yet other reefs, however, where the seastar had varying levels of impact on massive corals, modeling of coral population dynamics suggested that some coral populations could maintain themselves in the face of renewed outbreaks, while others would easily be driven to local extinction (Done 1987, 1988). These differences in vulnerability depended on the initial coral population structures and levels of damage, the abundance of different coral size (age) classes, the return frequency and intensity of the outbreaks, and model assumptions made about coral recruitment, growth, and survival between outbreaks. The results cast doubt on the view that massive coral populations dominated by large, old colonies could not have developed under historical outbreak regimes similar to those of the 1960s to 1980s (Done 1987, 1988; cf. Cameron, Endean, and DeVantier 1991a).

Elsewhere in the Indo-West Pacific, in the Ryukyu Islands and Guam, persistent suboutbreak populations of *A. planci* have caused shifts in coral community structure from assemblages dominated by preferred prey species (e.g., *Acropora* and *Montipora*) to assemblages dominated by nonpreferred prey, particularly massive corals (e.g., *Porites*, *Leptastrea*, and *Favia*) (Keesing 1992). On the GBR, by contrast, persistent suboutbreak populations of the seastar did not develop following the major outbreaks of the 1960s to 1980s. These various shifts in coral assemblage structure to differing levels of predation pressure indicate that, as with the aforementioned coral diversity, the community response to predation is not consistent, but rather varies in relation to a wide array of factors (cf. Porter 1972; Glynn 1974, 1990; Cameron et al. 1991a,b; Keesing 1992).

4.1.5 Geological Studies

Reef geologists have searched for the characteristic skeletal elements of *A. planci* in cores through reef sediments dating back some 3000 ybp (years before present). The studies found seastar skeletal remains distributed throughout the record and concluded that outbreaks had occurred previously (Frankel 1975, 1977; Walbran et al. 1989a,b). These findings were initially welcomed by some marine park managers as proof that outbreaks were natural events and thus did not require management intervention. This view was criticized by some ecologists and paleoecologists as being statistically invalid and inconclusive (Moran, Bradbury, and Reichelt 1985; Fabricius and Fabricius 1992; Keesing et al. 1992; Pandolfi 1992). Storms, bioturbation, and other taphonomic processes produced significant mixing and rearrangement of the seastar skeletal elements, adversely affecting the accuracy of age determinations from bulk skeletal samples. In a similar vein, Greenstein (1989, Chapter 2) found that taphonomic processes obliterated any sedimentological signature of the 1983–1984 mass mortality of the Caribbean sea urchin *Diadema antillarum*.

Several key questions remained unanswered by the geological studies. Did the seastar skeletal results represent the sedimentological signature of past major outbreaks, long-term accumulations of more stable seastar populations, or the well-mixed signature of the recent outbreaks distributed with those of past populations throughout the sedimentary record by bioturbation and storms? If indeed there were major past outbreaks, at what frequency did they occur? Ultimately it was shown that the geological studies could not answer these key questions of past seastar abundance or the frequency and intensity of past predation activity. Given the strikingly different ecological effects of different-sized seastar populations and return frequencies of outbreaks (Done 1987, 1988), these questions remain central to understanding the phenomenon.

4.1.6 Ecological Effects and the Detection of Past Outbreaks

Different-sized seastar populations produce different ecological effects and leave different historical signals of their presence across multiple spatial and temporal scales. Small seastar populations on the GBR (<1 seastar ha^{-1}) tend to prey sublethally and selectively on preferred coral species, particularly fast-growing, opportunistic branching corals, somewhat analogous to selective "weeding" or logging in a forest. Coral assemblages tend to maintain their long-lived structural elements, including massive corals, with comparatively little change in coral population size and age structures. Recovery largely involves growth of surviving corals and "in-filling" of the more ephemeral coral taxa. Few scars are left on massive corals and the history of seastar activity is masked rapidly by recovery of the more ephemeral taxa.

By contrast, major outbreaks (10^2–10^3 seastars ha^{-1} on the GBR) tend to cause death of most coral colonies, with loss of a major proportion of hard-coral cover and diversity at the scales of communities and reefscapes. The major loss of coral biomass and ecological complexity and the narrowing of size and age structures of most coral taxa are more analogous to heavy clearing in forest understory. Here the large massive corals remain standing like rounded boulders on a hill slope, with multiple scars left as testament to the voracious feeding that rapidly destroyed the surrounding coral thickets. Recovery on these reefs is largely dependent on larval recruitment from upstream reefs and has usually been dominated by rapid-recruiting, fast-growing corals (mostly the preferred acroporid prey) following the 1960s' and 1980s' outbreaks, perhaps providing positive feedback in the future propagation of outbreaks. The heavily scarred, large massive corals provide longer-term, decadal-scale evidence of the outbreak events.

4.2 Massive Corals

On the GBR, massive corals are represented by over 100 species from 34 genera in 10 families of Scleractinia (Veron 1986; DeVantier 1995). Some of these corals can attain ages of several centuries, growing slowly (~4 to 20 mm yr^{-1}) to great size (>5 m in diameter and height) and weight (tens of tonnes). These corals are major reef-builders (Potts et al. 1985; Done and Potts 1992) and, because of their size and longevity, they can form the basis of complex reef communities (Cameron and Endean 1985; DeVantier and Endean 1988). On the GBR, massive corals in two genera, *Porites* and *Diploastrea*, regularly grow to very large size and age, and thus are relevant to the present study.

4.2.1 *Porites*

From the scleractinian family Poritidae, these corals are common in most reef biotopes of the GBR, from mainland fringing reefs to outer shelf platform reefs. There are at least six large massive species on the GBR, of which *Porites lutea*, *P. lobata*, *P. solida*, and *P. australensis* are the most common (Veron 1986). Small corallite sizes (~1 mm) and intra- and interspecific variability in corallite structural characters preclude consistent identification to the species level in the field, so these species were considered as a closely related species complex for the purposes of this study.

These corals are dioecious broadcast spawners: they produce sperm or eggs that are fertilized in the water column, develop into planula larvae, and are dispersed in currents for periods of days to weeks (Harrison and Wallace 1990). Reproductive success is enhanced through colony fragmentation, which produces clones of "satellite colonies" or ramets (Highsmith 1980; DeVantier and Endean 1989) that may outnumber sexually produced genets in local populations (Done and Potts 1992). These corals grow slowly (~4–22 mm yr^{-1} radially, Isdale 1981; Lough et al. 1999) and may be exceptionally long-lived. The oldest known living colony on the GBR was dated at ~700 years (Potts et al. 1985), although individual genotypes may be much older because of the propensity for fragmentation (Potts et al. 1985; DeVantier and Endean 1989; Done and Potts 1992).

Cores taken from *Porites* from near-shore and midcontinental shelf reefs of the central GBR exhibit distinctive luminescent banding patterns when exposed to ultraviolet light (Isdale 1984). The intensity of luminescence and width of the bands are highly correlated with historical episodes of flood runoff from coastal rivers (Isdale et al. 1998; Lough, Barnes, and McAllister 2002), allowing accurate dating of growth bands in these corals (see also Halley and Hudson, Chapter 6).

4.2.2 *Diploastrea heliopora*

From a monotypic genus in the family Faviidae, this species can also attain great size and age, growing radially at ~4 to 7 mm yr^{-1} (DeVantier 1995, unpublished data; D. Barnes, Australian Institute of Marine Science, personal communication). Like massive *Porites*, *Diploastrea* is a dioecious spawner (Harrison and Wallace 1990), but unlike *Porites*, colony fragmentation is not an important mode of reproduction as its skeleton is very dense. Colonies of *Diploastrea* are usually far less common than *Porites* (Cameron, Endean, and DeVantier 1991b), occasionally occurring in local patches of high abundance. For reasons not well understood, this species is one of the least preferred of prey of *A. planci* on the GBR, and it is often a conspicuous survivor (albeit sometimes partially eaten) in coral assemblages otherwise denuded by predation.

4.3 Injury, Scarring, and Tissue Regeneration in Massive Corals

Individual predation episodes by *A. planci* leave a distinctive circular feeding scar on the prey coral of approximately the same size as the seastar's oral disk (~5–20 cm diameter). The scar is left where living coral tissues have been digested from the colony surface by the seastar's extruded stomach, exposing the clean white skeleton prior to fouling by algae and other organisms (Fig. 4.1). When feeding activity is more intense, one or more seastars can leave multiple contiguous scars, ultimately causing the loss of all living tissues and death of the coral colony.

Massive corals can heal small injuries (<10 cm^2) in one to several years, but there is a marked decline in the corals' capacity for regeneration of tissues over larger injuries (Bak and Steward-van Es 1980; Meesters, Bos, and Gast 1992; DeVantier 1995; van Woesik 1998). In the case of *Porites*, this relation is influenced by the presence of tissue to ~10 mm below the skeletal surface (Lough et al. 1999). Where injury is superficial, this subsurface tissue can facilitate regeneration. Predation by

Figure 4.1. Predation by *Acanthaster planci* on *Porites*, Great Barrier Reef, Australia.

A. planci produces large injuries (~50–300 cm²) and generally destroys the subsurface tissues through digestion.

On *Porites* corals subjected to major levels of predation, surviving coral remnants are significantly associated with the presence of inquiline serpulid worms and pectinid scallops. The worms and scallops tend to deter predation on the coral surface in their immediate vicinity, leaving characteristic rings of living coral tissue surrounding the inquiline species (DeVantier, Reichelt, and Bradbury 1986; DeVantier and Endean 1988; see also Glynn 1990; Pratchett 2001).

With increasing time following predation, the isolated surviving rings or patches of living coral tissue continue to grow above the surrounding dead colony surfaces, often as separate subcolonies, producing distinctive morphological anomalies of miniature coral heads on the otherwise dead colony surfaces (Fig. 4.2). Over periods of years to decades following predation, such surviving remnant patches may fuse and obscure the intervening dead coral surface, or they may remain as isolated clone-mates on the parent (Fig. 4.3).

(A)

(B)

Figure 4.2. Morphological anomalies on *Porites*. (**A, B**) Green Island Reef, showing scarred surfaces, remnant surviving coral, and the steps from the scarred surfaces to the living surface. Arrows indicate decades in which scarring occurred.

(C)

(D)

Figure 4.2. (*continued*) (**C**) Kepulauan Seribu, Indonesia. Here, the first scarred surface with the seastar corresponds to disturbance in the 1960s, and the earlier scarred surface with the urchins corresponds to disturbance in the 1930s to 1940s. (**D**) Green Island Reef. Formation of patch reef on large, massive coral killed by seastar predation in the 1980s.

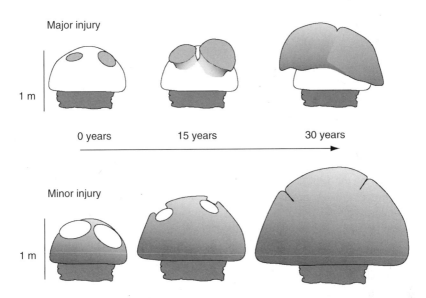

Figure 4.3. Schematic representation of growth of scarred corals following predation, showing scarring, with remnants fusing and failing to fuse over periods of years to decades. Remnant fusion is also visible in Fig. 4.2A.

The probability of survival and ultimate fusion of remnants depends on their initial size and distance apart (i.e., on the sizes of the intervening dead patches), and on the outcomes of interspecific interactions with organisms that colonize the adjacent dead surfaces. In some cases, herbivorous fish graze the dead surfaces, removing turf algae and other colonizing taxa (Edmunds 2000; T. Done personal observation). This tends to keep the scarred surface free of recruitment and thus readily visible over long periods (decades). When the remnants do not fuse, an irregular surface of one, two, or more steps or levels is produced from the dead scarred surface to the top of the living remnant coral head(s), herein referred to as steps or step-heights.

The activities of grazing fishes notwithstanding, some partially dead corals are prone to secondary interactions of overgrowth by biota that successfully colonize the dead surfaces (Endean, Cameron, and DeVantier 1988). This process ultimately leads to formation of patch reefs nucleated by the original massive coral (Done and DeVantier 1990; Fig. 4.2D). With increasing time since disturbance, bioerosion around the base of remnants may contribute to their fragmentation from the parent. With fouling and overgrowth, remnant fragmentation on the one hand, or fusion on the other, reduces the likelihood of finding good examples of old scarring.

4.4 Other Causes of Scarring

Factors other than seastar predation cause injury and death of massive corals. These include predation by fish and snails; damage from tropical storms; bleaching associated with fluctuations in temperature, insolation, and/or salinity; sedimentation;

diseases; and overgrowth by other taxa, particularly sponges and soft corals (Moran 1986; Glynn 1990, 1993; Lang and Chornesky 1990; Rogers 1990; Bythell, Gladfelter, and Bythell 1993; Turner 1994; DeVantier et al. 1997; Edmunds 2000). For most of these other forms of coral mortality on the GBR, patterns of injury and scarring are initially distinct from those caused by seastar predation (DeVantier 1995; T. Done personal observation), although such differences become less clear-cut with increasing time following the disturbance.

During the period of this study (and of prior scientific observations on the GBR since the 1920s), causes other than seastar outbreaks produced relatively minor levels of scarring to massive corals in the central GBR. For example, intense tropical storms during the 1980s to 1990s (Cyclones Winifred in 1986 and Joy in 1990) and the major coral bleaching event of 1998 had little impact on massive corals on the vast majority of midcontinental shelf reefs in the central GBR. Surveys on reefs in the direct path of Cyclone Winifred (1986) indicated that most large massive corals had been little affected (although some colonies had been toppled), despite heavy impacts that severely fragmented other corals, notably beds of branching *Acropora* (Done et al. 1986; DeVantier 1995).

Bleaching events in 1982, 1994, and 1998 also had relatively minor effects on massive corals in comparison with taxa of other growth forms, but they did cause scarring (mostly on colony tops) and death of colonies on the worst affected reefs (Fisk and Done 1985; DeVantier et al. 1997; Marshall and Baird 2000; L. DeVantier and T. Done personal observation). Notably, there is no evidence of such intense mass bleaching events prior to the 1980s (Fisk and Done 1985; Hoegh-Guldberg 1999; Lough 2000), nor have there been any reported major outbreaks of predatory *Drupella* snails on the GBR, like those of the Ningaloo Reef tract of northwestern Australia (Turner 1994). Thus predation during major outbreaks of *A. planci* remains the major known cause of broad-scale scarring on massive corals on the GBR, particularly prior to the recent mass bleaching events of 1998 and 2002. From these observations, we reasoned that intense predation on large, old corals prior to known seastar outbreaks of the 1960s may have left morphological evidence in earlier scarring patterns. We thus searched for old scarring on massive corals that preceded the first reported seastar outbreak of the 1960s (also see Crown-of-Thorns Study 1987; Done and Potts 1992; Musso 1994; DeVantier 1995).

4.5 Field Surveys

Approximately 1600 large colonies of *Porites* and *Diploastrea* were surveyed in meandering swims using SCUBA on 12 reefs of the central GBR from 1985 to 1991 (Table 4.1). The reefs were distributed along a roughly NW–SE axis over ~500 km of the central GBR, in the path of the major longshore East Australian Current. The reefs were chosen because there was some information on their histories of seastar outbreaks since the 1960s. Ten of the reefs were known to have been affected by seastar outbreaks once or twice since the 1960s

Table 4.1. Reefs surveyed for scarring on massive corals on the central GBR from 1985 to 1991. Listing is in chronological order of outbreaks and corresponds approximately to increasing latitude in the central GBR. Lack of field surveys, particularly during the 1960s' to 1970s' outbreak series, means that absence of recorded outbreaks cannot be taken as definitive proof that outbreaks did not occur

Reef	Type	Known outbreak history	Survey years
Green Island	Midshelf platform	1962–1966, 1980–1982	1985, 1991
Normanby Is.	Inner fringing	1965–1967	1986
Feather	Midshelf platform	1966–1968, 1981–1983	1985
Potter	Outer shelf platform	1982–1985	1985, 1987
Beaver	Midshelf platform	1966–1968, 1982–1985	1987
Brook Is.	Inner fringing	—	1988
Myrmidon	Outer shelf platform	—	1987
Rib	Midshelf platform	1967–1969, 1983–1985	1985
John Brewer	Midshelf platform	1969–1971, 1983–1986	1985, 1987
Grub	Outer shelf platform	1984–1987	1986
Wheeler	Midshelf platform	1985–1987	1985
Holbourne Is.	Inner fringing	1985–1987	1988

(Crown-of-Thorns Study 1987). Two of the reefs were thought to have been unaffected by seastar outbreaks since the 1960s (Table 4.1).

Because of time limits and the patchy and clumped distributions of massive corals on these reefs, field surveys were conducted as swim-searches rather than quantitative belt transects. Although levels of search effort were roughly similar on each reef, the density and abundance of massive corals varied markedly among reefs. Follow-up surveys were conducted on three of the reefs (Green Island, Potter, and John Brewer Reefs) in areas not visited during the initial surveys (Table 4.1).

4.5.1 Step-Height Measurements of Scar Remnants

More than 2000 scar step heights were measured with a purpose-built plastic measuring device consisting of flat base and top plates connected by a measuring scale bar or fiberglass surveyor's tape (Crown-of-Thorns Study 1987). For each step, the base of the measuring device was placed on the scarred surface immediately adjacent to the living coral remnant, with the top of the device placed on the apex of the remnant and the measurement taken parallel to the axis of growth of the remnant. Differences in height between the dead surface and top of adjacent living surface were measured to the nearest centimeter, normal to the dead and live surfaces.

These step-height measurements provided an estimate of the height of growth of the surviving remnant above the dead surface. Some corals had several steps associated with different previous disturbance episodes (Fig. 4.2). For corals with two or more distinct dead surfaces, the sum of each of the individual step-height measurements was used to determine height of growth to the live surface from the earliest dead surface. Bumpy coral surfaces, parallax error, and difficulties of working underwater caused measurement errors. These errors increased with remnant height, such that for small remnants (<10 cm height of growth above adjacent dead surface), errors were of the order of ± 0.5 cm; for medium-sized remnants (~50 cm step height), errors were of the order of ± 2 cm; and for large remnants (~100 cm step height), errors were of the order of ± 5 cm.

4.5.2 Coral Coring for Growth Rates and Scar Ages

For *Porites*, ages of scarred surfaces and average annual growth rates were determined by drilling 127 short (~20 cm) cores from scarred coral surfaces along the axis of coral growth with a SCUBA-powered pneumatic drill. The short cores were compared to longer standard reference cores (Isdale 1984). The scar cores were sectioned longitudinally, and the luminescent banding patterns (representing ca. 10–20 years of growth prior to scarring) were compared with the longer reference cores that had been dated previously using densitometry and luminescence (Fig. 4.4). For most cores from scarred surfaces, it was possible to determine the year of scarring by matching the luminescent banding patterns with those of the reference cores, although some cores from old scarred surfaces could not be dated because bioerosion and additional luminescence prevented a match.

For five reefs with sufficient cores, regressions of the date of the dead *Porites* surfaces (from the cores) and the step height of growth of remnants above the scarred surfaces were calculated. The regressions between the step heights and the ages of the scarred surfaces were significant ($P < 0.05$, r^2 range 0.73–0.91 for different reefs, Table 4.2). This allowed calculation of average coral growth rates on each reef, which were used to provide estimates of the ages of scarred surfaces from the step-height measurements of *Porites* corals without cores (also see Crown-of-Thorns Study 1987). For reefs without sufficient cores, *Porites* radial growth rates were assumed to range between 7 and 12 mm yr^{-1}. For *D. heliopora*, high skeletal density precluded coring, and growth rates were assumed to range between 4 and 7 mm yr^{-1} radial expansion (DeVantier 1995, unpublished data; D. Barnes, Australian Institute of Marine Science, personal communication).

The significant correlations and use of "average" coral growth rates notwithstanding, inter- and intracolony variation in the timing of predation and remnant growth constituted an additional source of error in the scar-age estimates for corals without cores. Predation associated with an outbreak may range over a 4- to 6-year period. This is because a moderate seastar population can be present on a reef for several years prior to the detection of an outbreak, the outbreak may last for several years (Table 4.1), and a residual seastar population may remain for several years after the population has declined from outbreak density (Endean 1976; Moran 1986; Endean, Cameron, and DeVantier 1988). With differences in growth rate of ~4 to 20 mm yr^{-1} among different *Porites* corals (Isdale 1981; Lough et al. 1999), this can result in a wide spread in remnant step heights. Even among remnants on a single coral head growing from the same scarred surface, differences in initial remnant size and subsequent growth can produce different

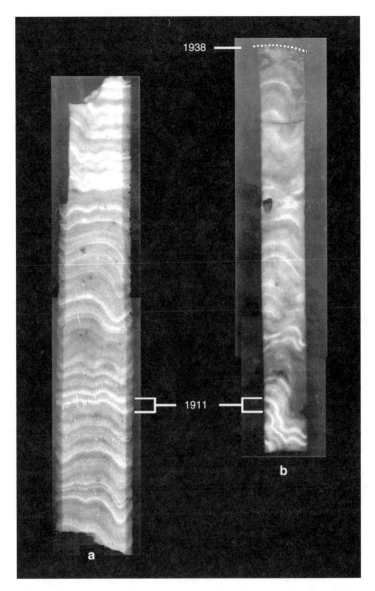

Figure 4.4. Comparison of luminescent banding patterns on thick reference coral core (**a**, left) and thinner scar core (**b**, right) from Normanby Island Reef. The intense double luminescent band near the base of both core sections was dated at 1910 to 1911. The dead, scarred surface of the small scar core was dated at 1938.

step heights among remnants (L. DeVantier personal observation; see below). As with step-height measurements, this error increases with time since disturbance, and we therefore assigned decadal periods rather than years to disturbances detected from scarring without supporting core dates.

Table 4.2. Regressions between height of remnant growth above scarred surfaces and age of scarred surfaces (from cores) on *Porites* from five reefs of the central GBR. Regressions were obtained using Bartlett's three-group method, allowing for error in both dependent and independent variables (Sokal and Rohlf 1969). All were significant at $P < 0.05$. Mean radial coral growth rates are also listed (modified from Crown-of-Thorns Study 1987)

Reef name and year of survey	Sample size (no. cores)	r^2	Radial growth (mm yr^{-1})
Rib 1985	21	0.91	8.9
Green Island 1985	29	0.89	7.0
Feather 1985	14	0.78	10.2
John Brewer 1985	15	0.78	10.9
Potter 1985	24	0.73	11.7

4.6 Coral Scar Patterns: The 1980s' Outbreaks

On the reefs known to have hosted *A. planci* outbreaks in the 1980s, there were large numbers of recently scarred colonies, particularly of *Porites* (Fig. 4.5A–M), with distinct peaks in scarring generally consistent with the timing of the 1980s' outbreaks. The recent scarring represented the first scar step on most massive corals, and for *Porites* the scar peaks were well developed on all of the outbreak-affected reefs. For *Diploastrea*, recent peaks in scarring were also apparent on Green Island, Feather, Potter, Rib and John Brewer Reefs. There were very few *Diploastrea* colonies at the sites on Beaver, Holbourne Island, and Grub Reefs, and thus the signal of the 1980s' outbreaks was not apparent. On Wheeler Reef there was no evidence of scarring in the 1980s, but some earlier scarring was evident (Fig. 4.6F; see below).

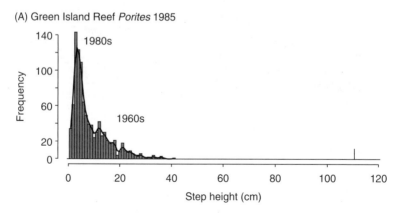

(A) Green Island Reef *Porites* 1985

Figure 4.5. Frequency bar graphs of scar levels on massive *Porites* corals at (**A**) Green Island Reef 1985.

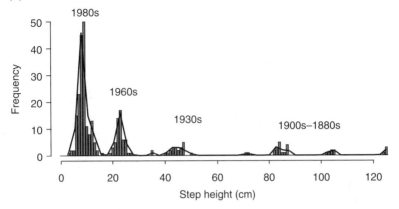

(B) Green Island Reef *Porites* 1991

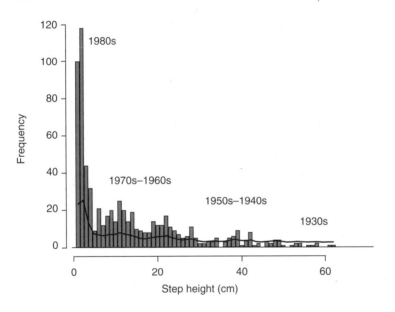

(C) Potter Reef *Porites* 1985

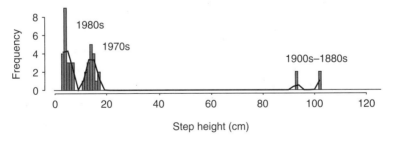

(D) Potter Reef *Porites* 1987

Figure 4.5. (*continued*) (**B**) Green Island Reef 1991. (**C**) Potter Reef 1985. (**D**) Potter Reef 1987.

(E) Normanby Island Reef *Porites* 1986

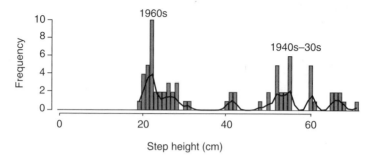

(F) Feather Reef *Porites* 1985

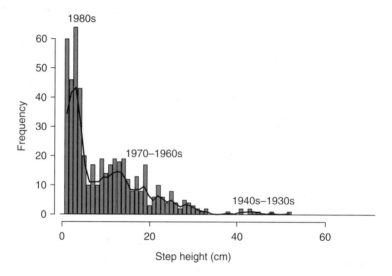

(G) Beaver Reef *Porites* 1987

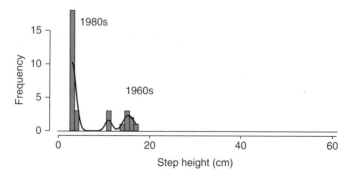

Figure 4.5. (*continued*) (**E**) Normanby Island Reef 1986. (**F**) Feather Reef 1985. (**G**) Beaver Reef 1987.

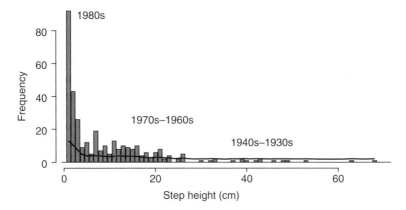

(H) Rib Reef *Porites* 1985

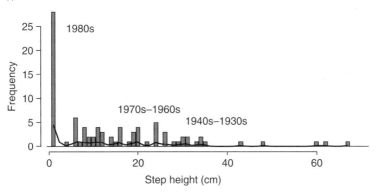

(I) John Brewer Reef *Porites* 1985

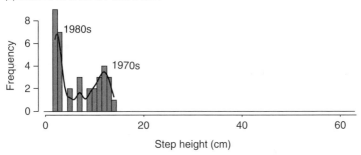

(J) John Brewer Reef *Porites* 1987

Figure 4.5. (*continued*) (**H**) Rib Reef 1985. (**I**) John Brewer Reef 1985. (**J**) John Brewer Reef 1987.

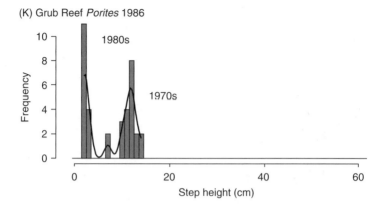

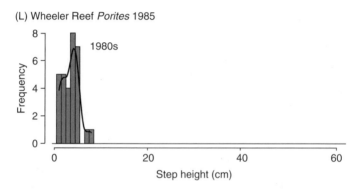

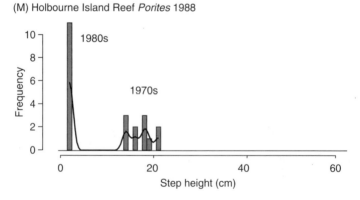

Figure 4.5. (*continued*) (**K**) Grub Reef 1986. (**L**) Wheeler Reef 1985. (**M**) Holbourne Island Reef 1988. *X*-axis: step height in cm, *Y*-axis: number of measured colonies. Dates of scarring were determined from scar step heights and coral cores.

Massive corals on the inner-shelf Brook Islands Reef, with no known seastar outbreaks at time of survey (Table 4.1), generally were intact or showed only minor injury and scarring. There was little evidence of any major mortality, either recent or past, on most corals, and there were very few old scars. Indeed, the great size and pristine condition of many colonies indicated that the corals had been unaffected by major disturbance for many years (DeVantier and Endean 1989). Similarly, at the outer-shelf Myrmidon Reef the great majority of massive corals were in good condition, with the very few scars present corresponding in etiology (being mostly restricted to colony tops) and timing with a bleaching event in 1982. This bleaching episode, among the first reported for the GBR (Fisk and Done 1985), had comparatively little effect on massive corals, but it caused major mortality in some taxa of other growth forms on this reef. The high levels of recent scarring on outbreak-affected reefs, and low levels on reefs known not to have been affected by outbreaks in the 1980s, provided strong support for *Acanthaster* predation being the major cause of the scarring, consistent with the observational evidence (Fig. 4.1).

4.7 Coral Scar Patterns: The 1960s' to 1970s' Outbreaks

For *Porites*, scarring attributable to the 1960s' to 1970s' outbreaks was found at Green Island, Feather, Rib, John Brewer, Normanby Island, and Beaver Reefs, consistent with the historical record, and also on Grub, Potter, and Holbourne Island Reefs, for which no record existed (Fig. 4.5A–M, Table 4.1). For *Diploastrea*, scarring consistent with the 1960s' to 1970s' outbreaks was found on Green Island, Feather, Rib, and John Brewer Reefs, and also on Potter and Wheeler Reefs, for which there was no outbreak record (Fig. 4.6A–F). The scarring on massive corals of both genera at Potter, Grub, Wheeler, and Holbourne Island Reefs suggests that seastar outbreaks did occur there during the 1960 to 1970s, as on most other reefs in the vicinity, but those outbreaks were not recorded at the time.

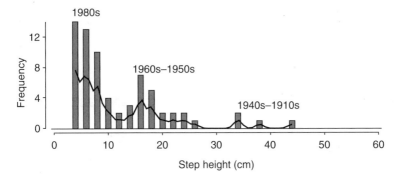

(A) Green Island Reef *Diploastrea* 1985

Figure 4.6. Frequency bar-graphs of scar levels on *Diploastrea heliopora* in 1985 at (**A**) Green Island Reef.

(B) Feather Reef *Diploastrea* 1985

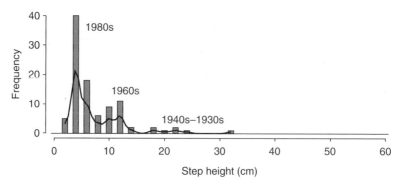

(C) Potter Reef *Diploastrea* 1985

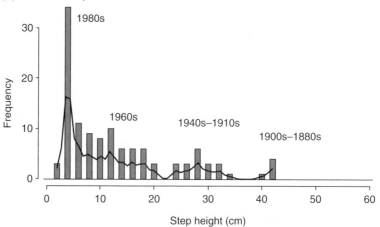

(D) Rib Reef *Diploastrea* 1985

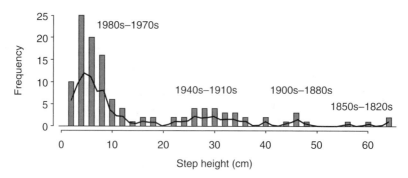

Figure 4.6. (*continued*) (**B**) Feather Reef. (**C**) Potter Reef. (**D**) Rib Reef.

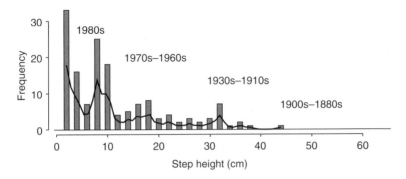

(E) John Brewer Reef *Diploastrea* 1985

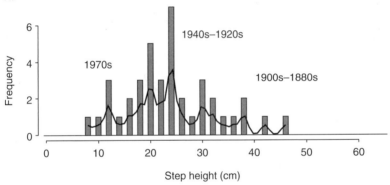

(F) Wheeler Reef *Diploastrea* 1985

Figure 4.6. (*continued*) (**E**) John Brewer Reef. (**F**) Wheeler Reef. *X*-axis: step height in cm, *Y*-axis: number of measured colonies. Approximate dates of scarring were determined from scar step heights.

There is considerable uncertainty in the historical record of outbreaks, particularly for the 1960s' to 1970s' outbreaks (Endean and Stablum 1975; Moran, Bradbury, and Reichelt 1988). Dedicated surveys for the crown-of-thorns seastar were in their infancy on the GBR and elsewhere at that time and consisted mostly of "spot-check" dives rather than the more comprehensive assessments of entire reef slopes using manta-tow (English, Wilkinson, and Baker 1997). Even today less than 5% of GBR reefs are surveyed or monitored on a regular basis (Oliver et al. 1995; Sweatman 1997; Sweatman et al. 1998).

On most of our survey reefs, scarring attributable to the 1960s' to 1970s' outbreaks was recognizable as the second step on the scarred corals (Fig. 4.2). However, on Normanby Island Reef, unaffected by an outbreak in the 1980s, the 1960s' scarring formed the first step (Table 4.1, Fig. 4.5E). Here, remnant heights from the step ranging from 18 to 33 cm dated to the 1965 to 1967 outbreak in cores from six colonies, demonstrating the substantial variability in heights of remnant growth that can be associated with the same disturbance event.

There was also a substantial reduction in the number of scars attributable to the 1960s' to 1970s' outbreaks in comparison with those from the 1980s' outbreaks (Figs. 4.5, 4.6). Scarring attributable to the 1980s' outbreaks accounted for more than half the scars measured on these reefs, whereas scars attributable to the 1960s' to 1970s' outbreaks accounted for most of the remainder. This is consistent with the historical evidence of increasing predation on massive corals in the 1980s' outbreaks (Endean and Cameron 1990). The reduction is also attributable in part to masking of old scars through fouling, and overgrowth, fragmentation, and fusion of remnants (Figs. 4.2, 4.3).

4.8 Evidence for Seastar Outbreaks Prior to the 1960s

For *Porites* and/or *Diploastrea*, scarring consistent with disturbance prior to the 1960s was apparent on 7 of the 12 reefs surveyed, although the signal of disturbance was far less well-defined than for the 1960s' to 1970s' or 1980s' outbreaks. Far fewer corals had prior scarring, the error margins associated with step-height measurements increased, and some of the coral cores from old dead coral surfaces were uninterpretable.

4.8.1 Green Island Reef

Porites corals on the southern part of this reef had the oldest and best-preserved coral-scar record (Fig. 4.5B). Some of the largest *Porites* corals had up to five different scar steps, indicating repeated episodes of disturbance, with one scar core dating back to the 1880s. The first two steps generally corresponded to the 1980s' and 1960s' outbreaks and were found on many corals around much of the reef. The third step, represented by a small peak in *Porites* scarring (40–50 cm step height), corresponded to disturbance in the 1930s to 1940s (Figs. 4.2, 4.5B), the fourth step (73–85 cm) to disturbance in the 1880s to 1900s, and the fifth step (100–125 cm) to disturbance in the 1850s to 1900s. Only two cores were interpretable from the early scar surfaces (50-cm step height, 1936; 102-cm step height, 1881). These earlier scar steps were not as widely distributed around the reef as those of the 1960s and 1980s, being concentrated on corals near the southern end of the reef (Fig. 4.5B). For *Diploastrea*, three distinct steps were apparent, the first two corresponding to the 1980s' and 1960s' outbreaks and the third step (heights of 34–44 cm) to disturbance between the 1880s and 1930s (Fig. 4.6A, assuming radial growth rates of 4–7 mm yr^{-1}).

4.8.2 Potter Reef

Massive *Porites* on this reef also had a long history of scarring over the past century. Four steps were present on some of the larger corals, the first two steps corresponding to the 1980s' and 1970s' outbreaks (step heights from 11 to 28 cm). The third step (heights of 38–48 cm) corresponded to disturbance in the 1930s to

1950s (Fig. 4.5C). Earlier scarring, detected during the follow-up survey (1987, Table 4.1, Fig. 4.5D), corresponded to disturbance around 1890–1910. No cores from the earliest scarred surfaces were interpretable. For *Diploastrea*, three distinct steps were apparent, the first and second corresponding to the 1980s' and 1970s' outbreaks and the third, ranging in step height from 24 to 32 cm, to disturbance during the 1900s to 1940s. Earliest scarring (42 cm) was consistent with disturbance in the 1880s to 1920s (Fig. 4.6C).

4.8.3 Normanby Island Reef

Large *Porites* colonies on this reef exhibited two distinct steps, the first corresponding with the outbreak of the mid-1960s and the second being represented by three small peaks in scarring from 41 to 72 cm (Fig. 4.5E). Cores from this oldest scar-level on three colonies dated to 1936 to 1938. As with the wide range in step heights associated with the 1960s' outbreak on this reef, the range associated with the second step was largely attributable to differences in growth rate among remnants. No scarred *Diploastrea* were found in the site at this reef.

4.8.4 Feather Reef

Three distinct steps were apparent on some of the larger *Porites* corals, the first and second corresponding to the 1980s' and 1970s' outbreaks and the third (heights of 40–45 cm) corresponding to disturbance in the 1930s to 1940s (Fig. 4.5F). No cores from the earliest scarred surfaces were interpretable. For *Diploastrea*, three steps were also present on some corals, with the first and second corresponding to the known outbreaks and the third (heights of 18–24 cm) to disturbance during the 1920s to 1940s (Fig. 4.6B).

4.8.5 Beaver Reef

No scarring prior to the 1960s' to 1970s' outbreak was found at the site on this reef (Fig. 4.5G).

4.8.6 Rib Reef

Here some of the large scarred *Porites* corals had three steps, the first and second corresponding to the 1980s' and 1960s' to 1970s' outbreaks. There was a wide spread of earlier scarring on *Porites* colonies, but there were no distinct peaks prior to the 1980s. The third step (40–70 cm) corresponded to disturbance(s) between the 1910s to 1940s (Fig. 4.5H). No cores from these scarred surfaces were interpretable. For *Diploastrea*, there was a long record of disturbance, with up to four steps on some corals. The first and second steps did not aggregate into clear height peaks. Rather, there was a spread of scars from ~1 to 12 cm step height in the period since the 1970s' outbreaks (Fig. 4.6D). An earlier scar peak (26–30 cm, third step) corresponded to disturbance between the 1910s and 1940s and matched the third step on Feather and Potter Reefs. Several earlier

Diploastrea scars (step heights of 44–48 and 55–65 cm) corresponded to disturbance(s) from the 1820s to 1900s (Fig. 4.6D).

4.8.7 John Brewer Reef

Here three steps were present on some of the larger *Porites* colonies, with the first two matching the known seastar outbreaks of the 1980s and 1960s to 1970s and the third (43–60 cm heights) corresponding to disturbance in the 1930s to 1940s (Fig. 4.5I). Earliest scarring (77 cm below live surface) indicated prior disturbance around the beginning of the 20th century. The follow-up survey in 1987 found small numbers of *Porites* scars corresponding closely to the 1970s' and 1980s' outbreaks (Fig. 4.5J). For *Diploastrea*, some colonies had up to four steps, the first two attributable to the known outbreaks, the third step (18 cm height peak) to disturbance in the 1940s to 1960s, and the fourth (32 cm peak) to disturbance in the 1900s to 1930s (Fig. 4.6E). As with several of the other reefs, the oldest scarring on *Diploastrea* (44 cm) corresponded to disturbance in the 1870s to 1920s.

4.8.8 Grub Reef

No scarring prior to the 1960s' to 1970s' outbreak was found at the site on this reef (Fig. 4.5K).

4.8.9 Wheeler Reef

No scarring corresponding to disturbance prior to the 1980s was found on the *Porites* colonies surveyed (Fig. 4.5L). However, a wide spread of earlier scarring was apparent in *Diploastrea*, with two steps on some corals. Here there was no record in *Diploastrea* scars from the 1980s' outbreak, which was underway at time of survey. The first step (peak at 12 cm) corresponded to disturbance in the 1960s, the second step (20–30 cm height) to disturbance in the 1910s to 1940s, and the earliest scarring (to 46 cm height) with disturbance in the 1870s to 1920s (Fig. 4.6F), again consistent with earliest scarring on several other reefs.

4.8.10 Holbourne Island Reef

No scarring prior to the 1960s' to 1970s' outbreak was found at the site on this reef (Fig. 4.5M) or on Brook Islands or Myrmidon Reefs (not illustrated).

To summarize:

1. The known seastar outbreaks of the 1960s to 1970s and 1980s were well preserved in the coral-scar record as distinct morphological anomalies (first two scar steps on corals on most reefs).
2. Similar scarring on other reefs for which there was no historical record for the 1960s to 1970s suggested that outbreaks had occurred there also. The 1980s'

and 1960s' to 1970s' signals were well preserved in corals around much of the perimeters on most reefs, being found in follow-up surveys on three reefs.

3. There were major declines in the numbers of scars with increasing time since disturbance, from the 1980s' outbreaks to the 1960s' to 1970s' outbreaks and prior to the 1960s, both within and among reefs.
4. Scarring prior to the 1960s was apparent on 7 of the 12 reefs surveyed, with some large, massive corals exhibiting three to five scar steps.
5. Peaks in scarring on *Porites* and/or *Diploastrea* corresponded to disturbance on the seven reefs in the 1920s to 1940s (supported by *Porites* core dates for the 1930s).
6. Earliest scarring on five reefs corresponded to disturbance from the 1820s to 1920s (with one core date of 1881 from Green Island Reef).
7. The scar record suggests that there may have been seastar outbreak series on the GBR prior to the 1960s, most notably in the 1930s to 1940s.
8. The scar record also suggests that the outbreaks have increased in frequency and geographic spread in recent decades.

4.9 Outbreak Chronology

The three outbreak series of the 1960s to 1970s, 1980s, and 1990s to 2000s propagated sequentially throughout the central GBR tract from north to south (Table 4.1), with a return period at many individual reefs of ~15 to 17 years. Green Island Reef was affected prior to Normanby Island, Feather, Potter, Rib, John Brewer, Wheeler, and Holbourne Island reefs in each outbreak series (Table 4.1). This sequential spread of outbreaks is largely related to the dispersal in longshore currents of seastar larvae from spawning parent populations on upstream source reefs to downstream sink reefs (Dight, James, and Bode 1988; Reichelt, Bradbury, and Moran 1990).

To hindcast past outbreak activity, we assumed that any earlier outbreak series had similar temporal and spatial patterns as the known outbreaks since the 1960s, recurring on individual reefs at intervals of 15 to17 years and spreading sequentially through the reef tract (Table 4.3). Comparison of the scar and core records from the above seven reefs does indicate earlier patterns of disturbance roughly consistent in timing with the hindcast outbreaks, particularly during the 1930s to 1940s. Scar step-height measurements and dates from cores (1936 to 1938) both support the hindcasts, although it is not possible categorically to assign this scarring to past seastar outbreaks. Nonetheless, the broad geographic spread of scarring in the 1930s to 1940s, distributed over some 300 km of the central GBR on seven of our survey reefs, is more similar to that of recent seastar outbreaks than to other known forms of disturbance, including large tropical storms (Lourensz 1981), major flooding events, and/or elevated sea temperatures (Hoegh-Guldberg 1999; Lough 2000; Marshall and Baird 2000).

Oral histories from other areas of the Indo-Pacific include knowledge of crown-of-thorns seastar activity during the 1930s in Indonesia (Mortensen 1931; Fig. 4.2C), the Philippines (Domantay and Roxas 1938), Palau (Hayashi 1938; see also Dana 1970), and American Samoa (Birkeland 1981). The concurrence of

Table 4.3. Hindcast of seastar outbreak history for 10 reefs of the GBR, based on timing and return periods of known outbreaks since the 1960s. Dates of known outbreaks and dates from coral cores (*) are listed. Hindcast outbreak periods with coral scar evidence from *Porites* and/or *Diploastrea* are signified P/D

Reef	1995–2002	1979–1988	1962–1972	1946–1956	1930–1940	1850–1910s
Green Is.	Small to moderate population 1992-98	1979–1981* P/D	1962 1966* P/D		1936* P/D	1881* P/D
Normanby Is.	No outbreak	No outbreak	1965–1967* P		1936–1938* P	
Feather	1999-00	1981–1983* P/D	1966–1968* P/D		P/D	
Beaver	1999-00	1982–1985 P	1965–1967 P			
Potter	1999-00	1983–1985* P/D	1967–1971* P		P/D	P/D
Rib	2000-02	1983–1985* P/D	1967–1969* P/D		P/D	D
John Brewer		1984–1986* P/D	1969–1971* P/D		P/D	P/D
Wheeler		1985–1988* P	D		D	D
Grub		1985–1987 P	1969–1970 P			
Holbourne Is.		1987–1988 P	1970–1974 P			

seastar activity in the 1930s at widespread sites throughout the Indo-Pacific is consistent with the subsequent concurrence of later outbreak series. For example the 1960s' outbreaks co-occurred on the GBR and parts of Polynesia in the Southern Hemisphere, and in the Ryukyu Islands, Guam, and other parts of Micronesia in the Northern Hemisphere (Moran 1986).

This Indo-Pacific-wide concurrence of outbreaks is suggestive of ocean-basin scale synchrony, perhaps triggered through El Niño–Southern Oscillation (ENSO)-related changes in rainfall, runoff, and temperature that may enhance larval survival (Zann 2000). There were large ENSO anomalies in the 1870s to 1880s, 1900s, and 1920s to 1930s, and they have occurred more commonly since the 1960s (Lough 2000). The lack of accurate dates for outbreak initiation makes precise correlation unachievable, however.

4.10 The 1990s' Outbreaks and Future Prognosis

Another series of seastar outbreaks has been occurring on the GBR since the mid-1990s, ~15 to 17 years after the previous outbreaks (Table 4.3), and coincident with reestablishment of coral cover to levels approximating the preoutbreak levels of the 1960s and 1980s (i.e., >30 – 50%). On the shallow slopes of some reefs, cover returned to these levels in ~10 years, dominated by fast-growing acroporid table and branching corals (Done et al. 1988; Oliver et al. 1995), preferred prey of the seastar. Other reefs (e.g., Green Island Reef), however, have not recovered their coral cover since the 1980s' outbreaks (Sweatman 1997; Sweatman et al. 1998; Seymour and Bradbury 1999). It will take considerably longer than 15 to 17 years for some deeper reef-slope communities, composed of more diverse coral assemblages than their shallow counterparts, to recover (T. Done and L. DeVantier unpublished data). The three outbreak series since the 1960s are recurring at a faster rate than complete coral recovery (Seymour and Bradbury 1999). This does not prove that the outbreaks are related to human activities, but it implies that the synecology of the central GBR continues to favor their propagation.

It is clear from the work conducted to date that there is little likelihood of answering the questions of novelty and human causation of outbreaks. One reason for this is that it is not possible to conduct definitive "before-and-after, control-and-impact" type studies, as the GBR, like most other reef systems, is not in a pristine state. The GBR, particularly its central third, has been subjected to increasing human pressures from fishing, agricultural and urban runoff, and other developments since at least the beginning of the 20th century (Done et al. 1997; see also Brown 1987; Wilkinson 1992; Ginsburg 1994; see Jackson 1997 for the global perspective).

If the recurrent waves of seastar outbreaks are anomalies rather than natural features of reef ecology, they may provide an example of large and catastrophic events arising from circumstances that appear unrelated or insignificant at the time but cascade into later prominence. One example of this phenomenon has occurred on reefs of the Caribbean Sea since the 1980s. Chronic overfishing of herbivorous fishes, mass mortality of herbivorous sea urchins, impacts from several hurricanes, coral diseases, and other factors have caused dramatic changes in ecosystem function, manifested as large and apparently persistent shifts in coral and algal cover (Hughes 1994; Jackson 1997; Aronson and Precht 2001).

4.11 Causes of Seastar Outbreaks

It is highly unlikely that outbreaks always arise from a single cause; rather, a variety of predisposing factors are likely to act in their initiation and propagation. These may include relaxation of predation pressure on juvenile, subadult, and adult seastars, which allows greater adult maturation (Endean and Cameron 1990; Ormond et al. 1990) and promotes spawning aggregations and high fertilization success (Babcock and Mundy 1992). Additional factors include enhancement of larval survivorship through coastal runoff, reducing sea-surface salinity to levels preferred by the seastar larvae and increasing their food supply through enrichment by land-derived nutrients (Birkeland 1982; Lucas 1982; Brodie 1992; Brodie et al. 2005).

As noted above, the annual propagation of secondary outbreaks through the central GBR since the 1960s (1963–1979, 1981–1992, 1997–2005 and continuing) indicates that environmental conditions and levels of nutrition are now usually

adequate for survival of seastar larvae once primary outbreaks have been initiated. Whether this was so prior to the agricultural, urban, and industrial development of the Queensland coast remains contentious, although there has been a
marked, fourfold increase in nutrient supply to the central GBR Lagoon since
European settlement of the catchments (Bell and Elmetri 1995; Brodie 1995;
Pulseford 1996; Brodie et al. 2001, 2005; Furnas and Mitchell 2001; McCulloch
et al. 2003).

Development, growth, and survival of the seastar larvae increase almost ten-
fold with doubled concentrations of large phytoplankton (Okaji 1996; Ayukai,
Okaji, and Lucas 1997; Okaji, Ayukai, and Lucas 1997). Concentrations of large
phytoplankton on the inshore central GBR shelf in the summer wet season, when
A. planci larvae develop, are double those of other places and times, suggesting
that frequent *A. planci* outbreaks on the GBR may result from increased terrigenous nutrient input (Brodie et al. 2005).

Mathematical models also suggest that increased larval nutrition (chlorophyll
levels) can lead to increased frequency and severity of seastar outbreaks on the
GBR (G. De'ath personal communication). Other models suggest that following
high seastar larval settlement there can be qualitative differences in the type of
resultant outbreak episode, as governed by predation (Bradbury, van der Laan, and
Macdonald 1990; van der Laan and Hogeweg 1992; Bradbury and Seymour 1997).
Modeled high levels of fish predation on the seastars can restrict their activity to
isolated, sporadic outbreaks, whereas intermediate to low levels of predation allow
initiation of a "travelling wave" or "global pulse" of outbreaks over hundreds of
kilometers. These results appear consistent with a model of self-organized criticality (Bak, Tang, and Wiesenfeld 1987, 1988; Aronson and Plotnick 1998),
wherein at a particular critical point, initially small-scale effects of sporadic outbreaks on individual reefs cascade through many reefs in the system.

Attribution of a single cause for seastar outbreaks (e.g., increased nutrients or
release from predation), whilst offering worthwhile insights at local scale, has not
proven convincing when considered in the context of the entire Indo-Pacific. For
example, while increased chlorophyll levels resulting from enhanced nutrients in
river runoff may contribute to increased frequency and severity of seastar outbreaks
on the GBR (Brodie et al. 2005), recent and apparently unprecedented major
seastar outbreaks in the northern Red Sea may not be attributable to this cause.
There and elsewhere (e.g., isolated, open ocean atolls such as Cocos-Keeling and
Europa, Indian Ocean), other putative causes, including reduction in predation
pressure, may also be of importance. Complex, multiple causality and threshold
dynamics are characteristic of many ecological systems (Aronson and Plotnick
1998), and *Acanthaster* outbreaks appear to be a case in point.

There are good reasons to consider seastar outbreaks as natural. There are also
good reasons for concern that what were once isolated and sporadic events are
now more widespread and frequent, having become tightly entrained in the
dynamics of Indo-Pacific reef communities. The strengthening signal of outbreaks
in recent decades in scar patterns on massive corals (Figs. 4.5, 4.6) suggests that
reefs of the GBR have become increasingly predisposed to them in the period

since major human alteration to river catchments and fish stocks. We consider two spatial scales to be of key importance in this regard: the broad scale of the central GBR and the fine scale on Green Island Reef.

On the broad scale, reefs of the central section of the GBR have been affected the worst by seastar outbreaks since the 1960s. This central section, particularly the area around Green Island in the north to Holbourne Island and the Whitsundays in the south, lies adjacent to most of the GBR's major coastal population centers (Cairns, Townsville, Bowen, and Mackay). Since the mid- to late 1800s these centers have supported much of the fisheries development on the GBR. They have also served as the nuclei of extensive development of coastal and hinterland agriculture, with related modification of river catchments that include many of the major rivers discharging into the GBR Lagoon. The central section is also under the regular oceanographic influence of the East Australian Current, providing a reliable dispersal medium for seastar larvae through the reef tract, fostering the regular propagation of travelling waves of outbreaks, and indeed of coral larvae initiating recovery.

On the fine scale, Green Island Reef formed the apparent epicenter of outbreaks in the 1960s and 1980s. Green Island receives nutrient-enriched waters (Udy et al. 1999). It is located in the flood plumes of the Russell-Mulgrave, Johnstone, and (to a lesser extent) Burdekin Rivers, which are focused onto this midshelf reef through topographic steering by Cape Grafton to the south (Devlin et al. 2001). A local tourist resort also discharged sewage directly onto the reef for several decades, from at least as early as the 1950s through the late 20th century, further enhancing local nutrient supply. Green Island is the closest reef to the major fishing port of Cairns and has been subject to human use since the end of the 19th century. Situated in the region where the East Australian Current bifurcates into southerly and northerly streams, Green Island is well positioned as a source reef for dispersal of larvae to other reefs of the central GBR.

Given more than a century of human use, why were there no historical reports of seastar outbreaks on Green Island Reef (or indeed other reefs) prior to the 1960s, even though it has been populated and visited since the 1900s? The coral-scar record suggests that at the fine scale as well as the broad, prior seastar activity was less widespread than during the outbreaks since the 1960s. On Green Island, most scarred corals in the vicinity of the coral cay, where human activity was (and is) focused, exhibited one or two scar steps attributable to the 1980s' and 1960s' outbreaks (Fig. 4.5A). Most prior scarring, however, was found on corals at the opposite (southern) end of the reef to the cay (Figs. 4.2, 4.5B).

The buildup of seastar populations on Green Island (and perhaps adjacent reefs) probably formed a primary reproductive source for the wider spread of outbreaks through the central GBR in the 1930s, 1960s, and 1980s, and possibly as far back as the early to mid-1800s. In our view this has been fostered through reduced levels of piscine and gastropod predation, favoring adult seastar maturation, spawning aggregation and high fertilization success. Subsequent larval survival of the seastars has been favored by the enhanced nutrient levels locally on Green Island Reef and more broadly in the central lagoon of the GBR.

Green Island Reef did not support a major seastar outbreak during the 1990s series, which appears to have spread from reefs farther north (Sweatman 1997; Sweatman et al. 1998). Following the 1980s' outbreaks, coral cover on Green Island has not returned to pre-outbreak levels. Recovery has been set back by, among other factors, a series of cyclones and ongoing predation on corals by a small to moderate seastar population present on the reef since the early to mid-1990s (L. DeVantier personal observation; Sweatman et al. 1998; Table 4.3). In this respect, the coral–seastar dynamic on Green Island in the 1990s appeared to resemble that on some reefs of Okinawa in the Ryukyu Islands. There, seastar populations have remained at moderate densities following the outbreaks that occurred from the late 1950s (the first reported) to the early 1980s. The persistent seastar population on Okinawa selectively preys on preferred corals, maintaining coral cover at a low level (Yamaguchi 1986; Keesing 1992; R. van Woesik personal communication) and apparently represents a chronic, postoutbreak phase in the seastar–coral community interaction.

We do not argue that seastar outbreaks are a novel phenomenon of the latter part of the 20th century. Rather, it is likely that there have been sporadic outbreaks over the evolutionary history of the seastar. It does appear, however, that the frequency and geographic spread of outbreaks have increased during the 20th century. In retrospect, earlier polarization of the controversy into the natural versus anthropogenic dichotomy, with their respective corollaries of "do nothing" versus "control," now appears simplistic (Sapp 1999). The seastar is well-suited to major population fluctuations (Moore 1978), yet the initiation, propagation, and ecological effect of outbreaks may well be linked with human activities (Cameron 1977; Cameron and Endean 1982; Ormond et al. 1990; Keesing et al. 1992). In a world where humans seek desired outcomes in natural resource systems, it does not necessarily follow that all natural disturbances should be left unchecked, nor that all anthropogenic ones require intervention (see also Precht et al. 2005).

Of more concern than outbreak novelty per se is that the recent decimation of reef corals by seastars is now exacerbated in many reef regions by destructive fishing and overfishing, pollution, sedimentation, and bleaching, among a litany of other impacts. With this increasing array of problems presently besetting reefs, continuing seastar outbreaks are causing widespread losses of biological diversity and ecological complexity. The synergistic and cumulative effects of these impacts are exceeding any intrinsic resilience that reef communities have against individual impacts, with concomitant long-term changes in their biological structure and ecological function (cf. Nyström et al. 2000).

For example, shallow fore-reef slopes of the central GBR exhibited characteristic zonation and community resilience following catastrophic disturbance by the seastars in the 1960s to 1980s (Done 1982, 1992a,b; Done et al. 1997). The community dominants (tabular and branching *Acropora* spp.) exhibited rapid, high recruitment from upstream reefs, further strengthened by their rapid growth and propensity to overtop other corals (T.J. Done, L. DeVantier, E. Turak, R. Van Woesik, and D. Fisk unpublished data). However, this resilience is now highly threatened by the recent addition of a third major disturbance regime, unprecedented bleaching events in 1998 and 2002 (Berkelmans and Oliver 1999; Lough 2000) that have severely reduced diversity and abundance of corals in very shallow waters (Marshall and Baird 2000; L. DeVantier personal observation). These were previously refugia from seastar predation and hence sources of larval supply for replenishment (DeVantier et al. 2004).

4.12 Risk Management in a Complex "Natural" System

Centuries of increasing development of coastlines and hinterlands, and human use of coral reefs for fisheries, tourism, and other industries mean that the distribution, variety, and intensity of human impacts are greater now than ever before. Impacts from seastar outbreaks and the wide array of other problems facing coral reefs continue to escalate. It has become increasingly clear since the 1960s that the combined effects of these impacts are causing widespread degradation of coral reef ecosystems globally (see, e.g., Brown 1987; Richmond 1993; Jackson 1997; Gardner et al. 2003; Pandolfi et al. 2003). The risks that these disturbances pose to the ecological integrity of the reefs themselves, as well as to reef-based industries, present enormous challenges for future sustainable management. To meet these challenges successfully will require appropriate and dedicated initiatives at local community, governmental, and intergovernmental levels, based on an improved understanding of reef ecology. It was with this goal in mind that the present contribution was written.

4.13 Summary

Major population outbreaks of the coral-feeding crown-of-thorns seastar *A. planci* were unknown to science prior to the 1960s, but they have subsequently caused massive damage to coral assemblages on many Indo-Pacific coral reefs. The scale of damage initially raised fears of the widespread loss of Indo-Pacific reef ecosystems and extinction of reef-building corals. This crown-of-thorns "phenomenon" has caused major schisms in the scientific community and proven an intractable problem for management. The controversy initially centered on whether outbreaks were natural events, forming a normal part of coral reef community dynamics, or were novel events caused by human activities. More recently, attention has focused on possible increases in frequency and intensity of outbreaks linked to human activities on and adjacent to reefs. There have been several previous attempts to address these questions. We have briefly reviewed these studies and described another approach: the search for evidence of past seastar outbreaks in scar patterns on large, old, massive coral heads on the Great Barrier Reef.

Intense seastar predation on massive corals, some of which are several centuries old, leaves characteristic feeding scars that can remain visible for decades to centuries. These old coral scar patterns provide inferential evidence of seastar outbreaks prior to the 1960s, most notably in the 1930s to 1940s, increasing in frequency and geographic spread in recent decades coincident with increasing human use and impact on the GBR. The evidence suggests that outbreaks are neither completely natural nor entirely human-induced. Rather, what appear to have been isolated and sporadic events in the past are now more widespread and frequent. The earlier polarization of the controversy into the natural versus anthropogenic dichotomy now appears simplistic. The seastar is apparently adapted to major population fluctuations, yet the initiation, propagation, and ecological effect of outbreaks may all be linked with human activities.

The third series of seastar outbreaks since the 1960s is presently occurring at widespread locations throughout the Indo-Pacific. Of more concern than novelty per se is that the recent decimation of corals by the seastars is now exacerbated in many regions by a litany of other impacts, including destructive fishing and overfishing, pollution, sedimentation, and heat-induced bleaching. With the increasing array of other problems besetting reefs, the continuing outbreaks are causing widespread losses of coral cover, biological diversity, and ecological complexity. The synergistic and cumulative effects of these impacts are exceeding any intrinsic resilience that reef communities have against individual impacts, with concomitant long-term changes in their biological structure and ecological function. Collectively, these impacts degrade reef attributes valued most highly by humans: biodiversity, high coral cover and structural complexity, and fish stocks. Managing coral reefs in a way that retains these values presents a major challenge for present and future human generations.

Acknowledgments. We gratefully acknowledge the personnel of the Australian Institute of Marine Science Crown-of-thorns Study 1985, particularly Mathew Wilce. Peter Isdale provided access to the AIMS *Porites* reference core collection, and with Bruce Parker assisted in comparison and interpretation of cores. We thank Barbara Musso for provision of additional core data, Glenn De'ath for statistical advice, and Katharina Fabricius, Richard Aronson, Janice Lough, and two anonymous reviewers for critical comments on earlier drafts of this manuscript. We also thank Richard Aronson for encouraging us to publish this work. This research was funded by the Australian Institute of Marine Science and Australian Cooperative Research Centres Program.

We dedicate this chapter to the memory of Dr. Robert Endean and Dr. Ann Cameron. Bob and Ann were among the first to recognize the importance of crown-of-thorns seastar outbreaks in coral reef ecology. Together they worked tirelessly in raising awareness and fostering research on causes and effects of the outbreaks on the Great Barrier Reef and elsewhere.

References

Aronson, R.B., and R.E. Plotnick. 1998. Scale-independent interpretations of macroevolutionary dynamics. In *Biodiversity Dynamics Turnover of Populations, Taxa and Communities*, eds. M.L. McKinney and J.A. Drake, 430–450. New York: Columbia University Press.

Aronson, R.B., and W.F. Precht. 2001. Evolutionary paleoecology of Caribbean coral reefs. In *Evolutionary Paleoecology: The Ecological Context of Macroevolutionary Change*, eds. W.D. Allmon and D.J. Bottjer, 171–233. New York: Columbia University Press.

Ayukai, T., K. Okaji, and J.S. Lucas. 1997. Food limitation in the growth and development of crown-of-thorns starfish larvae in the Great Barrier Reef. *Proc. Eighth Int. Coral Reef Symp.* 1:621–626.

Babcock, R., and C. Mundy. 1992. Seasonal changes in fertility and fecundity in *Acanthaster planci*. In *The Possible Causes and Consequences of Outbreaks of the Crown-of-Thorns Starfish*. ed. B. Lassig. Great Barrier Reef Marine Park Authority Workshop Series.

Bak, P., C. Tang, and K. Wiesenfeld. 1987. Self-organized criticality: An explanation of 1/f noise. *Phys. Rev. Lett.* 59:381–384.

Bak, P., C. Tang, and K. Wiesenfeld. 1988. Self-organized criticality. *Phys. Rev.* 38:364–374.

Bak, R.P.M., and Y. Steward-van Es. 1980. Regeneration of superficial damage in the scleractinian corals *Agaricia agaricites f. purpurea* and *Porites astreoides*. *Bull. Mar. Sci.* 30:883–87.

Bell, P.R.F. 1991. Status of eutrophication in the Great Barrier Reef Lagoon. *Mar. Pollut. Bull.* 23:89–93.

Bell, P.R.F., and I. Elmetri. 1995. Ecological indicators of large-scale eutrophication in the Great Barrier Reef lagoon. *Ambio* 24:208–215.

Berkelmans, R., and J.K. Oliver. 1999. Large-scale bleaching of corals on the Great Barrier Reef. *Coral Reefs* 18:55–60.

Birkeland, C. 1981. *Acanthaster* in the cultures of high islands. *Atoll Res. Bull.* 255:55–58.

Birkeland, C. 1982. Terrestrial runoff as a cause of outbreaks of *Acanthaster planci* (Echinodermata, Asteroidea). *Mar. Biol.* 69:175–85.

Birkeland, C., and J.S. Lucas. 1990. *Acanthaster planci: Major Management Problem of Coral Reefs*. Boston: CRC Press.

Bradbury, R.H., and R. Seymour. 1997. Waiting for COTS. *Proc. Eighth Int. Coral Reef Symp.* 2:1357–1362.

Bradbury, R.H., J.D. van der Laan, and B. Macdonald. 1990. Modeling the effects of predation and dispersal on the generation of waves of starfish outbreaks. *Math. Comput. Model.* 13:61–67.

Brodie, J.E. 1992. Enhancement of larval and juvenile survival and recruitment in *Acanthaster planci* from the effects of terrestrial run-off: A review. *Aust. J. Mar. Freshwater Res.* 43:539–554.

Brodie, J. 1995. The problem of nutrients and eutrophication in the Australian marine environment. In *The State of the Marine Environment Report for Australia. Technical Annex 2*, eds. L. Zann and D. Sutton, 1–29. Canberra: Department of the Environment, Sport and Territories.

Brodie, J., C. Christie, M. Devlin, D. Haynes, S. Morris, M. Ramsay, J. Waterhouse, and H. Yorkston. 2001. Catchment management and the Great Barrier Reef. *Water Sci. Tech.* 43:203–211.

Brodie, J., K. Fabricius, G. De'ath, and K. Okaji. 2005. Are increased nutrient inputs responsible for more outbreaks of crown-of-thorns starfish? An appraisal of the evidence. *Mar. Pollut. Bull.* 51:266–278.

Brown, B.E. 1987. Worldwide death of corals—Natural cyclical events or man-made pollution? *Mar. Pollut. Bull.* 18:9–13.

Bythell, J.C., E.H. Gladfelter, and M. Bythell. 1993. Chronic and catastrophic natural mortality of three common Caribbean corals. *Coral Reefs* 12:143–152.

Cameron, A.M. 1977. *Acanthaster* and coral reefs: Population outbreaks of a rare and specialised carnivore in a complex high-diversity system. *Proc. Third Int. Coral Reef Symp.* 1:193–199.

Cameron, A.M., and R. Endean. 1982. Renewed population outbreaks of a rare and specialized carnivore (the starfish *Acanthaster planci*) in a complex high-diversity system (the Great Barrier Reef). *Proc. Fourth Int. Coral Reef Symp.* 2:593–596.

Cameron, A.M., and R. Endean. 1985. Do long-lived species structure coral reef ecosystems? *Proc. Fifth Int. Coral Reef Congr.* 6:211–215.

Cameron, A.M., R. Endean, and L.M. DeVantier. 1991a. Predation on massive corals: Are devastating population outbreaks of *Acanthaster planci* novel events? *Mar. Ecol. Prog. Ser.* 75:251–258.

Cameron, A.M., R. Endean, and L.M. DeVantier. 1991b. The effects of *Acanthaster planci* predation on two species of massive coral. *Hydrobiologia* 216/217:257–262.

Chesher, R.H. 1969. Destruction of Pacific corals by the sea star *Acanthaster planci*. *Science* 165:280–283.

Colgan, M.W. 1982. Succession and recovery of a coral reef after predation by *Acanthaster planci*. *Proc. Fourth Int. Coral Reef Symp.* 2:333–338.

Colgan, M.W. 1987. Coral reef recovery on Guam (Micronesia) after catastrophic predation by *Acanthaster planci*. *Ecology* 68:1592–1605.

Connell, J. 1978. Diversity in tropical rainforests and coral reefs. *Science* 199:1302–1310.

Crown-of-Thorns Study 1987. *An assessment of the Distribution and Effects of Acanthaster planci (L.) on the Great Barrier Reef*, 13 volumes. Townsville: Australian Institute of Marine Science. Volume 2, *Massive Coral Study*.

Dana, T.F. 1970. *Acanthaster*: A rarity in the past? *Science* 169:894.

De'ath, G., and P.J. Moran. 1998. Factors affecting the behaviour of crown-of-thorns starfish (*Acanthaster planci* L.) on the Great Barrier Reef: 2: Feeding preferences. *J. Exp. Mar. Biol. Ecol.* 220:107–126.

DeVantier, L.M. 1995. The structure of assemblages of massive corals on the central Great Barrier Reef: An assessment of the effects of predation by *Acanthaster planci*. PhD thesis, University of Queensland.

DeVantier, L., G. De'ath, R. Klaus, S. Al-Moghrabi, M. Abdulaziz, G.B. Reinicke, and C. Cheung. 2004. Reef-building corals and coral communities of the Socotra Islands, Yemen: A zoogeographic 'crossroads' in the Arabian Sea. *Fauna Arabia* 20:117–168.

DeVantier, L.M., and R. Endean. 1988. The scallop *Pedum spondyloideum* mitigates the effects of *Acanthaster planci* predation on the host coral *Porites*: Host defence facilitated by exaptation? *Mar. Ecol. Prog. Ser.* 47:293–301.

DeVantier, L.M., and R. Endean. 1989. Observations of colony fission following ledge formation in massive reef corals of the genus *Porites*. *Mar. Ecol. Prog. Ser.* 58:191–195.

DeVantier, L.M., R.E. Reichelt, and R.H. Bradbury. 1986. Does *Spirobranchus giganteus* protect host *Porites* from predation by *Acanthaster planci*: Predator pressure as a mechanism of coevolution? *Mar. Ecol. Prog. Ser.* 32:307–310.

DeVantier, L.M., E. Turak, T.J. Done, and J. Davidson. 1997. The effects of Cyclone Sadie on coral communities of nearshore reefs in the central Great Barrier Reef. In *Cyclone Sadie Flood Plumes in the Great Barrier Reef Lagoon: Composition and Consequences*, ed. A. Steven. Townsville: Great Barrier Reef Marine Park Authority Workshop Series 22:65–88.

Devlin, M., J. Waterhouse, J. Taylor, and J. Brodie. 2001. *Flood Plumes in the Great Barrier Reef: Spatial and Temporal Patterns in Composition and Distribution*. Townsville: Great Barrier Reef Marine Park Authority Research Publication 68.

Dight, I.J., M.K. James, and L. Bode. 1988. Models of larval dispersal within the central Great Barrier Reef: Patterns of connectivity and their implications for species distributions. *Proc. Sixth Int. Coral Reef Symp.* 3:217–223.

Domantay, J.S., and H.A. Roxas. 1938. The littoral Asteroidea of Port Galera Bay and adjacent waters. *Philipp. J. Sci.* 65:203–238.

Done, T.J. 1982. Patterns in the distribution of coral communities across the central Great Barrier Reef. *Coral Reefs* 1:95–107.

Done, T.J. 1985. Effects of two *Acanthaster* outbreaks on coral community structure: The meaning of devastation. *Proc. Fifth Int. Coral Reef Congr.* 5:313–320.

Done, T.J. 1987. Simulation of the effects of *Acanthaster planci* on the population structure of massive corals in the genus *Porites*: Evidence of population resilience? *Coral Reefs* 6:75–90.

Done, T.J. 1988. Simulation of recovery of pre-disturbance size structure in populations of *Porites* spp. damaged by the crown-of-thorns starfish *Acanthaster planci*. *Mar. Biol.* 100:51–61.

Done, T.J. 1992a. Constancy and change in some Great Barrier Reef coral communities: 1980–1990. *Am. Zool.* 32:655–662.

Done, T.J. 1992b. Phase-shifts in coral reef communities and the ecological significance. *Hydrobiologia* 247:121–132.

Done, T.J., and L.M. DeVantier. 1990. Fundamental change in coral community structure at Green Island. *Coral Reefs* 9:166.

Done, T.J., L.M. DeVantier, E. Turak, L. McCook, and K. Fabricius. 1997. Decadal changes in community structure in Great Barrier Reef coral reefs. In *State of the GBR World Heritage Area Workshop*, eds. D. Wachenfeld and J. Oliver. Great Barrier Reef Marine Park Authority Workshop Series 23:97–108.

Done, T.J., P.J. Moran, and L.M. DeVantier. 1986. Cyclone Winifred—Observations of some ecological and geomorphological effects. In *Workshop on the Offshore Effects of Cyclone Winifred*, ed. I.M. Dutton. Great Barrier Reef Marine Park Authority Workshop Series 7:50–52.

Done, T.J., K. Osborne, and K.F. Navin. 1988. Recovery of corals post-*Acanthaster*: Progress and prospects. *Proc. Sixth Int. Coral Reef Symp.* 2:137–143.

Done, T.J., and D.C. Potts. 1992. Influences of habitat and natural disturbances on contributions of massive *Porites* corals to reef communities. *Mar. Biol.* 114:479–493.

Edmunds, P. 2000. Recruitment of scleractinians onto the skeletons of corals killed by black band disease. *Coral Reefs* 19:69–74.

Endean, R. 1976. Destruction and recovery of coral communities. In *Biology and Geology of Coral Reefs, Volume 3, Biology 2*, eds. O.A. Jones and R. Endean. New York: Academic Press.

Endean, R., and A.M. Cameron. 1990. *Acanthaster planci* population outbreaks. In *Coral Reefs (Ecosystems of the World 25)*, ed. Z. Dubinsky, 419–437. Amsterdam: Elsevier Science.

Endean, R., A.M. Cameron, and L.M. DeVantier. 1988. *Acanthaster planci* predation on massive corals: The myth of rapid recovery of devastated reefs. *Proc. Sixth Int. Coral Reef Symp.* 2:143–148.

Endean, R., and W. Stablum. 1975. Population explosions of *Acanthaster planci* and associated destruction of the hard coral cover of reefs of the Great Barrier Reef. *Environ. Cons.* 2:247–256.

English, S., C. Wilkinson, and V. Baker. 1997. *Survey Manual for Tropical Marine Resources 2nd Edition*. Australian Institute of Marine Science.

Fabricius, K.E., and F.H. Fabricius. 1992. Re-assessment of ossicle frequency patterns in sediment cores: Rate of sedimentation related to *Acanthaster planci. Coral Reefs* 11:109–114.

Fisk, D., and T.J. Done. 1985. Taxonomic and bathymetric patterns of bleaching in corals, Myrmidon Reef (Queensland). *Proc. Fifth Int. Coral Reef Congr.* 6:149–154.

Frankel, E. 1975. *Acanthaster* in the past: Evidence from the G.B.R. *Proceedings Crown-of-Thorns Starfish Seminar*, 159–165. Canberra: Australian Government Publishing Service.

Frankel, E. 1977. Evidence from the Great Barrier Reef of ancient *Acanthaster* aggregations. *Atoll Res. Bull.* 220:75–86.

Furnas, M. 2003. *Catchments and Corals: Terrestrial Runoff to the Great Barrier Reef.* Townsville: Australian Institute of Marine Science and CRC Reef.

Furnas, M., and A. Mitchell. 2001. Runoff of terrestrial sediment and nutrients into the Great Barrier Reef World Heritage Area. In *Oceanographic Processes of Coral Reefs*, ed. E. Wolanski. Boston: CRC Press.

Gardner, T.A., I.M. Côté, J.A. Gill, A. Grant, and A.R. Watkinson. 2003. Long-term region-wide declines in Caribbean corals. *Science* 301:958–960.

Ginsburg, R. (compiler). 1994. *Proceedings of the Colloquium on Global Aspects of Coral Reefs: Health, Hazards and History.* Rosenstiel School of Marine and Atmospheric Science, University of Miami.

Glynn, P.W. 1973. *Acanthaster*: Effect on coral reef growth in Panama. *Science* 180: 504–506.

Glynn, P.W. 1974. The impact of *Acanthaster* on corals and coral reefs in the eastern Pacific. *Environ. Cons.* 1:295–304.

Glynn, P.W. 1982. *Acanthaster* population regulation by a shrimp and a worm. *Proc. Fourth Int. Coral Reef Symp.* 2:607–612.

Glynn, P.W. 1990. Feeding ecology of selected coral reef macro-consumers: Patterns and effects on coral community structure. In *Coral Reefs (Ecosystems of the World 25)*, ed. Z. Dubinsky, 365–400. Amsterdam: Elsevier Science.

Glynn, P.W. 1993. Coral reef bleaching: Ecological perspectives. *Coral Reefs* 12:1–17.

Greenstein, B.J. 1989. Mass mortality of the West-Indian echinoid *Diadema antillarum* (Echinodermata: Echinoidea): A natural experiment in taphonomy. *Palaios* 4:487–492.

Harrison, P.L., and C.C. Wallace. 1990. Reproduction, dispersal and recruitment of scleractinian corals. In *Coral Reefs (Ecosystems of the World 25)*, ed. Z. Dubinsky, 133–207. Amsterdam: Elsevier Science.

Hayashi, R. 1938. Sea stars of the Caroline Islands. *Palao Trop. Biol. Station Stud.* 1:417–446.

Highsmith, R.C. 1980. Passive colonization and asexual colony multiplication in the massive coral *Porites lutea* Milne Edwards. *J. Exp. Mar. Biol. Ecol.* 47:55–67.

Hoegh-Guldberg, O. 1999. Climate change, coral bleaching and the future of the world's coral reefs. *Mar. Freshwater Res.* 50:839–866.

Hughes, T.P. 1994. Catastrophes, phase shifts and large scale degradation of a Caribbean coral reef. *Science* 265:1547–1551.

Isdale, P.J. 1981. Geographical variation in the growth rate of the hermatypic coral *Porites* in the Great Barrier Reef Province, Australia. PhD Thesis, James Cook University of North Queensland.

Isdale, P.J. 1984. Fluorescent bands in massive corals record centuries of coastal rainfall. *Nature* 310:578–579.

Isdale, P.J., B.J. Stewart, K.S. Tickle, and J.M. Lough. 1998. Palaeohydrological variation in a tropical river catchment: A reconstruction using fluorescent bands in corals of the Great Barrier Reef, Australia. *Holocene* 8:1–8.

Jackson, J.B.C. 1997. Reefs since Columbus. *Coral Reefs* 16:S23–S32.

Keesing, J.K. 1992. Influence of persistent sub-infestation density *Acanthaster planci* (L.) and high density *Echinometra mathaei* (de Blainville) populations on coral reef community structure in Okinawa, Japan. *Proc. Seventh Int. Coral Reef Symp.* 2:769–779.

Keesing, J.K., R.H. Bradbury, L.M. DeVantier, M. Riddle, and G. De' ath. 1992. The geological evidence for recurring outbreaks of the crown-of-thorns starfish: A reassessment from an ecological perspective. *Coral Reefs* 11:79–85.

Keesing, J.K., and A.R. Halford. 1992. Importance of post-settlement processes for the population dynamics of *Acanthaster planci* (L.). *Aust. J. Mar. Freshwater Res.* 43: 635–651.

Keesing, J.K., and J.S. Lucus. 1992. Field measurements of feeding and movement rates of the crown-of-thorns starfish *Acanthaster planci* (L.). *J. Exp. Mar. Biol. Ecol.* 156:89–104.

Lang, J.C., and E.A. Chornesky. 1990. Competition between reef corals—A review of mechanisms and effects. In *Coral Reefs (Ecosystems of the World 25)*, ed. Z. Dubinsky, 209–252. Amsterdam: Elsevier Science.

Lough, J. M. 2000. Perspectives on global change and coral bleaching: 1997–98 sea surface temperatures at local and global scales. *JAMSTEC Int. Coral Reef Symp., Tokyo* 215–229.

Lough, J.M., D.J. Barnes, M.J. Devereaux, B.J. Tobin, and S. Tobin. 1999. Variability in growth characteristics of massive *Porites* on the Great Barrier Reef. Cooperative Research Centre for the Great Barrier Reef Technical Report No. 28.

Lough, J.M., D.J. Barnes, and F.A. McAllister. 2002. Luminescence lines in corals from the Great Barrier Reef provide spatial and temporal records of reefs affected by land runoff. *Coral Reefs* 21:333–343.

Lourensz, R.S. 1981. *Tropical cyclones in the Australian Region July 1909 to June 1980.* Canberra: Australian Government Printing Service.

Lucas, J.S. 1982. Quantitative studies of feeding and nutrition during larval development of the coral reef asteroid *Acanthaster planci* (L.). *J. Exp. Mar. Biol. Ecol.* 65:173–194.

Lucas, J.S., W.J. Nash, and M. Nishida. 1985. Aspects of the evolution of *Acanthaster planci*. *Proc. Fifth Int. Coral Reef Congr.* 5:327–332.

Marshall, P.A., and A.H. Baird. 2000. Bleaching of corals on the Great Barrier Reef: Differential susceptibilities among taxa. *Coral Reefs* 19:155–163.

McCallum, H.I. 1987. Predator regulation of *Acanthaster planci*. *J. Theor. Biol.* 127:207–220.

McCallum, H.I., R. Endean, and A.M. Cameron. 1989. Sublethal damage to *Acanthaster planci* as an index of predation pressure. *Mar. Ecol. Prog. Ser.* 56:29–36.

McCulloch, M., S. Fallon, T. Wyndham, R. Hendy, J. Lough, and D. Barnes. 2003. Coral record of increased sediment flux to the inner Great Barrier Reef since European settlement. *Nature* 421:727–730.

Meesters, E.H., A. Bos, and G.J. Gast. 1992. Effects of sedimentation and lesion position on coral tissue regeneration. *Proc. Seventh Int. Coral Reef Symp.* 2:671–678.

Moore, R.J. 1978. Is *Acanthaster planci* an r-strategist? *Nature* 271:56–57.

Moran, P.J. 1986. The *Acanthaster* phenomenon. *Oceanogr. Mar. Biol. Ann. Rev.* 24:379–480.

Moran, P.J., R.H. Bradbury, and R.E. Reichelt. 1985. Assessment of the geological evidence for previous *Acanthaster* outbreaks. *Coral Reefs* 4:235–238.

Moran, P.J., R.H. Bradbury, and R.E. Reichelt. 1988. Distribution of recent outbreaks of the crown-of-thorns starfish (*Acanthaster planci*) along the Great Barrier Reef: 1985–1986. *Coral Reefs* 7:125–137.

Mortensen, T.H. 1931. Contributions to the study of the developmental and larval forms of echinoderms I and II. *K. Dan. Vidensk. Selsk. Skr. Nat. Math. Afd.* 9, 4(1):1–39.

Musso, B. 1994. Internal bioerosion of in situ living and dead corals on the Great Barrier Reef. PhD Thesis, James Cook University of North Queensland.

Nyström, M., C. Folke, and F. Moberg. 2000. Coral reef disturbance and resilience in a human-dominated environment. *Trends Ecol. Evol.* 15:413–417.

Okaji, K. 1996. Feeding ecology in the early life stages of the crown-of-thorns starfish, *Acanthaster planci* (L.). PhD thesis, James Cook University, Townsville, Australia.

Okaji, K., T. Ayukai, and J. Lucas. 1997. Selective feeding by larvae of the crown-of-thorns starfish, *Acanthaster planci* (L.). *Coral Reefs* 16:47–50.

Oliver, J., G. De'ath, T.J. Done, D. McB. Williams, M. Furnas, and P. Moran. 1995. *Long-term Monitoring of the Great Barrier Reef, Status Report no. 1*. Townsville: Australian Institute of Marine Science.

Olson, R.R. 1987. In situ culturing as a test of the larval starvation hypothesis for the crown-of-thorns starfish, *Acanthaster planci*. *Limnol. Oceanogr.* 32:895–904.

Ormond, R.F.G., R.H. Bradbury, S. Bainbridge, K. Fabricius, J.K. Keesing, L.M. DeVantier, P. Medlay, and A.D.L. Steven. 1990. Test of a model of regulation of crown-of-thorns starfish by fish predators. In Acanthaster *and the Coral Reef: A Theoretical Perspective,* ed. R.H. Bradbury, 180–207. *Lecture Notes in Biomathematics* 88. New York: Springer-Verlag.

Ormond, R.F.G., and A.C. Campbell. 1974. Formation and breakdown of *Acanthaster planci* aggregations in the Red Sea. *Proc. Second Int. Coral Reef Symp.* 1:595–619.

Pandolfi, J. 1992. A palaeobiological examination of the geological evidence of recurring outbreaks of the crown-of-thorns starfish, *Acanthaster planci*. *Coral Reefs* 11:87–94.

Pandolfi, J.M., R.H. Bradbury, E. Sala, T.P. Hughes, K.A. Bjorndal, R.G. Cooke, D. McArdle, L. McClenachan, M.J.H. Newman, G. Paredes, R.R. Warner, and J.B.C. Jackson. 2003. Global trajectories of the long-term decline of coral reef ecosystems. *Science* 301:955–958.

Porter, J.W. 1972. Predation by *Acanthaster* and its effects on coral species diversity. *Am. Nat.* 106:487–491.

Potts, D.C. 1981. Crown of thorns starfish: Man-induced pest or natural phenomenon? In *The Ecology of Pests*, eds. R.L. Kitching and R.E. Jones, 55–86. Melbourne: Commonwealth Scientific and Industrial Research Organization.

Potts, D.C., T.J. Done, P.J. Isdale, and D.A. Fisk. 1985. Dominance of a coral community by the genus *Porites* (Scleractinia). *Mar. Ecol. Prog. Ser.* 23:79–84.

Pratchett, M. 2001. Influence of coral symbionts on feeding preferences of crown-of-thorns starfish *Acanthaster planci* in the Western Pacific. *Mar. Ecol. Prog. Ser.* 214:11–119.

Precht, W.F., R.B. Aronson, S.L. Miller, B.D. Keller, and B. Causey. 2005. The folly of coral restoration following natural disturbances in the Florida Keys National Marine Sanctuary. *Ecol. Restor.* 23:24–28.

Pulseford, J.S. 1996. *Historical Nutrient Usage in Coastal Queensland River Catchments Adjacent to the Great Barrier Reef Marine Park*. Townsville: Great Barrier Reef Marine Park Authority Research Publication No. 40.

Randall, J.E. 1972. Chemical pollution in the sea and the crown-of-thorns starfish (*Acanthaster planci*). *Biotropica* 4:132–144.

Reichelt, R.E., R.H. Bradbury, and P.J. Moran. 1990. Distribution of *Acanthaster planci* outbreaks on the Great Barrier Reef between 1966 and 1989. *Coral Reefs* 9:97–103.

Richmond, R.H. 1993. Coral reefs: Present problems and future concerns resulting from anthropogenic disturbance. *Am. Zool.* 33:524–536.

Rogers, C.S. 1990. Responses of coral reefs and reef organisms to sedimentation. *Mar. Ecol. Prog. Ser.* 62:185–202.

Sapp, J. 1999. *What is Natural? Coral Reef Crisis.* New York: Oxford University Press.

Seymour, R.M., and R.H. Bradbury. 1999. Lengthening reef recovery times from crown-of-thorns outbreaks signal systemic degradation of the Great Barrier Reef. *Mar. Ecol. Prog. Ser.* 176:1–10.

Sokal, R.R., and F.J. Rohlf. 1969. *Biometry: The Principles and Practice of Statistics in Biological Research.* San Francisco: W.H. Freeman.

Sweatman, H. 1996. Commercial fishes as predators of adult *Acanthaster planci. Proc. Eighth Int. Coral Reef Symp.* 1:617–620.

Sweatman, H. (ed.). 1997. *Long-term Monitoring of the Great Barrier Reef. Status Report Number 2, 1997.* Townsville: Australian Institute of Marine Science.

Sweatman, H., D. Bass, A. Cheal, G. Coleman, I. Miller, R. Ninio, K. Osborne, W. Oxley, D. Ryan, A. Thompson, and P. Tomkins. 1998. *Long-term Monitoring of the Great Barrier Reef. Status Report Number 3, 1998.* Townsville: Australian Institute of Marine Science.

Turner, S.J. 1994. The biology and population outbreaks of the corallivorous gastropod *Drupella* on Indo-Pacific reefs. *Oceanogr. Mar. Biol. Ann. Rev.* 32:461–530.

Udy, J.W., W.C. Dennison, W.J. Lee Long, and L.J. McKenzie. 1999. Responses of seagrass to nutrients in the Great Barrier Reef, Australia. *Mar. Ecol. Prog. Ser.* 185:257–271.

van der Laan, J.D., and P. Hogeweg. 1992. Waves of crown-of-thorns starfish outbreaks—Where do they come from? *Coral Reefs* 11:207–213.

van Woesik, R. 1998. Lesion healing on massive *Porites* spp. corals. *Mar. Ecol. Prog. Ser.* 164:213–220.

Veron, J.E.N. 1986. *Corals of Australia and the Indo-Pacific.* Sydney: Angus and Robertson.

Veron, J.E.N. 1995. *Corals in Space and Time: The Biogeography and Evolution of the Scleractinia.* Sydney: University of New South Wales Press.

Veron, J.E.N., and R. Kelly. 1988. Species stability in reef corals of Papua New Guinea and the Indo-Pacific. *Assoc. Aust. Palaeontol. Mem.* 6:1–69.

Vine, P.J. 1973. Crown of thorns (*Acanthaster planci*) plagues: The natural causes theory. *Atoll Res. Bull.* 166:1–10.

Walbran, P.D., R.A. Henderson, A.J.T. Jull, and M.J. Head. 1989a. Evidence from sediments of long-term *Acanthaster planci* predation on corals of the Great Barrier Reef. *Science* 245:847–850.

Walbran, P.D., R.A. Henderson, J.W. Faithful, H.A. Polach, R.G. Sparks, G. Wallace, and D.C. Lowe. 1989b. Crown-of-thorns starfish outbreaks on the Great Barrier Reef: A geological perspective based upon the sediment record. *Coral Reefs* 8:67–78.

Wilkinson, C.R. 1992. Coral reefs of the world are facing widespread devastation: Can we prevent this through sustainable management practices? *Proc. Seventh Int. Coral Reef Symp.* 1:11–21.

Yamaguchi, M. 1986. *Acanthaster planci* infestations on reefs and coral assemblages in Japan: A retrospective analysis of control efforts. *Coral Reefs* 5:23–30.

Zann, L.P. 2000. Status of crown-of-thorns starfish in the Indian Ocean. In *Coral Reefs of the Indian Ocean,* eds. T. McClanahan, C. Sheppard, and D. Obura. New York: Oxford University Press.

5. Influence of Terrigenous Runoff on Offshore Coral Reefs: An Example from the Flower Garden Banks, Gulf of Mexico

Kenneth J.P. Deslarzes and Alexis Lugo-Fernández

5.1 Introduction

Direct anthropogenic impacts, chronic disturbances, and subtle indirect impacts can severely stress coral reef ecosystems. Overfishing, destructive fishing techniques, mechanical damages (e.g., anchoring, ship groundings, mining, and blasting), chronic and extreme sedimentation associated with land use changes, polluted runoff (e.g., inorganic nutrients, organic carbon, insecticides and herbicides, and pathogens), and the introduction of invasive species are among the established list of human insults on reefs (Brown and Howard 1985; Hughes 1994; Jackson 1997; Jackson et al. 2001; Szmant 2002; Turgeon et al. 2002; McCulloch et al. 2003). An evolving list of seemingly uncontrollable impacts includes diseases and plagues, coral bleaching, and effects of global climate change (Lessios 1988; Glynn 1993; Hughes 1994; Peters et al. 1997; Done 1999; Hoegh-Guldberg 1999; Kleypas, McManus, and Meñez 1999; Pittock 1999; Aronson et al. 2000; Aronson and Precht 2001a; Cervino et al. 2001; Kleypas, Buddemeier, and Gattuso 2001; Wellington et al. 2001; Aronson et al. 2002; Bruckner 2002; Harvell et al. 2002; Szmant 2002; Gardner et al. 2003; Hughes et al. 2003). The accelerated degradation of many coral reefs and their inability to recover from disturbances could be the result of synergistic and compounded effects of the increased frequency and/or intensity of natural and human-induced impacts (Knowlton 2001; McClanahan, Polunin, and Done 2002; Hughes et al. 2003). Although some reefs may be recovering or remain in good condition (e.g., Glynn et al. 2000; Edmunds and

Carpenter 2001; Causey et al. 2002; Cho and Woodley 2002), predictions of disaster on coral reefs make preservation and restoration an urgent and global matter (Bellwood et al. 2004). Efforts to preserve and restore reefs will be worthwhile only if regional environmental conditions that support coral reefs are maintained. To best serve science-based coral reef and land-based management, dedicated surveys of coral reef biotas should be complemented by the study and monitoring of regional and site-specific environmental parameters.

Excessive sedimentation and nutrification are probably the most harmful sources of impacts on coral reefs in nearshore environments. Sedimentation and nutrification can cause decreased coral cover, growth, and recruitment; shifts in the local composition of corals from framework-building species (*Montastraea* and *Acropora*) to more opportunistic species (*Agaricia* and *Porites*); and the local shift from a coral-dominated to an alga-dominated community (Dodge, Aller, and Thompson 1974; Rogers 1990; Cortés 1993; McCook 1999; Sullivan-Sealy and Bustamante 1999; Aronson et al. 2002; Causey et al. 2002; Szmant 2002; Turgeon et al. 2002; Gardner et al. 2003; McCulloch et al. 2003; Nugues and Roberts 2003a,b; Wolanski et al. 2003; Aronson et al. 2004). The scientific consensus is that remote oceanic reefs (atolls, barrier reefs, banks) are not affected by nearshore terrigenous runoff (i.e., sedimentation and nutrification) (Szmant 2002). Rather, because of their remoteness coral degradation on offshore reefs has been attributed primarily to bleaching and mortality as a result of global climate change (Hoegh-Guldberg 1999; Wilkinson 2000; Mumby et al. 2001; Sheppard 2003). Yet, considering the connectivity of remote reefs with nearshore environments (e.g., Cappo and Kelley 2001; King et al. 2001; Andréfouët et al. 2002; McCulloch et al. 2003), terrigenous runoff has the potential of being or becoming a regional stressor. In particular, land-use changes since the 1950s have generated an increased supply of sediments and nutrients to river runoff (e.g., Rabalais et al. 1996; McCulloch et al. 2003). Reefs in the downstream path of river plumes have been exposed to increasing amounts of sediments and nutrients. Furthermore, above-average river discharge rates since the 1950s and a predicted 15% global increase in river runoff during the next century have and will accentuate the delivery and propagation of the increasingly abundant sediments and nutrients (Cole 2003; McCulloch et al. 2003; Wolanski et al. 2003; Manabe et al. 2004).

Corals exposed to river runoff and rainfall, and to upwelled water, contain discrete growth increments (bands or lines) that fluoresce when excited by ultraviolet (UV) radiation (Isdale 1984; Boto and Isdale 1985; Smith et al. 1989; Fang and Chou 1992; Tudhope et al. 1996; Barnes and Taylor 2001). The fluorescence is believed to be caused either by organic compounds incorporated into the coral skeleton (Boto and Isdale 1985; Susic, Boto, and Isdale 1991; Ramseyer et al. 1997) or by a change in skeletal architecture (increased number and variation in size of holes) caused by decreased salinity (Barnes and Taylor 2001). While fluorescent bands in corals of the Great Barrier Reef are observed mainly in colonies sampled within 20 km of the shoreline, corals growing in locations more distant from shore also include fluorescent bands (e.g., Scoffin, Tudhope, and Brown 1989; Tudhope et al. 1996; Barnes and Taylor 2001). Considering the significant impacts caused by terrigenous runoff on coral reefs, and the predicted increase in river discharge, tracking fluorescence in corals is a critical environmental monitoring strategy to gauge the progression of riverine influence. Coral reefs currently influenced by river runoff could become exposed to increasing amounts of organic and inorganic substances and lower salinity. Offshore locations, which have thus far yielded coral samples exhibiting faintly fluorescing bands or lines, could soon contain more intense fluorescence. Coral communities in offshore locations thus far exposed to dilute river runoff will probably have to adapt to changing environmental characteristics.

In addition to the effects of local runoff and river discharge, reefs of the Caribbean region are exposed to the seasonal runoff via the Amazon and Orinoco Rivers (Muller-Karger et al. 1989), which are, respectively, the world's first and third largest rivers (mean discharge rates are 6642 and 1129 km^3 yr^{-1}, Dai and Trenberth 2002). The seasonal Orinoco River plume reaches Puerto Rico before dissipating as it drifts westward (Muller-Karger et al. 1989). This plume is a vehicle for nitrogen transport that promotes carbon fixation (Müller-Karger et al. 1989). The recurrence and potential increase in the discharge volume and sediment/nutrient content of the Orinoco plume may influence the condition of reefs of the Antilles island arc from Venezuela to Puerto Rico. Farther north in the western Atlantic region, two other large rivers—the Usumacinta River (mean discharge rate is 89 km^3 yr^{-1}; Dai and Trenberth 2002) and the Mississippi River (mean discharge rate is 610 km^3 yr^{-1}; Dai and Trenberth 2002)—discharge their runoff into the Gulf of Mexico. Monthly changes in river runoff rates were recorded in midshelf corals of the southwestern Gulf of Mexico (Beaver et al. 1996). Surveys of coral reefs in northern and eastern Caribbean, and the southwestern Gulf of Mexico report the deterioration of fringing reefs, which is attributed to multiple local stresses and to coral bleaching and diseases (Lang et al. 1998; Aronson and Precht 2001b; Linton et al. 2002; Hoetjes et al. 2002; Gardner et al. 2003). Many of the reefs in the Veracruz, Mexico, area have been decimated by the compounding effects of increased and polluted river runoff, bleaching, and diseases (Lang et al. 1998). True coral reefs in the northwestern Gulf of Mexico (NWGOM) are limited to two large, submerged banks, the Flower Garden Banks (FGB), which are located 190 km offshore from Texas on the Louisiana–Texas shelf edge (Fig. 5.1). The condition of the FGB reefs has not changed since the 1970s and they remain in good condition with coral cover exceeding 40% (Gittings 1998; Dokken et al. 2003; Aronson et al. 2005). It is believed that their geographical remoteness, their shelf-edge location (150 m surrounding water depth), water depth on the reefs (>15 m), their exposure to deep-water circulation, a high-energy and dilute environment, and enforced protective legislation all contribute to maintaining the good condition of these reefs (McGrail 1983; Gittings 1998; Lugo-Fernández 1998; Lang, Deslarzes, and Schmahl 2001).

Examination of slabbed cores from *Montastraea faveolata* colonies (150 cm long and 8 cm in diameter) collected in 1990 at the FGB revealed recurrent fluorescent bands throughout the cores (Fig. 5.2). These bands were typically wide, occurred within the low-density nonsummer skeletal growth, and overlapped the high-density summer growth. This intriguing observation led us to hypothesize

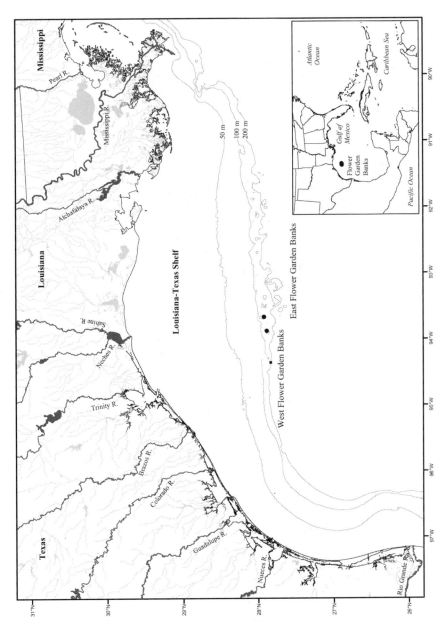

Figure 5.1. Map of the Flower Garden Banks, northwest Gulf of Mexico (■ - Mooring).

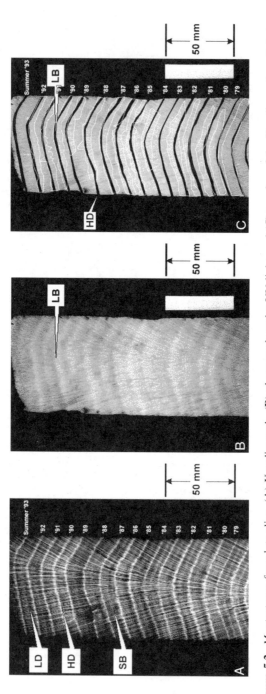

Figure 5.2. *Montastraea faveolata* slice. (**A**) X-radiograph, (**B**) photograph under UV light, and (**C**) overlay of the X-radiograph and photograph taken under UV light with outlines of high-density growth bands and fluorescence bands: LD, low-density growth band; HD, high-density growth band; SB, stress band; LB, fluorescence band.

that the fluorescence was due to (1) the incorporation of organic matter originating either from river runoff of upwelled water (Boto and Isdale 1985; Tudhope et al. 1996); (2) changes in skeletal structure resulting from exposure to low salinity (Barnes and Taylor 2001); or (3) a combination of the effects of low salinity and incorporation of organic matter. To test these hypotheses, we examined relevant environmental and oceanographic data of the NWGOM. We were particularly interested in the implications of the regional discharge of the Mississippi and Atchafalaya Rivers and Texas rivers (combined mean discharge: 20,117 m^3 s^{-1}; Dinnel and Bratkovich 1993; Nowlin et al. 1998) (Fig. 5.1). The Mississippi–Atchafalaya River system exhibits a seasonal pattern of discharge with peak flux of 30,000 m^3 s^{-1} in spring and low of 5000 m^3 s^{-1} in fall (Murray 1997). Nowlin et al. (2000) reported that low surface salinities (<33 psu), presumably derived from the Mississippi–Atchafalaya Rivers, have occurred over the Louisiana–Texas (LATEX) shelf edge since 1935. Also, Dodge and Lang (1983) found a negative correlation between the discharge of the Atchafalaya River and coral growth at the FGB, suggesting a correlation between freshwater discharge and the physiology of corals at the FGB. From 1989 to 1992, we observed recurrent episodes at the FGB of discolored water containing suspended matter, similar to that found in nearshore areas. For these reasons, we believed there might be a connection between the fluorescence we observed in the coral slabs and the low-salinity and associated discolored water events at the FGB.

5.2 Materials and Methods

5.2.1 Study Area

The FGB are two topographic highs (204 km^2) caused by salt diapirism (Rezak, Bright, and McGrail 1985). They are 18 km apart and located at the shelf edge, near the widest part of the LATEX shelf (190 km off Texas; 200 km off Louisiana; Fig. 5.1). The banks on which the FGB sit rise from depths of 100 to 150 m to within 17 m of the sea surface. Lush reefs of hermatypic corals (~2.7 km^2) cap the upper 48 m of the FGB (Bright et al. 1984; Gittings et al. 1992; Lugo-Fernández et al. 2001). Corals are found from the bank crest down to 52 m. The FGB coral reefs are unique in the NWGOM, hosting 21 coral species, >250 invertebrate species, and 257 reef fish species (Bright et al. 1984; Rezak, Bright, and McGrail 1985; Gittings et al. 1992; Pattengill 1998). The 17- to 25-m depth range has an estimated average of 46% coral cover, mainly composed of *M. faveolata*, *M. franksi*, *M. annularis*, *M. cavernosa*, *Diploria strigosa*, *Porites astreoides*, and *Colpophyllia natans* (Gittings et al. 1992). Coral cover appears to increase within the 25- to 30-m depth range.

The FGB are submerged, shelf-edge, Holocene reefs (Rezak 1977), but they differ from typical Caribbean shelf-edge reefs in several ways: (1) the FGB reefs are alive and active; (2) they occur at depths where Caribbean inner-shelf reefs are found (shallower than Caribbean submerged reefs; Macintyre 1972); and (3) they protrude into the mixed layer, away from the nepheloid layer generally present near the seafloor (McGrail 1983). The FGB reefs are, however, similar to Caribbean shelf-edge reefs in that both exist near deep water (Lugo-Fernández 1998).

5.2.2 Data Sources

Patterns of accretionary growth and fluorescent banding were assessed using a 4-mm-thick slice of a live *M. faveolata* head collected in the summer of 1994 at the FGB. The slice was X-radiographed to reveal skeletal growth following similar methods used by Knutson, Buddemeier, and Smith (1972) and photographed under UV light to reveal fluorescent bands (UV light: 625 nm; photograph: 200 ASA Kodak color print; 3.2 f; 3-s exposure). To investigate the timing and causes of fluorescent bands in *M. faveolata*, we began by qualitatively comparing the locations of fluorescent and growth bands in the coral skeleton. Relative positions of growth and fluorescent bands were compared by visually outlining the high-density and fluorescent bands before overlaying the coral slice X-radiograph and photograph taken under UV light.

We examined temperature, salinity, solar radiation, chlorophyll a, and nutrients to explore possible relationships with the coral's fluorescent bands. These data came from oceanographic studies conducted in the NWGOM and at the FGB (McGrail 1983; SAIC 1989; CSA 1996; Nowlin et al. 1998; Dokken et al. 2003). A 30-month time series of temperature and salinity came from "Mooring 8," located 30 to 48 km west of the FGB (Fig. 5.1), which operated from 1992 to 1994 (Nowlin et al. 1998). These data provided the most comprehensive regional environmental information. Hydrographic data (vertical profiles of salinity, nutrients, and solar radiation) came from stations within half a degree circle of the FGB and were collected during six cruises conducted in November and July/August 1992, 1993, and 1994 by Nowlin et al. (1998). Dokken et al. (2003) measured photosynthetically active radiation (PAR) on the crests of the FGB bank at approximately 20 m from 1997 to 1999. Additionally, PAR measurements along a cross-shelf transect near 92° W by SAIC (1989) were employed to examine its cross-shelf variation. We limited our analysis of PAR measurements to the first 20 m to make our results compatible with those of Dokken et al. (2003) and with Secchi disk measurements by Nowlin et al. (1998). Using deeper PAR profiles (e.g., 40 or 60 m) changes the absolute values but the trends remain unchanged. Nutrient and chlorophyll a data were from Rezak, Bright, and McGrail (1985), Kennicutt (1995), and Nowlin et al. (1998). Vertical profiles of chlorophyll a down to 100 m came from seven cruises and the nutrient profiles came from 10 cruises over the LATEX shelf (Nowlin et al. 1998).

The climatological and oceanographic data examined here contain strong seasonal signals, reflecting the Northern Gulf of Mexico (NGOM) environment. There are two seasons in the NGOM: winter (November through March) and summer (May through September). Two transitional months (April and October) exhibit large interannual variations (FAMU 1988). The LATEX shelf circulation, mainly wind-driven, also reflects two seasons: summer (June through September) and nonsummer (November through April), with May and October as transitions

(Nowlin et al. 1998). Moreover, while the river discharge cycle is in phase with the climate cycle, the LATEX shelf circulation cycle determines its ultimate distribution.

5.3 Results

5.3.1 Temperature

The annual seawater temperature profiles for 1990 to 1997 on the crests of the FGB displayed strong seasonal variation: roughly seven months of seawater warming (including two-and-a-half months of high temperatures) and five months of seawater cooling (Lugo-Fernández 1998; Fig. 5.3A). Water temperature on top of the banks was 22 to 23 °C in January, cooled to an annual minimum of 18 to 19 °C in February, and peaked at approximately 28 to 30 °C from late June to mid-September. Water temperature cooled beginning in late September. Maximum/minimum temperatures at the top of the banks varied ±2 °C or more from the mean. Furthermore, interannual variability was significant. The high-density growth bands observed in *M. faveolata* were probably laid down during peak water temperatures in mid-July to late September (Highsmith 1979).

Peculiar features of these annual temperature series are recurrent 1 to 4 °C drops in July and August (Fig. 5.4). The temperature record on top of the banks showed that the largest average temperature drop during the summer took place in July (Lugo-Fernández 1998). Consistent with our findings, temperature meas- urements at Stetson Bank, located approximately 58 km northwest of the East Flower Garden Bank, showed similar temperature drops during the summer (G.S. Boland unpublished data). Such temperature drops were probably caused by cooler water that moved through the FGB and Stetson Bank. Because of the tim- ing of these temperature drops, we speculate that the cooler water in question originated from the nearshore and was formed during the spring (high freshwater content, low temperature).

The annual water temperature cycle also included the effects of the seasonal shelf circulation and deep-water eddies (Nowlin et al. 1998). Deeper in the water column (100 and 190 m), temperature variations decreased and the timing of the minimum/maximum temperatures shifted in time (Fig. 5.3A). This vertical tem- perature variation produced a mixed layer, the thickness of which varied season- ally. The minimum vertical thickness (~20 m) occurred in April and the maximum (~70 m) in December to January (McGrail 1983).

5.3.2 Salinity

Fortnightly salinity averages from 1992 to 1994 at Mooring 8 ranged from 30.4 to 37.4 psu at a 13-m depth, and 36.0 to 36.5 psu at 100- and 190-m depths (Fig. 5.3B). At 13 m, salinity consistently decreased from a high value of approxi- mately 35.5 to 36 psu in December to 30.5 to 32.5 psu in late April to July. These low salinities represent a mixture of 12 to 17% freshwater and 83 to 88% seawater,

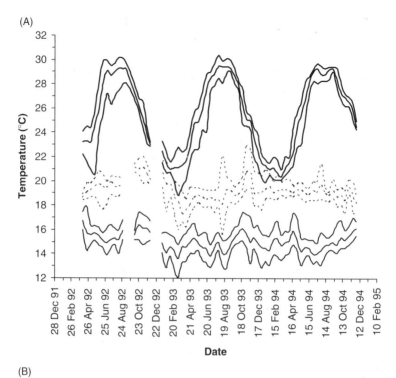

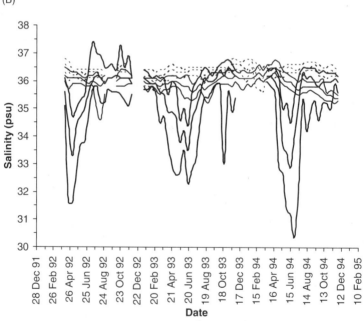

Figure 5.3. Maximum, average, and minimum seawater temperature and salinity 30–48 km west of the FGB: (**A**) Fortnightly temperature and (**B**) salinity time series measured at LATEX Mooring 8 at 13 m (thick solid line), 100 m (stippled line), and 190 m (thin solid line) below MSL. Data from Nowlin et al. (1998).

(A)

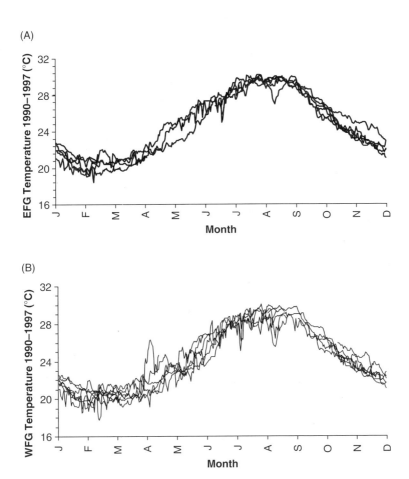

(B)

Figure 5.4. Average daily seawater temperature on the crest of (**A**) the East Flower Garden Bank (EFG) and (**B**) West Flower Garden Bank (WFG) from 1990 to 1997. SR Gittings and KJP Deslarzes unpublished data.

assuming a base salinity of 37 psu (Dinnel and Wiseman 1986). Salinity at 100 and 190 m, however, remained between approximately 35 and 36.5 psu throughout the year with little seasonal variability. McGrail (1983) also observed low-salinity water at the FGB (30–31 psu). CSA (1996) recorded low salinity near the surface and 1 m above the reefs in summer. In contrast to Dodge and Lang (1983), McGrail (1983) believed that the low-salinity water did not reach down to the FGB reefs. Nevertheless, data from Nowlin et al. (1998) collected at 13 m (Mooring 8) and those of CSA (1996) imply that low-salinity water does come in contact with FGB reefs on a seasonal basis.

The timing of near-surface salinity minima at the FGB showed little temporal variation, always occurring between late April and July. It is noteworthy that the phase relationships between fortnightly averages of temperature and salinity vary over time. Low salinities observed at the FGB tend to occur before or at the onset of summer maximum temperatures (Fig. 5.3). However, the minimum salinities coincide with the temperature drops in July noted in Fig. 5.4. Furthermore, comparing the summer surface salinity minimum with the discharge from the previous year of the Mississippi and Atchafalaya Rivers suggests a negative correlation between these two parameters.

To explore a potential correlation between summer sea-surface salinity (S) and river discharge (RD), we gathered all salinity (S) data collected at the FGB, same-year mean river discharge [RD(t)], and previous-year mean river discharge [RD(t−1)] (Table 5.1). The Pearson correlation coefficient between S and RD(t) was −0.30 (95% confidence interval: −0.79 to 0.47), and that between S and RD(t−1) was −0.46 (95% confidence interval: −0.85 to 0.35). As expected from our small sample ($N = 8$) these correlations are not statistically significant; however, they do suggest that river freshwater is important when explaining the salinity variations at the FGB. A probable scenario of same-year and preceding-year river runoff influence on salinity of LATEX shelf waters is as follows. The annual discharge of the Atchafalaya and Mississippi Rivers reaches a maximum in April, has a long period of low discharge from August to November, and concludes with a small, secondary discharge peak in December to January (Nowlin et al. 1998). During the April discharge peak, freshwater is confined mainly to the inner shelf, traveling westward for over 350 km until reaching Texas. Previous-year or "old" mixed river water and seawater is found on the midshelf (Dinnel and Wiseman 1986). By summer, some of the same-year mix of river water and seawater moves to the midshelf and by late fall it has moved east and south (Dinnel and Wiseman 1986). Murray (1997) observed same-year low-salinity water in an approximately 40-km-wide band trending southwestward off Galveston, Texas. Farther west and south, many local rivers discharging onto the narrow shelf off Texas add more freshwater (particularly in October) to the same-year midshelf river–seawater mix (Dinnel and Wiseman 1986; Nowlin et al. 1998).

We believe that the midshelf river–seawater mix (same year and previous year) reaches the FGB. The area near the Atchafalaya River probably makes the greatest contribution of river–seawater mix that reaches the FGB since it is the major

Table 5.1. Summer sea-surface salinity (S), same-year river discharge [RD(t)], and previous-year discharge [RD(t-1)] at the FGB from 1979 to 1995. Salinity is expressed in psu; river discharge is expressed in $m^3 s^{-1}$

Year	Surface salinity (S)	Mean river discharge [RD(t)]	Previous-year mean river discharge [RD(t-1)]
1979	31.50	1.73×10^9	1.15×10^9
1981	35.60	8.89×10^8	1.07×10^9
1989	33.50	1.40×10^9	9.17×10^8
1991	35.50	1.58×10^9	1.50×10^9
1992	31.50	1.15×10^9	1.58×10^9
1993	32.20	1.89×10^9	1.15×10^9
1994	30.20	1.36×10^9	1.89×10^9
1995	33.20	1.25×10^9	1.36×10^9
Pearson correlation, r		−0.30	−0.46

and closest source of freshwater. During April, the upper-layer water transport at the shelf edge is offshore (Current and Wiseman 2000), which can help draw water to the FGB. During the fall, the shelf circulation turns westward, advecting most of the river and seawater mix westward until it reaches Texas, where additional freshwater is added to the existing river–seawater mix. The fall (September to November) shelf-edge transport is onshore and coupled with the easterly flow could preclude freshwater from reaching the FGB. During the same time, however, the river–seawater mix located off Texas turns offshore and eastward along the shelf edge where it can reach the FGB. Furthermore, low-salinity water can also come as water parcels that break from the nearshore, low-salinity plume in summer, as observed in Australia (Wolanski and Jones 1981; Wolanski and van Senden 1983). Nowlin et al. (2000) also reported that low-salinity surface water (<33 psu) derived from the Mississippi and Atchafalaya Rivers is transported to the NWGOM shelf edge. Such low-salinity water cannot come from the offshore or deep Gulf because there the upper-layer (0–250 m) salinity ranges from 36.4 to 36.5 psu year-round (Nowlin et al. 2000). Low-salinity water is probably not a direct river input because the river plume (Mississippi and Atchafalaya Rivers) does not extend to the FGB (Walker 1996).

5.3.3 Light Attenuation

We estimated an average light attenuation coefficient (k), which represents the combined effects of pure water, suspended particulate matter (SPM), and dissolved organic matter (DOM) on photosynthetic active radiation (PAR) (Parsons, Takahashi, and Hargrave 1984). The minimum average attenuation coefficient in clear seawater is $k = 0.0384$ m^{-1}. This minimum is different from the specific attenuation coefficient at a given wavelength which can be less than 0.03 m^{-1} (Parsons, Takahashi, and Hargrave 1984). The in situ PAR measurements from Dokken et al. (2003) revealed episodes of increased attenuation from January to March, June through August, and October through December (attenuation episodes lasted 10–30 days). Dokken et al. (2003) reported values of k ranging from about 0.065 m^{-1} in late spring (May) to near 0.13 m^{-1} by summer (July). The episodic nature of the increased light attenuation at the FGB suggests that it is caused by water parcels containing high amounts of SPM or DOM.

In May, June, July, and October of 1989 to 1997 one of us (KJPD) witnessed murky, green-brown or discolored waters at the FGB (upper 10 m, occasionally extending down onto the crests of the banks). Dokken et al. (2003) reported similar water discolorations. Log entries on dive charters visiting the FGB from May 1988 to August 1997 contain numerous observations of "green, blue-green, brackish water, and terrible visibility" (Rinn Boats, Inc., Freeport, Texas, unpublished report). Discolored waters at the FGB occur near the surface and are discontinuous. Such discoloration could be caused by entrainment of materials from the near-bottom nepheloid layer around the FGB (McGrail 1985), but this seems unlikely since the nepheloid layer, with a thickness of ~20 m, is restricted to the near-bottom at depths of approximately 100 m. Entrainment of materials through 80 m of water seems unlikely in this environment.

Near the LATEX shelf edge, SPM, river-derived yellow substances, and water properties are documented causes for light attenuation (SAIC 1989). Hence, light attenuation measurements at the FGB can be used as indicators of river influence. SPM near the NGOM shelf edge have exhibited large interannual variations related mostly to river discharge characteristics (Nowlin et al. 1998). SPM observations at the shelf edge near the FGB range from 0.15 to 2.47 mg L^{-1} (SAIC 1989; Nowlin et al. 1998).

To investigate further the nature of light attenuation at the FGB, we sorted the vertical profiles of PAR into winter and summer profiles and fitted them with Beer's law, $I = I_0 \exp(-kz)$, to estimate an average attenuation (k). The shape of the PAR vertical profiles suggests that the FGB down to 20 m are generally part of near-surface optical layers; thus, using the first 20 m insures consistency in our results and reflects the near-surface conditions influencing FGB corals. Two estimates of k estimated from the regressions were discarded because they were smaller than the physical minimum of 0.0384 m^{-1}. The remaining k-values ranged from 0.0445 to 0.1237 m^{-1}, which is similar to the range reported by Dokken et al. (2003). Estimates of k sorted by season were plotted versus salinity (S) and transmissivity (VXMISS) (Fig. 5.5A,B) and statistics computed for each group. The summer average k was 0.0833 m^{-1} (range 0.0445 to 0.1237 m^{-1}; standard deviation 0.025 m^{-1}) and the winter average k was 0.0840 m^{-1} (range 0.0711–0.1021 m^{-1}; standard deviation 0.0077 m^{-1}). The k-distribution was bimodal in summer but not in winter (Fig. 5.5). The winter k-distribution (Fig. 5.5A) consisted of one cluster of near uniform values associated with high salinities. In summer (Fig. 5.5B) the k-distribution formed two clusters, one with $k > 0.07$ m^{-1}, associated mainly with low salinities, and a second cluster with $k \sim 0.05$ to 0.07 m^{-1}, associated with high salinities. We conducted a multiple regression of the form:

$$k = aS + b\text{VXMISS} + c\text{Season} + d\text{SPM} + \beta + \varepsilon$$

where S = salinity; VXMISS = transmissivity; SPM = suspended particulate matter; a, b, c, d, and β are coefficients; and ε is the error. The regression analysis, $r^2 = 0.63$, $df = 22$, was significant and explained about 63% of the k-variability. The β-constant was significant ($P = 0.0007$), salinity was significant ($P = 0.001$), season was significant ($P = 0.010$), and transmissivity was marginally significant ($P = 0.040$). SPM was not significant ($P = 0.412$). The regression equation is:

$$k = -0.0087 \cdot S - 0.1068 \cdot \text{VXMISS} + 0.0171 \cdot \text{Season} + 0.8324 + \varepsilon$$

Negative values represent decreasing values of k with increasing salinity and transmissivity, which means saltier and clearer waters attenuate light less. The multiple regression without SPM yielded essentially the same results. A regression of k with salinity explained about 42% of the k-variability. These results suggested plotting k versus season and salinity (Fig. 5.6), which reveal the two clusters in summer associated with different salinities, as well as more uniform k-values associated with high salinities in winter.

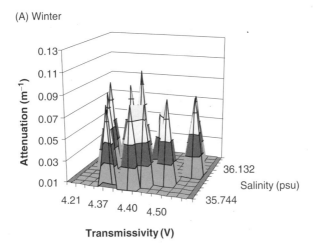

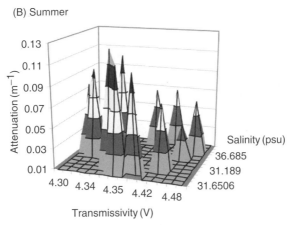

Figure 5.5. Average seawater light attenuation coefficient, k, near the FGB as a function of salinity and transmissivity in (**A**) winter and (**B**) summer. Data from Nowlin et al. (1998).

We also computed light attenuation coefficients (k_s) from Secchi disk depths (D_s) collected near the shelf edge (SAIC 1989; Nowlin et al. 1998) using the relationship $k_s = 1.7/D_s$ (Parsons, Takahashi, and Hargrave 1984). The k_s monthly averages varied between 0.04 and 0.1 m^{-1} and attained highest values in summer (Fig. 5.7). These values of k_s agreed with light attenuation coefficients in Dokken et al. (2003) and those estimated from our analysis of PAR. A multiple regression analysis similar to the one described above showed that k_s values were unrelated to salinity, transmissivity, and SPM. However, there was a marginally significant relationship with season ($P = 0.0468$). This result suggested that the time of the year affected the Secchi disk depth. However, a larger sample is needed to corroborate this finding.

PAR measurements along a cross-shelf transect near 92° W (SAIC 1989) were used to examine the variation of k versus cross-shelf water depth, with water depth serving as a proxy for distance from shore. Nearshore k values were very

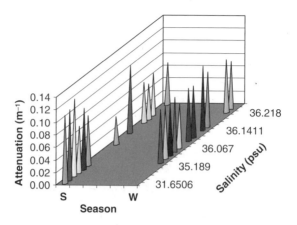

Figure 5.6. Distribution of average seawater light attenuation coefficient, k, as function of season and salinity near the FGB. Data from Nowlin et al. (1998).

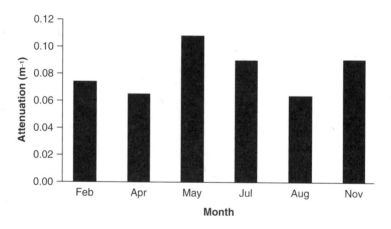

Figure 5.7. Average seawater light attenuation coefficient, k, measured by Secchi disk on the northern Gulf of Mexico shelf edge at 92°W. Data from SAIC (1989).

high ($k > 0.4$ m^{-1}). At depths of 40 to 50 m, k decreased to approximately 0.2 m^{-1}, and at the shelf edge, k was approximately 0.1 m^{-1}. Højerslev and Aarup (2002) explained the scatter of k as mainly due to SPM (detritus), which decreased from nearshore to offshore. The yellow substance, although well mixed in the upper layer, reflected the annual cycle of the river discharge in nearshore areas. Optically, deep oceanic-type waters of the NGOM generally occur near the shelf edge, which implies relatively clear waters (small k values). Because of the high input of yellow substances by the rivers, nearshore waters in the NGOM are of a yellow to greenish color and become blue in the offshore direction. Thus, water parcels of yellowish color observed at the FGB must have originated nearshore and been transported to the FGB by shelf circulation.

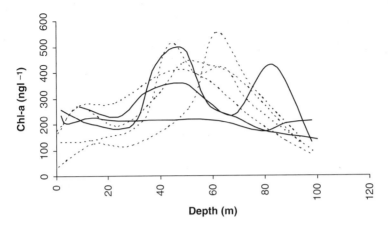

Figure 5.8. Vertical profiles of chlorophyll a acquired near the FGB from 1992 to 1994: stippled line, summer; solid line, winter. Data from Nowlin et al. (1998).

5.3.4 Chlorophyll a

Vertical profiles of chlorophyll a on the LATEX shelf sorted by winter and summer showed a chlorophyll a maximum at a 40-m water depth (Fig. 5.8). However, within the upper 30 m there was great seasonal and annual variability. The winter chlorophyll a values clustered around 200 to 300 ng L^{-1}, but summer chlorophyll a values showed greater scatter, 100 to 300 ng L^{-1}. Two summer chlorophyll a profiles overlapped the winter profiles, but the maxima of two other summer chlorophyll a profiles were less than 150 ng L^{-1}. Nowlin et al. (1998) found a high correlation between chlorophyll a and river discharge within the inner shelf. Furthermore, the distribution of chlorophyll a was closely related to the low-frequency circulation of the shelf (Nowlin et al. 1998). The yearly variation of chlorophyll a on the LATEX shelf was related to the magnitude of freshwater on the shelf (Nowlin et al. 1998). Thus, one might expect to see a seasonal signal near the FGB since the amount of freshwater varies seasonally. Time series data rather than snapshots of chlorophyll a would probably be best suited to determine the absence/presence of a seasonal signal at the FGB.

5.3.5 Nutrients

We examined silicate, phosphate, nitrite, and nitrate data that Nowlin et al. (1998) collected from 1992 to 1994 during ten cruises near the FGB. Nutrients were depleted in water shallower than 50 m but attained maximum concentrations below 100 m (Fig. 5.9). There was no seasonal or annual variability within the nutrient data except for silicate, which showed some variability over the entire depth range without any apparent trend. Kennicutt (1995) and Rezak, Bright, and McGrail (1983) found similar nutrient values at the FGB: nutrient depletion in superficial waters and increased nutrient concentrations below 50 m. Low nutrient levels at the shelf edge occurred along the entire NWGOM shelf (Nowlin et al. 1998).

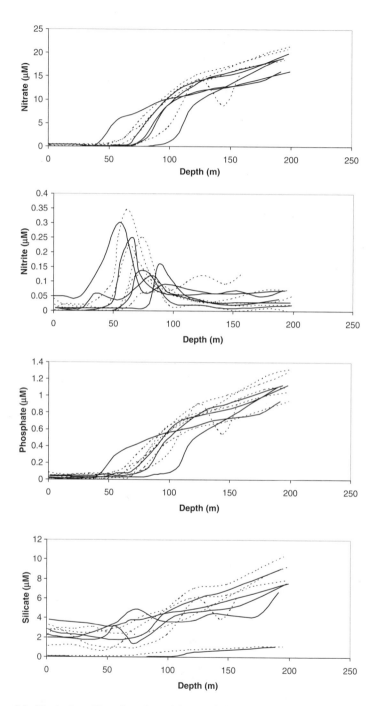

Figure 5.9. Vertical profiles of nutrients (nitrate, nitrite, phosphate, silicate) acquired near the FGB from 1992 to 1994: stippled line, summer; solid line, winter. Data from Nowlin et al. (1998).

5.3.6 Coral X-Radiography and Fluorescence

The X-radiograph of *M. faveolata* showed a succession of paired narrow, high-density and broad, low-density accretionary growth bands. Following Highsmith (1979) we interpret the high/low-density bands as representing, respectively, summer and nonsummer skeletal growth accreted from 1978 to 1994 (Fig. 5.2A). The uppermost high-density band represented the 1994 summer growth. The broad, low-density bands frequently contained thin, high-density stress bands located approximately halfway between successive high-density bands (Fig. 5.2A). Because such stress bands are laid down during the nonsummer growth season at the FGB, they may coincide with winter seawater temperature minima (~18 °C, Fig. 5.4A), similar to what Hudson (1977) found in *M. annularis* at a Florida Keys reef. We believe these stress bands are not associated with salinity changes because winter salinities around the FGB are near their normal open-water values (~36 psu).

Under UV light, the *M. faveolata* slice revealed annually recurrent fluorescent bands (Fig. 5.2B). By superimposing the X-radiograph positive and the photograph of the coral slab taken under UV light, we found that from 1983 to 1994 one fluorescent band occurred before each summer high-density band (Fig. 5.2C). Each fluorescent band was approximately twice as wide as the nearest high-density band and covered roughly a third of the low-density band. From 1979 to 1982, however, the yearly fluorescent band succeeded the summer high-density band while remaining adjacent to it. Thus, from 1979 to 1982 the fluorescent bands appeared to be incorporated in the coral skeleton after the annual seawater temperature maximum (August), whereas from 1983 to 1994 the bands were incorporated before the summer temperature maximum (probably June to August). In both intervals, though, the fluorescent bands extended over low-density skeleton.

5.4 Discussion

The existing environmental data on the NWGOM, and in particular data on nutrients (nitrate, nitrite, phosphate, and silicate), chlorophyll a, seawater temperature and salinity, PAR, and discharge of the Mississippi–Atchafalaya Rivers, are essential to decipher the influence of regional terrigenous runoff on corals of the FGB.

Nutrient and chlorophyll a levels at or near the FGB were low and did not show seasonal variations. Low nutrient values of nutrients are typical of the oligotrophic waters surrounding offshore coral reefs (Sorokin 1995). The uniform temporal pattern of chlorophyll a suggests that in situ algal blooms have not been occurring on a seasonal basis. This lack of seasonal variation was surprising, because we had hypothesized that river water and its nutrients reach the FGB and could drive phytoplankton production as observed elsewhere (e.g., Heil 1996). Low nutrients and the absence of seasonal variation implies that if river water is reaching the FGB it must be "old" water which had its nutrients depleted before reaching the FGB and is incapable of inducing in situ phytoplankton blooms. Low chlorophyll a values in the water column promote coral growth.

Temperature data from the FGB showed the expected seasonal cycle of cooling during the winter and warming in the summer. There were, however, consistent temperature drops (1–4 °C) during the summer warming. Salinity was high (~35–36 psu) most of the year, except from April to July when it consistently dropped 2 to 3 psu (Fig. 5.3B). Co-occurring measurements of declining seawater temperature and low salinity during the summer identified the relatively cool and freshwater masses (32–34 psu; ~27 °C; 12%–17% freshwater mixed with 83%–88% seawater) flowing through the FGB area. The most likely source of these transient water masses was a mix of warm Gulf water and river water derived from snowmelt runoff that had yet to be fully warmed by the summer heat. This fresh but cooler river–seawater mix found at the FGB was an "aged" river–seawater mass as indicated by its low nutrient contents. Finding river water at the FGB and at 1 m above the crest of the bank, as the low salinities suggest, differs from McGrail's (1983) early hydrographic assessment at the FGB. Similarly, Jaap (1984) reported that a mix of Mississippi River and open Gulf water occasionally reaches the Florida Keys, with salinities of 32 to 34 psu.

Light, a key ecological control of reef-building corals (Stoddart 1969; Veron 1995; Kleypas, McManus, and Menez 1999), is attenuated by SPM, DOM, and water (Parsons, Takahashi, and Hargrave 1984). Estimates of k at the FGB varied from 0.0445 to 0.1237 m^{-1}. A regression analysis showed that k correlated significantly with season, salinity, and transmissivity at the 95% confidence level. However, SPM did not significantly contribute to light attenuation at the FGB, in agreement with the low chlorophyll a values observed. There were events of high attenuation of light during the summer but not in winter (Fig. 5.6). The relationship of light attenuation and season most likely reflected changes in the inclination of the sun or a covariation with river discharge. Transmissivity, which is an indicator of the water column turbidity level, also affected light attenuation. The significant negative correlation between k and salinity meant that low salinity at the FGB was associated with high attenuation of light. Low seawater transmissivity associated with low salinity at the FGB signaled the presence of river-derived, colored DOM. Højerslev and Aarup (2002) found a similar negative correlation between salinity and light attenuation at the UV-B wavelengths over this shelf. They interpreted the k–salinity relationship in terms of the input of colored DOM from the Mississippi and Atchafalaya Rivers into the LATEX shelf. Recall that low chlorophyll a values precluded high optical attenuation through phytoplankton blooms. Del Castillo et al. (2000) observed a similar situation over the northeastern Florida shelf, where a mix of saltier Gulf water and river water with high DOM travels from nearshore to the outer shelf. A further relevant aspect of light attenuation data as it applies to the FGB was the discontinuous distribution of k (Fig. 5.5), which was in agreement with Dokken et al. (2003). This suggested that high attenuation of light occurred when water parcels of low temperature and salinity, containing colored DOM, moved through the FGB. These events of high attenuation of light lasted a finite time ($\leq$30 days) as observed by Dokken et al. (2003).

River runoff in the NGOM is a major source of freshwater input into the LATEX shelf, since the mean seasonal river discharge is more than four times greater than the mean rain input (Etter 1996). River discharge or runoff is either event-like or follows a regular pattern depending on the catchment basin. A large catchment basin (e.g., the Mississippi and Atchafalaya Rivers, draining nearly 42% of the area of United States) will act as a spatial and temporal integrator of freshwater, such that the extended and regular runoff can leave strong signals on the reefs because of its temporal and spatial repeatability. Rivers with small catchment basins such as small islands, or the small rivers of south Texas or northeastern Australia (Byron 1993; Nowlin et al. 1998; King et al. 2001) exhibit event-type runoff. The effects of event-type river runoff on corals depend on the magnitude of the freshwater flooding, locations of reefs, and the shelf circulation (King et al. 2001). However, because they tend to occur at any time, their signals in the reef are more difficult to detect. River runoff induces large salinity variations and transports sediments, nutrients, pollutants, and potential pathogens that may directly or indirectly affect corals (Bruckner, Bruckner, and Williams 1997; ISRS 1999; Bruckner 2002). Thus salinity at the FGB should display the strongest signal of river runoff.

Low salinities at the FGB correlated with same-year and previous-year river discharge from the Mississippi and Atchafalaya Rivers, a result that is in agreement with Dinnel and Wiseman (1986). It appears that an "aged" mix of seawater and Mississippi and Atchafalaya River water recurrently reaches the FGB during the summer (June through September). This is the probable source of fluorescence in corals of the FGB. Evidence supporting this link includes the following. First, seawater characterized by low concentrations of nutrients and chlorophyll a, low salinity, and low temperature that is found at the FGB in summer can only be the product of the mixing of seawater and freshwater. Second, a probable regional source of abundant freshwater is the discharge of the Mississippi and Atchafalaya Rivers. Third, the elevated optical attenuation coefficients that occur in summer at the FGB can only be explained by colored DOM originating from regional river discharge. Fourth, low salinity is known to occur over the shelf edge in the NWGOM and its occurrence can be explained by circulation on the LATEX shelf (Nowlin et al. 2000). Finally, Dodge and Lang (1983) found a negative correlation between river discharge and coral accretionary growth at the FGB.

Andréfouët et al. (2002) found a similar link in the western Caribbean between outer shelf reefs and nearshore river discharge. However, the nature of the linkage is different. In the Caribbean the link seems to occur during high river discharge years and through a direct contact with the river plume. Over the LATEX shelf the contact is annual and through water parcels transported to the FGB from the nearshore environment, because the actual river plume does not reach the shelf edge (Walker 1996).

Oceanographic processes control the fate of river discharge, regardless of its nature (King et al. 2001), and control its effects on corals (Wolanski and Jones 1981; Wolanski and van Senden 1983; Kitheka 1996). Prolonged exposure to freshwater is known to impact corals adversely, yet many reefs and their corals are exposed to salinities as low as 25 psu (Stoddart 1969; Coles and Jokiel 1992; Kleypas et al. 1999). Modern coral reefs are located between 30°N and 30°S (e.g., Veron 1995; Spalding et al. 2001) in areas subject to high precipitation especially in the Indo-Pacific region and experience a wide range of salinities (Coles and Jokiel 1992; NASA 2002). Most coral reefs, except for atolls, exist near coastlines or on insular or continental shelves, where they are exposed directly to freshwater through rainfall and/or runoff. Freshwater input from rainfall on reefs is mostly intermittent and occurs following torrential rainfalls or tropical storms (Coles and Jokiel 1992). Such direct input or large salinity fluctuations in the upper ocean can be harmful to corals found in shallow water (e.g., reef flats) (Stoddart 1969; Glynn 1973; Jaap 1984; Coles and Jokiel 1992). While one could be tempted to assume that rainfall is not relevant at the FGB because the reefs are located offshore and at ~18 m depth, the salinity data suggest otherwise.

Seawater temperature, salinity, light attenuation, and oceanographic data from the LATEX shelf and the FGB show that a mix of river and seawater reaches the FGB during the summer. The recurring low-salinity events at the FGB may affect corals in several ways. Low salinity values similar to those observed at the FGB are known to reduce fertilization rates of coral gametes (Richmond 1994) but cause little harm to coral planulae (Edmonson 1946). Salinities as low as observed at the FGB can make some corals become heterotrophic immediately before the spawning season (Moberg et al. 1997). This may be happening at the FGB since mass coral spawning takes place a month or so after the end of the low-salinity interval (Hagman, Gittings, and Deslarzes 1998). In some cases, exposure to low salinity can increase the sensitivity of corals to high seawater temperature and their susceptibility to bleaching (Coles and Jokiel 1978; Brown 1997), and this can be important because of the close timing of both types of events at the FGB. Further investigation is needed at the FGB to link the intensity of bleaching with preceding exposure to unusual river runoff (Hagman and Gittings 1992).

The low salinities certainly are enough to induce the recurrent changes in skeletal growth of *M. faveolata* seen as fluorescence bands under ultraviolet light (Barnes and Taylor 2001). We lack the fluorescence intensity measurements necessary to make statistical analysis of the river discharge record and fluorescence bands and to establish a direct link. However, even if we could, there is not an adequate causal mechanism to explain how the low-salinity water induces the fluorescence bands on coral skeletons (Barnes et al. 2003). Regardless, these results are contrary to the findings of Lough et al. (2002) who claim that only corals close to shore and exposed to dilute seawater can exhibit fluorescing bands. However, our results, along with those of Barnes and Taylor (2001) and Andréfouët et al. (2002), show that the delivery of diluted seawater that induces skeletal changes can explain fluorescent bands in offshore reef corals as well as other terrigenous influences on such reefs (including pollutants such as herbicides; Means and McMillin 1997).

5.4.1 Transport of River Runoff to the FGB

The northern GOM shelf circulation pattern (Cochrane and Kelly 1986; Nowlin et al. 1998) corroborates the pattern in the hydrographic data, demonstrating the transport of inner-shelf river–seawater mix and yellow substances to the shelf edge and FGB. From September through April, circulation is from Louisiana to Texas on the inner shelf, but in the opposite direction at the shelf edge. From May to August, the entire shelf circulation runs from Texas to Louisiana. Drifters released on the NWGOM shelf during 1993 to 1994 confirmed this shelf circulation pattern (Fig 5.10; Niiler, Johnson, and Baturin 1997; Lugo-Fernández et al. 2001). Drifters released over the middle and inner shelf (October 1993 to May 1994)

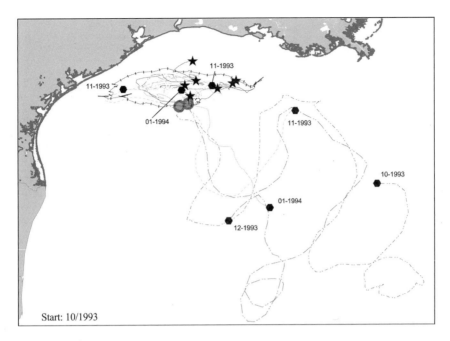

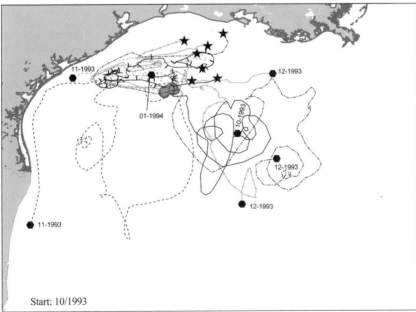

Figure 5.10. Trajectories of drifters released over the LATEX shelf between October 1993 and July 1994 that crossed the FGB: Stars and hexagons, beginnings and ends of the drifter trajectories, respectively. Data from Niiler, Johnson, and Baturin (1997).

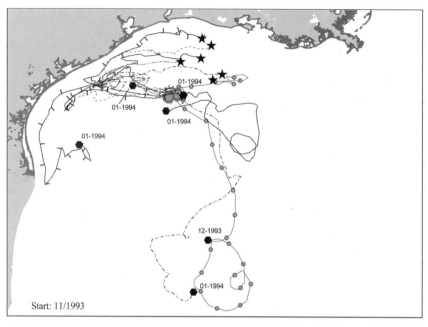

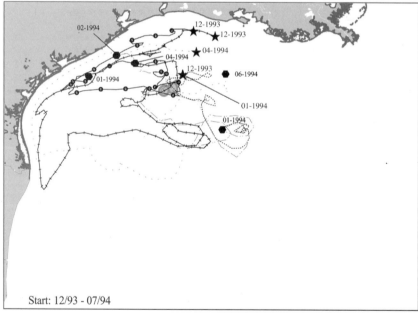

Figure 5.10. (*continued*)

south of the Atchafalaya River moved west until reaching near Texas, then turned east and moved along the shelf edge where they reached the FGB (Fig. 5.10). One drifter released during the peak river discharge in the spring traveled to the FGB in approximately two months, arriving in June (Fig. 5.10). This corroborates the above-mentioned correlation between low-salinity at the FGB and same-year peak river discharge. Drifters released in October farther east, near the Mississippi Delta, followed a similar trajectory and probably explain the correlation between low salinity at the FGB and previous-year peak river discharge. Drifters released at the FGB in August first moved eastward and then returned to near the FGB in about a month (Lugo-Fernández et al. 2001), helping to explain the event-like distribution of high light attenuation observed at the FGB.

The relationship of patterns of river discharge, mean shelf circulation, and cross-shelf transport point to three successive events that in all probability explain the link between river discharge and the FGB, including the formation of fluorescence bands in *M. faveolata* (Fig. 5.11). The timing of these events may vary from year to year. First, starting with the January-to-May period, inner-shelf currents are westward, while shelf-edge currents are eastward. Prevailing winds are westward and favorable to downwelling, thus keeping the inner-shelf water close to the coast. Cold fronts at time scales of approximately 7 to 10 days push water offshore and disrupt the nearshore current. During these cold-front episodes, water transport at the shelf edge in the upper 50 m is offshore (Current and Wiseman 2000). River discharge in the NWGOM is minimal until April when the spring flood occurs. Being less dense than ambient Gulf water, river water moves toward Texas near the coast as a wind-driven buoyant plume. This plume, with salinities of 10 to 20 psu, covers a distance of 350 km and extends about 50 km offshore (Murray 1997). Physical and biological processes mix the plume and deplete it of nutrients. This pattern explains why drifters released in October reached the FGB during the winter (Fig. 5.10).

Second, in May and June, wind direction is reversed. This reversal is followed by a reversal of the ocean current field from a westerly to an easterly flow in the inner shelf, while flow on the outer shelf remains eastward. This transports the river–seawater mix to the east across the entire shelf, creating a band of waters with salinities in the range of 27 to 30 psu south of Galveston and extending southeastward (Murray 1997). The remaining and declining discharges of regional rivers are transported east of their respective rivers. During this time, mixed river–seawater also reaches the FGB, as indicated by low salinities (Nowlin et al. 1998). This low-salinity water probably represents water parcels that break off the 27- to 30-psu band of water south of Galveston and are transported offshore to the FGB by upwelling-favorable winds and offshore eddies. Wolanski and van Senden (1983) observed a similar process in Australia. Year-to-year variations in the timing of salinity and light attenuation at the FGB are linked to variations in circulation, the volume of peak river runoff, and the amount of organic/inorganic substances transported during peak river discharge. Salinity variations at the FGB can also be caused by the short-term wind-induced sloshing of the river–seawater mix about the shelf edge.

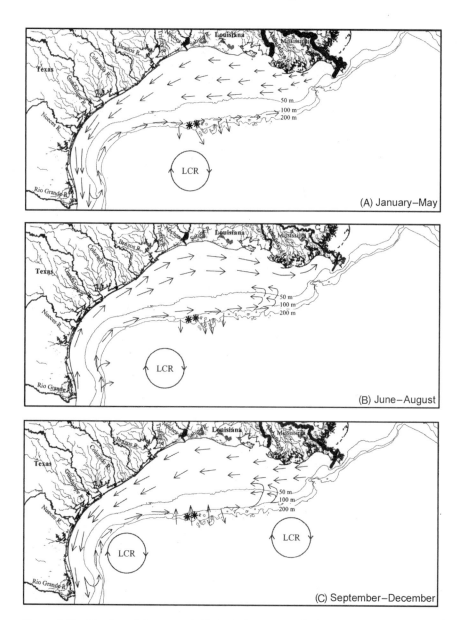

Figure 5.11. Sketches of the mean shelf circulation patterns in the northwestern GOM and Mississippi and Atchafalaya Rivers runoff from (**A**) January to May, (**B**) June to August, and (**C**) September to December: asterisks = locations of the FGB; LCR = loop current ring.

Third, in September to December, the wind field is reversed again and currents turn westward, probably advecting some of the mixed river–seawater toward Texas in the inner-shelf region. This transport may occur as a mixture of parcels of river–seawater, each having different physical/biological properties. Observations in Australia and numerical models of shelf surface salinity in the NWGOM exhibit a great degree of structure (i.e., parcels of different salinities), rather than smooth variations of salinity (Wolanski and van Senden 1983; Jim Herring, Dynalysis of Princeton, personal communication 2001). Some of the river–seawater parcels probably cross the FGB again, as shown by drifters tracked in the NWGOM (Fig. 5.10). Furthermore, drifter tracks show that after reaching the inshore Texas region, where additional mixing with discharges from local rivers takes place, the river–seawater mix can turn along the shelf edge toward the offshore and reach the FGB once again. This helps explain the correlation of previous-year river discharge and low salinity at the FGB.

The succession of these three events explains how mixed river–seawater and associated yellow substances originating from the inner shelf can reach the crests of the FGB on a seasonal basis. The low salinity and increased light attenuation associated with the mixed river–seawater that traverses the crests of the FGB probably induce skeletal variations in *M. faveolata* that are seen as fluorescent bands under UV light. Advected or upwelled nutrients are unlikely sources of fluorescence in *M. faveolata* at the FGB simply because the nutrients and pigment levels (chlorophyll a) never seem to be present in unusual concentrations at or near the FGB. Upwelling as a transport mechanism of nutrients, pigments, or sediments from the nepheloid layer onto the crests of the FGB should be ruled out because orographic effects around the banks are weak (vertical displacements of only 10 m; McGrail 1983). However, we cannot rule out upwelling induced by solitons (internal waves) as observed in the Florida Keys (Leichter et al. 1998) since the FGB are exposed to deepwater where such internal waves can be generated.

Embedded in our analysis of the transport of mixed river–seawater to and through the FGB are year-to-year transport variations caused by fluctuations in river discharge. The timing of the deposition of fluorescent bands in our coral slice from 1979 to 1982 may reflect temporal fluctuations of freshwater input into the LATEX shelf. Indeed, the fluorescent bands for those years succeeded the deposition of the summer high-density band. This seemingly unusual position of the fluorescence bands may either be an artifact produced during the overlapping of the coral slice X-radiograph and the photograph taken under UV light, or it may signal a different timing of river–seawater mix flowing through the FGB. Should the peak river discharge occur late in the spring (mid-May or later), the river–seawater mix would not have enough time to reach Texas, much less the FGB, because of the water circulation reversal in June (Fig. 5.11). But in September–October the circulation reverses once again to the west. The river–seawater mix could thus reach the FGB late in the year (September–December), following the deposition of the high-density growth band in *M. faveolata*. Furthermore, such late freshwater input into the LATEX shelf could be recorded as fluorescence bands in corals at the FGB should it be suddenly augmented by unseasonable and heavy river discharges resulting from torrential rains of tropical depressions or hurricanes (e.g., the Texas floods of October 1994; those from Tropical Storm Amelia, 1 to 4 August 1978; and those from Hurricane Norma, 11 to 13 October 1981). These exploratory analyses of historical data identify a possible correlation between the fluorescence in corals at the FGB and the volume of riverine input and highlight the importance of the timing of freshwater input into the LATEX shelf.

5.5 Summary and Conclusions

The geographic distribution of coral reefs suggests that corals tolerate a relatively brief exposure on the order of hours to days to unusually low salinity (<15 psu; Coles and Jokiel 1992). Freshwater input has a positive effect when it helps regulate salinity by offsetting the effects of evaporation. That same freshwater can affect corals negatively when associated with increased sedimentation, increased turbidity, nutrients, pollutants, and pathogens (Coles and Jokiel 1992; Richmond 1994; Veron 1995; Kleypas 1996; Peters et al. 1997; Kleypas, McManus, and Meñez 1999; Bruckner 2002; Lipp et al. 2002; Szmant 2002; McCulloch et al. 2003; Wolanski et al. 2003).

The study of environmental parameters and regional oceanographic/climatic processes can contribute significantly to our understanding of the changing conditions that may impact reefs. The FGB have gained Sanctuary status since 1992 (Public Law 102-251; 106 Stat. 66) because of their uniqueness. The in-depth study of environmental characteristics of the FGB revealed that from April through September, coral reef resources of the FGB are recurrently exposed to nearshore processes and the terrigenous runoff of the Mississippi and Atchafalaya Rivers and Texas regional rivers. This complicates management strategies to protect the FGB coral reefs from human disturbances (Gittings, Bright, and Hagman 1994; Gittings 1998). The FGB appear today in much better condition than many reefs of the Caribbean (Gardner et al. 2003) and seem safe from impacts that have degraded large expanses of reefs (Aronson et al. 2005). In fact, the presence of abundant, large, and healthy massive reef framework-building coral species at the FGB (*M. faveolata, M. annularis, M. franksi, P. astreoides, M. cavernosa, and Siderastrea siderea*) is in sharp contrast to their loss on other western Atlantic reefs (Gardner et al. 2003). The absence of branching corals (*Acropora palmata, Acropora cervicornis*) as framework builders at the FGB probably helped in maintaining stable coral populations. The FGB were able to avoid the catastrophic loss of acroporids to white-band disease and the weakening of their structure (Aronson and Precht 2001a,b; Aronson et al. 2005).

The focus of this study was the unexpected presence of recurrent fluorescent bands in slices of *M. faveolata* collected at the offshore Flower Garden Banks reefs (northwest Gulf of Mexico). The apparent geographical remoteness of the FGB and the absence of upwelling at these reefs seemingly preclude expected sources of fluorescence that influence corals, including low salinity and riverborne or upwelled organic matter. To fully examine potential causes of fluorescence in corals at the FGB, we surveyed all relevant environmental and regional physical oceanographic data. We found that the fluorescence is probably linked to the interrelationship of regional river discharge, Louisiana–Texas (LATEX) shelf circulation, and cross-shelf water transport. Supporting evidence includes the co-occurrence of declines in temperature and salinity (1 to 2 °C and ~5 psu below ambient levels) that occur in the summer (April–August), high light atten-uation coefficients that are significantly correlated with low salinity and season, and the absence of either chlorophyll a or nutrients.

The environmental significance of the input of freshwater on coral reefs is often overlooked (Coles and Jokiel 1992). In the case of the FGB, recurrent fresh-water input is probably of great importance, as these geographically isolated reefs are in fact physically linked to nearshore processes and regional river runoff and therefore exposed to sedimentation, turbidity, and waterborne pollutants and pathogens. Although we could not establish a definitive connection between the fluorescing bands observed in corals at the FGB and regional river runoff, future studies may find that the exposure to low salinity is a probable source of fluores-cence (our second working hypothesis). A multidisciplinary approach will be necessary to test our hypotheses and to uncover the source of fluorescence. The detailed study of the relationship of barium to calcium (Ba/Ca) molar ratios in corals of the FGB, the mix of seawater and river water reaching the FGB, and the discharge of the Mississippi and Atchafalaya Rivers would probably be a useful approach (McCulloch et al. 2003). The timing of changes in the skeletal structure of corals should not only be compared with pulses of river runoff but also with other environmental variables, including potentially stressful and recurring events such as low temperature and the passage of hurricanes (NGDC 2000).

Acknowledgments. We are grateful for support from the US Department of the Interior, Minerals Management Service, Gulf of Mexico OCS Region (USDOI/MMS), and Geo-Marine, Inc. (GMI) during the preparation of this man-uscript. Bill Precht and Bob Halley provided us with insightful review comments that helped improve our manuscript. We are especially grateful for Rich Aronson's patience and thoughtful guidance. Peter Gehring (GMI), Tara Montgomery (USDOI/MMS), and Ronnie Roller (GMI) are thanked for their help in preparing the figures, Niall Slowey (Texas A&M University, Department of Oceanography) for providing the coral slab and slab X-radiograph, and Matt Howard (Texas A&M University, Department of Oceanography) for providing us with river-discharge data. The opinions expressed are our own and do not reflect the opinions or policies of the USDOI/MMS or GMI.

References

Andréfouët, S., P.J. Mumby, M. McField, C. Hu, and F.E. Muller-Karger. 2002. Revisiting coral reef connectivity. *Coral Reefs* 21:43–48.

Aronson, R.B., I.G. Macintyre, C.M. Wapnick, and M.W. O'Neil. 2004. Phase shifts, alter-native states, and the unprecedented convergence of two reef systems. *Ecology* 85(7): 1876–1891.

Aronson, R.B., and W.F. Precht. 2001a. White-band disease and the changing face of Caribbean coral reefs. *Hydrobiologia* 460:25–38.

Aronson, R.B., and W.F. Precht. 2001b. Evolutionary paleoecology of Caribbean coral reefs. In *Evolutionary Paleoecology: The Ecological Context of Macroevolutionary Change*, eds. W.D. Allmon and D.J. Bottjer, 171–233. New York: Columbia University Press.

Aronson, R.B., W.F. Precht, I.G. Macintyre, and T.J.T. Murdoch. 2000. Coral bleach-out in Belize. *Nature* 405:36.

Aronson, R.B., W.F. Precht, T.J.T. Murdoch, and M.L. Robbart. 2005. Long-term persistence of coral assemblages on the Flower Garden Banks, northwestern Gulf of Mexico: Implications for science and management. *Gulf Mex. Sci.* 23:84–94.

Aronson, R.B., W.F. Precht, M.A. Toscano, and K.H. Koltes. 2002. The 1998 bleaching event and its aftermath on a coral reef in Belize. *Mar. Biol.* 141:435–447.

Barnes, D.J., and R.B. Taylor. 2001. On the nature and causes of luminescent lines and bands in coral skeletons. *Coral Reefs* 19:221–230.

Barnes, D.J., R.B. Taylor, and J.M. Lough. 2003. Measurement of luminescence in coral skeletons. *J. Exp. Biol. Ecol.* 295:91–106.

Beaver, C.R., K.J.P. Deslarzes, J.H. Hudson, and J.W. Tunnell. 1996. Fluorescent banding in reef corals as evidence of increased runoff onto the Veracruz coral reef complex. *Proc. Eighth Int. Coral Reef Symp., Panama*, Abstract p. 14.

Bellwood, D.R., T.P. Hughes, C. Folke, and M. Nyström. 2004. Confronting the coral reef crisis. *Nature* 429:827–833.

Boto, K., and P. Isdale. 1985. Fluorescent bands in massive corals result from terrestrial fulvic acid inputs to nearshore zone. *Nature* 315:396–397.

Bright, T.J., G.P. Kraemer, G.A. Minnery, and S.T. Viada. 1984. Hermatypes of the Flower Garden Banks, northwestern Gulf of Mexico: A comparison to other western Atlantic reefs. *Bull. Mar. Sci.* 34:461–476.

Brown, B.E. 1997. Coral bleaching: Causes and consequences. *Coral Reefs* 16:S129–S138.

Brown, B.E., and L.S. Howard. 1985. Assessing the effects of "stress" on reef corals. *Adv. Mar. Biol.* 22:1–63.

Bruckner, A.W. 2002. Priorities for effective management of coral diseases. *NOAA Technical Memorandum NMFS-OPR-22*. Silver Spring: US Department of Commerce, National Oceanic and Atmospheric Administration, National Marine Fisheries Service.

Bruckner, A.W., R.J. Bruckner, and E.H. Williams. 1997. Spread of black-band disease epizootic through the coral reef system in St. Ann's Bay, Jamaica. *Bull. Mar. Sci.* 61:918–928.

Byron, G. 1993. Impacts of cyclone-induced floods on fringing reefs: Case study of Keppel Bay, Queensland, Australia. In *Proceedings of the Colloquium on Global Aspects of Coral Reefs: Health, Hazards and History*, compiler R.N. Ginsburg, 8–14. Miami: University of Miami.

Cappo, M., and Kelley, R. 2001. Connectivity in the Great Barrier Reef World Heritage Area—An overview of pathways and processes. In *Oceanographic Processes on Coral Reefs: Physical and Biological Links in the Great Barrier Reef*, ed. E. Wolanski, 161–187. Boca Raton: CRC Press.

Causey, B., J. Delaney, E. Diaz, D. Dodge, J. Garcia, J. Higgins, B. Keller, R. Kelty, W. Jaap, C. Matos, G. Schmahl, C. Rogers, M. Miller, and D. Turgeon. 2002. Status of coral reefs in the U.S. Caribbean and Gulf of Mexico: Florida, Texas, Puerto Rico, US Virgin Islands, Navassa. In *Status of Coral Reefs of the World: 2002*, ed. C. Wilkinson, 251–276. Townsville: Australian Institute of Marine Science.

Cervino, J., T.J. Goreau, L. Nagelkerken, G.W. Smith, and R. Hayes. 2001. Yellow band and dark spot syndromes in Caribbean corals: Distribution, rate of spread, cytology, and effects on abundance and division rate of zooxanthellae. *Hydrobiologia* 460:53–63.

Cho, L.L., and J.D. Woodley. 2002. Recovery of reefs at Discovery Bay, Jamaica and the role of *Diadema antillarum. Proc. Ninth Int. Coral Reef Symp., Bali, Indonesia* 1:331–337.

Cochrane, J.D., and F.J. Kelly. 1986. Low-frequency circulation on the Texas-Louisiana continental shelf. *J. Geophys. Res.* 91:10,645–10,659.

Cole, J. 2003. Dishing the dirt on coral reefs. *Nature* 421:705–706.

Coles, S.L., and P.L. Jokiel. 1978. Synergistic effects of temperature, salinity, and light on the hermatypic coral Montipora verrucoasa. *Mar. Biol.* 49:187–195.

Coles, S.L., and P.L. Jokiel. 1992. Effects of salinity on coral reefs. In *Pollution in Tropical Aquatic Systems*, eds. D.W. Connell and D.W. Hawker, 147–166. Boca Raton: CRC Press.

Cortés, J. 1993. A reef under siltation stress: A decade of degradation. In *Proceedings of the Colloquium on Global Aspects of Coral Reefs: Health, Hazards, and History*, compiler R.N. Ginsburg, 240–246. Miami: Rosenstiel School of Marine and Atmospheric Science, University of Miami.

CSA (Continental Shelf Associates). 1996. Long-term monitoring at the East and West Flower Garden Banks. *OCS Study MMS 96-0046*. New Orleans: US Department of the Interior, Minerals Management Service.

Current, C.L., and W.J. Wiseman, Jr. 2000. Dynamic height and seawater transport across the Louisiana-Texas shelf break. *OCS Report/MMS 2000-045*. New Orleans: US Department of the Interior, Minerals Management Service.

Dai, A.D., and K.E. Trenberth. 2002. Estimates of freshwater discharge from continents: Latitudinal and seasonal variations. *J. Hydrometeorol.* 3:660–687.

Del Castillo, C.E., F. Gilbes, P.G. Coble, and F.E. Muller-Karger. 2000. On the dispersal of riverine colored dissolved organic matter over the west Florida shelf. *Limnol. Oceanogr.* 45:1425–1432.

Dinnel, S.P., and A. Bratkovich. 1993. Water discharge, nitrate concentration and nitrate flux in the lower Mississippi River. *J. Mar. Syst.* 4:315–326.

Dinnel, S.P., and W.J. Wiseman, Jr. 1986. Fresh water on the Louisiana and Texas shelf. *Cont. Shelf Res.* 6:765–784.

Dodge, R.E., R.C. Aller, and J. Thompson. 1974. Coral growth related to resuspension of bottom sediments. *Nature* 247:574–577.

Dodge, R.E., and J.C. Lang. 1983. Environmental correlates of hermatypic coral (Montastrea annularis) growth on the East Flower Garden Bank, northwest Gulf of Mexico. *Limnol. Oceanogr.* 28:228–240.

Dokken, Q.R., I.R. MacDonald, J.W. Tunnell, Jr., T. Wade, K. Withers, S.J. Dilworth, T.W. Bates, C.R. Beaver, and C.M. Rigaud. 2003. Long-term monitoring at the East and West Flower Garden Banks National Marine Sanctuary, 1998–2001. *OCS Study MMS 2003-031*. New Orleans: US Department of the Interior, Minerals Management Service.

Done, T.J. 1999. Coral community adaptability to environmental change at the scales of regions, reefs and reef zones. *Am. Zool.* 39:66–79.

Edmonson, C.H. 1946. Behavior of coral planulae under altered saline and thermal conditions. *Occ. Pap. Bishop Mus.* 18:283–304.

Edmunds, P.J., and R.C. Carpenter. 2001. Recovery of Diadema antillarum reduces macroalgal cover and increases abundance of juvenile corals on a Caribbean reef. *Proc. Natl. Acad. Sci. USA* 98:5067–5071.

Etter, P.C. 1996. Heat and freshwater budgets of the Texas-Louisiana shelf April 1992 through November 1994. *Technical Report No. 96-1-T*. College Station: Texas A&M University.

Fallon, S.J., M.T. McCulloch, and C. Alibert. 2003. Examining water temperature proxies in *Porites* corals from the Great Barrier Reef: A cross-shelf comparison. *Coral Reefs* 22:389–404.

FAMU (Florida A&M University). 1988. Meteorological database and synthesis for the Gulf of Mexico. *OCS Study MMS 88-0064*. New Orleans: US Department of the Interior, Minerals Management Service.

Fang, L.S., and Y.C. Chou. 1992. Concentration of fulvic acid in the growth bands of hermatypic corals in relation to local precipitation. *Coral Reefs* 11:187–191.

Gardner, T.A, I.M. Cote, J.A. Gill, A. Grant, and A.R. Watkinson. 2003. Long-term region-wide declines in Caribbean corals. *Science* 301:958–960.

Gittings, S.R. 1998. Reef community stability on the Flower Garden Banks, northwest Gulf of Mexico. *Gulf Mex. Sci.* 16:161–169.

Gittings, S.R., T.J. Bright, and D.K. Hagman. 1994. Protection and monitoring of reefs on the Flower Garden Banks. In *Proceedings of the Colloquium on Global Aspects of Coral Reefs: Health, Hazards and History*, compiler R.N. Ginsburg, 181–187. Miami: University of Miami.

Gittings, S.R., K.J.P. Deslarzes, D.K. Hagman, and G.S. Boland. 1992. Reef coral populations and growth on the Flower Garden Banks, northwest Gulf of Mexico. *Proc. Seventh Int. Coral Reef Symp., Mangilao* 1:90–96.

Glynn, P.W. 1973. Ecology of a Caribbean coral reef. The Porites reef-flat biotope: Part I. Meteorology and hydrography. *Mar. Biol.* 20:297–318.

Glynn, P.W. 1993. Coral reef bleaching: Ecological perspectives. *Coral Reefs* 12(1):1–17.

Glynn, P.W., S.B. Colley, J.H. Ting, J.L. Mate, and H.M. Guzman. 2000. Reef coral reproduction in the eastern Pacific; Costa Rica, Panama and Galapagos Islands (Ecuador). IV. Agariciidae, recruitment and recovery of *Pavona varians* and *Pavona* sp.a. *Mar. Biol.* 136:785–805.

Hagman, D.K., and S.R. Gittings. 1992. Coral bleaching on high latitude reefs at the Flower Garden Banks, NW Gulf of Mexico. *Proc. Seventh Int. Coral Reef Symp., Guam* 1:38–43.

Hagman, D.K., S.R. Gittings, and K.J.P. Deslarzes. 1998. Timing, species participation, and environmental factors influencing annual mass spawning at the Flower Garden Banks (northwest Gulf of Mexico). *Gulf Mex. Sci.* 16:170–179.

Harvell, C.D., C.E. Mitchell, J.R. Ward, S. Altizer, A.P. Dobson, R.S. Ostfeld, and M.D. Samuel. 2002. Climate warming and disease risks for terrestrial and marine biota. *Science* 296:2158–2162.

Heil, C.A. 1996. The influence of dissolved humic material (humic, fulvic, and hydrophilic acids) on marine phytoplankton. Doctoral dissertation, University of Rhode Island.

Hendy, E.J., J.M. Lough, and M.K. Gagan. 2003. Historical mortality in massive *Porites* from the central Great Barrier Reef, Australia: Evidence for past environmental stress? *Coral Reefs* 22:207–215.

Highsmith, R.C. 1979. Coral growth rates and environmental control on density banding. *J. Exp. Mar. Biol. Ecol.* 37:105–125.

Hoegh-Guldberg, O. 1999. Climate change, coral bleaching and the future of the world's coral reefs. *Mar. Freshwater Res.* 50:839–866.

Hoetjes, P., A. Lum Kong, R. Juman, A. Miller, M. Miller, K. De Meyer, and A. Smith. 2002. Status of coral reefs in the eastern Caribbean : The OECS, Trinidad and Tobago, Barbados and the Netherlands Antilles. In *Status of Coral Reefs of the World: 2002*, ed. C. Wilkinson, 325–342. Townsville: Australian Institute of Marine Science.

Højerslev, N.K., and T. Aarup. 2002. Optical measurements on the Louisiana shelf off the Mississippi River. *Estuarine Coastal Shelf Sci.* 55:599–611.

Hudson, J.H. 1977. Long-term bioerosion rates on a Florida reef: A new method. *Proc. Third Int. Coral Reef Symp., Miami* 2:492–497.

Hughes, T.P. 1994. Catastrophes, phase shifts, and large-scale degradation of a Caribbean coral reef. *Science* 265:1547–1551.

Hughes, T.P., A.H. Baird, D.R. Bellwood, M. Card, S.R. Connolly, C. Folke, R. Grosberg, O. Hoegh-Guldberg, J.B.C. Jackson, J. Kleypas, J.M. Lough, P. Marshall, M. Nyström, S.R. Palumbi, J.M. Pandolfi, B. Rosen, and J. Roughgarden. 2003. Climate change, human impacts, and the resilience of corals. *Science* 301:929–933.

Isdale, P. 1984. Fluorescent bands in massive corals record centuries of coastal rainfall. *Nature* 310:578–579.

Isdale, P. 1995. Coral rain gauges: The proxy fluorescence record in massive corals—a discussion paper. Paleoclimate and environmental variability in the Austral-Asian transect during the past 2000 years. *Proc. 1995 Nagoya IGBP-PAGES/PEP II Symp.* 51–59.

ISRS (International Society for Reef Studies). 1999. ISRS statement on diseases on coral reefs. *Reef Encounter* 25:24–26.

Jaap, W.C. 1984. The ecology of the south Florida coral reefs: A community profile. FWS/OBS-82/08. Washington, DC: US Fish and Wildlife Service.

Jackson, J.B.C. 1997. Reefs since Columbus. *Coral Reefs* 16:S23–S32.

Jackson, J.B.C., M.X. Kirby, W.H. Berger, K.A. Bjorndal, L.W. Botsford, B.J. Bourque, R.H. Bradbury, R. Cooke, J. Erlandson, J.A. Estes, T.P. Hughes, S. Kidwell, C.B. Lange, H.S. Lenihan, J.M. Pandolfi, C.H. Peterson, R.S. Steneck, M.J. Tegner, and R.B. Warner. 2001. Historical overfishing and the recent collapse of coastal ecosystems. *Science* 293:629–638.

Kennicutt, M.C., II. 1995. Gulf of Mexico Offshore Operations Monitoring Experiment (GOOMEX), Phase I: sublethal response to contaminant exposure—Volume I: Final Report. *OCS Study MMS 95-0045*. New Orleans: US Department of the Interior, Minerals Management Service.

King, B., F. McAllister, E. Wolanski, T. Done, and S. Spagnol. 2001. River plume dynamics in the Central Great Barrier Reef. In *Oceanographic Processes on Coral Reefs: Physical and Biological Links in the Great Barrier Reef*, ed. E. Wolanski, 145–155. Boca Raton: CRC Press.

Kitheka, J.U. 1996. Water circulation and coastal trapping of brackish water in a tropical mangrove-dominated bay in Kenya. *Limnol. Oceanogr.* 41:169–176.

Kleypas, J.A. 1996. Coral reef development under naturally turbid conditions: Fringing reef near Broad Sound Australia. *Coral Reefs* 15:153–167.

Kleypas, J.A., R.W. Buddemeier, D. Archer, J.P. Gattuso, C. Langdon, and B.N. Opdyke. 1999. Geochemical consequences of increased atmospheric carbon dioxide on coral reefs. *Science* 284:118–120.

Kleypas, J.A., R.W. Buddemeier, and J.P. Gattuso. 2001. The future of coral reefs in an age of global change. *Int. J. Earth Sci. (Geol. Rundsch.)* 90:426–437.

Kleypas, J.A., J.W. McManus, and L.A.B. Meñez. 1999. Environmental limits to coral reef development: Where do we draw the line? *Am. Zool.* 39:146–159.

Knowlton, N. 2001. The future of coral reefs. *Proc. Natl. Acad. Sci. USA* 98:5419–5425.

Knutson, D.W., R.W. Buddemeier, and S.V. Smith. 1972. Coral chronometers: Seasonal growth bands in reef corals. *Science* 177:270–272.

Lang, J., P. Alcolado, J.P. Carricart-Ganivet, M. Chiappone, A. Curran, P. Dustan, G. Gaudian, F. Geraldes, S. Gittings, R. Smith, W. Tunnell, and J. Wiener. 1998. Status of coral reefs in the northern areas of the wider Caribbean. In *Status of Coral Reefs of the World: 1998*, ed. C. Wilkinson, 123–134. Townsville: Australian Institute of Marine Science.

Lang, J.C., K.J.P. Deslarzes, and G.P. Schmahl. 2001. The Flower Garden Banks: Remarkable reefs in the NW Gulf of Mexico. *Coral Reefs* 20:126.

Leichter, J.J., G. Shellenbarger, S.J. Genovese, and S.R. Wing. 1998. Breaking internal waves on a Florida (USA) coral reef: A plankton pump at work? *Mar. Ecol. Prog. Ser.* 166:83–97.

Lee, T.N., and E. Williams. 1999. Mean distribution and seasonal variability of coastal currents and temperature in the Florida Keys with implications for larval recruitment. *Bull. Mar. Sci.* 64:35–56.

Lesly, R.A., and M.L. Coleman. 2003. Record of natural and anthropogenic changes in reef environments (Barbados West Indies) using laser ablation ICP-MS and sclerochronology on coral cores. *Coral Reefs* 22:416–426.

Lessios, H.A. 1988. Mass mortality of *Diadema antillarum* in the Caribbean: What have we learned? *Ann. Rev. Ecol. Syst.* 19:371–393.

Linton, D., R. Smith, P. Alcolado, C. Hanson, P. Edwards, R. Estrada, T. Fisher, R. Gomez Fernandez, F. Geraldes, C. McCoy, D. Vaughan, V. Voegeli, G. Warner, and J. Wiener. 2002. Status of coral reefs in the northern Caribbean and Atlantic node of the GCRMN. In *Status of Coral Reefs of the World: 2002*, ed. C. Wilkinson, 277–302. Townsville: Australian Institute of Marine Science.

Lipp, E.K., J.L. Jarrel, D.W. Griffin, J. Lukaisk, J. Jaeukiewiez, and J.B. Rose. 2002. Preliminary evidence for human fecal contamination in corals of the Florida Keys, USA. *Mar. Pollut. Bull.* 44:666–670.

Lough, J.M., D.J. Barnes, and F.A. McAllister. 2002. Luminescent lines in corals from the Great Barrier Reef provide spatial and temporal records of reefs affected by land runoff. *Coral Reefs* 21:333–343.

Lugo-Fernández, A. 1998. Ecological implications of hydrography and circulation to the Flower Garden Banks, northwest Gulf of Mexico. *Gulf Mex. Sci.* 2:144–160.

Lugo-Fernández, A., K.J.P. Deslarzes, J.M. Price, G.S. Boland, and M.V. Morin. 2001. Inferring probable dispersal of Flower Garden Banks coral larvae (Gulf of Mexico) using observed and simulated drifter trajectories. *Cont. Shelf Res.* 21:47–67.

Macintyre, I.G. 1972. Submerged reefs of eastern Caribbean. *AAPG Bull.* 56:720–738.

Manabe, S., R.T. Wetherald, P.C.D. Milly, T.L. Delworth, and R.J. Stouufer. 2004. Century-scale change in water availability: CO_2-quadrupling experiment. *Climatic Change* 64:59–76.

McClanahan, T., N. Polunin, and T. Done. 2002. Ecological states and resilience of coral reefs. *Conservation Ecology* 6(2):18 http://www.consecol.org/vol16/iss2/art18

McCook, L.J. 1999. Macroalgae, nutrients and phase shifts in coral reefs: Scientific issues and management consequences for the Great Barrier Reef. *Coral Reefs* 18:357–367.

McCulloch, M., S. Fallon, T. Wyndham, E. Hendy, J. Lough, and D. Barnes. 2003. Coral record of increased sediment flux to the inner Great Barrier Reef since European settlement. *Nature* 421:727–730.

McGrail, D.W. 1983. Flow, boundary layers, and suspended sediment at the Flower Garden Banks. In: *Reefs and Banks of the Northwestern Gulf of Mexico: Their Geological, Biological, and Physical Dynamics*, eds. R. Rezak, T.J. Bright, and D.W. McGrail, 141–230. Technical Report No. 83-1-T. New Orleans: US Department of the Interior, Minerals Management Service.

McGrail, D.W. 1985. Currents and suspended sediments at the East Flower Garden Bank. In *The Flower Gardens: A Compendium of Information*, eds. T.J. Bright, D.W. McGrail, and R. Rezak, 91–103. MMS OCS No. 85-0024. New Orleans: US Department of the Interior, Minerals Management Service.

Means, J.C., and D.J. McMillin. 1997. Pollutant chemistry. In *An Observational Study of the Mississippi-Atchafalaya Coastal Plume*, ed. S.P. Murray, 365–496. OCS Study MMS 98-0040. New Orleans: US Department of the Interior, Minerals Management Service.

Moberg, F., M. Nystrom, N. Kautsky, M. Tedengren, and P. Jarayabhand. 1997. Effects of reduced salinity on the rates of photosynthesis and respiration in the hermatypic corals *Porites lutea* and *Pocillopora damicornis. Mar. Ecol. Prog. Ser.* 157:53–59.

Muller-Karger, F.E., C.R. McClain, T.R. Fisher, W.E. Esais, and R. Varela. 1989. Pigment distribution in the Caribbean Sea: Observations from space. *Prog. Oceanogr.* 23:23–64.

Mumby, P.J., J.R.M. Chisholm, A.J. Edwards, C.D. Clark, E.B. Roark, S. Andréfouët, and J. Jaubert. 2001. Unprecedented bleaching-induced mortality in *Porites* spp. at Rangiroa Atoll, French Polynesia. *Mar. Biol.* 139:183–189.

Murray, S.P. 1997. An observational study of the Mississippi-Atchafalaya coastal plume: OCS Study MMS 98-0040. New Orleans: US Department of the Interior, Minerals Management Service.

NASA (National Aeronautics and Space Administration). 2002. Earth Observatory. Data and Images. Rainfall. Accessed May 2003. http://earthobservatory.nasa.gov/Observatory/Datasets/rainfall.gpcp.html

NGDC (National Geophysical Data Center). 2000. Preliminary Summary of workshop on Atlantic basin paleohurricane reconstructions from high resolution records. Accessed October 2004. http://www.ngdc.noaa.gov/paleo/hurricane/conference.html

Niiler, P.P., W.R. Johnson, and N. Baturin, 1997. *Surface Current and Lagrangian-Drift Program*. New Orleans: US Department of the Interior.

Nowlin, W.D., A.E. Jochens, R.O. Reid, and S.F. DiMarco. 1998. Texas–Louisiana shelf circulation and transport process study: Synthesis report LATEX A. Vols. I and II. OCS Study MMS 98-0035 and MMS 98-0036. New Orleans: US Department of the Interior, Minerals Management Service.

Nowlin, W.D., A.E. Jochens, S.F. DiMarco, and R.O. Reid. 2000. Physical oceanography. In Deepwater Gulf of Mexico Environmental and Socioeconomic Data Search and Synthesis, Volume 1, Narrative Report, ed. Continental Shelf Associates. OCS Study MMS 2000-049. New Orleans: US Department of the Interior, Minerals Management Service.

Nugues, M.M., and C.M. Roberts. 2003a. Partial mortality in massive reef corals as an indicator of sediment stress on coral reefs. *Mar. Pollut. Bull.* 46:314–323.

Nugues, M.M., and C.M. Roberts. 2003b. Coral morality and interaction with algae in relation to sedimentation. *Coral Reefs* 22:507–516.

Parsons, T.R., M. Takahashi, and B. Hargrave. 1984. *Biological Oceanographic Processes*. New York: Pergamon Press.

Pattengill, C.V. 1998. Structure and persistence of the reef fish assemblages of the Flower Garden Banks National Marine Sanctuary. PhD dissertation, Texas A&M University.

Peters, E.C., N.J. Gassman, J.C. Firman, R.H. Richmond, and E.A. Power. 1997. Ecotoxicology of tropical marine ecosystems. *Environ. Toxicol. Chem.* 16:12–40.

Pittock, B.A. 1999. Coral reefs and environmental change: Adaptation to what? *Am. Zool.* 39:10–29.

Porter, J.W., S.K. Lewis, and K.G. Porter. 1999. The effects of multiple stressors on the Florida Keys coral reef ecosystem: A landscape hypothesis and a physiological test. *Limnol. Oceanogr.* 44:941–949.

Rabalais, N.N., R.E. Turner, D. Justic, Q. Dortch, W.L. Wiseman Jr., and B.K. Sen Gupta. 1996. Nutrient changes in the Mississippi River and system responses on the adjacent continental shelf. *Estuaries* 19(2B):386–407.

Ramseyer, K., T.M. Miano, V. D'Orazio, A. Wildberger, T. Wagner, and J. Geister. 1997. Nature and origin of organic matter in carbonates from speleothems, marine cements and coral skeletons. *Org. Geochem.* 26:361–378.

Rezak, R. 1977. West Flower Garden Bank, Gulf of Mexico. *Stud. Geol.* 4:27–35.

Rezak, R., T.J. Bright, and D.W. McGrail. 1983. Reefs and banks of the northwestern Gulf of Mexico: Their geological, biological, and physical dynamics. Final Technical Report No. 83-1-T. New Orleans: US Department of the Interior, Minerals Management Service.

Rezak, R., T.J. Bright, and D.W. McGrail. 1985. *Reefs and Banks of the Northwestern Gulf of Mexico: Their Geological, Biological, and Physical Dynamics*. New York: Wiley.

Rezak, R., S.R. Gittings, and T.J. Bright. 1990. Biotic assemblages and ecological controls on reefs and banks of the northwest Gulf of Mexico. *Am. Zool.* 30:23–35.

Richmond, R.H. 1994. Effects of coastal runoff on coral reproduction. In *Proceedings of the Colloquium on Global Aspects of Coral Reefs: Health, Hazards and History*, compiler R.N. Ginsburg, 360–364. Miami: University of Miami.

Rogers, C. 1990. Responses of coral reefs and reef organisms to sedimentation. *Mar. Ecol. Prog. Ser.* 62:185–202.

SAIC (Science Applications International Corporation). 1989. Gulf of Mexico physical oceanography program. OCS Report/MMS 89-0068. New Orleans: US Department of the Interior, Minerals Management Service.

Scoffin, T.P., A.W. Tudhope, and B.E. Brown. 1989. Fluorescent and skeletal density banding in Porites lutea from Papua New Guinea and Indonesia. *Coral Reefs* 7:169–178.

Sheppard, C.R.C. 2003. Predicted recurrences of mass coral mortality in the Indian Ocean. *Nature* 425:294–297.

Smith, T.J., III, J.H. Hudson, M.B. Robblee, G.V.N. Powell, and P.J. Isdale. 1989. Freshwater flow from the Everglades to Florida Bay: A historical reconstruction based on fluorescent banding in the coral *Solenastrea bournoni. Bull. Mar. Sci.* 44:274–282.

Sorokin, Y.I. 1995. *Coral Reef Ecology*. Berlin: Springer.

Spalding, M.D., C. Ravilious, and E.P. Green. 2001. *World Atlas of Coral Reefs*. Berkeley: University of California Press.

Stoddart, D.R. 1969. Ecology and morphology of recent coral reefs. *Biol. Rev.* 44:433–498.

Sullivan-Sealy, K., and G. Bustamante. 1999. Setting geographic priorities for marine conservation in Latin America and the Caribbean. The Nature Conservancy, Arlington, Virginia.

Susic, M., K. Boto, and P. Isdale. 1991. Fluorescent humic acid bands in coral skeletons originate from terrestrial runoff. *Mar. Chem.* 33:91–104.

Szmant, A.M. 2002. Nutrient enrichment on coral reefs: Is it a major cause of coral reef decline? *Estuaries* 25:743–766.

Tudhope, A.W., D.W. Lea, G.B. Shimmield, C.P. Chilcott, and S. Head. 1996. Monsoon climate and Arabian Sea coastal upwelling recorded in massive corals from southern Oman. *Palaios* 11:347–361.

Turgeon, D.D., R.G. Asch, B.D. Causey, R.E. Dodge, W. Jaap, K. Banks, J. Delaney, B.D. Keller, R. Speiler, C.A. Matos, J.R. Garcia, E. Diaz, D. Catanzaro, C.S. Rogers, Z. Hillis-Starr, R. Nemeth, M. Taylor, G.P. Schmahl, M.W. Miller, D.A. Gulko, J.E. Maragos, A.M. Friedlander, C.L. Hunter, R.S. Brainard, P. Craig, R.H. Richmond, G. Davis, J. Starmer, M. Trianni, P. Houk, C.E. Birkeland, E.Y. Golbuu, J. Gutierrez, N. Idechong, G. Paulay, A. Tafileichig, and N. Vander Velde. 2002. The State of Coral Reef Ecosystems of the United States and Pacific Freely Associated States: 2002. Silver Spring: National Oceanic and Atmospheric Administration/National Ocean Service/National Centers for Ocean Coastal Science.

Veron, J.E.N. 1995. *Corals in Space and Time: The Biogeography and Evolution of the Scleractinia*. Ithaca: Cornell University Press.

Walker, N.D. 1996. Satellite assessment of Mississippi River plume variability: Causes and predictability. *Remote Sen. Environ.* 58:21–25.

Wellington, G.M., P.W. Glynn, A.E. Strong, S.A. Navarrete, E. Wieters, and D. Hubbard. 2001. Crisis on coral reefs linked to climate change. *EOS* 82:1–7.

Wilkinson, C. 2000. The 1997–98 mass coral bleaching and mortality event: 2 years on. In *Status of Coral Reefs of the World: 2000*, ed. C. Wilkinson, 21–34. Townsville: Australian Institute of Marine Science.

Wolanski, E., and M. Jones. 1981. Physical properties of Great Barrier Reef lagoon waters near Townsville. I. Effects of Burdekin River floods. *Aust. J. Mar. Freshwater Res.* 32:305–319.

Wolanski, E., R. Richmond, L. McCook, and H. Sweatman. 2003. Mud, marine snow and coral reefs. *Am. Sci.* 91:44–51.

Wolanski, E., and D. van Senden. 1983. Mixing of Burdekin River flood waters in the Great Barrier Reef. *Aust. J. Mar. Freshwater Res.* 34:49–63.

6. Fidelity of Annual Growth in *Montastraea faveolata* and the Recentness of Coral Bleaching in Florida

Robert B. Halley and J. Harold Hudson

6.1 Introduction

Annual growth bands in corals represent an important source of high-resolution (subannual) data for global change studies. Coral skeletal records have been a source of paleoclimatic, paleoceanographic, and paleoenvironmental data for more than two decades (Weber et al. 1976; Druffel 1982; Dodge and Lang 1983; Shen, Boyle, and Lea 1987; Smith et al. 1989; Dunbar and Cole 1992; Quinn, Crowley, and Taylor 1998). Important aspects of coral growth and degradation have come from analyses of coral records from this century (Hudson et al. 1994). More recently, sclerochronology is finding new applications to regional and local environmental issues over time scales of a few years to decades (Halley et al. 1994; Swart et al. 1996). Accurate chronologies are fundamental to these analyses, but only a few coral species construct skeletons with appropriate geometry, architecture, and density variation for exact temporal documentation. Even those species with appropriate external morphology may contain irregular annual growth bands or hiatuses caused by environmental disturbances or disease (Fig. 6.1).

Corals from the *Montastraea annularis* species complex are commonly used for sclerochronology in Florida and the Caribbean. The morphotype with the most regular annual density bands has been identified as *Montastraea faveolata* (Knowlton et al. 1992; Weil and Knowlton 1994). Typically, an annual growth band is a couplet consisting of a thin high-density (HD) band formed in late summer and a thicker low-density (LD) band formed during the rest of the year

(A)   (B)

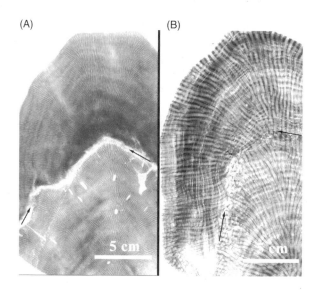

Figure 6.1. X-radiographs of coral colonies exhibiting poorly developed coral banding and hiatuses in two species of coral. (**A**) *Porites lobata* from Guam exhibiting indistinct annual density bands and a hiatus (arrows) indicating that most of the coral died and was subsequently overgrown. (**B**) *Solenastrea bournoni* from Florida Bay exhibiting distinct but irregular density banding with hiatus (arrows) indicating severe disruption of growth.

(Hudson et al. 1976; Dodge et al. 1992). Although there is still some discussion of this assignment (Szmant et al. 1997), we use the new nomenclature here because the morphology and skeletal architecture of this coral are distinct. *M. faveolata* is a particularly robust species for sclerochronology because its skeleton exhibits exceptionally well-defined density bands in X-radiographs. Prior to its redesignation by Knowlton et al. (1992), the growth form of *M. faveolata* appears to have been the sibling species of choice for sclerochronology. Descriptions of *M. annularis* and X-radiographs widely used for sclerochronology in the Caribbean appear to belong to *M. faveolata* (Leder, Szmant, and Swart 1991; Dodge et al. 1992).

This chapter documents the remarkable consistency of density banding in these corals, using cross correlation, during more than a century of skeletal growth from 12 coral colonies. The data presented here are a selection from 30 coral cores documented by Tao et al. (1999). Photographs of the 30 colonies of *M. faveolata*, all from Biscayne National Park, are also available in Tao et al. (1999) and provide the reader with a collection exhibiting the range of morphologies of this species.

6.1.1 Bleaching and Skeletal Growth

Coral bleaching is a common response to stress that can disrupt skeletal growth and has been widely associated with increased water temperature and UV penetration on reefs (Brown and Ogden 1993). Although scattered reports of coral bleaching appeared before the 1980s, the phenomenon has been widely reported during the past decade (Williams and Bunkley-Williams 1990, 1991). Coral bleaching disrupts metabolism and reproduction, diminishes coral growth, and, if persistent, can lead to death of the coral colony (Glynn 1983; Brown 1990). Goreau and Macfarlane (1990) note that *M. annularis* practically ceased calcification during the 1987 to 1988 bleaching event in Jamaica. Jokiel and Coles (1990) report that summer temperatures 1 to 2 °C above normal are sufficient to induce prolonged bleaching for the Indo-Pacific, but most colonies recover. Fitt et al. (2001) review the complicated relationships between temperature, light, and bleaching mechanisms and emphasize that a temperature for bleaching is difficult to define. To date in the Florida Keys, repeated episodes of bleaching have been predominantly in the sublethal range. However, it is possible that bleaching events have left the colonies more susceptible to coral diseases that have become widespread in recent years (Bruckner and Bruckner 1997; Richardson 1998).

Leder, Szmant, and Swart (1991) describe the skeletal effects of bleaching in *M. faveolata* and report significant reduction of skeletal extension in five of six corals that survived a severe bleaching event in south Florida during the summer and fall of 1987. One colony that bleached for 11 to 12 months failed to produce the low-density portion of the annual band for that coral growth year. Leder, Szmant, and Swart (1991) conclude that severe stress can result in the loss of a year's growth from the skeletal record of corals. They used the isotopic composition of the skeleton to determine the correct chronology and identify the gap in skeletal growth. Identification of this gap suggests that similar hiatuses may record previous bleaching events and that growth disruptions could compromise the accuracy of coral band counting. Crossdating using coral band fluorescence provides an additional mechanism for identification of such growth gaps.

6.1.2 Crossdating

Coral density bands, like tree rings, are most commonly dated by counting annual growth bands backward from the living surface. Dendrochronologists crossdate individual tree samples to verify assignments of exact calendar years (Schweingruber 1989). Typically, crossdating utilizes annual sequences of over 50 years and relies on common patterns of variation that can be recognized among a group of trees. Crossdating can identify missing years or false rings that may be confused with annual growth. The technique assumes that a missing year or a false ring for a given year is not present in all trees.

Most coral studies have relied solely on back-counting for coral chronology (Druffel 1982; Petzold 1984; McConnaughey 1989; Aharon 1991; Guzmán, Jackson, and Weil 1991; Winter, Goenaga, and Maul 1991; Dunbar et al. 1994; Quinn, Crowley, and Taylor 1998). In many corals, density banding is unclear and other methods must be used to discern annual growth increments, for example isotopic profiles (Cole, Fairbanks, and Shen 1993). As a result, correlation by back-counting alone becomes progressively uncertain with greater age.

Crossdating of coral band sequences has rarely been attempted, perhaps because samples of coral cores are much more difficult to obtain than tree ring cores. Also, coral cores often represent only a few decades of growth, limiting the cores that are available for intercolony comparison. In addition, coral skeletal architecture is much coarser than the cellular structure of wood. Even large-diameter coral cores may inadequately represent the continuity of banding sequences over an entire coral head. Hudson et al. (1976) crossdated coral colonies in much the same way as dendrochronologists crossdate trees, recognizing patterns of variation in skeletal banding. That study examined seven cores spanning 32 years taken from an area of reef approximately 1 km². Cold-water-induced high-density bands (stress bands) that form in the low-density portion of the annual density couplet provided characteristic markers and patterns of variation that could be correlated between corals.

Other methods used to verify back-counted age assignments have included staining (Leder et al. 1996), repeated sampling (Hudson et al. 1976; Swart et al. 1996), and measuring radioactive tracers from weapons testing and natural sources (Knutson, Buddemeier, and Smith 1972; Dodge and Thompson 1974; Buddemeier and Kinzie 1975). These methods provide verification of growth banding for a few years to a few decades. Thorium-230 ages provide approximate verification of band counting and suggest accuracy of ±5% over a time scale of two centuries (Edwards, Taylor, and Wasserburg 1988).

6.1.3 Origin of Fluorescence

Yellow-green fluorescent bands are preserved in the coral skeleton and result from fulvic acid introduced to coastal reefs during periods of high runoff (Isdale 1984; Boto and Isdale 1985; Susik, Boto, and Isdale 1991). In tropical and subtropical settings, concentrations of fulvic acids in fresh water may be an order of magnitude greater than concentrations in oligotrophic seawater (Zepp and Schlotzhauer 1981; Harvey et al. 1983; Ertel et al. 1986). The so-called "black-water" rivers of the southeastern United States and mangrove swamps contain particularly high concentrations of fulvic acid (Averett et al. 1987). Growing corals incorporate these dissolved organic acids into their skeletons, where they may remain stable for hundreds of thousands of years (Klein et al. 1990).

Fluorescent bands also occur in the skeletons of corals far removed from the influence of major terrestrial runoff (Guillaume 1995). In the open ocean, fulvic and humic acids form as degradation products of phytoplankton and are known as "yellow substance" (Carder et al. 1989). The concentration of yellow substance is presumably related to primary productivity, but Bricaud, Morel, and Prieur (1981) have shown that there is little concentration of yellow substance in upwelling areas. In all cases the amount of yellow substance in the open ocean is several orders of magnitude less than in coastal waters. Perhaps fluorescent bands in atoll corals are the result of degradation products of zooxanthellae being incorporated into the skeleton, but that process remains speculative.

Barnes and Taylor (2001) found luminescence in corals from the Great Barrier Reef to be predominantly caused by variations in skeletal architecture. They believe that annual fluorescent banding is an expression of the annual skeletal density-banding pattern. In particular, they found that fluorescent bands are almost always associated with the low-density regions of annual density bands. This latter finding is substantially different from the occurrences of fluorescence in Biscayne National Park, which are predominantly in the dense portion of the annual couplet.

Although the sources of fluorescence discussed above suggest that there are several possible mechanisms capable of generating fluorescent banding, we argue below that most of the fluorescence observed here is caused by the incorporation of land-derived substances during periods of high freshwater runoff. A similar origin has been reconfirmed for previously described fluorescent banding in corals in the Great Barrier Reef (Lough, Barnes, and McAllister 2003).

6.2 Methods

Between 16 July and 4 October 1986, core samples from 38 colonies of *M. faveolata* were collected from eight patch reefs in Biscayne National Park (Fig. 6.2) following procedures described by Hudson (1981) and Hudson et al. (1994). The cores were taken from large (>1 m) colonies distributed across the Atlantic shelf 5 km east to west and 25 km north to south, in anticipation of broad influence from Biscayne Bay on the reef tract. The sites range from outer shelf (site 6) where the influence from the Florida Current is greatest, to inshore (site 8) at the mouth of Biscayne Bay. This distribution of sites captures natural variability typical of near-shore ocean water across the shelf (Shinn 1966; Hudson 1981; Hudson et al. 1991). During the summer of 1986 there was no evidence of bleaching or disease among large colonies of *M. faveolata,* and no selection was made based on the health of the colonies. X-radiographs were digitally scanned and positive images prepared for analyses. A preliminary chronology was determined by back-counting high-density bands from the living surface. Growth rates were measured and are reported in Hudson et al. (1994).

Long coral cores (>1 m) are typically not continuous and fracture during sampling. This requires reassembly and correlation across breaks. Fractures that intersect density bands at a high angle are simple to bridge. Horizontal breaks, however, may allow the cored portion of the coral to rotate on top of the next portion, potentially grinding away some skeleton. Horizontal fractures can introduce uncertainty in the dating assignments, but they may not be as serious as they appear in X-radiographs. Whole cores may often be easily matched to solve fracture breaks. Whole cores can be rotated to approximate original positions more easily than individual slabs, which may be cut on different planes (Fig. 6.3).

Coral slabs were placed in a dark box and illuminated with two 35-cm short-wave ultraviolet lamps (peak intensity: 1100 μW cm^{-2} of 254 nm at 30 cm) and imaged with a black-and-white video camera using a yellow No. 2 lens filter. Image data were collected at a resolution of 0.2 mm/pixel, background corrected, standardized, enhanced, and reduced to a convenient working scale using commonly available image analysis software. The lamps were placed at 45° angles to

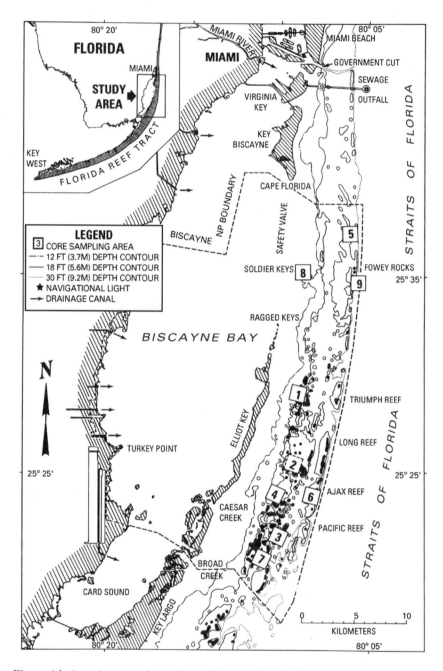

Figure 6.2. Location map of core sites in Biscayne National Park, southeastern Florida.

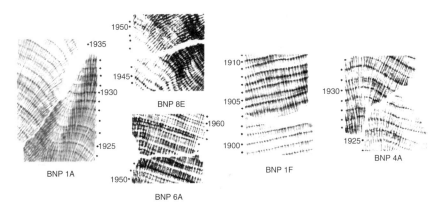

Figure 6.3. Examples of fractures in coral cores: BNP 1A, 8E, and 6A, fractures occur at a high angle to density bands, so tracing bands across the break is relatively simple; BNP 1F, spinning has rounded the core ends; BNP 4A, fracture is complex.

the plane of the coral slab, one on each side, and 10 cm apart. Assigning dates to the fluorescent images was accomplished by digitally manipulating the images. The positions of fluorescent bands were traced on a template that was superimposed on the X-radiograph using image-processing software. Slight changes in scale were corrected by matching the outlines of the two images. Dates were then assigned to fluorescent bands by noting the date of the density band underlying the tracing. Semiquantitative measures of fluorescence were developed using image brightness (on a scale of 0 = white, to 255 = black) as a proxy. Initial interpretations of fluorescent images were reported in Halley and Hudson (1992). The complete set of digital images (X-radiographs and fluorescence images), together with age assignments and site photographs, are available in Tao et al. (1999).

6.3 Results and Discussion

6.3.1 Fluorescence and Runoff

Three modes of fluorescence are apparent in the Biscayne core set (Fig. 6.4). The most distinctive mode occurs as aperiodic, yellow-green, moderately fluorescing, thin (1–3 mm) bands, on a more dully fluorescing background. A second mode appears as annual fluorescent banding, with fluorescent skeletal material occurring in every density couplet for several decades of growth. A third fluorescent pattern is mottled, with no apparent correspondence between growth bands and fluorescence. Only the aperiodic fluorescent banding is used for crossdating.

About 95% of the aperiodic fluorescent bands occur between 1878 and 1965 in the cores from Biscayne National Park. Distinct bands are rare prior to 1880, although several occur in the 1840s. Skeletal growth after 1965 rarely exhibits aperiodic fluorescent bands. The mottled fluorescence pattern is most common

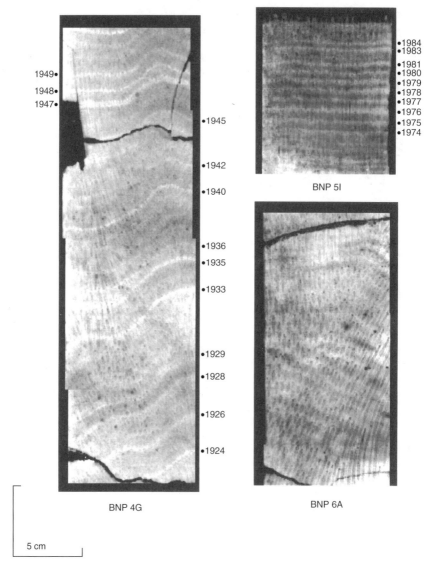

Figure 6.4. Examples of three modes of fluorescent bands in Biscayne National Park corals: BNP 4G, distinct aperiodic bands; BNP 5I, annual fluorescent bands; BNP 6A, mottled, indistinct fluorescent bands.

between 1965 and 1986, the period during which most freshwater flow had been diverted from Biscayne Bay.

Aperiodic fluorescent bands occur in about 40% of the annual growth intervals formed during the past 100 years and, when present, they correspond to the intervals of dense skeletal growth that form in the late summer. Fluorescent intensity

Table 6.1. Years for which fluorescent markers were used to crossdate coral cores from Biscayne National Park

1871	1875	1876	1878	1880	1884	1891	1893
1894	1899	1900	1901	1904	1905	1908	1909
1910	1915	1918	1922	1924	1926	1928	1929
1933	1935	1936	1940	1946	1947	1948	1949
		1952	1953	1959	1963	1967	

varies among horizons within a core, between cores, and between sites. A total of 37 distinct fluorescent horizons could be correlated between three or more cores (Table 6.1). Fifteen fluorescent bands may be visually correlated between more than half of the 38 cores from Biscayne National Park (Fig. 6.5). These fluorescent horizons are used to crossdate cores distributed over 100 km^2 of the inshore reef.

Rainfall in south Florida is highly seasonal (Duever et al. 1994). The area receives about half its precipitation (~1250 mm) from June through September. Runoff is also highly seasonal, but volume is modulated by the Biscayne Aquifer, which is highly porous and permeable Quaternary limestone with significant storage capacity (Fish and Stewart 1991). During late summer, runoff into Biscayne Bay produces distinct near-shore water masses that move south and east along the near-shore zone of the northern Florida Keys and about halfway across the reef tract (Lee 1986). Eventually these water masses disperse and are entrained into the northward-flowing Florida Current (Lee and Williams 1999).

Prior to the 1920s, runoff occurred by wet-season sheet-flooding from the Everglades, with dispersed point sources at the Miami River and through topographic depressions through the Atlantic Coastal Ridge known as transverse glades. Between 1920 and the 1950s, seasonal runoff increasingly reached the coast via drainage canals that were dug to drain wetlands and control flooding (Parker, Ferguson, and Love 1955). Since the 1950s, runoff has been controlled by water management practices in Florida and floodwaters are directed to the Atlantic Ocean and Gulf of Mexico north of Biscayne National Park. Between 1887 and 1960, water that historically flowed to South Florida has been diverted to the Atlantic and Gulf of Mexico through drainage canals at the latitude of Lake Okeechobee, approximately 120 km to the north.

Many of the brightly fluorescing bands correspond to years of flooding induced by tropical cyclones during the period 1871 to 1947. A correlation between fluorescence and historical changes in runoff from the Miami area reinforces the interpretation that fluorescent organic acids are derived from the mainland. Figure 6.6 compares flow of the Tamiami Canal, where it enters the Miami River, to the semiquantitative measure of fluorescent brightness. Although the correlation is not strong, periods of high runoff during 1947 to 1949 and 1952 to 1954 correlate with aperiodic fluorescent bands. Several prominent and widespread fluorescent bands occur during years of well-documented flooding in south Florida such as 1926 and 1928 (Huser 1989). Several of these floods were caused by hurricanes that also caused wind damage and widespread defoliation, introducing large inputs of organic matter into surface water. Both volume of runoff

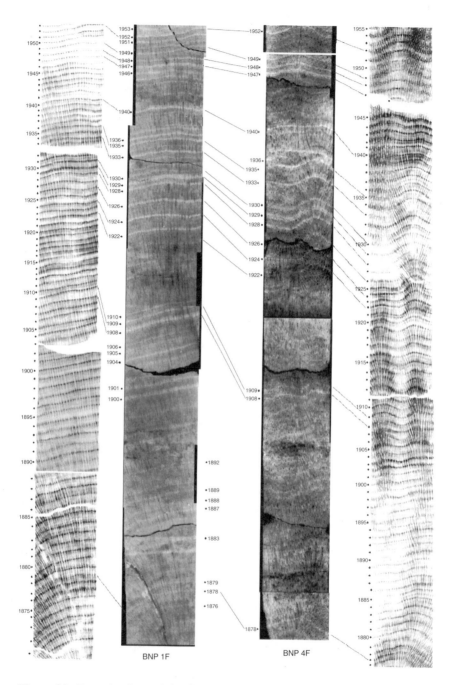

Figure 6.5. Example of crossdating between two coral cores, BNP 1F and BNP 4F. Sites are about 10 km apart. Correspondence of fluorescent band dates indicates complete fidelity of annual banding during the growth period 1878 to 1986.

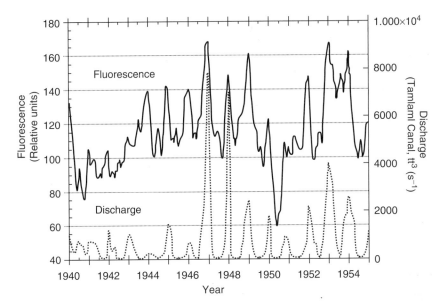

Figure 6.6. Time series of relative fluorescence from core BNP 1F (solid line) compared with discharge from the Tamiami Canal at the Miami River (dashed line).

and concentration of organic acids may increase after hurricanes. Flow from the Miami River during floods accounts for only a portion of the runoff to Biscayne Bay, and the contribution from other sources, such as traverse glades or drainage canals, may account for the low correlation. Nevertheless, the general correspondence supports a linkage between runoff and fluorescence.

Although more than half of the 38 cores taken for the growth rate study (Hudson et al. 1994) exhibit some aperiodic fluorescence, this feature alone is insufficient for crossdating. The following criteria were employed to select the cores used to evaluate the fidelity of annual growth:

1. Growth period of a century or more
2. Aperiodicity of fluorescent banding
3. Continuity between core sections
4. Geographic distribution

Twelve cores matched the criteria above and were used for further analyses (Table 6.2).

6.3.2 Crossdating

Dates were assigned to each of the fluorescent bands using chronologies from density banding, and the dated fluorescent bands were compared between cores. Distinctive groupings of bands were used to verify consistency of age assignments in these cores. For example, a particularly obvious pattern is the every-other-year

Table 6.2. Portions of cores used for skeletal density and fluorescence comparisons. Core locations are shown in Fig. 6.2

Location	Core	Interval of record	Length (years)
1	E	1870–1986	116
1	F	1870–1986	116
3	A	1870–1986	116
3	B	1877–1986	109
4	A	1870–1986	116
4	B	1870–1986	116
4	D	1870–1986	116
4	E	1870–1986	116
4	F	1870–1986	116
4	G	1880–1986	106
7	D	1870–1986	116
8	G	1870–1986	116
			Total 1375

pattern of 1922, 1924, 1926, and 1928 (Fig. 6.7). Other distinctive groupings of fluorescent bands include a couplet in 1908 and 1909 and the triplet of 1947, 1948, and 1949. Cores containing growth lapses would result in age assignments to fluorescent bands that would not correspond to those of other cores. Surprisingly, in this group of cores cumulatively representing 1375 annual growth bands, no missing high- or low-density bands were identified. In all cores, each year of growth was represented by a distinct annual density couplet. Annual growth represented by a single high-density band, as described for bleached corals (Leder, Szmant, and Swart 1991), was not found. The observed consistency of growth indicates that either (1) significant hiatuses (or extra bands) occur at the same intervals in all 12 cores spanning the entire study area or (2) *M. faveolata* colonies recorded uninterrupted growth for 108 years.

By noting portions of cores that did not correspond, we hoped to narrow the search and identify hiatuses in skeletal growth. Although it may, intuitively, seem reasonable that some environmental perturbations could result in widespread growth hiatuses among corals, direct observation suggests otherwise. *M. faveolata* is an environmentally tolerant species, as shown by transplant studies (Hudson 1981), and the species is widespread throughout the Caribbean. Biscayne National Park, where periodic cold air outbreaks severely damage reef corals, is near the northern latitudinal limit for this species (Hudson et al. 1976; Roberts et al. 1982). Also, during the last two decades reports of coral bleaching during the summer months have become more common (Brown 1990; Brown and Ogden 1993). However, the tolerance of individual colonies to environmental extremes at any one site is highly variable as evidenced by their individual responses to low- and high-temperature extremes (Hudson 1981; Brown and Ogden 1993). In the Florida Keys, most corals survive episodes of extreme cold or warm water. Typically, some colonies appear unaffected while their neighbors may be severely affected. X-radiographs reveal that the majority of coral skeletons continue to exhibit

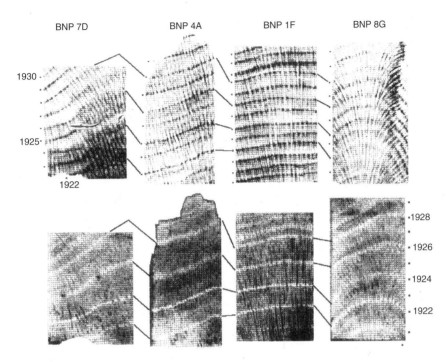

Figure 6.7. Correlation of fluorescent horizons between portions of four cores. Note that in Fig. 6.2, sites 7 and 8 are 25 km apart. Top: digitized X-radiographs illustrating annual dense bands dated following standard methods, with dense bands for 1922, 1925, and 1930 identified. Bottom: fluorescent images with tie-lines indicating years 1922, 1924, 1926, and 1928. These figures also illustrate that fluorescent bands occur in the high-density portions of annual density bands.

annual density patterns even after severe environmental insult (Hudson 1981; Leder, Szmant, and Swart 1991). Thus, it is unlikely that the same interval or intervals are missing from all the cores in this data set. Rather, we interpret these consistent correlations (Fig. 6.7) as strong support for exceptionally reliable annual growth in this species. As a consequence, we find no error in the age assignments to annual growth increments.

The fidelity of annual growth, as revealed in the cores studied here, indicates that severe bleaching events, such as those that affected Florida Keys corals in 1987, were rare events prior to the 1980s. Future bleaching events should be carefully monitored for skeletal effects in corals from the Florida Keys. If bleaching commonly results in loss of growth bands in large colonies, then bleaching is the result of processes that have not occurred on the reef tract during the record of growth in the coral colonies used in this study. If bleaching persists until mortality, then some of the oldest features of the reefscape will be lost, requiring many centuries to regrow.

6.4 Summary and Conclusions

Correlation with runoff and proximity to shore argue that fluorescent banding in *M. faveolata* from Biscayne National Park is the result of periods of high freshwater flow to the coast. The occurrence of frequent aperiodic fluorescent bands provides an independent set of markers that can be used for crosscorrelation with density bands to verify dates of growth and establish the general reliability of chronologies based on back-counting from the growing surface. In a set of 12 cores from large colonies, representing 1375 years of skeletal growth, no missing growth intervals were identified for the period 1878 to 1986. Although some growth intervals exhibited irregularities attributable to the natural disturbances, seriously reduced annual growth or absence of seasonal skeletal accretion did not occur. These observations support the assumption of past researchers that counting annual density couplets in *M. faveolata* produces reliable chronologies. The remarkable regularity of annual skeletal growth also argues for the absence of severe bleaching during the past century in this region. Conversely, the fidelity of annual growth suggests widespread coral bleaching is a phenomenon of the past two decades. The century of growth examined here appears to be in contrast with post-1986 growth when bleaching has become common. The recentness of bleaching supports the hypothesis of unprecedented environmental stress beginning in the late 20th century.

Acknowledgements. The authors express their gratitude to those who have worked with them during the collections and preparation of the material for the work reported here: Yucong Tao, Leanne M. Roulier, Karen M. Higgins, Marshall L. Hayes, and Jack Kindinger. Richard Curry of Biscayne National Park provided logistical and administrative support. We appreciate USGS St. Petersburg Center for Coastal Geology for archiving the coral cores since 1988. The USGS Coastal and Marine Geology and Climate Change Programs provided funding for this project. The authors appreciate constructive reviews of earlier versions of this manuscript by Gene Shinn and Kim Yates of the US Geological Survey, Dennis Hubbard, William Precht, and several anonymous reviewers.

References

Aharon, P. 1991. Recorders of reef environmental histories: Stable isotopes in corals, giant clams and calcareous algae. *Coral Reefs* 10:71–90.

Averett, R.C., J.A. Leenheer, D.M. McKnight, and K.A. Thorn. 1987. Humic substances in the Suwannee River, Georgia: Interactions, properties, and proposed structures. U.S. Geol. Survey Open File Rep. 87-557, Washington, DC.

Barnes, D.J., and R.B. Taylor. 2001. On the nature and causes of luminescent lines and bands in coral skeletons. *Coral Reefs* 19:221–230.

Boto, K., and P. Isdale. 1985. Fluorescent bands in massive corals result from terrestrial fulvic acid inputs to nearshore zone. *Nature* 315:396–397.

Bricaud, A., A. Morel, and L. Prieur. 1981. Absorption by dissolved organic matter in the sea (yellow substance) in the UV and visible domains. *Limnol. Oceanogr.* 26:43–53.

Brown, B.E. (ed.). 1990. Coral Bleaching. *Coral Reefs* 8:155–232.

Brown, B.E., and J.C. Ogden. 1993. Coral bleaching. *Sci. Am.* 268:64–70.

Bruckner, A.W., and R.J. Bruckner. 1997. Emerging infections on the reefs, *Nat. Hist.* 106:48.

Buddemeier, R.W., and R.A. Kinzie. 1975. The chronometric reliability of contemporary corals. In *Growth Rhythms and the History of the Earth's Rotation*, eds. G.D. Rosenberg and S.K. Runcorn, 135–147. New York: John Wiley & Sons.

Carder, K.L, R.G. Steward, G.R. Harvey, and P.B. Ortner. 1989. Marine humic and fulvic acids: Their effects on remote sensing of ocean chlorophyll. *Limnol. Oceanogr.* 34:68–81.

Cole, J.E., R.G. Fairbanks, and G.T. Shen. 1993. Recent variability in the Southern Oscillation: Isotopic results from a Tarawa Atoll coral. *Science* 260:1790–1793.

Dodge, R.E., and J.C. Lang. 1983. Environmental correlates of hermatypic coral (*Montastraea annularis*) growth on the East Flower Gardens Bank, Northwest Gulf of Mexico. *Limnol. Oceanogr.* 28:228–240.

Dodge, R.E., A.M. Szmant, R. Garcia, P.K. Swart, A. Forester, and J.J. Leder. 1992. Skeletal structural basis of density banding in the reef coral *Montastraea annularis*. *Proc. Seventh Int. Coral Reef Symp., Guam* 186–195.

Dodge, R.E., and J. Thompson. 1974. The natural radiochemical and growth records in contemporary hermatypic corals from the Atlantic and Caribbean. *Earth Planet. Sci. Lett.* 23:313–322.

Druffel, E.M. 1982. Banded corals: Changes in oceanic carbon-14 during the little ice age. *Science* 218:13–19.

Duever, M.J., J.F. Meeder, L.C. Meeder, and J.M. McCollom. 1994. The climate of south Florida and its role in shaping the Everglades ecosystem. In *Everglades. The Ecosystem and Its Restoration*, eds. S.M. Davis and J.C. Ogden, 225–248. Delray Beach, FL: St. Lucie Press.

Dunbar, R.B., and J.E. Cole. 1992. Coral records of ocean–atmosphere variability. NOAA Climate and Global Change Program Special Report No. 10.

Dunbar, R.B., G.M. Wellington, M.W. Cogan, and P.W. Glynn. 1994. Eastern Pacific sea surface temperature since 1600 A.D.: The $\delta^{18}O$ record of climate variability in Galapagos corals. *Paleoceanography* 9:291–315.

Edwards, R.L., F.W. Taylor, and G.J. Wasserburg. 1988. The natural radiochemical and growth records in contemporary hermatypic corals from the Atlantic and Caribbean. *Earth Planet. Sci. Lett.* 90:371–381.

Ertel, J.R., J.I. Hedges, A.H. Devol, J.E. Reichey, and M.N.G. Ribeiro. 1986. Dissolved humic substances of the Amazon River system. *Limnol. Oceanogr.* 31:739–754.

Fish, J.E., and M. Stewart. 1991. Hydrogeology of the surficial aquifer system. USGS Water-Investigations Report 90-4108. Florida: Dade Co.

Fitt, W.R., B.E. Brown, M.E. Warner, and R.P. Dunne. 2001. Coral bleaching: Interpretation of thermal tolerance limits and thermal thresholds in tropical corals. *Coral Reefs* 20:51–65.

Glynn, P. 1983. Extensive bleaching and death of reef corals on the Pacific coast of Panama. *Environ. Conserv.* 10:149–154.

Goreau, T.J., and A.H. Macfarlane. 1990. Reduced growth rate of *Montastraea annularis* following the 1987–1988 coral bleaching event. *Coral Reefs* 8:211–215.

Guillaume, M. 1995. Fluorescent growth bands in corals from Mururoa Atoll. In *Coral Reefs in the Past, Present and Future*, eds. B. Lathuiliere and J. Geister. Publications du Service Geologique du Luxumbourg 29:152.

Guzmán, H.M., J.B.C. Jackson, and E. Weil. 1991. Short term ecological consequences of a major oil spill on Panamanian subtidal reef corals. *Coral Reefs* 10:1–12.

Halley, R.B., and J.H. Hudson. 1992. Fluorescent growth bands in corals from south Florida and their environmental interpretation. *Proc. Seventh Int. Coral Reef Symp., Guam* 222.

Halley, R.B., P.K. Swart, R.E. Dodge, and J.H. Hudson. 1994. Decade-scale trend in seawater salinity revealed through $\delta^{18}O$ analysis of *Montastraea annularis* annual growth bands. *Bull. Mar. Sci.* 54:670–678.

Harvey, G.R., D.A. Boran, L.A. Chesal, and J.M. Tokar. 1983. The structure of marine fulvic and humic acids: Symposium on marine chemistry. *Mar. Chem.* 12:119–132.

Hudson, J.H. 1981. Response of *Montastraea annularis* to environmental change in the Florida Keys. *Proc. Fourth Int. Coral Reef Symp.* 2:233–240.

Hudson, J.H., R.B. Halley, A.J. Joseph, B.H. Lidz, and B. Schroeder. 1991. Long-term thermograph records from the Upper Florida Keys. USGS Open File Report 91-344.

Hudson, J.H., K.J. Hanson, R.B. Halley, and J.K. Kindinger. 1994. Environmental implications of growth rate changes in *Montastraea annularis*: Biscayne National Park, Florida. *Bull. Mar. Sci.* 54:647–669.

Hudson, J.H., E.A. Shinn, R.B. Halley, and B.H. Lidz. 1976. Sclerochronology—a tool for interpreting past environments. *Geology* 4:361–364.

Huser, T. 1989. *Into the Fifth Decade*. South Florida Water Management District, West Palm Beach, FL.

Isdale, P. 1984. Fluorescent bands in massive corals record centuries of coastal rainfall. *Nature* 310:578–579.

Jokiel, P.L., and S.L. Coles. 1990. Response of Hawaiian and other Indo-Pacific reef corals to elevated temperature. *Coral Reefs* 8:155–162.

Klein, R., Y. Loya, G. Gvirtzman, P.J. Isdale, and M. Susic. 1990. Seasonal rainfall in the Sinai Desert during the late Quaternary inferred from fluorescent bands in fossil corals. *Nature* 345:145–147.

Knowlton, N., E. Weil, L.E. Weigt, and H.M. Guzman. 1992. Sibling species in *Montastraea annularis*, coral bleaching, and the coral climate record. *Science* 255:330–333.

Knutson, D.W., R.W. Buddemeier, and S.V. Smith. 1972. Coral chronometers; seasonal growth bands in reef corals. *Science* 177:270–272.

Leder, J.J., A.M. Szmant, and P.K. Swart. 1991. The effect of prolonged "bleaching" on skeletal banding and the isotopic composition of *Montastraea annularis*. Preliminary observations. *Coral Reefs* 10:19–27.

Leder, J.J., P.K. Swart, A.M. Szmant, and R.E. Dodge. 1996. The origin of variation in the isotopic record of scleractinian corals: Oxygen. *Geochim. Cosmochim. Acta* 60:2857–2870.

Lee, T.N. 1986. Coastal circulation in the Key Largo Coral Reef Sanctuary. Physics of shallow estuaries and bays. *Lecture Notes in Coastal and Estuarine Studies* 16:178–198.

Lee, T.N., and E. Williams. 1999. Mean distribution and seasonal variability of coastal currents and temperature in the Florida Keys with implication for larval recruitment. *Bull. Mar. Sci.* 64:35–56.

Lough, J.M., D.J. Barnes, and F.A. McAllister. 2003. Luminescent lines in corals from the Great Barrier Reef provide spatial and temporal records of reef affected by land runoff. *Coral Reefs* 21:333–343.

McConnaughey, T. 1989. ^{13}C and ^{18}O isotopic disequilibrium in biological carbonates; I, Patterns. *Geochim. Cosmochim. Acta* 53:151–162.

Parker, G.G., G.E. Ferguson, and S.K. Love. 1955. Water Resources of South Florida. USGS Water-Supply Paper 1255.

Petzold, J. 1984. Growth rhythms recorded in stable isotopes and density bands in the reef coral *Porites lobata* (Cebu, Philippines). *Coral Reefs* 3:87–90.

Quinn, T.M., T.J. Crowley, and F.W. Taylor. 1998. A multicentury stable isotope record from a New Caledonia coral; interannual and decadal sea surface temperature variability in the Southwest Pacific since 1657 A.D. *Paleoceanography* 13:412–426.

Richardson, L.L. 1998. Coral diseases: What is really known? *Trends Ecol. Evol.* 13:438–443.

Roberts, H.H., L.J. Rouse, Jr., N.D. Walker, and J. H. Hudson. 1982. Cold-water stress in Florida Bay and the northern Bahamas—a product of winter cold-air outbreaks. *J. Sediment. Petrol.* 52:145–155.

Schweingruber, F.H. 1989. *Tree Rings*. Dordrecht, Holland: Kluwer Academic Publishers.

Shen, G.T., E.A. Boyle, and D.W. Lea. 1987. Cadmium in corals as a tracer of historical upwelling and industrial fallout. *Nature* 328:794–796.

Shinn, E.A. 1966. Coral growth rate, an environmental indicator. *J. Paleontol.* 40:233–240.

Smith, T.J., J. H. Hudson, M.B. Robblee, G.V.N. Powell, and P.J. Isdale. 1989. Freshwater flow from the Everglades to Florida Bay: A historical reconstruction based on fluorescent banding in the coral bands in the coral *Solenastrea bournoni*. *Bull. Mar. Sci.* 44:274–282.

Susic, M., K. Boto, and P. Isdale. 1991. Fluorescent humic acid bands in coral skeletons originate from terrestrial runoff. *Mar. Chem.* 33:91–104.

Swart, P.K., G.F. Healy, R.E. Dodge, P. Kramer, J. H. Hudson, R.B. Halley, and M.B. Robblee. 1996. The stable oxygen and carbon isotopic record from a coral growing in Florida Bay: A 160 year record of climatic and anthropogenic influence. *Palaeogeogr. Palaeoclimatol. Palaeoecol.* 123:219–237.

Szmant, A.M., E. Weil, M.W. Miller, and D.E. Colon. 1997. Hybridization within the species complex of the scleractinian coral *Montastraea annularis*. *Mar. Biol.* 129:561–572.

Tao, Y., L.M. Roulier, K.M. Higgins, M.L. Hayes, R.B. Halley, and J. H. Hudson. 1999. Digitized images of coral growth bands, fluorescent growth bands and their age assignments from cores of *Montastraea annularis*, Biscayne National Park, Upper Florida Keys. USGS Open-File Report 99–340.

Weber, J.N., P. Deines, P.H. Weber, and P.A. Baker. 1976. Depth related changes in the $^{13}C/^{12}C$ ratio of skeletal carbonate deposited by the Caribbean reef-frame building coral *Montastraea annularis*; further implications of a model for stable isotope fractionation by corals. *Geochim. Cosmochim. Acta* 40:31–39.

Weil, E., and N. Knowlton. 1994. A multi-character analysis of the Caribbean coral *Montastraea annularis* (Ellis and Solander, 1786) and its two sibling species, *M. faveolata* (Ellis and Solander, 1786) and *M. franksi* (Gregory, 1895). *Bull. Mar. Sci.* 55:151–175.

Williams, E.H., and L. Bunkley-Williams. 1990. The worldwide bleaching cycle and related sources of coral mortality. *Atoll Res. Bull.* 335:1–71.

Williams, E.H., Jr., and L. Bunkley-Williams. 1991. Coral reef bleaching alert; discussion. *Nature* 346:225.

Winter, A., C. Goenaga, and G.A. Maul. 1991. Carbon and oxygen isotope time series from an 18-year Caribbean reef coral. *J. Geophys. Res.* 96:16,673–16,678.

Zepp, R.G., and P.F. Schlotzhauer. 1981. Comparison of photochemical behavior of various humic substances in water: III. Spectroscopic properties of humic substances. *Chemosphere* 10:479–486.

Part III.   Patterns of Reef Development and Their Implications

7. Demise, Regeneration, and Survival of Some Western Atlantic Reefs During the Holocene Transgression

Ian G. Macintyre

7.1 Introduction

In the past 18,000 years, the coral reefs of the western Atlantic have experienced catastrophic stresses owing to a huge rise in sea level. The melting of late Pleistocene ice sheets during this period caused the world's seas to rise more than 100 m (Fig. 7.1). At times, surges in the meltwater runoff pushed sea level up as much as 25 to 30 m within a span of 1000 year (Fairbanks 1989). Through a review of some well-documented instances of the responses of coral reefs to changing sea levels, this chapter describes how supposedly fragile coral reef communities responded to and survived the postglacial rising seas.

Tropical corals, also known as hermatypic or zooxanthellate corals, form their skeletons with the help of symbiotic photosynthetic zooxanthellae (Goreau 1961a,b). Calcification occurs at such high rates that it enables these corals to form massive reef complexes, with one important restriction: because zooxanthellae are unable to function without sufficient light, active accumulation of these reef-building corals is limited to depths of about 20 m or less (James and Macintyre 1985). Rising seas therefore pose a serious threat to their survival.

In response to such conditions, some coral reefs are able to increase their accretion and thereby remain within the depth zone required for their accumulation. Others founder, perhaps hanging on until the rate of sea level decreases. Still others become stranded in deep water and perish. Think of a group of joggers (Fig. 7.2), with sea level leading the pack: reefs with growth rates that equal the rise in sea

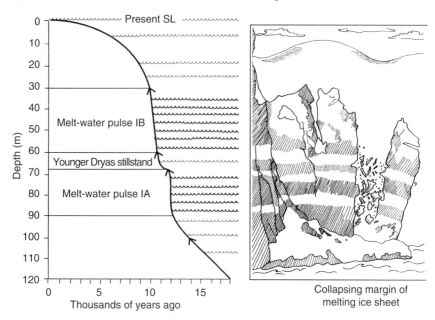

Figure 7.1. The rising seas of the Holocene Transgression caused by the melting of Pleistocene ice sheets as depicted on the right. This sea-level curve is based on data from Fairbanks (1989) and Lighty, Macintyre, and Stuckenrath (1982). Note that close spacing of sea-level positions represents periods of rapidly rising sea levels.

Figure 7.2. Schematic showing how a group of joggers can illustrate the response of coral reefs to rising sea levels. The lead jogger (in dark shorts) is "Sea Level" and right next to him there is "Keep-Up" reef with a growth rate equal to sea-level rise. He is followed by "Catch-Up" reef who is approaching the other two joggers with a growth rate that exceeds sea-level rise. Finally, at the rear, there is "Give-Up" reef in a state of little or no growth. Reproduced from Neumann and Macintyre (1985) with permission from the Ecole Pratique des Hautes Etudes, Université de Perpignan.

level are able to keep up, some lag behind but then rebound and catch up through energetic growth that surpasses the rise, but some fail to recover and give up (Neumann and Macintyre 1985).

Important clues to the behavior of Keep Up, Catch Up, and Give Up reefs lie not only in the history of coral-reef accumulation and the global rise in post-glacial seas, but also in two local factors: the topography of the sea floor and the frequency of destructive storms. Many of these details can be gleaned from the geological record preserved within the dead and living reefs of the western Atlantic, making it possible to document the best world record of reef development over the last 18,000 years.

As defined by Curray (1965) and Stoddart (1969), in this chapter the Holocene Epoch is extended back to approximately 18,000 ybp, which corresponds to the period of rising sea levels that resulted from the postglacial melting of the late Pleistocene ice sheets. This global rise in sea level has been termed the "Holocene Transgression" (Curray 1965, p. 725). In this chapter, the Holocene Transgression of sea level is considered to include all postglacial rises in sea level.

7.2 Holocene History of Western Atlantic Coral Reefs

7.2.1 Relict Reefs on the Outer Slopes

Information on the history of reefs in relation to sea level has been mounting ever since the discovery of relict submerged reefs on insular or continental slopes in various parts of the world. One such structure, described as an "old drowned reef of early Wurm age," parallels the north coast of Jamaica near Oracabessa for a distance of about 10 km and occurs at a depth of about 70 m (Goreau 1961b, p. 11). This feature appears to be very similar to a narrow 70-m ridge with 10-m relief that parallels the west coast of Barbados for a distance of 20 km (Macintyre 1967), as well as other "reef-like bodies" at depths of 80 to 120 m off western Guiana (Nota 1958, p. 18). More recently, submerged barrier reef platforms "backed by relict lagoons averaging 70 m depth" have been found on the outer edge of Australia's Great Barrier Reef (Harris and Davies 1989, p. 98). These and other features, such as small notches or terraces, relict oolitic dunes, and littoral sands (see Macintyre et al. 1991), are all indicative of a stillstand in sea level during the Holocene at depths of 75 to 90 m.

Until a decade ago, the notion that these were relict reefs established on deep slopes was considered highly speculative because the only evidence available to support it was the topography of the seafloor. This all changed in late 1988 when offshore drilling techniques were used to recover cores from relict reefs at depths of over 100 m below sea level off the south coast of Barbados (Fairbanks 1989). Now, for the first time, there was indisputable evidence that shallow-water reefs dating back 18,000 years (as measured by radiocarbon methods) had once flourished on insular slopes down to depths of 124 m (Fairbanks 1989; Bard et al. 1990).

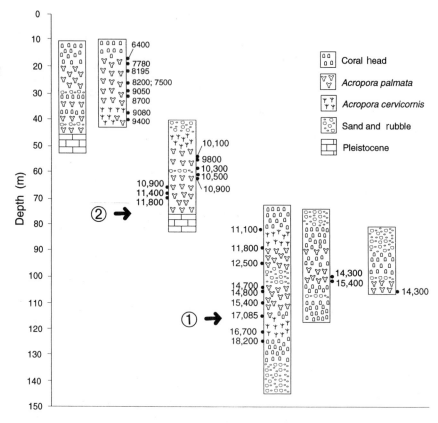

Figure 7.3. Reef facies and radiocarbon dates in a series of deep cores collected off the south coast of Barbados. Arrows point to the base of sections through relict shallow-water *Acropora palmata* reefs at approximate depths and dates of (1) 17,000 ybp at 115 m; (2) 12,000 ybp at 80 m. After Fairbanks (1989).

A graphic summary of some of Fairbanks' (1989) core-log data (Fig. 7.3) indicates a 25-m section of *Acropora palmata* framework ranging in age from 12,000 to 10,000 ybp and established on an erosional Pleistocene surface at a depth of about 80 m below present sea level. This erosional surface occurs at about the same depth as the relict reef ridge off the west coast of Barbados, which suggests that the two features probably formed sometime during the same time interval, most likely in the period of the Younger Dryas stillstand (Macintyre et al. 1991). A marked reduction in the rise of sea level at that time would have permitted these relict slope reefs to flourish for a brief interval (Fig. 7.1).

The present-day community on the surface of the relict reef off the west coast of Barbados consists of a rich cover of sponges, algae (particularly crustose corallines), and a few scattered corals (Macintyre et al. 1991). This feature is clearly no longer an actively forming reef, but a Give Up reef that was stranded on this slope by the rapidly rising seas of Melt-Water Pulse IB (Fig. 7.1). Evidence of an even earlier Give Up relict reef has been found in the cores collected off the south coast of Barbados (Fig.7.3). They contain sections of a shallow-water *Acropora palmata* framework dating back to 17,000 ybp that suggest the reef system died out about 12,000 ybp, when it was stranded by the rapidly rising seas associated with Melt-Water Pulse IA (Fig. 7.1).

7.2.3 Shelf-Edge Relict Reefs

Another important phase in the history of western Atlantic coral reefs began with the construction of reefs at the outer edges of insular and continental shelves about 9000 to 10,000 ybp. Today these reefs have a much richer live coral cover than those on the deeper slopes, although they lie at depths that are too great for active framework accumulation. These, too, are stranded Give Up reefs, and they are a common feature in the western Atlantic, particularly in the eastern Caribbean (Macintyre 1972). Although Fairbanks (1989) does not show bottom topography, it appears that his latest 20-m-thick *A. palmata* section, which ranges in age from 9400 to 6400 ybp at present-day depths of 20 to 40 m, is a shelf-edge Give Up reef. Excellent examples of such Give Up reefs are seen in the geological records of shelf-edge reefs in St. Croix and southeastern Florida, revealing a history of shallow-water growth and then demise.

Shelf-edge reef off the southeast coast of St. Croix. A complex ridge system occurs along the shelf-edge (approximately 15–20 m) of the northeastern, eastern, and southern coasts of St. Croix. Its surface community today is dominated by deeper-water, mostly massive corals (*Montastraea annularis, Diploria labyrinthiformis, Porites astreoides,* and *Colpophyllia natans*), octocorals, and sponges (Adey et al. 1977). An exposure in the Hess Refinery ship channel, on the south coast, indicates that much of the shelf-edge relief is related to a Pleistocene limestone rim (Adey et al. 1977).

A strikingly different feature present in this area, however, is a narrow outer ridge, generally less than 100 m wide, with a relief of 5 m, and occurring in water depths of 14 to 20 m. A 6.2-m core hole drilled into this ridge produced evidence to indicate that it was once a flourishing *Acropora palmata* reef. The lower half of the core section contained freshly preserved *A. palmata* with crusts of the shallow-water coralline alga *Porolithon pachydermum* (Fig. 7.4). A sample from the lower *A. palmata* section gave a radiocarbon date of 9075 ± 70 ybp. This material was quite different from the upper 3 m of the core, which consisted of a well-cemented, deeper-water *Montastraea/Diploria* coral head facies (Fig. 7.4). A sample from near the base of this upper section yielded a radiocarbon date of 6945 ± 70 ybp. This *Montastraea/Diploria* facies was the only reef facies recovered from two shoreward core holes (Fig. 7.4) and represents a reef community similar to that found on the present surfaces of the St. Croix ridges. Radiocarbon dates obtained from samples recovered in these shoreward core holes ranged in age from 6500 to 5000 ybp.

When the position of the rising seas of the Holocene Transgression is compared with the time *Acropora palmata* framework construction terminated, the

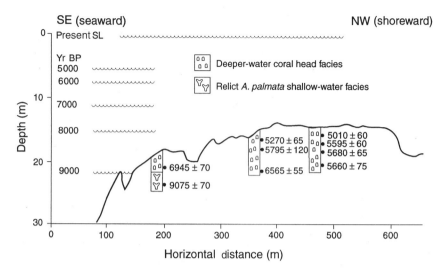

Figure 7.4. Transect of core holes drilled across the shelf edge off the southeast coast of St. Croix. Note the relict shallow-water *Acropora palmata* facies at the base of the outermost hole and the relationship of radiocarbon dates of cores to positions of sea level as indicated by the Lighty, Macintyre, and Stuckenrath (1982) sea-level curve. The demise of the *A. palmata* community coincides with the flooding of the insular shelf about 8000 to 9000 years ago. After Adey et al. (1977) and Macintyre (1988).

latter coincides with the initial flooding of the southern shelf off St. Croix (Fig. 7.4). A marked increase in noncarbonate debris, including illite, kaolinite/chlorite, quartz, amphibole, and feldspars, trapped in submarine cements in the coral-head facies (Adey et al. 1977) indicate turbid and high-nutrient conditions caused by the erosion of soil and lagoon sediments on the flooded shelf—in other words, conditions of stress that exceeded the tolerance level of most corals. The radiocarbon date obtained from the coral-head facies suggests that conditions did not improve until about 7000 years ago, but by then water at the shelf edge would have been too deep for an active *A. palmata* community to become reestablished (Adey et al. 1977; Macintyre 1988).

It seems, then, that a shallow-water *Acropora palmata* reef flourished on the shelf edge off the south coast of St. Croix 9000 years ago but eventually succumbed to stress caused by rising sea level and subsequent input of sediment-laden waters caused by the erosion of soil and lagoon sediments when waters began flooding the 5-km-wide shelf. The sediment-sensitive *A. palmata* was clearly unable to recover in time to allow this fast-accumulating shallow-water reef assemblage to become reestablished before water depth at the shelf edge became too deep to permit this coral to flourish and form a framework. Shelf-edge reefs throughout much of the eastern Caribbean appear to have followed this course of development, since relict Give Up reefs are a characteristic feature of the region (Macintyre 1972).

Relict shelf-edge reef off the southeastern coast of Florida. A relict reef ridge similar to the one off St. Croix extends along the shelf edge from Palm Beach to

Miami for a distance of at least 95 km (Duane and Meisburger 1969; Macintyre and Milliman 1970). It has a relief of about 10 m and occurs in water depths of 15 to 30 m (Macintyre and Milliman 1970; Lighty, Macintyre, and Stuckenrath 1978). This feature is adjacent to two inner ridges on the shelf, also running parallel to the shelf edge. All three ridges have been described as "reef lines" (Duane and Meisburger 1969). The outer ridge (at the shelf edge) shows little evidence of active reef-framework construction. Today it has a cover of deep reef fauna consisting mainly of massive corals, octocorals, sponges, antipatharians, and algae. The most common corals are *Montastraea cavernosa, Montastraea annularis, Meandrina meandrites, Agaricia agaricites, Dichocoenia stokesi, Stephanocoenia intersepta, Isophyllia sinuosa, Eusmilia fastigiata, Siderastrea siderea*, and *Madracis decactis* (Macintyre and Milliman 1970; Goldberg 1973).

Dredging associated with the placement of an offshore pipeline exposed a section of the shelf-edge reef near Hillsboro Inlet, about 40 km north of Miami (Lighty 1977; Lighty et al. 1978). A trench 3 m wide, 450 m long, and 11 m deep (Fig. 7.5) revealed what was once a typical Caribbean shallow-water *Acropora palmata* reef with thickets of *Acropora cervicornis* and massive corals in its back area, and massive corals and talus debris in the fore region. Radiocarbon dates of the *Acropora palmata* (Fig. 7.5) indicated that this reef flourished for a time starting at least 9500 years ago, and attained accumulation rates of 3.6 to 10.7 m/1000 years before dying out about 7000 years ago (Lighty, Macintyre, and Stuckenrath 1978).

This is another clear example of a Give Up reef. It was a fast-growing *Acropora palmata* community like the one off St. Croix, accumulating at the rate of at least 10 m/1000 years, which should have enabled it to keep up with the rising seas of the Holocene Transgression in the past 7000 years. Yet it did not. As in the example from St. Croix, the demise of this reef coincides with the flooding of large areas of the Florida continental shelf about 7000 years ago (Lighty, Macintyre, and Stuckenrath 1978), including areas to the south of this site. Increased turbidity and nutrients, caused by erosion of the shelf's sediment, may have exceeded the tolerance limit of this coral community as well (Lighty, Macintyre, and Stuckenrath 1978; Macintyre 1988).

An additional factor to consider in the Florida case is its present-day shallow-water temperatures, which can drop to 10 °C (Walford and Wicklund 1968). This is well below the minimum 16 to 18 °C required for active reef growth (James and Macintyre 1985) and is considerably cooler than the temperatures off St. Croix, which seldom fall below 25 °C (Adey et al. 1977). Lower water temperatures associated with shallow lagoon waters, as well as increases in turbidity and nutrients, could all have played a role in the death of the Florida shelf-edge reefs (Lighty, Macintyre, and Stuckenrath 1978; Macintyre 1988, Toscano and Lundberg 1998).

7.2.2 Shelf Reefs

Whereas outer-slope and shelf-edge reefs represent the early stages of Holocene reef history, those on continental or insular shelves represent the final stages of

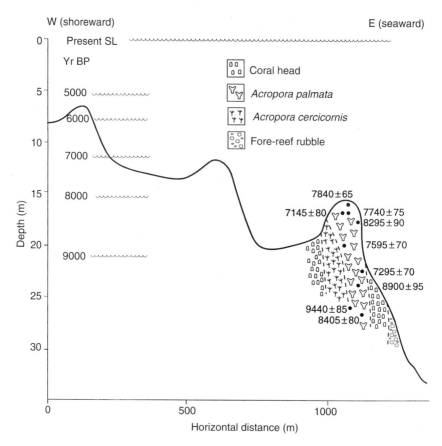

Figure 7.5. Transect of an exposed section across the shelf edge off the southeast coast of Florida near Hillsboro Inlet. Reef facies and accompanying radiocarbon dates, when compared with sea-level positions from the Lighty, Macintyre, and Stuckenrath sea-level curve, indicate that a typical Caribbean *Acropora palmata* reef flourished for at least 2000 years at the shelf edge and "gave up" about 7000 ybp when the inner shelf was flooded. After Lighty, Macintyre, and Stuckenrath (1978).

development. The shelf reefs of the western Atlantic are no more than about 7000 years old, and this short history has nothing to do with a sudden introduction of favorable conditions for reef growth caused by late postglacial warming as suggested by Newell (1959). Rather, it was governed by the depths and surface conditions of most of the shelves on which they became established. Most of these shelves were not flooded until about 7000 ybp, and even then vigorous reef growth did not begin in most cases until a suitable substrate was exposed following erosion of a soil or sediment cover.

Shelf reefs of the western Atlantic can be found in various stages of development. Some, like the Abaco reef in the Bahamas, are already in the Give Up stage, after being "shot in the back" by their own lagoon waters (Fig. 7.6), where major

Figure 7.6. Schematic illustrating how detrimental conditions can originate with a flooding of the back reef. These stress conditions, which can include increased turbidity, and increased or decreased temperature and salinity, can result in the demise of the bordering reef. Reproduced from Neumann and Macintyre (1985) with permission from the Ecole Pratique des Hautes Etudes, Université de Perpignan.

changes in temperature, salinity, and turbidity exceeded the tolerance limit of the bordering reefs (Neumann and Macintyre 1985, p. 107). Others, such as Alacran Reef, Mexico, are examples of Catch Up reefs: although at one point they lagged behind sea level, rapid framework accumulation enabled them to catch up. Still others, including Galeta Reef, Panama, are Keep Up reefs that have simply kept pace with the rise in sea level, now advancing more slowly than in earlier periods of the Holocene (Fig. 7.1). Yet another reef type exists in some areas (for example, the Bahamas) where water conditions have never been conducive to vigorous reef growth. These reefs are best described as veneers or scattered coral communities that have inherited their relief from the substrate.

A shallow Give Up reef: Abaco Bank, northern Bahamas. The bank-barrier reef on the eastern edge of Abaco Bank was originally thought to be a thriving community (Adey 1978). Though morphologically well developed, with a spur-and-groove system, this feature is now known to be almost entirely covered with brown fleshy algae, including *Turbinaria turbinata*, *Stypopodium zonale*, *Dictyota* sp., and *Sargassum polyceratium* (Fig. 7.7), along with a variety of octocorals and a few scattered corals (Lighty, Macintyre, and Neumann 1980). Thirteen core holes drilled into this reef complex, including one drilled into a spur, revealed that it was once a vigorous *Acropora palmata* reef-crest community, with massive corals (dominantly *Montastraea annularis*) in the back reef (Fig.7.8). Radiocarbon dates indicate that reef growth terminated about 3500 years ago, which was when

Figure 7.7. Alga-dominated community (*Turbinaria*, *Stypopodium*, *Dictyota*, and *Sargassum*) near the reef crest of the Abaco bank-barrier reef, northern Bahamas. There is no active reef accumulation on this Give Up reef, which has only scattered colonies of *Millepora complanata* (M) and *Diploria clivosa* (D).

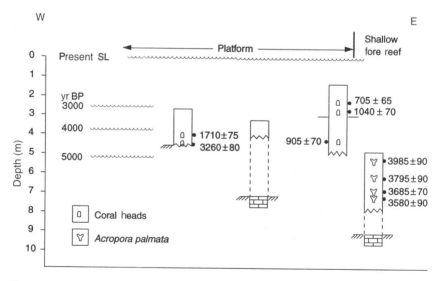

Figure 7.8. Transect of core holes drilled across the Abaco bank-barrier reef, northern Bahamas. The radiocarbon dates and types of coral dated are shown in each core. A comparison of these dated corals to the Lighty, Macintyre, and Stuckenrath (1982) sea-level curve positions indicates that the fore-reef *Acropora palmata* community gave up about 3500 ybp, which coincides with the flooding of the northern Bahamian platform. After Lighty, Macintyre, and Neumann (1980).

the extensive back-reef platform of the Little Bahama Bank was flooded by the rising seas of the Holocene Transgression (Lighty, Macintyre, and Neumann 1980).

The demise of the framework-constructing corals of this reef has been attributed to stress conditions caused by the off-bank flow of turbid and episodically cooled shallow platform waters (Lighty, Macintyre, and Neumann 1980). This suggestion is supported in part by the fact that patches of vigorous reef growth, including *Acropora palmata,* occur off the eastern coasts of some islands on Abaco Bank, where they are clearly protected from the detrimental shelf waters. The role of cold-water stress in the demise of these northern reefs has been particularly well illustrated by digital thermal infrared data from satellites and in situ water temperature measurements (Roberts et al. 1982). The results indicate that during a period of three successive cold fronts in January 1977, the shallow waters on Little Bahama Bank remained below 16 °C for a period of 7 to 8 days. As already mentioned, such temperatures are lethal for most tropical reef corals, particularly the more sensitive *Acropora palmata.* It can be concluded that this Bahama shallow-water reef gave up because it was unable to cope with the episodic runoff of ultracooled waters from the back-reef platform.

A Catch Up reef: Isla Perez, Alacran Reef Complex, Mexico. Isla Perez is the largest of five sand and gravel cays on a northwest-trending bank-barrier reef off the leeward southwest side of Mexico's Alacran Reef Complex. Alacran is on the Campeche Shelf, a submarine extension of the Yucatan Peninsula. One core hole, located on Isla Perez, was drilled into the bank-barrier and indicated a Holocene reef section of record thickness (33.6 m) on a Pleistocene substrate at a depth of 32.3 m below mean high sea level (Bonet 1967).

Reef development in this deeper-than-normal shelf setting apparently began with a coral-rubble and sand facies (Fig. 7.9). For a time, the accumulation rate approached no more than 2.5 m/1000 years (Macintyre, Burke, and Stuckenrath 1977), with the result that development soon lagged well behind the rising seas. About 6000 ybp, however, a well-developed coral-head facies became established in the area, which included an abundance of rapidly growing *Acropora cervicornis* that boosted the accumulation rate to 6 m/1000 years. The massive corals were gradually replaced, in large part by *Acropora cervicornis* forming an open framework at a vertical accumulation rate of 12 m/1000 years. At this rate of accumulation, the reef was eventually able to catch up with sea level (Macintyre, Burke, and Stuckenrath 1977). As the reef surface approached sea level, *Montastraea annularis, Porites astreoides,* and *Diploria* spp. increased, marking a transition to a reef-flat-rubble facies at depths of 7.1 to 2.3 m below present sea level. That facies, in turn, was replaced by a shallow rubble-and-sand facies (Macintyre, Burke, and Stuckenrath 1977). Trenches dug into the island indicated that it consisted of mostly *Acropora cervicornis* and *Porites porites* storm debris washed up above mean high sea level (Bonet 1967). In other words, this reef not only caught up with sea level, but surpassed it.

The successful catching up in this case was directly related to the ability of *Acropora cervicornis* to accumulate very rapidly in depths down to 20 m (Graus, Macintyre, and Herchenroder 1984, 1985; Graus and Macintyre 1989).

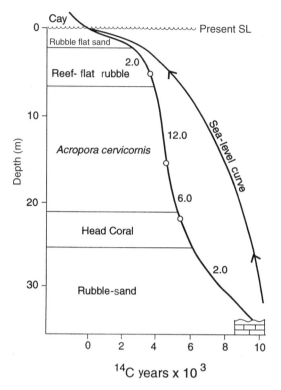

Figure 7.9. Schematic diagram illustrating the relationship between the Lighty, Macintyre, and Stuckenrath (1982) sea-level curve and the radiocarbon-dated reef facies recovered from a core hole drilled on Isla Perez, Alacran Reef Complex, Mexico. The reef accumulation curve, which connects the radiocarbon dates, has accompanying rates of accumulation of core intervals, and shows that the rapid accumulation of the *Acropora cervicornis* facies (up to 12 m/1000 years) has allowed this reef section to catch up with sea level. After Macintyre, Burke, and Stuckenrath (1977).

Where this community thrives in a well-protected setting and its corals are not damaged and transported by storms, impressive accumulations occur. Among other examples of *A. cervicornis* Catch Up reefs are the numerous atoll-like shoals in the shelf lagoon of the central section of the Belizean barrier reef (Purdy, Pusey, and Wantland 1975). Behind the barrier reef here, accumulations, mainly of *A. cervicornis,* have reached more than 13 m in thickness, and their age, as indicated by radiocarbon dates, extends to 3370 ± 80 ybp (Westphall and Ginsburg 1985; Westphall 1986). In this area, the final shoaling-up facies occurs at a depth of about 1 m and consists predominantly of a loose framework of *Porites divaricata* (Westphall 1986; Aronson and Precht 1997; Macintyre, Precht, and Aronson 2000) and marks a transition from the *A. cervicornis* catch-up stage to a *P. divaricata* keep-up stage (Aronson, Precht, and Macintyre 1998).

A Keep Up Reef: Galeta Point Reef, Panama. Galeta is a Caribbean fringing reef established off a promontory on the Panamanian coast 6 km east of the Panama Canal (Macintyre and Glynn 1976). It is characterized by a broad, shallow reef flat that consists of sand and rubble with a rich floral cover. The inner reef-flat areas are dominated by the red alga *Acanthophora spicifera* and the seagrass *Thalassia testudinum*; the more exposed outer edge supports dense growths of the fleshy red alga *Laurencia papillosa* and crustose coralline algae, with patches of the zoanthid *Zoanthus sociatus* (Meyer and Birkeland 1974).

Seaward of the reef flat, the hydrocoral *Millepora complanata* covers a crustose coralline pavement descending to about 1 m, where additional corals begin to appear, including *Siderastraea siderea, Porites astreoides, Porites furcata, Diploria* spp., *Favia fragum, Agaricia agaricites, Montastraea annularis,* and a few *Acropora palmata.* This scattered coral community becomes more dense beyond the dropoff (at a depth of 3–4 m) and consists mainly of *Porites furcata, Agaricia agaricites, Siderastrea siderea, Diploria* spp., and *Isophyllia sinuosa* (Macintyre and Glynn 1976). At a depth of about 6 m, coral abundance decreases markedly, and scattered colonies of *Siderastrea siderea* and platy *Agaricia* sp. cover a muddy sand and coral rubble bottom. These corals give way to fleshy algae at about 12 m, and the bottom levels off at a depth of 14 m. Although 36 coral species have been reported in this area (Porter 1972), there is little evidence of active reef-framework construction (Macintyre and Glynn 1976).

Clues to the history of this feature have been found in 13 core holes drilled along two transects that together extended across the entire reef, from the mangrove shoreline to the fore-reef slope (Fig. 7.10; see Macintyre and Glynn 1976). This reef has a maximum thickness of 14.3 m and was established on the uneven erosional surface of the mid-Miocene Gatún Formation. Six distinct reef facies could be identified in the cores. First, a fore-reef coral-head facies occurs at the outer edge of the reef, consisting dominantly of massive corals including *Dichocoenia stokesi, Siderastrea siderea, Stephanocoenia intersepta, Diploria strigosa, Porites astreoides,* and *Montastraea annularis.* These corals were commonly bored by sponges and molluscs and filled with a microcrystalline magnesium calcite submarine cement (Macintyre 1977). Second, an *Acropora palmata* facies forms most of the framework of this reef. It consists predominantly of well-preserved *Acropora palmata,* which for the most part exhibits corallite growth patterns indicating that the corals have not been overturned and were thus probably in growth position when cored (Fig. 7.11). Third, a reef-flat rubble facies

---

Figure 7.10. Stages of development of a Caribbean fringing reef off Galeta Point, Panama. This scenario is based on reef facies recovered from 13 core holes drilled along two transects across this reef (core holes shown in top profile). A comparison of 18 radiocarbon dates obtained from these cores and the sea-level positions indicated by the Lighty, Macintyre, and Stuckenrath (1982) sea-level curve, allows a detailed reconstruction of how this reef kept up with the late Holocene rising seas. After Macintyre and Glynn (1976) and Macintyre (1983).

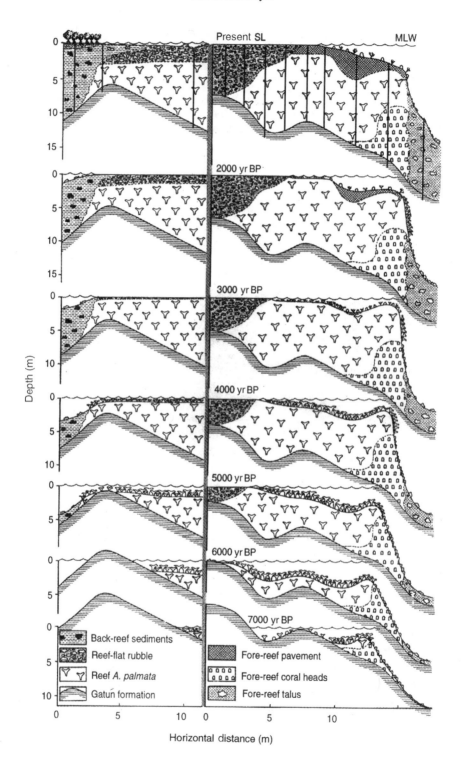

Present SL

MLW

2000 yr BP

3000 yr BP

4000 yr BP

5000 yr BP

6000 yr BP

7000 yr BP

Depth (m)

Back-reef sediments

Reef-flat rubble

Reef *A. palmata*

Gatún formation

Fore-reef pavement

Fore-reef coral heads

Fore-reef talus

Horizontal distance (m)

Figure 7.11. A core of *Acropora palmata* collected from the *A. palmata* facies from Galeta Reef, Panama. Note how the corallite growth patterns and surface encrustation of crustose coralline algae indicate that this colony was probably in growth position when cored. Reproduced from Macintyre and Glynn (1976) with permission from the American Association of Petroleum Geologists.

overlies the *A. palmata* facies and consists of a variety of water-worn, bored, and encrusted corals, including *Diploria clivosa*, *Agaricia agaricites*, *Millepora complanata*, *Siderastrea siderea*, *Acropora palmata*, *Porites astreoides*, *Porites furcata*, and *Diploria strigosa*. These corals occur in a *Halimeda*-rich sand matrix. Fourth in the sequence is a fore-reef pavement facies that is extensively lithified and contains mostly *Millepora complanata*, *Agaricia agaricites*, *Porites astreoides*, and crustose coralline algae. This is a well-indurated pavement limestone (Macintyre and Marshall 1988), in which most of the original skeletal components have been destroyed by multicyclic boring, filling, and cementation (Macintyre 1977). Fifth, a fore-reef talus facies can be found in an apron of sediment lying in front of the reef. This facies has no framework and consists mainly of gravel-size skeletal debris in a muddy sand matrix. The sixth and last is a back-reef sediment facies, another unconsolidated facies formed by coral rubble, including *Acropora cervicornis*, in a muddy sand matrix. The innermost limit of this facies grades up into a calcareous mangrove peat.

The various stages of development of these facies were postulated by relating 18 radiocarbon dates of core samples to a minimum sea-level curve (Lighty, Macintyre, and Stuckenrath 1982) for the tropical western Atlantic (Macintyre and Glynn 1976; Macintyre 1983). These stages of growth, shown in Fig. 7.10, can be summarized as follows:

7000 years ago. A small, typical Caribbean fringing reef, dominated by *Acropora palmata* with a fore reef of mixed coral heads, was established on the erosional surface of Miocene siltstone.

6000 years ago. The reef prograded shoreward, keeping up with the advancing seas of the Holocene Transgression. The reef topography still reflected some of the relief of the underlying erosional surface. Debris in the fore reef limited seaward reef growth.

5000 years ago. The Miocene siltstone substrate was no longer exposed and its relief was masked by the rapid accumulation of the *A. palmata* facies. Reef-flat rubble started to accumulate over an area of erosional high relief. Significant accumulations of back-reef lagoonal sediments made their first appearance.

4000 years ago. A thriving *A. palmata* fringing reef reached its peak of development. Reef-flat rubble continued to prograde over the *A. palmata* facies, which in turn continued to prograde over the fore-reef coral head facies.

3000 years ago. The reef had almost caught up with sea level and was now characterized by a wide rubble-covered reef flat. The *A. palmata* community was considerably reduced with the establishment of a pavement community in the shallow fore reef.

2000 years ago. Active framework-building had ceased. The shallow areas consisted of back-reef sediments, reef-flat rubble, and a heavily lithified shallow pavement. A little *A. palmata* survived on the upper section of a slope that was extensively covered by talus debris.

Present. A mature emergent reef is present, with mangroves and back-reef sediments prograding over the inner edge of an extensive rubble reef flat. The fore reef is dominated by a dense, well-cemented pavement and a thick talus cover that is protected from the erosive action of waves by a developing seaward bank-barrier reef.

The general absence of reef-framework accumulation, encroachment of mangroves over the lagoonal back-reef sediments, and extensive mud-rich sediments covering most of the fore-reef slope all indicate that this Panamanian Keep Up reef has fully caught up with sea level and is in a state of postclimactic degradation. The successful accumulation of the *A. palmata* community at this site is probably related to the fact that the area is well removed from the paths of severe hurricane activity (Neumann et al. 1987). Only one hurricane has reportedly "brushed Panama's northwestern coastline" in the last 120 years (Clifton et al. 1997). As mentioned earlier, the *A. palmata* cores in general appeared to be in growth position. Unlike so many other Caribbean fringing reefs that have little or no shallow reef flat because their framework has been destroyed by intermittent hurricanes, Galeta Reef has flourished through in situ framework construction and thereby has kept up with the advancing seas.

7.3 Discussion

The past three decades have brought to light a great deal of information on the Holocene history of western Atlantic coral reef development, as preserved within their framework. A particularly important finding was that coral-reef communities are not the fragile entities they were once thought to be. On the contrary, the internal record of coral reefs shows an impressive ability to survive under conditions of rapidly rising sea level. Furthermore, in the aftermath of rapid sea-level-rise events, many reefs went on to create structures having substantial topographic relief.

At the same time, the geological record reflects a story of survival in varying degrees. Some Give Up reefs, stranded on the continental and insular slopes by spectacular surges in the rise of sea level related to periods of rapid melting of late Pleistocene ice sheets, gave up the struggle, as did a number at the outer edges of continental and insular shelves. These shelf-edge reefs appear to have succumbed to stress conditions introduced by the runoff of sediment-rich, nutrient-rich, and intermittently cooled shallow inshore waters formed by the initial flooding of the shelf platforms. But there are also numerous examples of Catch Up and Keep Up reefs. These communities were able to flourish and build spectacular coral reefs on shelves in protected waters when the advance of the rising seas began to slow down. This impressive final stage of reef development should put to rest the notion that reefs are unable to survive profound changes in environmental conditions because of their fragility.

Western Atlantic coral reefs could face yet another serious threat to their survival if expected rates of sea-level rise in association with global warming are valid (Hoffman, Keyes, and Titus 1983; IPCC 1995). Experiments with a computer model designed to simulate the growth of Caribbean coral reefs indicate that by the year 2100, these reefs may begin to be submerged, even with a minimum projected sea-level rise of about 50 cm (Graus and Macintyre 1998). With deeper reef crests, more wave energy would propagate leeward and would resuspend back-reef sediments and erode shorelines, which would result once again in many Give Up reefs that would probably succumb to the outflow of detrimental waters produced in their lagoons. These would not be unlike the conditions that terminated framework construction in western Atlantic shelf-edge reefs, which were eventually succeeded by flourishing inshore shelf reefs. Similarly, computer simulations extended to the year 2500 suggest that conditions will improve and that new reef crests could become established shoreward of the previous submerged crests (Graus and Macintyre 1998).

In reviewing the geological record of western Atlantic coral reefs of the past 18,000 years, one can only marvel at the capacity of these reefs to withstand cataclysmic changes in environmental conditions. Although the early Holocene reefs on the outer slopes and shelf edges were terminated by rapid rises in sea level and inundation by deadly sediment-laden, nutrient-rich, or supercooled waters, with time coral survivors thrived on the shallow shelves. In the face of ever-present environmental stress of one kind or another, some or all of these reef-building communities may fall into decline. Yet their resilience, now amply demonstrated in the geological record, gives scientists every reason to believe that reef communities have the ability to regain their healthy status at or near their original site once suitable conditions return.

7.4 Summary

The internal framework of western Atlantic coral reefs provides excellent examples of patterns of reef development over the last 18,000 years when the postglacial seas rose over 100 m, occasionally in rapid bursts. The reefs did not always survive but the zooxanthellate corals did, and they reinitiated reef building. The overall history of western Atlantic coral reefs during these advancing seas is one of Give Up reefs on the outer slopes, stranded by rapid meltwater pulses; Give up reefs on the shelf edges, killed off by stress conditions caused by the flooding of the shelves; and finally Catch Up and Keep Up reefs on the shelves, but also Give Up reefs that have succumbed to exposure to the lethal waters that have originated in their shallow lagoons.

References

Adey, W.H. 1978. Coral reef morphogenesis: A multi-dimensional model. *Science* 202:831–837.

Adey, W.H., I.G. Macintyre, R. Stuckenrath, and R.F. Dill. 1977. Relict barrier reef system off St. Croix: Its implications with respect to late Cenozoic coral reef development in the western Atlantic. *Proc. Third Int. Coral Reef Symp., Miami* 2:15–21.

Aronson, R.B., and W.F. Precht. 1997. Stasis, biological disturbance, and community structure of a Holocene coral reef. *Paleobiology* 23:326–346.

Aronson, R.B., W.F. Precht, and I.G. Macintyre. 1998. Extrinsic control of species replacement on a Holocene reef in Belize: The role of coral disease. *Coral Reefs* 17:223–230.

Bard, E., B. Hamelin, R.G. Fairbanks, and A. Zindler. 1990. Calibration of the 14 °C timescale over the past 30,000 years using mass spectrometric U-Th ages from Barbados corals. *Nature* 345:405–410.

Bonet, F. 1967. Biogeologia subsuperficial del arrecife Alacranes, Yucatan. *Mex. Univ. Nac. Autonoma Inst. Geol. Bol.* 80.

Curray, J.R. 1965. Late Quaternary history, continental shelves of the United States. In *The Quaternary of the United States*, eds. H.E. Wright and D.G. Frey, 723–735. Princeton, NJ: Princeton University Press.

Clifton, K.E., K. Kim, and J.L. Wulff. 1997. A field guide to the reefs of Caribbean Panama with an emphasis on western San Blas. *Proc. Eighth Int. Coral Reef Symp., Panama* 1:167–184.

Duane, B.B., and E.P. Meisburger. 1969. Geomorphology and sediments of the nearshore continental shelf, Miami to Palm Beach, Florida. Washington, DC: U. S. Army Corps of Engineers Coastal Engineering Research Center Tech. Memo. 29.

Fairbanks, R.G. 1989. A 17,000-year glacio-eustatic sea-level record: Influence of glacial melting rates on the Younger Dryas event and deep-ocean circulation. *Nature* 342:637–642.

Goldberg, W.M. 1973. The ecology of the coral-octocoral communities off the southeast Florida coast: Geomorphology, species composition and zonation. *Bull. Mar. Sci.* 23:465–488.

Goreau, T.F. 1961a. Problems of growth and calcium deposition in corals. *Endeavour* 20:32–39.

Goreau, T.F. 1961b. Geological aspects of the structure of Jamaican coral reef communities. U. S. Navy Office of Naval Research, Biol. Branch, Final Progress Report.

Graus, R.R., and I.G. Macintyre. 1989. The zonation pattern of Caribbean coral reefs as controlled by wave and light energy input, bathymetric setting and reef morphology. *Coral Reefs* 8:9–18.

Graus, R.R., and I.G. Macintyre. 1998. Global warming and the future of Caribbean coral reefs. *Carbonates and Evaporites* 13:43–47.

Graus, R.R., I.G. Macintyre, and B.E. Herchenroder. 1984. Computer simulation of the reef zonation at Discovery Bay, Jamaica: Hurricane disruption and long-term physical oceanographic controls. *Coral Reefs* 3:59–68.

Graus, R.R., I.G. Macintyre, and B.E. Herchenroder. 1985. Computer simulation of the Holocene facies history of a Caribbean fringing reef (Galeta Point, Panama). *Proc. Fifth Int. Coral Reef Congr., Tahiti* 3:317–322.

Harris, P.T., and P.J. Davies. 1989. Submerged reefs and terraces on the shelf edge of the Great Barrier Reef, Australia. Morphology, occurrence and implications for reef evolution. *Coral Reefs* 8:87–98.

Hoffman, J.S., D. Keyes, and J.G. Titus. 1983. *Projecting Future Sea Level Rise.* Washington, DC: U.S. Government Printing Office.

IPCC. 1995. *Climate Change 1995: The Science of Climate Change.* New York: Cambridge University Press.

James, N.P., and I.G. Macintyre. 1985. Carbonate depositional environments. Modern and ancient. Part 1: Reefs. zonation, depositional facies, diagenesis. *Colo. Sch. Mines Q.* 80:1–70.

Lighty, R.G. 1977. Relict shelf-edge Holocene coral reef: Southeast coast of Florida. *Proc. Third Int. Coral Reef Symp., Miami* 2:215–221.

Lighty, R.G., I.G. Macintyre, and A.C. Neumann. 1980. Demise of a Holocene barrier-reef complex, northern Bahamas. *Geol. Soc. Am. Abstr. Prog.* 12:471.

Lighty, R.G., I.G. Macintyre, and R. Stuckenrath. 1978. Submerged early Holocene barrier reef south-east Florida shelf. *Nature* 276:59–60.

Lighty, R.G., I.G. Macintyre, and R. Stuckenrath. 1982. *Acropora palmata* reef framework: A reliable indicator of sea level in the western Atlantic for the past 10,000 years. *Coral Reefs* 1:125–130.

Macintyre, I.G. 1967. Submerged coral reefs, west coast of Barbados, West Indies. *Can. J. Earth Sci.* 4:461–474.

Macintyre, I.G. 1972. Submerged reefs of eastern Caribbean. *AAPG Bull.* 56:720–738.

Macintyre, I.G. 1977. Distribution of submarine cements in a modern Caribbean fringing reef, Galeta Point, Panama. *J. Sediment. Petrol.* 47:503–516.

Macintyre, I.G. 1983. Growth, depositional facies, and diagenesis of a modern bioherm, Galeta Point, Panama. In *Carbonate Buildups—A Core Workshop*, ed. P.M. Harris, 578–593. SEPM Core Workshop No. 4.

Macintyre, I.G. 1988. Modern coral reefs of western Atlantic: New geological perspective. *AAPG Bull.* 72:1360–1369.

Macintyre, I.G., R.B. Burke, and R. Stuckenrath. 1977. Thickest recorded Holocene reef section, Isla Perez core hole, Alacran Reef, Mexico. *Geology* 5:749–754.

Macintyre, I.G., and P.W. Glynn. 1976. Evolution of modern Caribbean fringing reef, Galeta Point, Panama. *AAPG Bull.* 60:1054–1072.

Macintyre, I.G., and J.F. Marshall. 1988. Submarine lithification in coral reefs: Some facts and misconceptions. *Proc. Sixth Int. Coral Reef Symp., Australia* 1:263–272.

Macintyre, I.G., and J.D. Milliman. 1970. Physiographic features on the outer shelf and slope, Atlantic continental margin, southeastern United States. *Geol. Soc. Am. Bull.* 81:2577–2598.

Macintyre, I.G., W.F. Precht, and R.B. Aronson. 2000. Origin of the Pelican Cays ponds, Belize. *Atoll Res. Bull.* 466:1–11.

Macintyre, I.G., K. Rutzler, J.N. Norris, K.P. Smith, S.D. Cairns, K.E. Bucher, and R.S. Steneck. 1991. An early Holocene reef in the western Atlantic: Submersible investigations of a deep relict reef off the west coast of Barbados, W.I. *Coral Reefs* 10:167–174.

Meyer, D.L., and C. Birkeland. 1974. Environmental sciences program marine studies— Galeta Point. In *1973 Environmental Monitoring and Baseline Data*, 129–253. Washington, DC: Smithsonian Institution Press.

Neumann, A.C., and I.G. Macintyre. 1985. Reef response to sea level rise: Keep-up, Catch-up, or Give-up. *Proc. Fifth Int. Coral Reef Congr., Tahiti* 3:105–110.

Neumann, C.J., B.R. Jarvinen, A.C. Pike, and J.D. Elms. 1987. *Tropical Cyclones of the North Atlantic Ocean, 1871–1986, Third Revision.* Ashville, NC: NOAA National Climatic Data Center.

Newell, N.D. 1959. The coral reefs. *Nat. Hist.* 68:119–131.

Nota, D.J.D. 1958. Sediments of the western Guiana shelf. In *Reports of the Orinoco shelf expedition. 2. Meded. Landbouwhogesch. Wageningen,* 58:1–98.

Porter, J.W. 1972. Ecology and species diversity of coral reefs on opposite sides of the Isthmus of Panama. *Biol. Soc. Wash. Bull.* 2:89–116.

Purdy, E.G., W.C. Pusey, and K.F. Wantland. 1975. Continental shelf of Belize-regional shelf attributes. In *Belize Shelf—Carbonate Sediments, Clastic Sediments, and Ecology,* eds. K.F. Wantland and W.C. Pusey, 1-52. Tulsa: American Association of Petroleum Geologists.

Roberts, H.H., L.J. Rouse, N.D. Walker, and J.H. Hudson. 1982. Cold-water stress in Florida Bay and northern Bahamas: A product of winter cold-air outbreaks. *J. Sediment. Petrol.* 52:145–155.

Stoddart, D.R. 1969. Ecology and morphology of recent coral reefs. *Biol. Rev.* 44:433–498.

Toscano, M.A., and J. Lundberg. 1998. Early Holocene sea-level record from submerged fossil reefs on the southeast Florida margin. *Geology* 26:255–258.

Walford, L.A., and R.I. Wicklund. 1968. Serial atlas of the marine environment: Folio 15, monthly sea temperature structure from the Florida Keys to Cape Cod: American Geographical Society, Plates 13, 14, and 15.

Westphall, M.J. 1986. Anatomy and history of a ringed reef complex, Belize, Central America. MS Thesis. Miami: University of Miami.

Westphall, M.J., and R.N. Ginsburg. 1984. Substrate control and taphonomy of Recent (Holocene) lagoon reefs from Belize, Central America. In *Advances in Reef Science,* eds. P.W. Glynn, A.M. Szmant-Froelich, and P.K. Swart, 135–136. Miami: Rosenstiel School of Marine and Atmospheric Science, University of Miami.

8. Broad-Scale Patterns in Pleistocene Coral Reef Communities from the Caribbean: Implications for Ecology and Management

John M. Pandolfi and Jeremy B.C. Jackson

8.1 Introduction

Ecologists are asking very important questions about the maintenance of the high species diversity so typical of tropical reef ecosystems. Some ecologists believe that species diversity in reef communities is influenced by differential recruitment to specific sites. Community structure, then, is dependent upon larval dispersal, or limits to dispersal, and the taxonomic composition of the local community should reflect the abundance and migration of larvae in the regional species pool. In this view, what biotic interaction exists imposes little influence on community structure. Local community structure might be similar over large spatial and temporal scales but only because the regional species pool is constant (Doherty and Williams 1988; Hubbell 1997). Conversely, local community structure might be highly variable over large spatial and temporal scales because the composition of the regional species pool is varying, or stochastic events interrupt or enhance the supply of larvae over space and time. An important corollary is that regional species diversity strongly influences local species diversity (Cornell and Lawton 1992; Cornell and Karlson 1996, 1997; Caley and Schluter 1997; Karlson and Cornell 1998; Karlson 1999; Karlson, Cornell, and Hughes 2004).

Other ecologists ascribe more importance to biotic interactions operating locally, such as competition, predation and herbivory, disease, symbiosis, and other density-dependent processes (Odum and Odum 1955). Here local community structure is more a function of species-specific adaptations to environment, niche breadth, and the interaction of biotic components than the relative abundance of larvae in the regional species pool. Hence, local communities in similar environments are similar not because of similar larval supply, but because of similar ecological processes which filter out other associations of taxa. That is, there is "limited membership" (Pandolfi 1996). Metapopulation models incorporate both competition and migration into community dynamics (Jackson, Budd, and Pandolfi 1996).

Still another group of ecologists ascribes community structure on coral reefs to disturbance (Connell 1978; Huston 1985; Tanner, Hughes, and Connell 1994; Connell 1997; Connell, Hughes, and Wallace 1997; Hughes and Connell 1999; Connell et al. 2004). Disturbance occurs at a variety of spatial and temporal scales and species abundances are in a perpetual state of recovery from these disturbances (Karlson and Hurd 1993). Important ecological processes occur between disturbance events, but the length of the intervals between disturbance events, and their intensity, can have dramatic effects on the composition and structure of local communities (Done 1988; Connell 1997; Hughes and Connell 1999).

Due to the present rampant degradation of coral reefs worldwide (Ginsburg 1994; Hughes 1994; Jackson 1997; Pandolfi et al. 2003), it is critical that we resolve the relative role of various processes in the maintenance of diversity on reefs. Unfortunately, many of the proposed mechanisms are either unstudied or essentially untestable on modern long-lived corals at the appropriate spatial and temporal scales. For example, until recently (Hughes et al. 1999; Murdoch and Aronson 1999; Bellwood and Hughes 2001), the distribution of species abundance patterns over broad spatial scales was unstudied in living reefs. A real possibility exists that there are no longer any living "pristine" reefs to study (Jackson 1997; Pandolfi et al. 2003, 2005). However, the global extent of Quaternary reef deposits worldwide provides a reliable database with which to test ecological predictions (Pandolfi 1996, 2001; Pandolfi and Jackson 1997, 2001) and begin to understand the long-term effects of habitat degradation in coral reefs (Jackson et al. 2001) over large spatial and temporal scales.

In this paper we present data on the community structure of Pleistocene reefs along a 2500-km transect across the southern Caribbean Sea (the "SOCA" transect). We use the results to evaluate contemporary studies of living reefs in the light of various ecological theories explaining species distribution patterns on reefs. We also discuss the implications of these results to management issues in living reefs. Our study indicates: (1) species abundance in reef coral communities is highly predictable at broad scales; (2) the lack of broad-scale analyses in living reefs precludes testing much of ecological theory that is relevant to the maintenance of diversity in coral reefs; and (3) Pleistocene baseline data, incapable of shifting, indicate that Caribbean coral reefs were very different 125 ka than they are today.

8.2 Patterns over Large Temporal Scales: The Fossil Record

Ecologists have recently turned to the Quaternary fossil record of coral reefs to examine ecological patterns over long time scales. Potential effects of humans on reef ecology were absent or insignificant on most reefs until the last few hundreds or thousands of years, so that it is possible to analyze "natural" distribution patterns of truly "pristine" coral reefs before human disturbance began. Fossil reefs generally accumulate in place, and their former biological inhabitants and physical environments can be determined. Community patterns can be documented over long time periods and large distances, often encompassing environmental conditions that are beyond values recorded by humans but within the range of projected global changes (Pandolfi 1999). Reef corals, in particular, record their ecological history especially well because they form large resistant skeletons that can be identified to the species level. Temporal resolution is on the order of hundreds to a couple of thousand years, so studies of reef-coral dynamics extend to a scale appropriate for their colony longevity, generation time, and other aspects of their biology and life history.

Of course, there are also limits to the degree to which the fossil record can contribute to community ecology. For example, manipulative experimentation is impossible, so mechanisms are often not forthcoming in studies of fossil assemblages. We attempt to minimize this limitation by: (1) using a large number of observations on patterns over a variety of spatial and temporal scales ("macroecology" of Brown 1995) and (2) testing competing predictions of hypotheses that assume the operation of different processes. The way in which the original communities preserve to become fossil reef deposits is also a major concern to those who study the fossil record. In a series of recent papers it has become apparent that there is a high degree of fidelity between the original coral life assemblages and both their adjacent death assemblages and their fossil counterparts in both the Caribbean and Indo-Pacific (Pandolfi and Minchin 1995; Greenstein and Curran 1997; Pandolfi and Greenstein 1997; Greenstein and Pandolfi 1997; Greenstein, Curran, and Pandolfi 1998; Greenstein, Pandolfi, and Curran 1998; Edinger, Pandolfi, and Kelley 2001; Greenstein, Chapter 2). Thus it appears as though a large amount of ecological information is preserved in Pleistocene reef deposits. For example, the ecological patterns we observed on Curaçao using fossil relative abundance data were very similar to those noted in living reefs from Curaçao and elsewhere in the Caribbean prior to the 1980s (Pandolfi and Jackson 2001).

The potential for understanding the community ecology of living coral reefs by looking at the recent past history of corals and other reef inhabitants is enormous at specific localities worldwide in the tropical sea. One of the main reasons for this lies in the sea level 125,000 years ago. Sea level was 2 to 6 m higher 125 ka than it is today (Chappell et al. 1996), resulting in a terrace 2 to 6 m above current sea level throughout the tropics preserving fossil coral reefs. Thus, Pleistocene reef deposits are distributed throughout the Caribbean, Indo-Pacific, and Red Sea. Along the coastal passive margin of East Africa, slight uplift of this terrace has resulted in an almost continuous Pleistocene sea cliff, up to 20 m in height, for hundreds of kilometers in Kenya, Tanzania, and Mozambique (Pandolfi personal observation 1998). The extent of Quaternary deposits is not limited to the fortuitous level of the sea at 125 ka. In many places throughout the world tectonic uplift caused by local volcanic or earthquake activity has interacted with global sea level changes to form series of raised reef terraces, each built during successive sea level rise. In the Caribbean, these kinds of terraces occur in Cuba (Iturralde-Vinent 1995), Haiti (Dodge et al. 1983), Barbados (Mesolella 1967), the Dominican Republic (Geister 1982), and Curaçao (de Buisonjé 1964, 1974; Pandolfi, Llewellyn, and Jackson 1999). They also occur in the Red Sea (Dollo 1990), Indonesia (Chappell and Veeh 1978), Papua New Guinea (Chappell 1974; Pandolfi and Chappell 1994), Henderson Island (Pandolfi 1995; Blake and Pandolfi 1997), Japan (Nakamori, Iryu, and Yamada 1995), and many other Indo-Pacific localities. Such places provide enormous insight into the ecological history of coral reefs over broad spatial and temporal scales.

Study of Quaternary reef coral assemblages has yielded a very stable view of reef communities over broad temporal scales. Jackson's (1992) semiquantitative analysis of Mesolella's (1968) Pleistocene data from Barbados provided the first tantalizing evidence for large-scale patterns in the recent past history of living coral reefs. Mesolella (1968) recognized that the recurrent patterns in species dominance and diversity that he found in the raised terraces of Barbados were very similar to those being described in the living reefs of Jamaica (Goreau 1959; Goreau and Wells 1967). Jackson (1992) used these data to suggest that reef coral communities reassembled after global sea level changes in similar ways throughout a 500-kyr interval. Pandolfi (1996) formally tested this notion in reef coral communities that had repeatedly reassembled on the Huon Peninsula, Papua New Guinea, finding that both species presence/absence and diversity were remarkably persistent through a 95-kyr interval (125–30 ka). Coral species composition and diversity were no different among nine different reef-building episodes, even though the communities varied spatially. Hubbell (1997) criticized the work as only focusing on the dominant species, suggesting that focus on the most dominant species will give the impression of persistence when the rare taxa are ignored. But there were no differences in any of the results whether or not rare species were included (Pandolfi 1996). Similar order in reef coral assemblages through space is also characteristic of Pleistocene assemblages from Curaçao (Pandolfi and Jackson 2001), and through time in Pleistocene assemblages from Barbados where separate analysis of both common and rare species gave similar results (Pandolfi 2000; Pandolfi and Jackson, 2006). This does not necessarily mean that coral assemblages do not vary latitudinally. For example, Pleistocene occurrences of *Acropora* spp. and additional warm-water corals occur well south of their modern range (Playford 1983; Kendrik, Wyrwoll, and Szabo 1991) and the extent to which these latitudinal range shifts along the coast of Western Australia impacted reef coral community membership is under study (Greenstein and Pandolfi 2004).

In contrast to reef coral assemblages, molluscan communities inhabiting reef environments on islands of the central tropical Pacific appear to have been continually rearranged as a result of the dynamic distributions of their component taxa in response to cyclic changes in sea level (Paulay and Spencer 1988; Paulay 1990, 1991; Kohn and Arua 1999). Taylor (1978) demonstrated that bivalve communities sampled from Late Pleistocene (Last Interglacial) limestones exposed on Aldabra Atoll were very different from both the modern fauna and older Pleistocene communities. He implicated changes in available habitat in response to sea-level cycles as a source for the faunal response observed. Changes in habitat availability were also suggested by Crame (1986) to explain the results of a comparative study of bivalve communities preserved in Late Pleistocene limestones along the Kenya coast and modern communities occurring offshore. Here, geographic range differences resulted in variable community composition among modern and Pleistocene molluscan faunas.

Recent studies of Holocene reef history indicate that unprecedented changes in modern assemblages have recently occurred throughout the Caribbean (Lewis 1984; Aronson and Precht 1997; Greenstein, Curran, and Pandolfi 1998; Pandolfi 2000; Aronson et al. 2002; Pandolfi et al. 2003). These studies point to the alarming realization that although multiple stable states may be part of the ecology of coral reefs (Hatcher 1984; Done 1992; Knowlton 1992; Hughes 1994), the actual stable states themselves may be changing. The duration, replacement rates, and effects of "stable" configurations in living reefs on their ultimate regional and global persistence are areas of reef ecology that demand immediate attention, and ones which cannot be addressed by simply studying living reefs over years or decades. In this chapter, we expand the spatial framework developed on Curaçao (Pandolfi and Jackson 2001) to include 2500 km of the Caribbean Sea in an attempt to: (1) demonstrate scale dependence in the variability of reef coral community composition and (2) place the catastrophic changes in Caribbean reefs (Hughes 1994; Gardner et al. 2003) in a historical perspective at multiple spatial scales. Specifically, we ask if there really is a predictable baseline for coral communities in the 125-ka reef that can serve as a comparison for pre- and post-1980s' Caribbean coral reefs.

8.3 Study Sites

We surveyed late Pleistocene (125-ka) reef-coral assemblages across a southern Caribbean (SOCA) transect extending from San Andrés (Colombia) in the west, to Curaçao (Netherlands Antilles) in the center, to Barbados (West Indies) in the eastern Caribbean Sea (Fig. 8.1). We compared species distribution patterns from the same shallow fore-reef environment, during a single reef-building episode, on the leeward side of the three Caribbean islands, focusing on predictability in species abundance patterns. We were interested in the degree to which species' relative abundance patterns varied over a broad geographic range at the same latitude from leeward shallow reefs. All three fossil reefs studied represent oceanic reefs well away from highly influential continental masses.

We also chose our three study sites to test the applicability of Geister's (1977, 1980) qualitative model's linking exposure to wave energy and coral distribution patterns within the Caribbean. Geister (1977) recognized six basic reef types that indicate, through the relative composition of the reef coral fauna, the degree of wave exposure within clearly delimited reef areas.

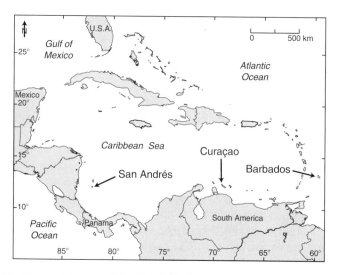

Figure 8.1. Geographic map of the Caribbean Sea showing the three sites for the southern Caribbean transect (SOCA). Pleistocene coral reefs were sampled at San Andrés, Curaçao, and Barbados.

8.3.1 San Andrés

The high island of San Andrés forms part of the Archipelago of San Andrés and Providencia and is comprised of a series of oceanic islands, atolls, and coral shoals lined up in a NNE direction (Fig. 8.1) (Geister 1973; Geister and Diaz 1997). The island occupies only 25 km². Wind direction is influenced by the regional NE trades, with winds primarily from the ENE with mean monthly velocities between 4 m s⁻¹ (May, September, October) and 7 m s⁻¹ (December, January, July). The climate has an average rainfall of about 1900 mm yr⁻¹ and average temperature of about 27 °C (with a 10 °C seasonal range). Sea surface temperatures average 28.5 °C with mean monthly values ranging between 26.8 °C during February and March and 30.2 °C during August and September (Geister and Diaz 1997). The reef complex surrounding San Andrés is about 18 km long and 10 km wide.

The Pleistocene Coral Reef Ecosystem

The San Luis Formation comprises two units, with a pre-125 ka limestone that is generally overlain by the 125-ka limestone (the Younger Low Terrace). Both units underlie the modern reef complex. San Andrés is fringed by the transgressive 125-ka reef complex, sandwiched between the living reef and the internal pre-125-ka and Tertiary rocks (Fig. 8.2A). Along the leeward west coast, the Pleistocene reef complex is represented as an emerged fringing reef. Exposures also occur along the north, east, and south coasts. Most of the fringing reef along the leeward coast comprised *Acropora cervicornis* (interpreted as a fossil reef flat; Geister 1975), but along the steep seaward margin of the reef terrace is an

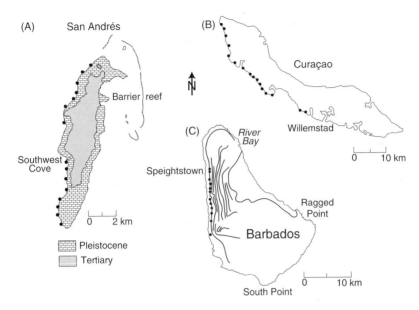

Figure 8.2. (**A**) Geographic map of San Andrés showing sampling localities of the 125-ka terrace of the Pleistocene San Luis Formation along the leeward coast. (**B**) Geographic map of Curaçao, Netherlands Antilles, showing sampling localities of the Pleistocene (125-ka) Hato Member of the Lower Terrace along the leeward coast. (**C**) Geographic map of Barbados showing Pleistocene (125-ka) sampling localities along the leeward coast. Abundant reef-terrace development occurs along the western and southern sides of the island (shown as parallel topographic contours).

association dominated by members of the *Montastraea "annularis"* species complex (interpreted as a shallow fore reef; Geister 1975). In contrast, there is little *A. cervicornis* living around the island today. Moreover, *Millepora* is common in the living windward reefs today, but is practically absent from the Pleistocene deposits. Geister (1975) attributed the difference in coral distribution patterns to the considerable difference in water depth above the two reef complexes, which resulted in different wave-exposure patterns. We sampled the shallow fore-reef Pleistocene deposits at 15 localities on the leeward side of the island (Fig. 8.2A).

8.3.2 Curaçao

Curaçao is a low-lying, arid, oceanic island in the Leeward Islands of the Dutch West Indies, approximately 60 km north of Venezuela (Fig. 8.1). The island is the largest of the group and is 61 km long and up to 14 km wide, with its long axis oriented SE to NW. Most of the NNE-facing coast is exposed to easterly trade winds with an average velocity of 8.1 m s^{-1} (Stienstra 1991), whereas the entire SSW coast experiences leeward conditions. The semiarid climate has an average rainfall of about 565 mm year^{-1} and average temperature of about 27 °C (De Palm 1985).

The mean tidal range is 0.3 m and the maximum range is 0.55 m (De Haan and Zaneveld 1959).

Turbidity increases markedly after heavy rainfalls (Van Duyl 1985). Direct impacts of tropical hurricanes are unusual (only two major hurricanes hit the coast-line between 1886 and 1980; Neumann et al. 1981), but hurricanes pass within 185 km (100 nautical miles) of the island on an average of once every four years (report Meteorological Service of the Netherlands Antilles 1981 cited in Van Duyl 1985). These storms are accompanied by westerly winds and generally bring much higher wave-energy conditions to the leeward coast, which may result in breakage of corals in otherwise normally sheltered reef habitats (de Buisonjé 1974; Bak 1975, 1977).

At Curaçao today, windward reef-crest environments show the highest wave energy, with typical wave heights from 2.0 to 3.5 m whereas leeward reef-crest environments have typical wave heights from 0.3 to 1.5 m (Van Duyl 1985). Both the leeward and windward sides of Curaçao are affected by the presence of Bonaire, located less than 50 km to the east/southeast. On the eastern, windward side of Curaçao, the position of Bonaire causes a significant reduction in wind and wave energy (reduced fetch) in the central and southern portion of the island, rela-tive to the northern portion. In addition, refracted water movement caused by large swells increases around the northern part of the island, which results in greater wave energy in the northern leeward coast than on the southern leeward coast (Van Duyl 1985). However, wind and wave exposure are more homogeneous along the lee-ward than windward coastlines. Living shallow-water reefs are largely confined to the leeward side and are fringing in nature (Bak 1977; Focke 1978; Van Duyl 1985).

The Pleistocene Coral Reef Ecosystem

During the Pleistocene, significant reef development occurred all around the island of Curaçao. Global sea level changes, coupled with regional tectonic uplift of the island, resulted in the formation of five geomorphological terraces com-posed of raised fossil reefs (Alexander 1961; de Buisonjé 1964, 1974; Herweijer and Focke 1978). The climatic conditions noted above have been favorable for the remarkable preservation of Curaçao's Quaternary fossil reef deposits. We studied the Lower Terrace, which is up to 600 m wide on the windward side of the island and up to 200 m wide along the leeward side (Fig. 8.2B). The Lower Terrace is exposed around the entire coast of the island, from 2 to 15 m above present sea level. Based on radiometric age dating and stratigraphic relationships, Herweijer and Focke (1978) correlated the uppermost unit of the Lower Terrace, the Hato Unit, with the 125-ka sea level interval (stage 5e of Emiliani 1966; see also Schubert and Szabo 1978), and tentatively suggested an age of 180 to 225 ka for the underlying unit of the Lower Terrace, the Cortelein Unit (i.e., stage 7). New radiometric age dates using high-resolution TIMS dating also give 125 ka as the age for the Hato Unit of the Lower Terrace (Pandolfi, Llewellyn, and Jackson 1999). Due to low uplift rates on Curaçao, the Lower Terrace preserves only very shallow-water reef environments, restricted to paleodepths <10 m (Pandolfi,

Llewellyn, and Jackson 1999). We sampled the reef crest at 16 localities on the leeward side of the island (Fig. 8.2B).

8.3.3 Barbados

Barbados is a high oceanic island about 150 km east of the Windward Islands of the Lesser Antilles (Fig. 8.1). The island is 32 km long and up to 23 km wide with its long axis oriented SE to NW and lies in the belt of NE trade winds. Barbados receives the NW-flowing Antilles–South Equatorial Current, and currents flowing along the leeward west coast are the product of large eddies. The predominantly subhumid to humid tropical climate has an average rainfall between 1100 and 1250 mm year^{-1} along the topographically low north and south coastal regions, increasing inland to a maximum of 1750 to 2120 mm year^{-1} over the higher more interior portions of the island. Peak rainfall occurs during the period from September to December. Average annual temperature ranges between 24 and 28 °C, with only slight seasonal variation (James, Stearn, and Harrison 1977). Sea surface temperatures measured off the west coast range between 29.5 °C in the fall to 26 °C in February (Sander and Stevens 1973) and salinity varies between 32 and 36 ppt. The low salinities are thought to be the result of masses of low-salinity water that drift northward from the Amazon and Orinoco River systems. Barbados has two high tides and two low tides daily, with a mean tidal range of 0.7 m and a diurnal range of 1.1 m (Lewis 1960a). Wave heights on the windward east coast are approximately 1.2 m on calm days and up to 4 m on rough days, whereas those on the leeward west coast only 0.2 m on calm days and up to 0.3 m on rough days (James, Stearn, and Harrison 1977).

Barbados is not usually subject to Caribbean hurricanes, because they generally pass to the north of the island. However, between 1651 and 1850, 11 hurricanes passed near the island and 3 were devastating (Bird, Richards, and Wong 1979). Three hurricanes in the last century (1895, 1955, and 1979) caused no recorded damage, but Hurricane Allen in 1980 caused considerable change in the Bellairs fringing reef (Mah and Stearn 1986).

The leeward west coast of Barbados is characterized by a well-developed but discontinuous fringing reef that is best developed around the headlands. The fringing reef occurs on a sloping shelf that may extend up to 300 m from the shore to a depth of around 10 m. Seaward of this shelf is a trough of mixed coral and sand composition with a depth of up to 30 m. Then the so-called "bank reef" comes up to a depth of around 15 m and is up to 100 m across.

The Pleistocene Coral Reef Ecosystem

Barbados preserves a spectacular series of Pleistocene raised terraces with excellently preserved coral reef assemblages occurring over a wide range of reef habitats (Mesolella 1967; Mesolella, Sealy, and Matthews 1970; Jackson 1992; Martindale 1992; Pandolfi 1999; Pandolfi, Lovelock, and Budd 2002) (Fig. 8.2C). The raised reef terraces range in age from >600 ka to 82 ka. Numerous authors have described the Pleistocene environments and reef development based on reef geomorphology and sediment characteristics (Mesolella, Sealy, and Matthews 1970), and epibionts (Martindale 1992; Perry 2001). Submerged fossil coral reefs have been extensively studied for sea-level change during the last deglaciation (Fairbanks 1989; Guilderson, Fairbanks, and Rubenstone 2001) and for refining age-dating methods (Bard et al. 1990). The magnitude and timing of Pleistocene global sea level fluctuations have been inferred from the uplifted reefs (Blanchon and Eisenhauer 2001), and thus the stratigraphy, radiometric age dating, and models for Pleistocene reef development are among the best in the world. Preservation of coral species is excellent, and the corals show clear depth zonation, providing an ecological frame of reference. Thus, the Barbados terraces are ideal for paleoecological study. We sampled the leeward reef crest at 12 localities on the leeward side of the island (Fig. 8.2C).

8.4 Methods

8.4.1 Sampling

Twelve to sixteen 40-m transects were laid in the Pleistocene shallow fore-reef environment along the leeward side of each of the three islands (Fig. 8.2). Transects were normally separated from each other by 500 to 1000 m. Each island was surveyed along 13 to 35 km of coastline (Table 8.1). Transects were sampled from either two or three sites per island, but here we only present results from the pooled analyses. Where corals were encountered, their orientation and degree of fragmentation were noted where possible, and they were identified to the lowest possible taxonomic level, usually to species. The length of transect that each coral colony intersected was recorded. Thus, the raw relative abundance data for corals from the transects was the total length of transect intercepted by each coral species. After the transects were recorded, a one-hour search for additional (rare) coral species not intercepted along the transect was made. The results of the raw presence/absence data set were similar to those of the abundance data set and are not discussed further. Comparison among quantitative and binary results from this study and other Pleistocene Caribbean studies is the subject of a separate contribution (Pandolfi 2001).

Table 8.1. Sampling data for Pleistocene coral communities from the three Caribbean islands spanning over 2500 km. Species richness is given for transects and parenthetically for transects plus a one-hour search

	Total number of 40-m transects	Length of island sampled (km)	Total length of transects (m)	Species richness
San Andrés	15	13.7	600	23 (32)
Curaçao	16	35.6	640	21 (30)
Barbados	12	13.1	480	17 (22)

8.4.2 Data Analysis

We used species-sampling curves to investigate whether our methodology adequately accommodated the diversity present at each island. The cumulative number of species encountered in each transect was plotted for each island. Species diversity per transect was compared using one-way ANOVA. Comparison of taxonomic composition among all possible pairs of transects was calculated using the Bray–Curtis (BC) dissimilarity coefficient (Bray and Curtis 1957). Abundance data were transformed to their square roots prior to the calculation to reduce the influence of occasional large abundance values for some taxa (Field, Clarke, and Warwick 1982). The transformed abundance values for each taxon were standardized. We used nonmetric multidimensional scaling (NMDS) as an ordination technique to provide a visual summary of the pattern of BC values among the samples. Analysis of similarities (ANOSIM; Clarke 1993; Clarke and Warwick 1994) was used to determine whether assemblages from different islands differed significantly in coral composition. Details of these methods can be found in Pandolfi and Minchin (1995), Pandolfi (1996), Jackson, Budd, and Pandolfi (1996), Pandolfi (2001), and Pandolfi and Jackson (2001). Within each island, BC similarity values were plotted against distance separating compared communities.

8.5 Results

We encountered 36 reef-coral species from 43 transects covering 1.72 km of fossil reef from the Pleistocene leeward shallow reef environment of the three islands (Table 8.1). Inspection of the species sampling curves showed that 5 to 7 transects (samples) were sufficient to capture greater than 90% of the coral species richness from San Andrés and 8 or 9 transects captured greater than 90% of the coral species richness from Curaçao and Barbados (Fig. 8.3). We found 23 species in

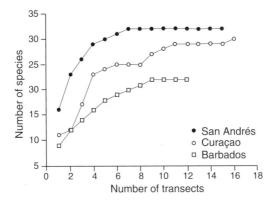

Figure 8.3. Species-sampling curves for the Pleistocene shallow leeward environments at San Andrés, Curaçao, and Barbados. Curves show the cumulative number of species encountered at each island, along the line transects with an additional one-hour search.

San Andrés, 21 species in Curaçao, and 17 species in Barbados along the transects in the leeward shallow reef environment, and 32, 30, and 22 species, respectively, when the one-hour search data were included (Table 8.1). Species diversity per transect was significantly different among the three islands ($F_{(2,42)} = 29.52$, $P < 0.00001$), with means of 16.9 for San Andrés, 11.6 for Curaçao, and 6.8 for Barbados.

The ordination showed that taxonomic composition of corals within the leeward shallow fore-reef environment varied among the three islands over the 2500-km southern Caribbean transect (Fig. 8.4). Assemblages from Barbados were the most distinctive, whereas those from San Andrés and Curaçao showed overlap. Nonetheless, the ANOSIM tests were significant both overall and for all three pairwise comparisons (Table 8.2), so there is clear geographic separation in community composition.

Comparison of Species among Islands

Assemblages from San Andrés are dominated by the extinct organ-pipe *Montastraea* (Pandolfi 1999; Pandolfi, Jackson, and Geister 2001; Pandolfi, Lovelock, and Budd 2002) and *Acropora cervicornis* (Fig. 8.5). These species, together with *Diploria strigosa*, *Diploria labyrinthiformis*, *Montastraea faveolata*, and *M. annularis* (sensu stricto) account for greater than 85% of the total composition of the San Andrés shallow leeward assemblages. Assemblages from Curaçao are also

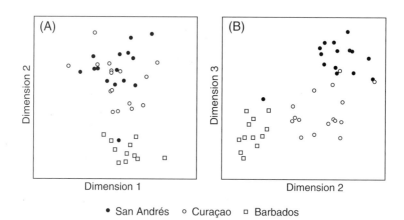

Figure 8.4. Nonmetric multidimensional scaling (NMDS) ordination of reef-coral species composition from Pleistocene leeward reef-crest communities along the SOCA transect at San Andrés, Curaçao, and Barbados. Note that Barbados plots well away from San Andrés and Curaçao. However, San Andrés and Curaçao are also distinct. ANOSIM statistical tests reveal that the separation of reef-coral assemblages from different islands in the ordination corresponds to significant differences in species composition (Table 8.2). The NMDS started with 20 random configurations and proceeded through 200 iterations for each of four dimensions. Plots are (**A**) dimensions 1 and 2 and (**B**) dimensions 2 and 3, from the three-dimensional analysis. The minimum stress value for the three-dimensional analysis was 0.08.

Table 8.2. Results of the ANOSIM test for significant differences in Pleistocene reef-coral species composition (relative abundance) within leeward reef-crest environments among the three Caribbean islands along the SOCA transect: San Andrés, Curaçao, and Barbados. R = ANOSIM test statistic

	R	P-value
Overall	0.73	<0.0001
San Andrés/Curaçao	0.56	<0.0001
San Andrés/Barbados	0.91	<0.0001
Curaçao/Barbados	0.72	<0.0001

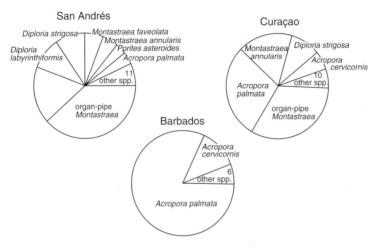

Figure 8.5. Coral species abundance patterns within the leeward shallow reef environment in Pleistocene reefs from San Andrés, Curaçao, and Barbados. The pie diagrams show differences in the relative abundances of the dominant reef-coral taxa among islands. Note the gradients in abundance for many of the dominant taxa, including *A. palmata*, organ-pipe *Montastraea*, and *Diploria*. However, the most abundant taxa are constant among islands throughout the Caribbean.

dominated by the extinct organ-pipe *Montastraea* along with *A. palmata* (Fig. 8.5). These species, together with *M. annularis* (s.s.), *D. strigosa*, and *A. cervicornis* account for greater than 98% of the total composition of the Curaçao shallow leeward assemblages. *A. palmata* and *A. cervicornis* dominate shallow leeward assemblages from Barbados (Fig. 8.5). These two species account for over 94% of the total composition of the Barbados shallow leeward assemblages. Thus, differences in coral species composition among islands are mainly due to the distribution and abundance of the dominant taxa: *A. palmata*, organ-pipe *Montastraea*, *M. annularis* (s.s.), *D. strigosa*, and *A. cervicornis*. There were 7 species that occurred in only one of the islands, and 10 species that were absent from only one of the islands (Table 8.3).

Table 8.3. Species occurrences at each of the three Caribbean islands in the Pleistocene leeward shallow reef. Seven species occurred on only one island, 10 species occurred on two islands, and 19 species occurred on all three islands

	San Andrés	Curaçao	Barbados	Number of islands
Acropora cervicornis	x	x	x	3
Acropora palmata	x	x	x	3
Agaricia agaricites	x	x	x	3
Agaricia crassa		x		1
Colpophyllia amaranthus	x	x		2
Colpophyllia breviserialis	x	x		2
Colpophyllia natans	x	x		2
Dendrogyra cylindricus	x	x	x	3
Dichocoenia stokesi			x	1
Diploria clivosa	x	x		2
Diploria labyrinthiformis	x	x	x	3
Diploria strigosa	x	x	x	3
Eusmilia fastigiata	x	x		2
Favia fragum	x	x	x	3
Isophyllastrea rigida	x		x	2
Isophyllia sinuosa	x			1
Madracis mirabilis	x	x		2
Meandrina meandrites	x	x	x	3
Millepora complanata		x		1
Montastraea annularis (sensu stricto)	x	x	x	3
Montastraea faveolata	x	x	x	3
Montastraea franksi		x		1
Montastraea-II *cavernosa-1*	x	x	x	3
Montastraea-II *cavernosa-2*	x			1
Montastraea -II *cavernosa-3*	x	x		2
Montastraea -II *cavernosa-4*	x	x		2
Mussa angulosa	x			1
organ-pipe *Montastraea*	x	x	x	3
Pocillopora cf. *palmata*	x	x	x	3
Porites astreoides	x	x	x	3
Porites furcata	x	x	x	3
Porites porites	x	x	x	3
sheet *Montastraea*	x	x	x	3
Siderastrea radians	x	x	x	3
Siderastrea siderea	x	x	x	3
Stephanocoenia intersepta	x		x	2
Total number of species	32	30	22	Grand total 36

Community Similarity with Distance

At the scale of tens of kilometers, plots of distance versus BC index values from all three islands were relatively flat and showed high levels of similarity in community composition over the large distances surveyed on each island (Fig. 8.6). There was no significant change in BC with distance along the 13-km coast we sampled at Barbados ($r^2 = -0.8$, $P = 0.48$) (Fig. 8.6C), nor along the

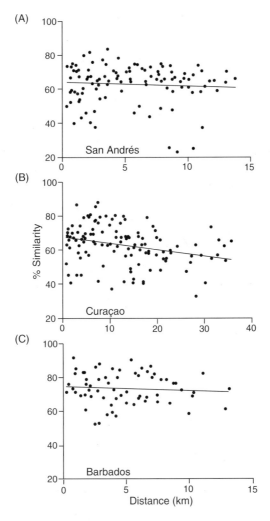

Figure 8.6. Plot of Bray–Curtis similarity values among transects within the leeward shallow reef as a function of distance between transects for Pleistocene reef coral communities. The species composition within all three reef islands remains relatively constant over very broad spatial scales. Notice difference in scale for Curaçao (**B**) as opposed to San Andrés (**A**) and Barbados (**C**).

14-km coast we sampled at San Andrés ($r^2 = -0.6$, $P = 0.55$) (Fig. 8.6A). However, along the longer 36-km coast we sampled at Curaçao, community similarity decreased slightly, but significantly, with distance ($r^2 = 0.34$, $P = 0.023$) (Fig. 8.6B). Mean similarity values were highest among Barbados assemblages (74%) and lower among San Andrés and Curaçao assemblages (63% for each island).

8.6 Discussion

8.6.1 Pleistocene Patterns of Community Structure

Geister (1977, 1980) formulated a qualitative model of coral community composition that related coral abundance patterns to wave energy in the Caribbean Sea. Geister's model (1977, 1980) predicts changes in *A. palmata*, and branching and massive corals, on the basis of the taxonomic composition of "calm"- versus "rough"-water reefs he observed both living in the Caribbean Sea and preserved as fossil deposits. As wave energy increases on the reef front, the shallow coral communities should be characterized by species of the *M. "annularis"* complex, then *Porites*, then *A. cervicornis*, then *D. strigosa/A. palmata*, then *Palythoa/Millepora*. Reef zones exposed to the highest wave energies should be characterized by Melobesieae (calcareous algae or an algal ridge).

One of the cornerstones of the model is the positive relationship between high wave energy and abundance of *A. palmata* in shallow water. *A. palmata* is a highly wave-resistant coral of the Caribbean usually confined to the top 5 m of water depth. Leeward Pleistocene assemblages from Barbados show almost monospecific dominance of *A. palmata*. Those from Curaçao show *A. palmata* as a co-dominant with the organ-pipe *Montastraea*, with abundant *M. annularis* (s.s.) and *D. strigosa*. Those from San Andrés show much lower abundance of *A. palmata* and are characterized by *A. cervicornis* and organ-pipe *Montastraea*. In contrast, *A. cervicornis* and organ-pipe *Montastraea* are more abundant in Curaçao and San Andrés than they are in Barbados. The organ-pipe *Montastraea* also lived in areas of high wave energy (Pandolfi 1999; Pandolfi and Jackson 2001; Pandolfi, Jackson, and Geister 2001). Barbados has the highest frequency of hurricanes and Curaçao the highest average wave energy, whereas San Andrés is outside the hurricane belt and experiences much lower wave energy than the islands to the east (Neumann et al. 1981; SeaWinds database 2005). Based on these modern meteorological data, our quantitative surveys of coral distribution patterns over broad spatial scales, and Geister's wave energy model, Barbados leeward crests had the highest, Curaçao intermediate, and San Andrés the least wave exposure at 125 ka. Coral assemblages appear to be highly predictable, varying in species abundance depending on wave exposure. These results corroborate earlier suggestions that corals are highly adapted to wave-energy regimes (Jackson 1991).

Even though the significant differences we found in Pleistocene coral species abundance among the three Caribbean islands (Table 8.2; Figs. 8.4, 8.5) can be directly related to varying wave-energy regimes, the group of taxa with the highest abundance remained constant among the islands (Fig. 8.5). The same species tended to dominate wherever they occurred in the Pleistocene leeward shallow waters of the southern Caribbean Sea. They included *A. palmata*, *A. cervicornis*, and species of *Diploria* and the *M. "annularis"* species complex (Fig. 8.5). Thus, these coral assemblages showed a high degree of similarity in their community structure.

However, regional differences in coral species composition may not always be related to wave energy. We found significant differences in species diversity among the three islands. This may represent a biogeographic overprint on assemblages that are largely controlled by local wave regime. For example, Liddell and Ohlhorst (1988) provided a preliminary comparison of coral taxonomic composition (species abundance patterns) across the Caribbean in 10- to 15-m- deep fore-reef slopes. Our application of nonmetric multidimensional scaling ordination to these early census data also shows clear biogeographic separation (Fig. 5.5 in Jackson, Budd, and Pandolfi 1996). In this case, the taxonomic composition of the reef corals at Antillean sites was distinct from that at mainland and more northern localities. Thus, fore-reef community composition at intermediate depths, at least, also varied over broad spatial scales in the Caribbean Sea. Geographical and exposure differences can be teased apart by analyzing the SOCA data using both Site and Island as factors, and this will be the subject of a future contribution.

At the intermediate scale of individual islands we found higher degrees of order. Coral community composition was related more to local environment than distance separating the assemblages (Fig. 8.6). Thus, within each island, the taxonomic composition of communities was markedly similar over broad geographic areas, indicating nonrandom species associations and communities comprised of species occurring in characteristic abundances. This community similarity was maintained with little decrease over distances up to 35 km. Regardless of the cause, these results suggest greater degrees of similarity in community composition at scales of tens of kilometers than hundreds to thousands of kilometers in the Pleistocene Caribbean Sea (Pandolfi 2002).

Community variability at the large and intermediate spatial scales shown in our Pleistocene SOCA transect contrasts with many studies of living coral reefs at small spatial scales where a much larger degree of variance in community structure has been observed (Sale 1977; Tanner, Hughes, and Connell 1994; Connell, Hughes, and Wallace 1997). Connell, Hughes, and Wallace (1997) attributed ecological differences to different disturbance regimes and histories at individual sites at Heron Island. Hurricanes can cause selective damage to those reefs and/or parts of reefs most heavily exposed to the advancing storm front (Woodley et al. 1981). Thus, damage to the sessile benthos can be highly patchy.

However, in both coral reefs (Done 1982; Ault and Johnson 1998a,b; Robertson 1996; Connell, Hughes, and Wallace 1997; Hughes et al. 1999; Bellwood and Hughes 2001; Pandolfi 2001, 2002; Pandolfi and Jackson 2001, this study) and forest communities (Terborgh, Foster, and Percy 1996; Clark and McLachlan 2003), workers who have enlarged their spatial scale of study usually have found more consistent patterns in community structure (but see Murdoch and Aronson 1999). Thus the high variability inherent at individual sites within reefs appears to decrease markedly as more area of the reef or more time is studied. The variation at the smallest scales may even be higher than the biogeographic differences we found, both in our transects and in Liddell and Ohlhorst's (1988) compilation. When these patterns are considered together, order in reef coral communities appears to be lowest at small spatial scales (Connell, Hughes, and Wallace 1997), highest at intermediate spatial scales (Pandolfi 1996, 1999; Pandolfi and Jackson 2001), and intermediate at the broadest spatial scales

(Jackson, Budd, and Pandolfi 1996, this study) within the same biogeographic province (Pandolfi 2002). Variance is lowest over scales of kilometers to tens of kilometers where we suggest that species niche dimensions (biotic interaction and environmental preferences) and principles of limiting similarity result in similar community structure (Pandolfi 2002). This is the scale at which we recognize certain species assemblages recurring from place to place and from time to time (i.e., ecotones; Goreau 1959; Goreau and Wells 1967; Geister 1977, 1980; Done 1982). Given certain environmental parameters, reef-coral communities within a reef, or among reefs that are close by, may resemble one another very closely. Moreover, the trade-off among niche-dimension processes and regional effects of the species pool (dispersal, recruitment, migration) probably resulted in the dominance of the same few coral species throughout the southern Caribbean in the Pleistocene and subtle differences in their exact community structure (composition and diversity), respectively. Trends in predictability are also apparent in studies of community structure over varying temporal scales: compare the persistent, long-term framework of Pandolfi (1996), Aronson and Precht (1997), and Greenstein, Curran, and Pandolfi (1998) to the more variable, short-term framework of Tanner, Hughes, and Connell (1994) and Connell, Hughes, and Wallace (1997).

We believe that one reason that causality has not been forthcoming in studies of coral reef ecology is because of the confusion engendered when scale is not explicitly incorporated into comparative discussions of research results. An example is Karlson and Cornell's (1998) criticism of Pandolfi's (1996) result documenting persistence in coral communities from Papua New Guinea (PNG) throughout 95 ka. They argued that too few species of *Acropora* had been encountered to make any generalizations about that portion of the coral fauna, so an important component of the fossil community had been missed. Many Indo-Pacific coral communities are composed of a disproportionate number of species of the dominant Indo-Pacific branching coral *Acropora* (up to 125 species; Wallace 1999). Although only 4 species of *Acropora* occurred in the Pleistocene surveys (Pandolfi 1996), a survey of 120 species of modern corals from the adjacent living reefs on the Huon Peninsula found 18 species of *Acropora* (Nakamori, Wallensky, and Campbell 1994). Only 4 of the 14 additional species could be classified as "abundant," and this occurred at only two sampling localities. To bolster their criticism, Karlson and Cornell (1998) referred to the Tomascik, van Woesik, and Mah (1996) study of colonization by what are usually quite rare species of *Acropora* immediately after a lava flow in the Banda Islands, Indonesia. However, this is an inappropriate comparison, because the Banda Islands community was in a very early stage of succession whereas the Pleistocene PNG communities were sampled at the very latest stage of their respective reef-building episodes. Thus, they represented the culmination of thousands of years of uninterrupted reef growth and development. This example illuminates the point that careful consideration of scale-dependent patterns and associated processes is necessary for comparisons of studies conducted over very different spatial or temporal scales.

8.6.2 Perspectives on Ecological Theory of Community Structure

Several ecological processes are purported to have some role in maintaining species diversity in highly variable local reef populations studied over limited temporal scales. We discuss our Pleistocene results in the light of current ecological concepts that are based mainly on studies of limited temporal or spatial scale in coral reefs.

Recruitment Limitation

"Supply-side ecology" implies that the spatial distribution and supply of planktonic larvae exerts control over the distribution and abundance of adults (Young 1987,1990; Grosberg and Levitan 1992). The "recruitment limitation" hypothesis (Doherty 1981) predicts that planktonic larval supply is the limiting factor that determines future population abundance (reviews in Caley et al. 1996; Hixon 1998). Variation in larval supply appears to be especially important for local variability in fish populations (*Australian Journal of Ecology* 1998, vol. 23). Aside from the problems associated with ambiguities in the use of the terms "recruitment" and "limitation," adult density dependence has only recently been incorporated into studies of larval dynamics (Ault and Johnson 1998a; Hughes et al. 1999). Studies conducted over longer time intervals and larger spatial scales have shown that adult abundance patterns are not always correlated with larval abundance, and that postsettlement processes can exert greater influence over adult community structure than larval supply (Robertson 1996; Ault and Johnson 1998a). Thus, our highly predictable Pleistocene coral communities at individual Caribbean islands (Pandolfi and Jackson 1997, 2001), over the entire Caribbean, and over long time scales (Jackson 1992; Pandolfi 1996, 1999, 2000, 2001; Pandolfi and Jackson 2006) indicate alternative or additional factors to coral larval supply. In this regard, two of the most prominent Caribbean coral species, *A. palmata* and *A. cervicornis*, are rarely observed to spawn (Bak and Engel 1979) and may have much reduced larval supplies, even though they were the most dominant taxa in Pleistocene and pre-1980s' Caribbean reefs.

Saturation

Since Ricklefs (1987) called for more regional and historical components to ecological studies and Cornell (1985a,b) outlined a methodology for relating local species richness to regional species richness, many workers have investigated this relationship using regression analyses (Cornell and Lawton 1992; Cornell and Karlson 1996, 1997; Caley and Schluter 1997; Karlson and Cornell 1998). When a positive linear relationship is found between local and regional species richness, communities are said to be unsaturated with species, meaning that they are open to invasion and the diversity of the regional species pool plays a major role in maintaining community structure. When curvilinear relationships are found, communities are saturated and biotic interactions and local environment are more important in structuring communities than the size of the regional species pool. General concerns that have been raised are the scale at which "local" assemblages are defined and the appropriateness of combining different groups of organisms (Westoby 1998) and/or biotas with different evolutionary histories (Hugueny et al. 1997) in a single study.

Our Pleistocene data caution against overinterpreting these kinds of analyses, aside from the obvious point that regression analysis cannot document causal relationships. In the Caribbean, both Pleistocene (this study and Pandolfi and Jackson 2001) and pre-1980s' (Goreau 1959; Goreau and Wells 1967) coral communities were dominated by one or a few species, even though a large number of species inhabited each environment. Thus, local richness may appear unsaturated when in reality a subset of the community is indeed saturated. In this case it may matter less to local community structure how diverse the regional species pool is than which particular species compose the species pool. Thus, "saturation" studies may have limited applicability to reef-coral communities in which only a small fraction of the species present is numerically abundant.

Dispersal Limitation

Hubbell's (1997, 2001) "unified theory of biogeography and relative species abundance" incorporates speciation and species relative abundance into MacArthur and Wilson's island biogeography theory. Species abundance patterns can be explained simply by stochastic events of migration, extinction, and speciation. For example, Hubbell (1995) stressed limitation to dispersal as a key factor in the unpredictability of rainforest communities at small scales. Hubbell's theory is elegant in its simplicity, has appealing explanatory power, and uses only a few very simple parameters with very little biological complexity. Unfortunately, nowhere does a central assumption of the theory hold: that all species are equal in their per capita fitness. This does not make the theory useless; on the contrary it enlightens us as to how some components of the processes that affect ecosystem dynamics operate. Our task is to evaluate the extent to which the "propensity" (sensu Ulanowicz 1997) for migration rates among populations to control species abundance patterns is affected by other ecological agents such as biotic interactions, variations in larval supply, and disturbance. We raise this discussion because Hubbell (1997) predicted constancy in time and space for populations that produce abundant larvae capable of long-distance transport. He used this theoretical argument to explain patterns of similarity in coral composition through time and space found in Papua New Guinea and Curaçao (Pandolfi and Jackson 1997). However, in Caribbean corals, locally dominant species do not appear to produce more or better larvae than locally rare taxa (Bak and Engel 1979), nor do they appear to be any more widespread in any tropical region (Pandolfi and Jackson 1997).

Metapopulation Dynamics

A fourth concept is metapopulation dynamics, in which dispersal capability and competitive ability are traded off such that their interaction determines species abundance patterns at any one patch (locality) (Hanski 1998). Surprisingly, there has been little effort to incorporate metapopulation models into coral reef ecology (but see Jackson, Budd, and Pandolfi 1996; Mumby 1999), despite the fact that these ecosystems display populations that are presumably highly dispersed and competitively interactive. Reef fish ecologists, in particular, raised several objections to the metapopulation perspective. First, field studies are conducted mainly at the level of local populations, so links between these studies and population dynamic theory can only be made at a small spatial scale (Caley et al. 1996). However, as broader spatial scales are considered in studies by more ecologists (Hughes et al. 1999), this problem should recede. Second, since larval life can be so long for marine invertebrates, virtually all recruitment is from elsewhere (Caley et al. 1996). For reef corals, however, there has been no relationship established between larval life and distance of transport; thus long larval life does not equate with recruitment from elsewhere. For example, although coral planulae may live for over 100 days in aquaria (Richmond 1987; Wilson and Harrison 1998), there still are no data to indicate how long they live in the field, how far they travel, or the relationship between larval abundance and distance from their source. Moreover, metapopulation models are being expanded to include those metapopulations displaying long-range dispersal (Doebeli and Ruxton 1998). And fish populations show a significant amount of larval retention on coral reefs (Jones et al. 1999; Swearer et al. 1999). Third, the "open" recruitment assumed for coral reef fish is contrary to metapopulation theory, in which subpopulations are assumed to be mostly closed, with only infrequent dispersal between patches (Caley et al. 1996; Hixon 1998). However, molecular genetic studies are showing that patterns of gene flow in many coral reef species do not correspond to present-day oceanic circulation patterns and instead point to highly pulsed dispersal, perhaps related to range expansion during interglacial periods (Benzie 1999). Moreover, spatially explicit metapopulation models are beginning to take into account dispersal distance among patches (Hanski 1998), even in the marine environment (Roughgarden and Iwasa 1986). The degree to which coral reefs are "open" to recruitment is an unresolved issue (Hughes et al. 1999; Jones et al. 1999; Swearer et al. 1999). Fourth, metapopulation theory is incapable of being applied to marine systems because marine populations are so persistent; hence, local extinction among patches is infrequent (Caley et al. 1996). This may be true over short time scales, but over longer time scales it may not. For example, Pleistocene global sea level fluctuations caused periodic elimination of reef substrates with local extinction. Because the sea level changes caused significant changes in oceanic circulation currents (Fleminger 1985), local extinction may well have occurred as habitat was gradually reduced. Local extinction in Caribbean reef corals also appears to be occurring today in two previously dominant species: *A. palmata* and *A. cervicornis* (Lewis 1984; Aronson and Precht 1997; Greenstein, Curran, and Pandolfi 1998). These extinctions are in part related to habitat degradation, one of the primary themes for studies of metapopulation dynamics (Hanski 1998).

Even though there are large differences between terrestrial and coral reef ecosystems, they may not be large enough to preclude consideration of metapopulation dynamics in coral reef ecology. The final evaluation of the role of metapopulation dynamics in coral reef ecology must await sufficient studies conducted at large spatial or temporal scales. However, Pandolfi (1999) used the predictions from a model by Nee and May (1992) to suggest that the Pleistocene extinction of two species of reef corals might have resulted from habitat reduction during the last glacial maximum, 18 ka. If linkages among disparate coral reef populations are important for maintaining local species diversity, then decreasing the amount of available habitat below a certain threshold might have deleterious effects for coral reefs (Pandolfi 1999). Thus it is important to consider the role of metapopulation dynamics in the explanation of persistence in time and space in Pleistocene Caribbean reef coral communities, and their subsequent collapse in the 1980s.

Biological Interactions

Biological interactions are not only ubiquitous in coral reef settings, they have also been shown to have major effects on the population density of individual species, and thus the community structure of many reefs. The principal causal agents of such effects are competition, predation, herbivory, symbiosis, and disease. Competition has been observed among corals (Lang and Chornesky 1990; Van Veghel 1994), corals and algae (Miller and Hay 1998), soft corals (Alino, Sammarco, and Coll 1992), fish (Robertson 1996; Steele 1998), and a host of other taxa. Both intense grazing by schooling herbivores and damselfish territoriality strongly influence the standing crop, productivity, and community structure of reef macroalgae and corals (Carpenter 1986, 1997; Hay et al. 1989; Hay 1997; Hixon 1997). For example, in the Gulf of Panamá, the omnivorous damselfish *Eupomacentrus acapulcoensis* facilitates zonation in the reef corals *Pocillopora* and *Pavona* (Wellington 1982). Where disease decimates or even eliminates dominant coral taxa, as with white-band disease in the acroporid corals (Gladfelter 1982), the very structure of the reef may be destroyed, and habitat loss may result in a decline in biodiversity and collapse of previous community structure (Richardson 1998). Our knowledge of symbiosis in coral reefs is very limited, but recent studies are showing immense complexity among symbionts and hosts that varies with environment (Rowan and Knowlton 1995; Rowan et al. 1997; Baker 2001; Baker et al. 2004; Rowan 2004; Sotka and Thacker 2005). Thus, interaction between host and symbiont may lead to zonation patterns among coral reef organisms.

An important question concerning biotic interactions among coral reef organisms is the degree to which such processes are important in maintaining the structure of coral reef communities. Whereas so much biotic interaction naturally invites study of deterministic processes in the maintenance of coral reef diversity, it is equally true that the processes are very complex and the outcomes difficult to predict based on biotic interactions alone. Many of the ecological mechanisms involved are species-specific and unsatisfying when attempting to use them as generalities in coral reef community ecology. However, the general, predictable patterns of spatial and temporal persistence in taxonomic composition found among the Pleistocene localities examined in this study are at least consistent with the idea of biotic interaction and local adaptation as playing a role in controlling coral reef community structure.

Disturbance

Because they predominate in tropical belts where severe storms can wreak environmental havoc in a very short time, coral reef ecosystems commonly suffer disturbance. Disturbance has interested community ecologists studying coral reefs ever since Connell's (1978) exposition of the intermediate disturbance hypothesis (IDH) (see review by Karlson and Hurd 1993). According to the IDH, coral reef communities are prevented from reaching an equilibrial state because disturbance occurs often enough to prevent competitive exclusion. Thus, reef communities are extremely dynamic, depending upon the disturbance regime to which they are exposed. Long-term studies using 1-m² quadrats on coral reefs from Heron Island on the Great Barrier Reef show dramatic changes in coral community structure at multiple spatial scales (Tanner, Hughes, and Connell 1994; Connell, Hughes, and Wallace 1997). Connell (1997) found that recovery from acute, short-term disturbance occurred much more commonly than from chronic, longer-term disturbances. Local patches can be altered by biological disturbances, most notably from the crown-of-thorns starfish (Moran 1986), but also from damselfish territories and spatial heterogeneity in herbivorous fish dynamics (Hixon 1997). And coral composition is extremely variable at the smallest scales (meters to hundreds of meters) where local disturbance regimes are highly unpredictable (Connell, Hughes, and Wallace 1997).

Using simulation models of data collected during the long-term study of Heron Island, Tanner, Hughes, and Connell (1994) found that the length of time required to reach a climax assemblage is around 20 years. Because 20 years was far greater than the amount of time between disturbances, Tanner, Hughes, and Connell (1994) favored nonequilibrium theories of coral reef communities, arguing that their shallow-water communities were nearly always in a state of recovery from the latest cyclone. However, the IDH was inadequate to explain their simulations because diversity remained very high long after a major disturbance; thus competitively inferior species were not eliminated as predicted by IDH. These authors found that climax assemblages were highly diverse and varied in composition from site to site. It must be asked whether these patterns are exactly those that one might predict from examining quadrats that are less than the size of many individual coral colonies on reefs. Although individual quadrats changed dramatically, we have no feel in this study for how whole communities within reef habitats might have changed. Our SOCA study using 40-m transects suggests much greater similarity in community composition both within and among islands than that observed and simulated from 1-m² quadrats on Heron Island.

In a subsequent paper, Connell, Hughes, and Wallace (1997) sought to address the spatial-scale issue in their long-term study. They found a large amount of spatial variability in coral abundance (and recruitment), with most occurring between habitats, not within them. They ascribed most of the variability in coral abundance to varying degrees of disturbance within and between the habitats they studied. Their study records change in coral abundance over a 27-year interval and provides valuable information on how coral abundance is affected by disturbance. We have also found within-habitat variability to be much less than between-habitat variability in Pleistocene coral reef communities in Curaçao (Pandolfi and Jackson 2001). In Curaçao, limited variability within habitats occurs in the absence of major hurricanes, which commonly pass to the north of the island. Thus, disturbance regime alone does not seem to explain high degrees of within-habitat order in coral community structure in both living and ancient reefs.

8.6.3 Management Implications

Perhaps the most important finding in our work is that Pleistocene species distribution patterns are dramatically different from present-day ones, but this difference was not as great 25 years ago as it is today. In fact, the dominant corals in the Pleistocene are the same taxa that dominated Caribbean sites until the early 1980s when the effects of overfishing forced major changes in the community structure of living reefs (Hughes 1994; Jackson et al. 2001; Pandolfi et al. 2003). Thus, the Pleistocene provides a very powerful baseline with which to evaluate reef degradation in living systems (Greenstein, Curran, and Pandolfi 1998; Pandolfi 2002). The fossil record provides a history of community ecology unfettered by human influence, and consequently provides goals for reef management (Pandolfi et al. 2005).

Extensive degradation of reef habitats in the Archipelago of San Andrés and Providencia has been attributed to bleaching, black-band and white-band disease, hurricanes, dredging, overfishing, sewage pollution, and shipping activities (Díaz, Garzón-Ferreira, and Zea 1992; Garzón-Ferreira and Kielman 1994; Zea et al. 1998). This has resulted in an overall decrease in coral cover over 50%, with the greatest effects occurring near sewage outfalls close to the shorelines, especially on the NW side of San Andrés (Geister and Diaz 1997). Consistent with many other Caribbean reefs, macroalgae in San Andrés have increased greatly during the past two decades, especially since the mass mortality of the sea urchin *Diadema antillarum* in 1983 (Lessios, Robertson, and Cubit 1984; Lessios 1988). Although Pleistocene and modern reef environments are somewhat different in topography and physiography (Geister 1975), one of the most dominant branching corals from the Pleistocene, *A. cervicornis*, is almost absent from living reefs. It seems clear that modern reefs have suffered a decline with respect to their Pleistocene counterparts.

The leeward reef at Curaçao has been extensively studied in a series of papers spanning almost 25 years (Bak 1975, 1977; Bak and Luckhurst 1980; Bak and Nieuwland 1995). Prior to the 1980s, these leeward reefs showed a diversity of reef habitats and, in many places, the common zonation pattern with depth observed throughout the Caribbean: *A. palmata*-dominated communities in the shallow-water reef crest, often with abundant head corals, followed downslope by *A. cervicornis*-dominated communities in the shallow fore-reef, and below a mixed head-coral assemblage in deeper fore-reef areas (Bak 1977; Van Duyl 1985). This zonation pattern is less distinct today, however, as degradation of the leeward reefs has proceeded (Bak and Nieuwland 1995). Pleistocene leeward reefs from Curaçao are dramatically different than living reefs are today, but similar to those that characterized the leeward coast prior to the 1980s. This is a clear case showing the utility of the Pleistocene as a baseline and shows that reef degradation during the past 20 years has proceeded at an alarming and seemingly unprecedented rate.

Jackson (1992) analyzed Mesolella's data on species abundance patterns from the Pleistocene of Barbados and found striking similarity through time and with the "classic" coral zonation patterns found at Discovery Bay in Jamaica. He concluded that the reef-coral communities from Barbados were stable throughout a 250,000-year interval. In the shallow leeward reef crest *A. palmata* was the most dominant coral. The high abundance of *A. palmata* in the Pleistocene of Barbados is also a feature of the Holocene deposits on the leeward side of the island. Lewis (1984) radiometrically dated coral death assemblages along the leeward coast and also noted relative abundance of the dominant coral species. His results show that *A. palmata* was still a major player along the leeward coastline from 2300 to 950 ybp.

In contrast to the Pleistocene and Holocene data, living leeward coast assemblages are depauperate in *A. palmata*. Tomascik and Sander (1987) censused reef-coral communities from seven localities along the leeward coast. Living *A. palmata* was found at only two of these localities, Greensleeves and Sandridge, and at each locality it is now rare (average relative coverage of 1.4 and 0.2%, respectively). Thus, *A. palmata* is significantly reduced in shallow water around Barbados leeward reefs, relative to their Holocene and Pleistocene counterparts. A similar situation occurs for its congener *A. cervicornis* in deeper water. Thus, the coral community composition throughout the Pleistocene fossil record on Barbados is strikingly different from that of post-1980s reef: the present is anomalous with respect to most of the Pleistocene history.

The best-studied reef on the leeward side of the island is the Bellairs fringing reef adjacent to the Bellairs Research Institute of McGill University in Holetown (Lewis 1960b; Stearn, Scoffin, and Martindale 1977). Significant changes occurred on the Bellairs reef due to Hurricane Allen in 1980 (Mah and Stearn 1986), but these reefs had already been severely degraded within 60 years of European colonization of the island when the creation of sugar plantations caused massive deforestation throughout the island (Lewis 1984). The unprecedented runoff, today including abundant fertilizers, has resulted in a severely degraded coral reef.

Woodley (1992) raised the intriguing hypothesis that perhaps the classic Caribbean zonation patterns that characterized Jamaica between 1940 and 1980 are anomalous because there were few or no hurricanes that affected the Jamaican reefs during this time. He reasoned that if the community dynamics of coral reefs are disturbance-driven, they should maintain high levels of species diversity. But, shallow-water Caribbean reefs prior to the 1980s were consistently composed of nearly monospecific stands of *A. palmata*.

Therefore, the low frequency of hurricanes during the past 40 years on the Caribbean may have led to anomalous community composition. Perhaps, then, we are worrying too much about an ecological concept that may well be the exception rather than the norm. However, the Pleistocene fossil record argues against such an interpretation. Nearly monospecific stands of *Acropora* have been a consistent feature of Caribbean reefs for the past 250 kyr and even longer (Pandolfi 2000). We cannot ascribe the luxuriant reefs observed between the 1940s and 1980s to historical accident: the classic pattern is the norm and extends throughout much of the Caribbean Pleistocene. We must view living reefs in the context of their Pleistocene counterparts and acknowledge that the past 25 years might have been the most degrading and disruptive to reef coral communities in their entire Caribbean history. Our task is to acknowledge the documentation of declining reefs (Porter and Meier 1992) and move on with remediating the damage we have done. Monitoring is fine, but only as a yardstick to measure the real progress that needs to be made in restoring coral reefs to health before it is too late (Jackson 1997; Pandolfi et al. 2005).

The results from paleoecological studies suggest that we know what to conserve and how to document deviation from historical trends (Pandolfi 1999; Pandolfi et al. 2003, 2005). We can judge the merits of conserving habitats and species on the basis of their past history. The Barbados example in particular shows the importance of the species *A. palmata* in the shallow reef-crest zone along its leeward coast for hundreds of thousands of years. Its present rarity should alert us into action. The Pleistocene fossil record of coral reefs offers a database to test for a shifting ecological baseline (Pauly et al. 1998); indeed it provides us with a very early baseline independent of human activity. Analysis of this database can provide a clear frame of reference to resource managers as to what they are trying to manage or conserve.

8.7 Summary

Three central tenets have occupied much of the discussion concerning the ecology of coral reefs: dispersal, biotic interaction, and disturbance. Dispersalist theories have stressed the role of larval supply, the size of the regional species pool, migration rates, and metapopulation dynamics. Numerous other studies document a myriad of biotic interactions that influence community structure on coral reefs, including competition, predation, herbivory, disease, and symbiosis. Disturbance regime is also considered a major controlling factor in the degree to which dispersal and biotic interaction can influence community structure (Connell et al. 2004). Recognition of the relative importance of each of these processes must await tests of their predictions at multiple spatial and temporal scales, which are so far largely lacking in studies of living reef ecosystems.

The fossil record provides a means of understanding coral community dynamics over longer time scales than traditionally have been used to argue for the relative influence of dispersal, biotic interactions, and disturbance. We discuss some of the major recent findings in the long-term ecological dynamics of coral reefs, and their relevance for understanding coral reef ecology. Comparison of results from the fossil record with those of living reefs suggests different patterns at different scales of observation. At small spatial and temporal scales, living reefs appear to show high degrees of variance in community structure. At intermediate spatial and temporal scales variance is sharply reduced, and at large temporal scales variance is intermediate. We suggest that this pattern is generated by the interaction of ecological processes and responses that vary at different spatial and temporal scales.

Pleistocene ecological data provide a fundamental baseline, incapable of shifting, for understanding natural variability in coral reefs and for putting the ecological effects of modern human habitat degradation into a historical context.

References

Alexander, C.S. 1961. The marine terraces of Aruba, Bonaire, and Curaçao, Netherlands Antilles. *Ann. Assoc. Am. Geogr.* 51:102–123.

Alino, P.M., P.W. Sammarco, and J.C. Coll. 1992. Competitive strategies in soft corals (Coelenterata, Octocorallia). IV. Environmentally induced reversals in competitive superiority. *Mar. Ecol. Prog. Ser.* 81:129–145.

Aronson, R.B., I.G. Macintyre, W.F. Precht, T.J.T. Murdoch, and C.M. Wapnick. 2002. The expanding scale of species turnover events on coral reefs in Belize. *Ecol. Monogr.* 72:233–249.

Aronson, R.B., and W.F. Precht. 1997. Stasis, biological disturbance, and community structure of a Holocene coral reef. *Paleobiology* 23:326–346.

Ault, T.R., and C.R. Johnson. 1998a. Spatially and temporally predictable fish communities on coral reefs. *Ecol. Monogr.* 68:25–50.

Ault, T.R., and C.R. Johnson. 1998b. Spatial variation in fish species richness on coral reefs: Habitat fragmentation and stochastic structuring processes. *Oikos* 82:354–364.

Bak, R.P.M. 1975. Ecological aspects of the distribution of reef corals in the Netherlands Antilles. *Bijdragin tot de Dierkunde* 45:181–190.

Bak, R.P.M. 1977. Coral reefs and their zonation in Netherlands Antilles. *AAPG Stud. Geol.* 4:3–16.

Bak, R.P.M., and M.S. Engel. 1979. Distribution, abundance and survival of juvenile hermatypic corals (Scleractinia) and the importance of life history strategies in the parent coral community. *Mar. Biol.* 54:341–352.

Bak, R.P.M., and B.E. Luckhurst. 1980. Constancy and change in coral reef habitats along depth gradients at Curaçao. *Oecologia* 47:145–155.

Bak, R.P.M., and G. Nieuwland. 1995. Long-term change in coral communities along depth gradients over leeward reefs in the Netherlands Antilles. *Bull. Mar. Sci.* 56:609–619.

Baker, A.C. 2001. Ecosystems: Reef corals bleach to survive change. *Nature* 411:765–766.

Baker, A.C., C.J. Starger, T.R. McClanahan, and P.W. Glynn. 2004. Corals adaptive response to climate change. *Nature* 430:741.

Bard, E., B. Hamelin, R.G. Fairbanks, and A. Zindler. 1990. Calibration of the ^{14}C timescale over the past 30,000 years using mass spectrometric U-Th ages from Barbados corals. *Nature* 345:405–410.

Bellwood, D.R., and T.P. Hughes. 2001. Regional-scale assembly rules and biodiversity of coral reefs. *Science* 292:1532–1534.

Benzie, J.A.H. 1999. Genetic structure of coral reef organisms: Ghosts of dispersal past. *Am. Zool.* 39:131–145.

Bird, J.B., A. Richards, and P.P. Wong. 1979. Coastal subsystems of western Barbados, West Indies. *Geogr. Ann. Ser. A Phys. Geogr.* 61:221–236.

Blake, S.G., and J.M. Pandolfi. 1997. Geology of selected islands of the Pitcairn Group, southern Polynesia. In *Geology and Hydrogeology of Carbonate Islands,* eds. H.L. Vacher and T. Quinn, 407–431. Developments in Sedimentology, Vol. 54.

Blanchon, P., and A. Eisenhauer. 2001. Multi-stage reef development on Barbados during the last interglaciation. *Quat. Sci. Rev.* 20:1093–1112.

Bray, J.R., and J.T. Curtis. 1957. An ordination of the upland forest communities of southern Wisconsin. *Ecol. Monogr.* 27:325–349.

Brown, J.H. 1995. *Macroecology*. Chicago: University of Chicago Press.

Caley, M.J., M.H. Carr, M.A. Hixon, T.P. Hughes, G.P. Jones, and B.A. Menge. 1996. Recruitment and the local dynamics of open marine populations. *Ann. Rev. Ecol. Syst.* 27:477–500.

Caley, M.J., and D. Schluter. 1997. The relationship between local and regional diversity. *Ecology* 78:70–80.

Carpenter, R.C. 1986. Partitioning herbivory and its effects on coral reef algal communities. *Ecol. Monogr.* 56:345–363.

Carpenter, R.C. 1997. Invertebrate predators and grazers. In *Life and Death of Coral Reefs,* ed. C. Birkeland, 198–229. New York: Chapman and Hall.

Chappell, J. 1974. Geology of coral terraces, Huon Peninsula, New Guinea: A study of Quaternary tectonic movements and sea-level changes. *Geol. Soc. Am. Bull.* 85:553–570.

Chappell, J., A. Omura, T. Esat, M. Mcculloch, J. Pandolfi, Y. Ota, and B. Pillans. 1996. Reconciliation of late Quaternary sea levels derived from coral terraces at Huon Peninsula with deep sea oxygen isotope records. *Earth Planet. Sci. Lett.* 141: 227–236.

Chappell, J., and H.H. Veeh. 1978. Late Quaternary tectonic movements and sea-level changes at Timor and Atauro Island. *Geol. Soc. Am. Bull.* 89:356–367.

Clark, J.S., and J.S. McLachlan. 2003. Stability of forest biodiversity. *Nature* 423:635–638.

Clarke, K.R. 1993. Non-parametric multivariate analyses of changes in community structure. *Aust. J. Ecol.* 18:117–143.

Clarke, K.R., and R.M. Warwick. 1994. Change in marine communities: An approach to statistical analysis and interpretation. National Environment Research Council, UK.

Connell, J.H. 1978. Diversity in tropical rain forests and coral reefs. *Science* 199:1302–1310.

Connell, J.H. 1997. Disturbance and recovery of coral assemblages. *Proc. Eighth Int. Coral Reef Symp., Panamá* 1:9–22.

Connell, J.H., T.P. Hughes, and C.C. Wallace. 1997. A 30-year study of coral abundance, recruitment, and disturbance at several scales in space and time. *Ecol. Monogr.* 67:461–488.

Connell, J.H., T.P. Hughes, C.C. Wallace, J.E. Tanner, K.E. Harms, and A.M. Kerr. 2004. A long-term study of competition and diversity of corals. *Ecol. Monogr.* 74:179–210.

Cornell, H.V. 1985a. Local and regional richness of cynipine gall wasps on California oaks. *Ecology* 66:1247–1260.

Cornell, H.V. 1985b. Species assemblages of cynipid gall wasps are not saturated. *Am. Nat.* 126:565–569.

Cornell, H.V., and R.H. Karlson. 1996. Species richness of reef-building corals determined by local and regional processes. *J. Anim. Ecol.* 65:233–241.

Cornell, H.V., and R.H. Karlson. 1997. Local and regional processes as controls of species richness. In *Spatial Ecology: The Roles of Space in Population Dynamics and Interspecific Interactions,* eds. D. Tilman and P. Kareiva, 250–268. Princeton: Princeton University Press.

Cornell, H.V., and J.H. Lawton. 1992. Species interactions, local and regional processes, and limits to the richness of ecological communities: A theoretical perspective. *J. Anim. Ecol.* 61:1–12.

Crame, J.A. 1986. Late Pleistocene molluscan assemblages from the coral reefs of the Kenya coast. *Coral Reefs* 4:183–196.

de Buisonjé, P.H. 1964. Marine terraces and sub-aeric sediments on the Netherlands Leeward Islands, Curaçao, Aruba and Bonaire as indications of Quaternary changes in sea level and climate. I and II. *Koninklijke Nedelandse Akademie van Wetenschappen Amsterdam, Series B* 67:60–79.

de Buisonjé, P.H. 1974. Neogene and Quaternary geology of Aruba, Curaçao and Bonaire (Netherlands Antilles). *Uitgaben Natuarwetenschap-pelijke Stud. Suriname Nede. Antillen (Utrecht)* 78.

De Haan, D., and J.S. Zaneveld. 1959. Some notes on tides in Annabaai harbour Curaçao, Netherlands Antilles. *Bull. Mar. Sci.* 9:224–236.

De Palm, J.P. (ed.). 1985. *Encyclopedie van de Nederlandse Antillen.* Zutphen: De Walburg Pers.

Díaz, J.M., J. Garzón-Ferreira, and S. Zea. 1992. Evaluación del estado actual del arrecife coralino de la Isla de San Andrés. Final Report Project, INVEMAR/CORPES, Santa Marta.

Dodge, R.E., R.G. Fairbanks, L.K. Benninger, and F. Maurrasse. 1983. Pleistocene sea levels from raised coral reefs of Haiti. *Science* 219:1423–1425.

Doebeli, M., and G.D. Ruxton. 1998. Stabilization through spatial pattern formation in metapopulations with long-range dispersal. *Proc. R. Soc. London Ser. B* 265:1325–1332.

Doherty, P.J. 1981. Coral reef fishes: Recruitment-limited assemblages? *Proc. Fourth Int. Coral Reef Symp., Manila* 2:465–470.

Doherty, P.J., and Williams, D.M. 1988. The replenishment of coral reef fish populations. *Oceanogr. Mar. Biol. Ann. Rev.* 32:487–551.

Dollo, W.C. 1990. Facies, fossil record, and age of Pleistocene reefs from the Red Sea (Saudi Arabia). *Facies* 22:1–46.

Done, T.J. 1982. Patterns in the distribution of coral communities across the central Great Barrier Reef. *Coral Reefs* 1:95–107.

Done, T.J. 1988. Simulation of recovery of pre-disturbance size structure in populations of *Porites* spp. damaged by the crown of thorns starfish *Acanthaster planci*. *Mar. Biol.* 100:51–61.

Done, T.J. 1992. Constancy and change in some Great Barrier Reef coral communities: 1980–1990. *Am. Zool.* 32:655–662.

Edinger, E.N., J.M. Pandolfi, and R. Kelley. 2001. Diversity and taxonomic composition of Holocene reefs and modern reef coral life and death assemblages in Madang Lagoon, Papua New Guinea. *Paleobiology* 27:669–694.

Emiliani, C. 1966. Paleotemperature analyses of the Caribbean cores P6304-8 and P6304-9 and a generalized temperature curve for the past 425,000 years. *J. Geol.* 74:109–126.

Fairbanks, R.G. 1989. A 17,000-year glacio-eustatic sea level record: Influence of glacial melting rates on the Younger Dryas event and deep ocean circulation. *Nature* 342:637–642.

Field, J.G., K.R. Clarke, and R.M. Warwick. 1982. A practical strategy for analysing multispecies distribution patterns. *Mar. Ecol. Prog. Ser.* 8:37–52.

Fleminger, A. 1985. The Pleistocene equatorial barrier between the Indian and Pacific Oceans and a likely cause for Wallace's Line. *UNESCO Tech. Pap. Mar. Sci.* 49:84–97.

Focke, J.W. 1978. Limestone cliff morphology and organism distribution on Curaçao (Netherlands Antilles). *Leidse Geol. Meded.* 51:131–150.

Gardner, T.A., I.M. Côté, J.A. Gill, A. Grant, and A.R. Watkinson. 2003. Long-term region-wide declines in Caribbean corals. *Science* 301:958–960.

Garzón-Ferreira, J., and M. Kielman. 1994. Extensive mortality of corals in the Colombian Caribbean during the last two decades. In *Proceedings of the Colloquium on Global Aspects of Coral Reefs: Health, Hazards and History, 1993*, ed. R.N. Ginsburg, 247–253. Miami: Rosenstiel School of Marine and Atmospheric Science, University of Miami.

Geister, J. 1973. Los arricifes de la Isla de San Andrés (Mar Caribe, Colombia). *Mittedungen Instituto Colombo-Alemán de Investigaciones Cientificas* 7:211–228.

Geister, J. 1975. Riffbau und geologische Entwicklungsgeschichte der Insel San Andrés (westliches Karibisches Meer, Kolumbien). *Stuttg. Beitr. Naturkd. Ser. B* 15:1–203.

Geister, J. 1977. The influence of wave exposure on the ecological zonation of Caribbean coral reefs. *Proc. Third Int. Coral Reef Symp., Miami* 1:23–29.

Geister, J. 1980. Calm-water reefs and rough-water reefs of the Caribbean Pleistocene. *Acta Palaeontol. Pol.* 25:541–556.

Geister, J. 1982. Pleistocene reef terrraces and coral environments at Santo Domingo and near Boca Chica, southern coast of the Dominican Republic. *Trans. Ninth Caribbean Geol. Conf.* 2:689–703.

Geister, J., and J. Diaz. 1997. A field guide to the oceanic barrier reefs and atolls of the southwestern Caribbean (archipelago of San Andres and Providencia, Colombia). *Proc. Eighth Int. Coral Reef Symp., Panamá* 1:235–262.

Ginsburg, R.N. (ed.). 1994. *Proceedings of the Colloquium on Global Aspects of Coral Reefs: Health, Hazards and History, 1993*. Miami: Rosenstiel School of Marine and Atmospheric Science, University of Miami.

Gladfelter, W.G. 1982. Whiteband disease in *Acropora palmata*: Implications for the structure and growth of shallow reefs. *Bull. Mar. Sci.* 32:639–643.

Goreau, T.F. 1959. The ecology of Jamaican coral reefs. I. Species composition and zonation. *Ecology* 40:67–90.

Goreau, T.F., and J. W. Wells. 1967. The shallow-water Scleractinia of Jamaica: Revised list of species and their vertical distribution range. *Bull. Mar. Sci.* 17:442–453.

Greenstein, B.J., and A. Curran. 1997. How much ecological information is preserved in fossil coral reefs and how reliable is it? *Proc. Eighth Int. Coral Reef Symp., Panama* 1:417–422.

Greenstein, B.J., H.A. Curran, and J.M. Pandolfi. 1998. Shifting ecological baselines and the demise of *Acropora cervicornis* in the western North Atlantic and Caribbean Province: A Pleistocene perspective. *Coral Reefs* 17:249–261.

Greenstein, B.J., and J.M. Pandolfi. 1997. Preservation of community structure in modern reef coral life and death assemblages of the Florida Keys: Implications for the Quaternary fossil record of coral reefs. *Bull. Mar. Sci.* 61:431–452.

Greenstein, B.J., and J.M. Pandolfi. 2004. A comparison of gradients in reef coral community composition between late Pleistocene and modern coral reefs of Western Australia. Geological Society of America Annual Meeting, Denver, November 7–10, 2004.

Greenstein, B.J., J.M. Pandolfi, and A. Curran. 1998. The completeness of the Pleistocene fossil record: Implications for stratigraphic adequacy. In *The Adequacy of the Fossil Record*, eds. S.K. Donovan and C.R.C. Paul, 75–109. London: John Wiley and Sons.

Grosberg, R.K., and D.R. Levitan. 1992. For adults only? Supply-side ecology and the history of larval biology. *Trends Ecol. Evol.* 7:130–133.

Guilderson, T.P., R.G. Fairbanks, and J.L. Rubenstone. 2001. Tropical Atlantic coral oxygen isotopes: Glacial-interglacial sea surface temperatures and climate change. *Mar. Geol.* 172:75–89.

Hanski, I. 1998. Metapopulation dynamics. *Nature* 396:41–49.

Hatcher, B.G. 1984. A maritime accident provides evidence for alternate stable states in benthic communties on coral reefs. *Coral Reefs* 3:199–204.

Hay, M.E. 1997. The ecology and evolution of seaweed–herbivore interactions on coral reefs. *Proc. Eighth Int. Coral Reef Symp.* 1:23–32.

Hay, M.E., J.R. Pawlik, J.E. Duffy, and W. Fenical. 1989. Seaweed–herbivore–predator interactions: Host-plant specialization reduces predation on small herbivores. *Oecologia* 81:418–427.

Herweijer, J.P., and J.W. Focke. 1978. Late Pleistocene depositional and denudational history of Aruba, Bonaire and Curaçao (Netherlands Antilles): *Geol. Mijnbouw* 57:177–187.

Hixon, M.A. 1997. Effects of reef fishes on corals and algae. In *Life and Death of Coral Reefs*, ed. C. Birkeland, 230–248. London: Chapman and Hall.

Hixon, M.A. 1998. Population dynamics of coral-reef fishes: Controversial concepts and hypotheses. *Aust. J. Ecol.* 23:192–201.

Hubbell, S.P. 1995. Towards a theory of biodiversity and biogeography on continuous landscapes. In *Preparing for Global Change: A Midwestern Perspective*, eds. G.R. Carmichael, G.E. Folk, and J.L. Schnoor, 173–201. Amsterdam: SPB Academic Publishing.

Hubbell, S.P. 1997. A unified theory of biogeography and relative species abundance and its application to tropical rain forests and coral reefs. *Proc. Eighth Int. Coral Reef Symp., Panama* 1:33–42.

Hubbell, S.P. 2001. *The Unified Neutral Theory of Biodiversity and Biogeography.* Monographs in Population Biology 32. Princeton, NJ: Princeton University Press.

Hughes, T.P. 1994. Catastrophes, phase shifts and large-scale degradation of a Caribbean coral reef. *Science* 265:1547–1551.

Hughes, T.P., A.H. Baird, E.A. Dinsdale, N. Moltschaniwskyj, M.S. Pratchett, J.E. Tanner, and B. Willis. 1999. Patterns of recruitment and abundance of corals along the Great Barrier Reef. *Nature* 397:59–63.

Hughes, T.P., and J.H. Connell. 1999. Multiple stresses on coral reefs. *Limnol. Oceanogr.* 44:932–940.

Hugueny, B., L.T. de Morais, S. Mérigoux, B. de Mérona, and D. Ponton. 1997. The relationship between local and regional species richness: Comparing biotas with different evolutionary histories. *Oikos* 80:583–587.

Huston, M.A. 1985. Patterns of species diversity on coral reefs. *Ann. Rev. Ecol. Syst.* 16:149–177.

Iturralde-Vinent, M. 1995. Sedimentary geology of western Cuba. Fieldguide, 1st SEPM Congress on Sedimentary Geology, St. Petersburg Beach, Florida, August 13–16, 1995.

Jackson, J.B.C. 1991. Adaptation and diversity of reef corals. *BioScience* 41:475–482.

Jackson, J.B.C. 1992. Pleistocene perspectives on coral reef community structure. *Am. Zool.* 32:719–731.

Jackson, J.B.C. 1997. Reefs since Colombus. *Coral Reefs* 16:S23–S32.

Jackson, J.B.C., A. Budd, and J.M. Pandolfi. 1996. The shifting balance of natural communities? In *Evolutionary Paleobiology: Essays in Honor of James W. Valentine*, eds. D. Jablonski, D.H. Erwin, and J.H. Lipps, 89–122. Chicago: University of Chicago Press.

Jackson, J.B.C., M.X. Kirby, W.H. Berger, K.A. Bjorndal, L.W. Botsford, B.J. Bourque, R. Bradbury, R. Cooke, J. Erlandson, J.A. Estes, T.P. Hughes, S. Kidwell, C.B. Lange, H.S. Lenihan, J.M. Pandolfi, C.H. Peterson, R.S. Steneck, M.J. Tegner, and R. Warner. 2001. Historical overfishing and the recent collapse of coastal ecosystems. *Science* 293:629–638.

James, N.P., C.W. Stearn, and R.S. Harrison. 1977. Field guidebook to modern and Pleistocene reef carbonates, Barbados, W.I. Third International Symposium on Coral Reefs. Miami: The Atlantic Reef Committee, University of Miami.

Jones, G.P., M.J. Milicich, M.J. Emslie, and C. Lunow. 1999. Self-recruitment in a coral reef fish population. *Nature* 402:802–804.

Karlson, R.H. 1999. *Dynamics of Coral Communities*. Population and Community Biology Series, Vol. 23. Dordrecht: Kluwer Academic Publishers.

Karlson, R.H., and H.V. Cornell. 1998. Scale-dependent variation in local vs. regional effects on coral species richness. *Ecol. Monogr.* 68:259–274.

Karlson, R.H., H.V. Cornell, and T.P. Hughes. 2004. Coral communities are regionally enriched along an oceanic biodiversity gradient. *Nature* 429:867–870.

Karlson, R.H., and L.E. Hurd. 1993. Disturbance, coral reef communities, and changing ecological paradigms. *Coral Reefs* 12:117–125.

Kendrik, G.W., K.-H. Wyrwoll, and B.J. Szabo. 1991. Pliocene-Pleistocene coastal events and history along the western margin of Australia. *Quat. Sci. Rev.* 10:419–439.

Knowlton, N. 1992. Thresholds and multiple stable states in coral reef community dynamics. *Am. Zool.* 32:674–682.

Kohn, A.J., and I. Arua. 1999. An early Pleistocene molluscan assemblage from Fiji: Gastropod faunal composition, paleoecology and biogeography. *Palaeogeogr. Palaeoclimatol. Palaeoecol.* 146:99–145.

Lang, J.C., and E.A. Chornesky. 1990. Competition between scleractinian reef corals—A review of mechanisms and effects. In *Coral Reefs: Ecosystems of the World 25*, ed. Z. Dubinsky, 209–252. Amsterdam: Elsevier.

Lessios, H.A. 1988. Mass mortality of *Diadema antillarum* in the Caribbean: What have we learned? *Ann. Rev. Ecol. Syst.* 19:371–393.

Lessios, H.A., D.R. Robertson, and J.D. Cubit. 1984. Spread of *Diadema* mass mortality through the Caribbean. *Science* 226:335–337.

Lewis, J.B. 1960a. The fauna of the rocky shores of Barbados, West Indies. *Can. J. Zool.* 38:391–435.

Lewis, J.B. 1960b. The coral reefs and coral communities of Barbados, West Indies. *Can. J. Zool.* 38:1133–1145.

Lewis, J.B. 1984. The *Acropora* inheritance: A reinterpretation of the development of fringing reefs in Barbados, West Indies. *Coral Reefs* 3:117–122.

Liddell, W.D., and S.L. Ohlhorst. 1988. Comparison of Western Atlantic coral reef communities. *Proc. Sixth Int. Coral Reef Symp., Australia* 3:281–286.

Mah, A.J., and C.W. Stearn. 1986. The effect of Hurricane Allen on the Bellairs fringing reef, Barbados. *Coral Reefs* 4:169–176.

Martindale, W. 1992. Calcified epibionts as palaeoecological tools: Examples from the Recent and Pleistocene reefs of Barbados. *Coral Reefs* 11:167–177.

Mesolella, K.J. 1967. Zonation of uplifted Pleistocene coral reefs on Barbados, West Indies. *Science* 156:638–640.

Mesolella, K.J. 1968. The uplifted reefs of Barbados: Physical stratigraphy, facies relationships and absolute chronology. Ph.D. dissertation, Brown University, Providence.

Mesolella, K.J., H.A. Sealy, and R.K. Matthews. 1970. Facies geometries within Pleistocene reefs of Barbados, West Indies. *AAPG Bull.* 54:1890–1917.

Miller, M.W., and M.E. Hay. 1998. Effects of fish predation and seaweed competition on the survival and growth of corals. *Oecologia* 113:231–238.

Moran, P.J. 1986. The *Acanthaster* phenomenon. *Oceanogr. Mar. Biol. Ann. Rev.* 24:379–480.

Mumby, P.J. 1999. Can Caribbean coral populations be modeled at metapopulation scales? *Mar. Ecol. Prog. Ser.* 180:275–288.

Murdoch, T.J.T., and R.B. Aronson. 1999. Scale-dependent spatial variability of coral assemblages along the Florida Reef Tract. *Coral Reefs* 18:341–351.

Nakamori, T., Y. Iryu, and T. Yamada. 1995. Development of coral reefs of the Ryukyu Islands (Southwest Japan, East China Sea) during Pleistocene sea-level change. In *West Pacific and Asian Carbonates*, eds. M. Tucker and R. Matsumoto. *Sediment. Geol.* 99:215–231.

Nakamori, T., E. Wallensky, and C. Campbell. 1994. Recent hermatypic coral assemblages at Huon Peninsula. In *Study on coral reef terraces of the Huon Peninsula, Papua New Guinea: Establishment of Quaternary sea level and tectonic history—A preliminary report on project 04041048 of the Monbusho International Research Program*, ed. Y. Ota, 111–116. Japan: Yokohama University.

Nee, S., and R.M. May. 1992. Dynamics of metapopulations: Habitat destruction and competitive coexistence. *J. Anim. Ecol.* 61:37–40.

Neumann, C.J., G.W. Cry, E.L. Caso, and B.R. Jarvinen. 1981. Tropical cyclones of the North Atlantic Ocean, 1871–1980. Ashville, NC: National Climatic Center.

Odum, H.T., and E.P. Odum. 1955. Trophic structure and productivity of a windward coral reef community at Eniwetok Atoll. *Ecol. Monogr.* 25:291–320.

Pandolfi, J.M. 1995. Geomorphology of the uplifted Pleistocene lagoon at Henderson Island, Pitcairn Island Group, South Central Pacific. *Biol. J. Linn. Soc.* 56: 63–77.

Pandolfi, J.M. 1996. Limited membership in Pleistocene reef coral assemblages from the Huon Peninsula, Papua New Guinea: Constancy during global change. *Paleobiology* 22:152–176.

Pandolfi, J.M. 1999. Response of Pleistocene coral reefs to environmental change over long temporal scales. *Am. Zool.* 39:113–130.

Pandolfi, J.M. 2000. Persistence in Caribbean coral communities over broad spatial and temporal scales. Presentation. *Ninth Int. Coral Reef Symp., Bali*, October 23–27, 2000.

Pandolfi, J.M. 2001. Taxonomic and numerical scales of analysis in paleoecological data sets: Examples from the Pleistocene of the Caribbean. *J. Paleontol.* 75:546–563.

Pandolfi, J.M. 2002. Coral reef ecology at multiple spatial and temporal scales. *Coral Reefs* 21:13–23.

Pandolfi, J.M., R.H. Bradbury, E. Sala, T.P. Hughes, K.A. Bjorndal, R.G. Cooke, D. Macardle, L. McClenahan, M.J.H. Newman, G. Paredes, R.R. Warner, and J.B.C. Jackson. 2003. Global trajectories of the long-term decline of coral reef ecosystems. *Science* 301:955–958.

Pandolfi, J.M., and J. Chappell. 1994. Stratigraphy and relative sea level changes at the Kanzarua and Bobongara sections, Huon Peninsula, Papua New Guinea. In *Study on coral reef terraces of the Huon Peninsula, Papua New Guinea: Establishment of Quaternary sea level and tectonic history—A preliminary report on project 04041048 of the Monbusho International Research Program*, ed. Y. Ota, 19–140. Japan: Yokohama University.

Pandolfi, J.M., and B.J. Greenstein. 1997. Preservation of community structure in death assemblages of deep water Caribbean reef corals. *Limnol. Oceanogr.* 42:1505–1516.

Pandolfi, J.M., and J.B.C. Jackson. 1997. The maintenance of diversity on coral reefs: Examples from the fossil record. *Proc. Eighth Int. Coral Reef Symp., Panama* 1:397–404.

Pandolfi, J.M., and J.B.C. Jackson. 2001. Community structure of Pleistocene coral reefs of Curaçao, Netherlands Antilles. *Ecol. Monogr.* 71:49–67.

Pandolfi, J.M., and J.B.C. Jackson. 2006. Ecological persistence interrupted in Caribbean coral reefs. *Ecol. Lett.* 9:818–826.

Pandolfi, J.M., J.B.C. Jackson, N. Baron, R.H. Bradbury, H.M. Guzman, T.P. Hughes, C.V. Kappel, F. Micheli, J.C. Ogden, H.P. Possingham, and E. Sala. 2005. Are US coral reefs on the slippery slope to slime? *Science* 307:1725–1726.

Pandolfi, J.M., J.B.C. Jackson, and J. Geister. 2001. Geologically sudden natural extinction of two widespread Late Pleistocene Caribbean reef corals. In *Process from Pattern in the Fossil Record*, eds. J.B.C. Jackson, S. Lidgard, and F.K. McKinney, 120–158. Chicago: University of Chicago Press.

Pandolfi, J.M., G. Llewellyn, and J.B.C. Jackson. 1999. Interpretation of ancient reef environments in paleoecological studies of community structure: Curaçao, Netherlands Antilles, Caribbean Sea. *Coral Reefs* 18:107–122.

Pandolfi, J.M., C.E. Lovelock, and A.F. Budd. 2002. Character release following extinction in a Caribbean reef coral species complex. *Evolution* 53:479–501.

Pandolfi, J.M., and P.R. Minchin. 1995. A comparison of taxonomic composition and diversity between reef coral life and dead assemblages in Madang Lagoon, Papua New Guinea. *Palaeogeogr. Palaeoclimatol. Palaeoecol.* 119:321–341.

Paulay, G. 1990. Effects of late Cenozoic sea-level fluctuations on the bivalve faunas of tropical oceanic islands. *Paleobiology* 16:415–434.

Paulay, G. 1991. Late Cenozoic sea level fluctuations and diversity and species composition of insular shallow water marine faunas. In *The Unity of Evolutionary Biology, Proceedings of the Fourth International Congress of Systematic and Evolutionary Biology, College Park, MD, 1990*, ed. E.C. Dudley, 184–193. Portland, OR: Dioscorides Press.

Paulay, G., and T. Spencer. 1988. Geomorphology, palaeoenvironments and faunal turnover, Henderson Island, S. E. Polynesia. *Proc. Sixth Int. Coral Reef Symp., Australia* 3:461–466.

Pauly, D., V. Christensen, J. Dalsgaard, R. Froese, and F. Torres, Jr. 1998. Fishing down marine food webs. *Science* 279:860–863.

Perry, C.T. 2001. Storm-induced coral rubble deposition: Pleistocene records of natural reef disturbance and community response. *Coral Reefs* 20:171–183.

Playford, P.E. 1983. Geological research on Rottnest Island. *J. R. Soc. West. Aust.* 66:10–15.

Porter, J.W., and O.W. Meier. 1992. Quantification of loss and change in Floridian reef coral populations. *Am. Zool.* 32:625–640.

Richardson, L.L. 1998. Coral diseases: What is really known? *Trends Ecol. Evol.* 13:438–443.

Richmond, R.H. 1987. Energetics, competency and long-distance dispersal of planula larvae of the coral *Pocillopora damicornis*. *Mar. Biol.* 93:527–533.

Ricklefs, R.E. 1987. Community diversity: Relative roles of local and regional processes. *Science* 235:167–171.

Robertson, D.R. 1996. Interspecific competition controls abundance and habitat use of territorial Caribbean damselfishes. *Ecology* 77:885–899.

Roughgarden, J., and Y. Iwasa. 1986. Dynamics of a metapopulation with space-limited subpopulations. *Theor. Pop. Biol.* 29:235–261.

Rowan, R. 2004. Thermal adaptation in reef coral symbionts. *Nature* 430:742.

Rowan, R., and N. Knowlton. 1995. Intraspecific diversity and ecological zonation in coral algal symbiosis. *Proc. Natl. Acad. Sci. USA* 92:2850–2853.

Rowan, R., N. Knowlton, A. Baker, and J. Jara. 1997. Landscape ecology of algal symbionts creates variation in episodes of coral bleaching. *Nature* 388:265–269.

Sale, P.F. 1977. Maintenance of high diversity in coral reef fish communities. *Am. Nat.* 111:337–359.

Sander, F. and D.M. Stevens. 1973. Organic productivity of inshore and offshore waters of Barbados: A study of the island mass effect. *Bull. Mar. Sci.* 23:771–792.

Schubert, C., and B.J. Szabo. 1978. Uranium-series ages of Pleistocene marine deposits on the islands of Curaçao and La Blanquilla, Caribbean Sea. *Geol. Mijnbouw* 57:325–332.

SeaWinds database, Remote Sensing Systems and the NASA Ocean Vector Winds Science Team. 2005. http://www.remss.com/qscat/qscat_browse.html

Sotka, E.E. and R.W. Thacker. 2005. Do some corals like it hot? *Trends Ecol. Evol.* 20:59–62.

Stearn, C.W., T.P. Scoffin, and W. Martindale. 1977. Calcium carbonate budget of a fringing reef on the west coast of Barbados. *Bull. Mar. Sci.* 27:479–510.

Steele, M.A. 1998. The relative importance of predation and competition in two reef fishes. *Oecologia* 115:222–232.

Stienstra, P. 1991. Sedimentary petrology, origin and mining history of the phosphate rocks of Klein Curaçao, Curaçao and Aruba, Netherlands West Indies. Publications Foundation for Scientific Research in the Caribbean Region No. 130.

Swearer, S.E., J.E. Caselle, D.W. Lea, and R.R. Warner. 1999. Larval retention and recruitment in an island population of a coral reef fish. *Nature* 402:799–802.

Tanner, J.E., T.P. Hughes, and J.H. Connell. 1994. Species coexistence, keystone species, and succession: A sensitivity analysis. *Ecology* 75:2204–2219.

Taylor, J.D. 1978. Faunal response to the instability of reef habitats: Pleistocene molluscan assemblages of Aldabra Atoll. *Palaeontology* 21:1–30.

Terborgh, J., R.B. Foster, and N.V. Percy. 1996. Tropical tree communities: A test of the nonequilibrium hypothesis. *Ecology* 77:561–567.

Tomascik, T.R., and F. Sander. 1987. Effects of eutrophication on reef-building corals. II. Structure of scleractinian coral communities on fringing reefs, Barbados, West Indies. *Mar. Biol.* 94:53–75.

Tomascik, T.R., R. van Woesik, and A.J. Mah. 1996. Rapid coral colonization of a recent lava flow following a volcanic eruption, Banda Islands, Indonesia. *Coral Reefs* 15: 169–175.

Ulanowicz, R.E. 1997. *Ecology, the Ascendent Perspective.* New York: Columbia University Press.

Van Duyl, F.C. 1985. Atlas of the living reefs of Curaçao and Bonaire (Netherlands Antilles). Foundation for Scientific Research in Surinam and the Netherlands Antilles 117, Utrecht.

Van Veghel, M.L.J. 1994. Polymorphism in the Caribbean reef building coral *Montastrea annularis.* Unpublished Ph.D. dissertation.

Wallace, C.C. 1999. *Staghorn Corals of the World: A Revision of the Genus Acropora.* CSIRO.

Wellington, G.M. 1982. Depth zonation in corals in the Gulf of Panamá: Control and facilitation by resident reef fishes. *Ecol. Monogr.* 52:223–241.

Westoby, M. 1998. The relationship between local and regional diversity: Comment. *Ecology* 79:1825–1827.

Wilson, J.R., and P.L. Harrison. 1998. Settlement-competency periods of larvae of three species of scleractinian corals. *Mar. Biol.* 131:339–345.

Woodley, J.D., 1992. The incidence of hurricanes on the north coast of Jamaica since 1870: Are the classic reef descriptions atypical? *Hydrobiologia* 247:133–138.

Woodley, J.D., E.A. Chornesky, P.A. Clifford, J.B.C. Jackson, L.S. Kaufman, N. Knowlton, J.C. Lang, M.P. Pearson, J.W. Porter, M.C. Rooney, K.W. Rylaarsdam, V.J. Tunnicliffe, C.W. Wahle, J.L. Wulff, A.S.G. Curtis, M.D. Dallmeyer, B.P. Jupp, M.A.R. Koehl, J. Neigel, and E.M. Sides. 1981. Hurricane Allen's impact on Jamaican coral reefs. *Science* 214:749–755.

Young, C.M. 1987. Novelty of "supply-side ecology". *Science* 235:415–416.

Young, C.M. 1990. Larval ecology of marine invertebrates: A sesquicentennial history. *Ophelia* 32:1–48.

Zea, S., J. Geister, J. Garzon-Ferreira, and J.M. Diaz. 1998. Biotic changes in the reef complex of San Andres Island (southwestern Caribbean Sea, Colombia) occurring over nearly three decades. *Atoll Res. Bull.* 456:1–30.

9. Ecological Shifts along the Florida Reef Tract: The Past as a Key to the Future

William F. Precht and Steven L. Miller

> As it is not in human record, but in natural history, that we are to look for the means of ascertaining what has already been . . . in order to be informed of operations which have been transacted in times past . . . or to events which are in time to happen.
>
> —James Hutton (1785)

9.1 Introduction

Does the Quaternary fossil record of Caribbean coral reefs provide an adequate baseline from which a variety of pressing issues and challenges facing modern reefs can be addressed? For more than a decade it has been recognized that the sedimentary and fossil record of Quaternary coral reefs has the potential to help decipher the role of history in the study of living reefs (Jackson 1991, 1992; Jackson, Budd, and Pandolfi 1996). Pandolfi and Jackson (Chapter 8) demonstrate that the predictable patterns of community membership and dominance of acroporid species in the Caribbean throughout the Pleistocene epoch allow for a baseline of pristine coral community composition before human exploitation. Thus, the Pleistocene fossil record of Caribbean reefs provides a clear frame of reference as to what to manage and conserve, and why. Or does it?

9.2 The Problem

The ecology of Caribbean and western Atlantic coral reefs has changed dramatically in recent decades and it is now believed that these reefs are in crisis (Rogers 1985; Wilkinson 1993; Ginsburg 1994; Brown 1997; Connell 1997; Eakin et al. 1997; Aronson and Precht 2001a; Gardner et al. 2003). This is especially true along the Florida reef tract (Shinn 1989; Ward 1990; Lidz 1997). But what defines the baseline from which we draw conclusions about change? Beginning in the 1950s with the work of the late Thomas F. Goreau and colleagues, it was recognized that typical Caribbean and western Atlantic reefs displayed a generalized zonation pattern with three common species of scleractinian corals as the primary builders of reef framework (Goreau 1959; Glynn 1973; Goreau and Goreau 1973; Kinzie 1973; Ginsburg and James 1974; Bak 1983; Jaap 1984; Hubbard 1988; Graus and Macintyre 1989; Aronson and Precht 2001b). These foundation species include *Acropora palmata*, *A. cervicornis*, and the *Montastraea annularis* species complex. The thickly branching elkhorn coral, *A. palmata*, was dominant at the reef crest and in the shallowest depths of the fore reef (0–5 m depth; Fig. 9.1A). The more thinly branching staghorn coral, *A. cervicornis*, was dominant at intermediate depths (~5–25 m) on exposed reefs, and it ranged into shallower habitats on more protected reefs (Adey and Burke 1977; Geister 1977; Hubbard 1988; Fig. 9.1B). The massive corals of the *M. annularis* species complex (Knowlton et al. 1992, 1997) were (and remain) common in a variety of reef habitats from <5 to >30 m. These *Montastraea* sibling species commonly exhibit intraspecific changes in morphology along depth gradients revealing broad zonational overlap (Fig. 9.1C). In addition, the branching coral *A. prolifera* was sometimes found at the interface between the *A. palmata* and *A. cervicornis* zones (Fig. 9.1D). The growth and form of this coral is intermediate between the other two Caribbean acroporids and it is most probably a hybrid (Vollmer and Palumbi 2002).

Similarly, *Acropora*-dominated zonation has been found in Pleistocene and Holocene fossil and subfossil reef deposits throughout the region. These patterns attest to the importance of *Acropora* spp. in time and space for the Caribbean (Mesolella 1967; Macintyre and Glynn 1976; James, Stern, and Harrison 1977; Geister 1983; Macintyre 1988; Precht and Hoyt 1991; Jackson 1992; Stemann and Johnson 1992; Hubbard, Gladfelter, and Bythell 1994; Greenstein, Harris, and Curran 1998; Aronson and Precht 2001b). However, starting in the late 1970s, and by the early 1980s, the coral zonation pattern dominated by the *Acropora* spp. had essentially disappeared on many, if not most, Caribbean reefs (Jackson 1991, 1992; Aronson and Precht 2001b,c). Disturbances of various types have been invoked to explain the changing face of Caribbean reefs over the last 25 years with coral mortality, especially mortality of the *Acropora* spp., being a major driving force in the transition (references in Aronson and Precht 2001b). At the same time, herbivorous fishes have been reduced on some Caribbean reefs by human exploitation, and the herbivorous echinoid *Diadema antillarum* experienced >95% mortality from disease throughout the region in 1983 to 1984 (Hay 1984; Lessios 1988). Coral mortality has in general been followed by the proliferation of fleshy and filamentous (noncoralline) macroalgae, because the reduced populations of these herbivores have not been able to keep pace with algal growth in the vast areas of space opened by the death of corals (Steneck 1994; Aronson and Precht 2000, 2001b; McCook, Wolanski, and Spagnol 2001; Williams and Polunin, 2001; Williams, Polunin, and Hendrick 2001). A question of great significance to scientists, managers, and policymakers is whether the recent changes are something new or part of a long-term pattern of repeated community shifts.

Figure 9.1. (A) Underwater photograph of typical in situ *A. palmata* forming reef flat prior to the mortality events of the past two decades. Upper horizontal surface of live coral is controlled by growth up to level of spring low tide. **(B)** Monospecific thicket of the branching coral *A. cervicornis* at intermediate depths on the fore reef.

(C)

(D)

Figure 9.1. (*continued*) (**C**) Three common sibling species of the *Montastraea annularis* complex showing species/habitat overlap including *M. franksi* (lower left), *M. faveolata* (upper left), and *M. annularis* (right). (**D**) Photograph of *A. prolifera* at the zonational interface between *A. palmata* and *A. cervicornis*.

There are a number of factors responsible for *Acropora* mortality, with white-band disease, temperature stress, predation, and hurricanes being some of the most significant at reducing populations both locally and regionally. Corals of the *M. annularis* species complex have also declined on some reefs throughout the region. Their mortality, however, was caused by factors different from those for the *Acropora* species (see Aronson and Precht 2001b). In Florida, it is well documented that *Acropora*-dominated communities are dynamic at the scale of individual reefs and they have been extremely volatile over the past century (Mayer 1902; Shinn 1976; Enos 1977; Davis 1982; Shinn et al. 1989; Jaap and Sargent 1994; Jaap 1998). At the scale of the entire reef tract, deteriorating conditions throughout the late Holocene led to contraction of acroporid reefs and the reef system as a whole. The current regional decline of acroporids, however, appears to be unprecedented in at least the last few thousand years (Aronson and Precht 1997; Aronson et al. 2002a) and possibly longer (Greenstein, Pandolfi, and Curran 1998; Pandolfi and Jackson 2001; but see Rogers et al. 2002; Shinn et al. 2003).

Assessing the novelty of the recent demise of acroporid corals requires a multidisciplinary approach that addresses multiple scales of space and time (Aronson 2001). Because coral reefs are both geologic and biologic entities, it should be possible to observe the effects of various disturbances in ecological time, detect historical changes in the paleoecological record, and deduce the multiscale processes behind those patterns. By comparing and contrasting the structure, anatomy, and biofacies patterns in reef-building episodes of the Pleistocene and Holocene with the living, modern coral reef community, we focused on the question, "How do past ecological responses of coral reef systems, recorded by the fossil record, aid in predicting future response to global change?"

Jackson (1992), Hunter and Jones (1996), Pandolfi (1999, 2002), Pandolfi and Jackson (1997, 2001), and Aronson and Precht (2001b) have shown that, almost without exception, Caribbean Pleistocene fossil-reef sections exhibit species composition and zonation similar to modern reefs at the same location (at least prior to the 1980s). Thus, Pleistocene reef-coral communities within the same environment are more distinct between reefs of the same age from different places than between reefs formed at different times at the same location. Jackson (1992), citing examples from Barbados and other Caribbean locations, concluded that acroporids dominated more or less continuously during the high sea-level stands for the past few hundred thousand years. This concordance of species composition shows that the present clearly is a key to the past. Pandolfi (2001, 2002) suggested that the Pleistocene data point to a high degree of order and predictability in Caribbean coral communities over broad spatial and temporal scales. An important result of these studies is that the recent regional demise of the acroporids in the Caribbean and western Atlantic may be without geological precedent and that present trends are not predicted from history, implicating humans as the vector for this change (Stokstad 2001).

There is one major exception to this general pattern, and that is between reefs of the last major interglacial and their Holocene to recent counterparts along the Florida reef tract. In the Pleistocene Key Largo Limestone (Marine Isotope Stage 5e; ~125 ka), *Acropora palmata* is absent and there is a paucity of *A. cervicornis*.

The Holocene architecture of reefs in the Florida Keys, however, is comprised primarily acroporids (Shinn 1963, 1980, 1984, 1988, 1989; Kissling 1977a; Lighty 1977; Shinn et al. 1981, 1989; Macintyre 1988; Lidz et al. 1997a; Precht et al. 2000). We contend that this temporal difference provides an important clue to understanding the present community configuration along the Florida reef tract. This difference through time also provides essential information to help predict future changes to Florida's reefs in response to environmental perturbations and global change, both natural and anthropogenic.

9.3 Major Causes of *Acropora* Mortality: The Roles of Temperature, Inimical Water, Disease, and Hurricanes

9.3.1 Temperature and Inimical Water

Vaughan (1914) recognized that coral reefs in the Florida Keys are at the northern latitudinal limit of extensive reef growth along the Americas (that is, reefs that form high-relief, three-dimensional, complex features constructed by scleractinian corals and coralline algae). Although zooxanthellate corals occur at higher latitudes on both coasts of Florida, only in the Florida Keys do corals construct nearly emergent reef systems (Jaap and Hallock 1990; Fig. 9.2). Temperature has

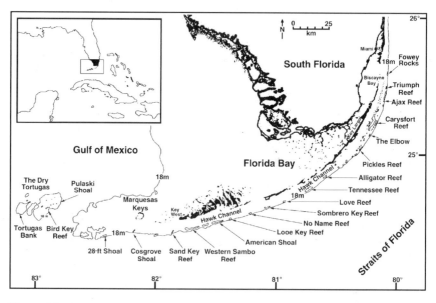

Figure 9.2. Index map of the Florida reef tract from the Dry Tortugas to Fowey Rocks. The named reefs on the map are the discontinuous, nearly emergent reefs that lie seaward of Pleistocene-age islands. Note the large tidal passes between Florida Bay and the open Atlantic in the middle Keys and the paucity of reefs in this area. Modified from Murdoch and Aronson (1999).

long been considered the main control on reef distribution (Dana 1843; Vaughan 1918, 1919a) with the optimum temperature for coral growth around 26 to 27 °C (Bosscher 1992). Cold-temperature tolerances are not well defined for corals but early experiments documented 16 °C as stressful to most corals and exposure to temperatures below 15 °C can result in their mortality (Mayer 1914, 1915). The present-day global distribution of coral reefs generally coincides with the 18 °C monthly minimum seawater isotherm (Kleypas, McManus, and Menez 1999; Kleypas, Buddemeier, and Gattuso 2001). South Florida lies between the 18 and 20 °C isotherm, and shallow coastal waters are especially susceptible to cooling by the passage of cold fronts (Jones 1977; Burns 1985; Walker, Rouse, and Huh 1987). On average, 30 to 40 cold fronts are recorded in south Florida every winter season (Warzeski 1977). Using patterns of generic scleractinian coral diversity, Porter and Tougas (2001) documented rapid faunal diminution northward along the east coast of Florida due primarily to cold-temperature limitations. Historically, Fowey Rocks was the northern extent of true modern reef growth and coral assemblages dominated by the acroporids (Vaughan 1914; Fig. 9.2). Well-developed assemblages of reef-building head corals and octocorals have been described for Broward and Palm Beach Counties (Raymond 1972; Goldberg 1973), and individual colonies of reef-building species have been found as far north as North Carolina (Macintyre and Pilkey 1969). Recently, seven areally limited areas comprising thickets of *A. cervicornis* were discovered in the waters off Broward County, ~50 km north of Fowey Rocks (Vargas-Ángel, Thomas, and Hoke 2003; Precht unpublished data). Although there has been much discussion as to the importance of these isolated coppices (Precht and Aronson 2004), being at the northern edge of acroporid survival and growth make them extremely vulnerable to natural as well as human disturbances.

Ginsburg and Shinn (1964) noted that the major reefs of south Florida occur seaward of islands of Pleistocene limestone. They speculated that the scarcity of thriving reefs opposite the large tidal passes of the Florida Keys is due to reduced water quality moving through the passes (see also Ginsburg and Shinn 1994). These waters are inimical to coral growth and hence to reef development. Newell et al. (1959) proposed a similar model for the depauperate nature of reefs adjacent to the Bahama Banks. Reefs in these areas are said to have been "shot in the back by their own lagoons" (Neumann and Macintyre 1985; Macintyre, Chapter 7).

Follow-up studies and natural abiotic disturbances have supported the inimical waters hypothesis (Marszalek et al. 1977; Shinn et al. 1989; Ginsburg and Shinn 1994; Shinn, Lidz, and Harris 1994; Smith 1994; Ginsburg, Gischler, and Kiene 2001; Cook et al. 2002; Lee et al. 2002; Smith and Pitts 2002). Specifically, Roberts et al. (1982) and Roberts, Wilson, and Lugo-Fernandez (1992) used satellite infrared spectral data and in situ water temperature measurements to identify plumes of cold (<16 °C), sediment-laden water flowing through the tidal passes for eight consecutive days in the winter of 1977, associated with the passage of three successive and unusually severe cold fronts (Fig. 9.3). The lowest measured water temperature in the middle Florida Keys during this period was 9 °C (Hudson 1981a). These prolonged cold temperatures proved lethal to hundreds

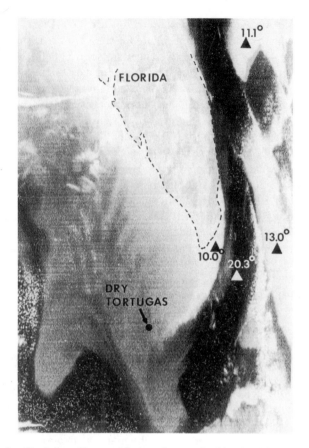

Figure 9.3. Satellite infrared image showing plumes of cold and turbid water flowing onto the reef tract in the lower Keys following the passage of numerous successive cold fronts in January 1977. Temperatures of inshore waters were ~10 °C cooler than waters in the Gulf Stream. Vast areas of acroporid-dominated reefs were killed during this event. Image courtesy of Harry Roberts.

of square kilometers of *A. palmata* and *A. cervicornis* (Davis 1982; Porter, Battey, and Smith 1982; Walker et al. 1982; Roberts, Rouse, and Walker 1983; Shinn 1989). In retrospect, this was not surprising, as coral translocation experiments moving *A. cervicornis* to inshore positions showed that the coral was killed outright when the water temperature fell to 13.3 °C (Shinn 1966, 1989). It has also been shown that *A. palmata* is equally sensitive to cold temperatures (Antonius 1977; Macintyre 1988; Shinn et al. 1989; Jaap and Sargent 1994).

In the Dry Tortugas, cold fronts have dramatically reduced acroporid populations at least twice in the last century (Jaap and Hallock 1990; Jaap and Sargent 1994). During the cold-water event of 1977 described above, the Dry Tortugas lost 96% of corals surveyed (Porter, Battey, and Smith 1982). Shinn (1984) specifically attributed the late Holocene reduction of living coral in the middle and lower Keys, especially *A. palmata* and *A. cervicornis*, to frequent exposure to cold water from the Gulf of Mexico during winter months. In the northern Keys, Burns (1985) described conditions east of Biscayne Bay as generally suboptimal for coral growth, due to cold-water flow from tidal passes out of the bay, with shallower-water sites affected more than deeper reefs. Cold fronts cause the loss of sensible and latent heat of the shallow bank and bay waters, resulting in advection and off-bank transport of superchilled water masses (Walker et al. 1982; Roberts, Rouse, and Walker 1983). Murdoch and Aronson (1999) showed that reefs adjacent to large tidal passes exhibited particularly low coral cover (e.g., Sombrero Key Reef), whereas reefs opposite large islands (e.g., American Shoal and Alligator Reef) had higher coral cover. However, all reefs in their study showed significantly diminished coral populations, especially the acroporids, when compared to historical observations. Lighty, Macintyre, and Neumann (1980) inferred that similar stress conditions introduced by the off-bank transport of turbid and episodically cooled waters led to the demise of the *Acropora*-dominated Abaco reef system in the Bahamas between 3000 and 4000 years ago. Following the mortality of the acroporids, the shallow Abaco reef system converted to an alternative community state dominated by macroalgae that persists to this day (Lighty 1981; Conrad Neumann personal communication).

The response of coral reefs to maximum temperature is less clear, but elevated water temperatures can be equally detrimental to corals (Mayer 1918). Bleaching, the loss of algal symbionts is a response to a number of potential stresses (Williams and Bunkley-Williams 1990). These stresses vary regionally and seasonally, and they may act singly or synergistically to cause corals to bleach (Fitt et al. 2001). The most obvious is temperature-induced stress. Corals are typically exposed during local summertime to temperatures near the upper limits of their thermal tolerances (Jokiel and Coles 1990; Glynn 1993; Hoegh-Guldberg 1999); therefore, coral reefs are often considered to be the ecosystems most threatened by global warming (Glynn 1996; Hoegh-Guldberg 1999; Kleypas et al. 1999; Walther et al. 2002). Field and laboratory studies have shown unequivocally that sustained, anomalously high summer water temperatures are associated with coral reef bleaching (Glynn and D'Croz 1990; Podestá and Glynn 1997, 2001) and, if temperatures are elevated above the average maximum for a prolonged period, many species die (Mayer 1918; Glynn 1983; Aronson et al. 2002b). Coral bleaching in response to anomalously high summer-season temperatures has become more frequent since the early 1980s (Glynn 1991, 1993; Hoegh-Guldberg 1999; Aronson et al. 2000; Lough 2000; Williams and Williams 2000; Kleypas, Buddemeier, and Gattuso 2001; Wellington et al. 2001) with the earliest known bleaching event in Florida having occurred in 1973 (Jaap 1979). Causey (2001) documented at least seven bleaching episodes along the Florida reef tract since then. The increased and widespread nature of these coral bleaching events over the past two decades is convincingly correlated with increases in maximum sea surface temperature (Kleypas, Buddemeier, and Gattuso 2001). Bleaching is not always fatal, and some bleaching episodes on reefs in Florida have been followed by recovery of most of the affected coral colonies (Jaap 1985; Porter et al. 1989;

Williams and Bunkley-Williams 1990; Lang et al. 1992; Fitt et al. 1993). Because
the *Acropora* species live in relatively shallow water, it is thought that they may
be tolerant of, or acclimatized to, warm summertime temperatures. Shinn (1966)
recognized that the growth for *A. cervicornis* was greatest when temperatures
ranged between 28 and 30 °C. However, Shinn (1966) described bleaching in
A. cervicornis when transplanted in water temperatures above 31 °C. Bleaching-
related mortality of *Acropora* has been observed in Costa Rica (Cortés et al.
1984), Belize (Kramer and Kramer 2000), and the Florida Keys and the Bahamas
(Causey et al. 1998; Miller et al. 2002; Precht unpublished data from the sum-
mer/fall of 1998).

The inimical-waters explanation of Ginsburg and Shinn (1964) may also be
rooted in pulses of high-temperature water flowing through the tidal passes (Porter,
Lewis, and Porter 1999). Shallow, lagoonal waters equilibrate rapidly with the
atmosphere relative to the more deeply mixed oceanic waters that front the reef
(Neumann and Macintyre 1985). Therefore, it is likely that at peak summertime
temperatures waters in excess of 32 °C are commonly discharged from the Gulf
of Mexico, Florida Bay, and Biscayne Bay onto the reefs of the Florida Keys
(Ginsburg and Shinn 1994; Ogden et al. 1994). Global warming will probably
exacerbate the situation (Hoegh-Guldberg 1999; Walther et al. 2002).

9.3.2 White-Band Disease

Although the effects of temperature are important at the scale of reef systems, it
is becoming apparent that for more than two decades white-band disease (WBD)
epizootics have been the primary cause of *Acropora* mortality over wide areas of
the Caribbean and western Atlantic (Bythell and Sheppard 1993; Aronson and
Precht 2001b,c), including Florida (Shinn 1989; Precht and Aronson 1997), with
losses in excess of 95% regionally (Precht et al. 2002). Robinson (1973), report-
ing on the reef condition of Buck Island National Monument in the U.S. Virgin
Islands, was the first to discuss mortality related to WBD. Subsequently, Gladfelter
(1982) recognized the devastating effects of WBD on coral reef community structure. Wells and Hanna (1992) noted anecdotally that acroporids from the Florida
reef tract had up to 96% of reef cover in places in 1981. By 1986 these corals had
succumbed to disease and were reduced to only 3% of the total reef cover. Jaap,
Halas, and Muller (1988) measured a 96% decline in *A. cervicornis* at Molasses
Reef in the Florida Keys National Marine Sanctuary (FKNMS) during this same
period. Gleason (1984) reported that by 1982 the acroporids had been extermi-
nated on a large portion of the Florida reef tract, supporting a case for their complete protection (Antonius 1994a,b). Quantitative evidence reveals that the
coverage of remnant *A. palmata* declined by 93% and *A. cervicornis* by 98%
between 1983 and 2000 at Looe Key (Miller et al. 2002). Photographic evidence
from Shinn shows that mass mortality of *A. cervicornis* at Grecian Rocks in
the upper FKNMS occurred during 1978 to 1979 with complete loss by 1983
(in Miller 2002). Before-and-after photographs in Ward (1990) also emphasize
the dramatic and devastating effects of coral death related to WBD in the

Florida Keys. These various reports suggest that the 1960s and early 1970s may represent a baseline for the status of "healthy" acroporid populations in Florida and the Caribbean (Kramer 2002).

Even after 20 years, the etiology of the WBD epizootic remains unknown, and recent reports reveal that there may be multiple varieties of the disease (or white syndromes) with differing characteristics and pathologies (Antonius 1981; Gladfelter 1982; Peters, Yevich, and Oprandy 1983; Peters and McCarty 1996; Peters 1997; Santavy and Peters 1997; Richardson 1998; Ritchie and Smith 1998; Richardson and Aronson 2002). WBD can generally be recognized as areas of bare skeleton, sometimes bordered by narrow bands of disintegrating, necrotic coral tissue, on otherwise healthy-looking, golden-brown *Acropora* branches. The disease spreads rapidly along the branches, usually from base to tip. The branches initially turn white and are quickly covered with a microalgal turf. The rapidly killed colonies are often left standing in growth position; in time they are broken and reduced to coral rubble by both physical and biological processes. Dead stands of *Acropora* are especially susceptible to breakage and transport during major storms (Hubbard et al. 1991). The resulting low-relief fields of *Acropora* rubble are then overgrown by replacement species, usually macroalgae (McClanahan et al. 1999). There is no association of WBD outbreaks with proximity to human influences; reefs both near and far from human population centers have been affected (Aronson and Precht 2001c). Of course, it could be argued that the Caribbean is so small that the entire region lies in close proximity to sources of anthropogenic stress (Hallock, Muller-Karger, and Halas 1993; Connell 1997; Roberts 1997; Miller and Crosby 1999; Jackson 2001; Andréfouët et al. 2002). Based on the results of coring Holocene reefs in Belize, Aronson and Precht (1997; see also Aronson et al. 2002a; Aronson and Ellner, Chapter 3) have shown that the regional mass mortality of the acroporids due to WBD was a novel event in at least the last 3 ky.

9.3.3 Hurricanes

As previously mentioned, local populations of *Acropora palmata* and *A. cervicornis* can be highly variable on a time scale of decades to centuries (Shinn et al. 1989; Jaap 1998). Both species experienced significant mortality from hurricanes and other severe storms in the past as they have recently (Glynn, Almodóvar, and González 1964; Ball, Shinn, and Stockman 1967; Hubbard 1988; Blanchon, Jones, and Kalbfleisch 1997). In fact, these studies and others have led to the opinion that hurricanes are a primary cause of present and past coral mortality, diversity, and distribution on Caribbean reefs (Connell 1978; Porter et al. 1981; Rogers 1993a; Aronson and Precht 1995; Blanchon 1997; and many others). Some areas, such as Costa Rica and Panama, receive virtually no hurricanes while others, including the Florida reef tract, suffer from regular hurricane damage (Ball, Shinn, and Stockman 1967; Perkins and Enos 1968; Neumann et al. 1993; Treml, Colgan, and Keevican 1997). Although hurricanes have been important at some localities (e.g., Stoddart 1963; Woodley et al. 1981; Graus, Macintyre, and

Herchenroder 1984; Hubbard et al. 1991; Lugo, Rogers, and Nixon 2000; Gardner et al. 2005), they do not explain recent patterns of coral mortality in much of the Caribbean region, including Florida.

Historical records of hurricanes in south Florida reliably date to 1871, with several trends of increased and decreased frequency apparent (Gentry 1984; Neumann et al. 1993). A large number of hurricanes struck the Florida Keys between 1910 and 1948, decreasing significantly over the next 50 years. In 1960 and 1965, Hurricanes Donna and Betsy struck the upper Keys but caused substantial damage to only a few reefs (Ball, Shinn, and Stockman 1967; Perkins and Enos 1968). Recovery was rapid after Hurricanes Donna and Betsy, with little evidence of storm effects present after one year and total recovery within five and three years, respectively (Shinn 1976). Two storms, Hurricane Andrew in 1992 and the March Storm of the century in 1993, caused different patterns of damage within the same reef system, suggesting that disturbance history is an important factor determining reef condition and resilience (Precht et al. 1993; Lirman and Fong 1997). Also, after Hurricane Andrew, 19 hard-bottom sites north of Miami revealed no pattern of damage relative to location, orientation, or depth (Blair, McIntosh, and Mostkoff 1994). In 1998, Hurricane Georges affected reefs of the lower and middle Keys. On the reef at Looe Key, the resulting damage included the near-elimination of remnant stands of *Acropora* spp. (Porter and Tougas 2001; Miller, Bourque, and Bohnsack 2002; WFP and SLM personal observations). Miller et al. (2002) commented that even in the upper Keys there was some mortality to remnant stands of *A. palmata* from this storm. Results from 12 reefs sampled before and six weeks after Hurricane Georges revealed that changes in the percent cover of hard corals varied among sites (Miller and Swanson 1999).

The effects of hurricanes in the Florida Keys over the last 50 years can be considered minimally important at the regional scale, but probably important at the scale of individual reefs (see Bythell, Hillis-Starr, and Rogers 2000). Significantly, the recent period of *Acropora* decline in Florida (the late 1970s to early 1980s) was coincident with an extended period of no major hurricanes. In addition, under all but the most extreme conditions, hurricanes can be favorable to the propagation and expansion of acroporid-dominated communities by asexual reproduction of storm-induced fragments (Shinn 1966, 1976; Gilmore and Hall 1976; Highsmith, Riggs, and D'Antonio 1980; Tunnicliffe 1981; Highsmith 1982; Fong and Lirman 1995; Lirman and Fong 1997; Lirman 2000; Precht et al. 2005a). It appears that, on a regional basis at least, corals suffer greater damage from chronic disturbances and stresses such as disease outbreaks (Shinn 1989; Bythell, Gladfelter, and Bythell 1993; Rogers 1993b; Bythell, Hillis-Starr, and Rogers 2000; Aronson and Precht 2001b,c; Precht et al. 2002).

9.3.4 Other Causes of *Acropora* Mortality

In Florida, Enos (1977) noted that on reefs close to Key West, spatial variability in reef vitality was random between reefs under similar physiographic and hydrodynamic settings. Specifically, he detailed a flourishing, healthy *A. palmata* reef assemblage at Sand Key reef, while at Rock Key about 2 km away the reef comprised totally dead in situ stands of *A. palmata*. What was responsible for the complete extirpation of *A. palmata* at Rock Key? In hindsight a disease epizootic or pest outbreak seems a likely cause. Enos (1977), commenting on these enigmatic observations made in the 1960s and early 1970s, stated, "The many dead reefs along the Florida shelf edges may attest to similar mass mortality rather than to creation of conditions continually unfavorable for reef growth." Unfortunately, we have neither premortality baselines for these reefs, nor before-during-and-after monitoring, limiting our ability to assign causality. At local scales, additional causes of acroporid mortality include cultural eutrophication (Weiss and Goddard 1977; Tomascik and Sander 1987; Bell and Tomascik 1994), predation by corallivores (Bak and van Eys 1975; Antonius 1977; Kaufman 1977; Tunnicliffe 1983; Knowlton, Lang, and Keller 1988, 1990; R. Bruckner, Bruckner, and Williams 1997; Miller 2001; Baums, Miller, and Szmant 2003), and sedimentation (Rogers 1983, 1990; Cortés 1994; Peters and McCarty 1996).

Jackson et al. (2001) and Jackson (2001) indicated that overgrowth by macroalgae following the mass mortality of *Diadema antillarum* was the main process responsible for the sudden and catastrophic mortality of the Caribbean corals during the 1980s. Based upon the timeline of disturbances in the Caribbean, however, this sequence of events is not possible. The rapid decline of *Acropora* in the late 1970s and early 1980s predated the mass die-off of the herbivorous urchin *D. antillarum* in 1983 to 1984, which in turn triggered dramatic increases in the abundance of the macroalgae on which they graze. While some mortality of head corals can be attributed to overgrowth by macroalgae, it is our contention that coral mortality itself was the crucial precursor to macroalgal dominance (discussed in Aronson and Precht 2001b; see also Lirman 2001). This is especially so for rapidly growing species like the *Acropora*, which can outcompete algae for space and light (Jackson 1991).

Understanding the causes of recent coral mortality as well as the resulting patterns of community variability along the Florida reef tract allows us to develop a better understanding of the processes and products preserved in their Quaternary fossil counterparts. Using analogy-based models, we will shuttle back and forth between the ancient and the modern to develop a picture of the changing face of Floridian reefs through time and space. Our comparative examples will, in turn, aid in predicting the future of Florida's reefs in an era of rapid ecological change and the impact of man.

9.4 The Quaternary History of Reef Building in the Florida Keys

9.4.1 The Pleistocene Reef System

The Key Largo Limestone was originally named by Sanford (1909). It crops out in an arcuate pattern, forming the present-day islands of the upper and middle Florida Keys. The exposure extends in a discontinuous fashion from Soldier Key in the northeast for approximately 170 km to the southwest, terminating at the Newfound Harbor Keys. The entire Pleistocene section is of variable composition and consists of at least five punctuated stratigraphic units (parasequences) separated by unconformity surfaces (Perkins 1977; Multer et al. 2002a). These units record the major Pleistocene high-stands of sea level on the south Florida Peninsula. Reef development in the earliest Pleistocene (Q1–Q2 units of Perkins) was restricted to the tops of a few bathymetric highs (Multer et al. 2002a). Reefs developed in earnest during MIS 9 (Q3 unit of Perkins) forming a bank-barrier reef system dominated by fused massive corals (Multer et al. 2002a). These Q3 bank-barrier reefs formed at the shelf break on a broad platform and provided the templates for late Pleistocene reef growth (Q4–Q5 units of Perkins). (Age determinations for the Q1–Q4 units and assignments of Marine Isotope Stages are based on amino acid diagenesis of fossil *Mercenaria* bivalves (Mitterer 1974).)

The unit that is the focus of this chapter is confined to the uppermost exposed portions of the Key Largo Limestone and was originally described by Agassiz (1896). These exposed sections are equivalent to the Q5 unit of Perkins (1977). The reefs were formed during the last major interglacial high sea-level stand (MIS 5e) and are equivalent to and transitional with the oolitic facies of the Miami Limestone (Coniglio and Harrison 1983; Kindinger 1986; Evans 1987; Shinn et al. 1989). Age determinations for this unit give an estimate of between 120 and 135 ka (Broecker and Thurber 1964; Osmund, Carpenter, and Windom 1964; Muhs et al. 1992; Fruijtier, Elliott, and Schlager 2000). Evidence for at least two separate sea-level peaks during MIS 5e have been documented throughout the Caribbean including south Florida (Muhs et al. 1992; Precht 1993; Neumann and Hearty 1996; White, Curran, and Wilson 1998; Fruijtier, Elliott, and Schlager 2000). The local maximum sea level during MIS 5e has been calculated at ~7 m above present (Randazzo and Halley 1997; Lidz 2000a,b) and is in close correlation with the elevation of 7.5 ± 1.5 m proposed by Cronin et al. (1981) for the U.S. Atlantic Coastal Plain. This also corresponds with an elevation of 6 ± 1 m measured along the north coast of Jamaica (Precht 1993). The maximum elevation of the Key Largo Limestone is presently found on Windley Key; the top surface of this formation is slightly less than 6 m above sea level (Halley, Vacher, and Shinn 1997). Where the elevated portions of the Key Largo Limestone are penetrated by mechanical methods (quarrying, canal cuts, marina excavations, mosquito ditches, building foundations, and core borings) they reveal the presence of a coral-rich, reefal limestone (Pasley 1972; Hodges 1977; Harrison, Cooper, and Coniglio 1984; Multer et al. 2002a).

All the coral species present in these limestones are found living today in the waters of the Florida reef tract (Stanley 1966). However, not all of the corals living today are present in the Key Largo Limestone (Hoffmeister et al. 1964). Specifically, there is an absence of *A. palmata*, a paucity of *A. cervicornis*, and dominance by an assemblage of *M. annularis, Diploria strigosa,* and *Porites astreoides.* Stanley (1966) commented that there is a fortuitous juxtaposition of these Pleistocene deposits with the recent, offering a special opportunity for application of the principle of uniformitarianism. Nevertheless, inferences drawn from comparison of modern and ancient reefs can give rise to misleading or incorrect interpretations, confounding our understanding of reef systems and hindering our ability to decipher the real story locked in the rocks (Wood 1999). Specifically, the time-averaged and incomplete nature of both the fossil and stratigraphic records may limit direct translation between the modern and ancient (Roy et al. 1996; Greenstein, Curran, and Pandolfi 1998; Pandolfi 2002; Shinn et al. 2003). Also, not all corals stand the same chance of preservation in the fossil record further confounding interpretations (Hubbard, Gladfelter, and Bythell 1994; Pandolfi 2002). Finally, having an understanding of paleogeographic setting is an important component to the application of uniformitarianism-based analogies. Interestingly, several different and contrasting explanations exist that describe the environment of formation of the exposed, coral-rich unit of the Pleistocene Key Largo Limestone. The most important aspect of the various explanations is reconciliation of the absence of *A. palmata* in the Pleistocene with an abundance of *A. palmata* in the Recent (the last 10 ka).

Six hypotheses have been developed for the Q5 unit (MIS 5e) of the Key Largo Limestone.

Hypothesis 1: The reefs grew in deeper water (water depths 6–12 m) dominated by *Montastraea annularis* and formed a broad and homogeneous coral plantation on the east Florida shelf but at some distance inside the shelf margin (Stanley 1966).

Hypothesis 2: The deposit represents an arcuate series of coalesced low-energy patch reefs, dominated by *M. annularis* and located shoreward of an active acroporid-dominated barrier reef on a broad carbonate shelf (Hoffmeister and Multer 1968; Hoffmeister 1974; Friedman 1977; Hodges 1977; Jones 1977; Multer 1977; Greenstein and Pandolfi 1997; Greenstein, Pandolfi, and Curran 1998).

Hypothesis 3: Linear concentrations of lagoonal back-reef corals grew along the edge of an elongate, submerged, antecedent high (Dodd, Hattin, and Liebe 1973).

Hypothesis 4: The shelf-margin sand-shoal complex known as White Bank, and associated inner-shelf patch reefs, are the modern-day analogy for the exposed Key Largo Limestone (Perkins 1977). Similar shelf-margin sand shoals and patch reefs migrated landward during Pleistocene (Q5) sea-level rise and did not require a seaward reef barrier for their inception. At sea-level maximum, the leading shallowest edge of the transgressive unit was preserved as an arcuate patch-reef and skeletal-sand complex.

Hypothesis 5: Numerous shallow-water patch reefs developed over an antecedent bathymetric high (the Northern Keys High), with associated skeletal grainstones, packstones, and wackestones. From this topographic high, the limestone dips steeply seaward, suggesting a sloping, ramp-type, carbonate shelf (Multer et al. 2002a,b).

Hypothesis 6: A concentric rim of shallow-water reefs dominated by *M. annularis* formed a bank-barrier complex. The bank-barrier complex separated a vast, shallow, carbonate platform dominated by bryozoan-rich skeletal packstones with localized lagoonal patch-reefs to the west (Gulf of Mexico), from a broad, coral-studded shelf to the east (Shinn et al. 1977, 1989; Shinn 1984, 1988; Harrison and Coniglio 1985; Textoris, Fuhr, and Merriam 1989; Halley, Vacher, and Shinn 1997).

Testing these hypotheses requires a careful evaluation of the chronostratigra-phy and biofacies in relation to the known sea-level history during Key Largo time (MIS 5e):

Hypothesis 1: Stanley (1966) described the exposed MIS 5e Key Largo reef limestones as having a deep-water (6- to 12-m) origin. He based this interpreta-tion on the absence of shallow-water *A. palmata* and an abundance of head corals, including large *M. annularis* colonies. Stanley (1966) emphasized that "the lack of zonation . . . and the apparent homogeneity of the entire Key Largo reef mass indicate that no significant energy gradient existed during reef growth." An excel-lent deeper-water modern counterpart exists in the *Diploria–Montastraea–Porites* zone on the reefs of the Flower Garden Banks in the Gulf of Mexico (Bright et al. 1984; Gittings et al. 1992). Recently, Lidz, Reich, and Shinn (2003) discussed the possibility that the "corals of the Key Largo Limestone prefer low-energy condi-tions and generally grow in water deeper than 5 m."

These deeper-water interpretations can be falsified based on the position of the exposed Key Largo Limestone relative to the known sea-level maximum. The maximum paleo-water depths for the corals can be calculated by taking the dif-ference between the maximum sea-level elevation (+7 m) and the elevation of the coral biofacies in question (+1–6 m). For Stanley's deeper-water (6–12 m depth) hypothesis to be correct, either neotectonic uplift or a MIS 5e high stand of at least 18 m above present sea level would be required. Furthermore, numerous corals exposed in outcrop are flat-topped or have developed "micro-atoll" mor-phology as a response to growth up to the mean level of spring low tide (Fig. 9.4; see Stoddart and Scoffin 1979). Based on this evidence, most corals in the Key Largo Limestone must have formed in waters ~1 to 5 m deep. These reefal deposits are also stratigraphically equivalent to and interfinger with the exposed oolitic shoal facies of the Miami Limestone negating the deeper-water interpreta-tion (Halley and Evans 1983; Kindinger 1986). The evidence for a relatively shal-low-water origin is clear, but the interpretation of shelf setting and physical environmental regimes is not.

Hypothesis 2: Hoffmeister and Multer (1968) postulated that the Key Largo Limestone represents a series of low-energy patch reefs that formed in a back-reef environment. They based this hypothesis on the high-energy conditions favorable for the growth of *A. palmata*. They argued that *A. palmata* is not a common biotic constituent on most modern-day patch reefs in the Florida Keys and that its absence in the Key Largo Limestone, therefore, demands a low-energy interpretation. This interpretation requires that a high-energy reef-crest community dominated by acroporids lie somewhere seaward on the outer margin of the Key Largo platform. They knew that to substantiate this interpretation they had to prove that this hypo-thetical outer reef existed (see Hoffmeister 1974). In the 1960s Hoffmeister and Multer drilled a series of deep cores throughout the Florida Keys to test and refine their facies model. Using a core boring drilled through the rear zone behind the crest of the present-day reef at Looe Key, they encountered what was believed to be an *A. palmata*-bearing Pleistocene limestone at ~17 m below present sea level (Multer and Hoffmeister 1977; also reported in Halley, Vacher, and Shinn 1997).

Figure 9.4. (**A**) *Montastraea annularis* colony in the Dry Tortugas. Coral is flat-topped in response to growth up to level of spring low tide. (**B**) Outcrop photograph of flat-topped *M. annularis* colony in growth position in Windley Key Quarry (Pleistocene, Florida Keys).

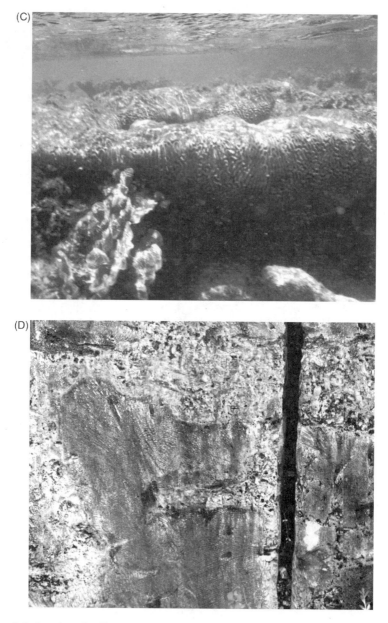

Figure 9.4. (*continued*) (**C**) Micro-atoll of living *Diploria strigosa* colony. This growth form is an excellent marker for spring low-tide levels. (**D**) Outcrop photograph of micro-atoll structure in *M. annularis* as a response of coral growth up to sea level, Windley Key Quarry. These paired photographs show the utility of the comparative approach in paleoecologic investigations.

This was the first time that *A. palmata* had ever been reported in the Pleistocene Key Largo Limestone, and at the time it suggested that an outer reef existed in approximately the same geographic position as the modern reef tract (Hoffmeister 1974; Multer and Hoffmeister 1977), some 7 km east of the exposed Pleistocene deposits of the Keys.

Although the hypothesis initially appears reasonable, the stratigraphic relationship between a shallow-water reef community with *A. palmata* at approximately −17 m and a known time-equivalent sea level of +7 m requires either active tectonic tilting of the platform and/or substantial erosion (~24 m) of the entire seaward platform margin. Hoffmeister and Multer (1968) preferred the latter interpretation although they indicated, "[T]here is a strong possibility that this was aided by a certain amount of down-tilting or faulting or both." However, the south Florida platform has been tectonically stable during the late Quaternary (Davis, Hine, and Shinn 1992; Lidz, Reich, and Shinn 2003), and although possible, erosion of up to 24 m of limestone in 120 ky is not likely. Actual erosion rates can be calculated by taking the difference between the maximum sea level at Key Largo time (+7 m) and the current preserved elevations of the exposed Key Largo Limestone (+1–6 m) throughout the Keys. Subtraction reveals no more than a few meters of loss in the last 120 kyr. Interestingly, Halley and Evans (1983) calculated the amount of surface dissolution to be around 2.64 m for the stratigraphically and mineralogically equivalent Miami Limestone during the same interval of subaerial exposure. In addition, the occurrence of *A. palmata* living at paleo-water depths in excess of 20 m cannot be reconciled with the known shallow-depth distribution and absolute water-depth limits of this species (Lighty, Macintyre, and Stuckenrath 1982).

Most important for testing the hypothesis are recently obtained dates of the Pleistocene *A. palmata* from the core samples. Dating revealed a Holocene age of 2.3 ka (see Multer et al. 2002a), falsifying this early hypothesis. Additional Pleistocene core samples obtained and dated by Multer et al. (2002a) lying directly beneath the Holocene reef tract are not time-equivalent with the exposed limestones of the Florida Keys. Instead, they are part of younger Pleistocene MIS 5c and 5a outlier reefs of Lidz et al. (1991; Fig. 9.5), and these are discussed in more detail later in this chapter.

Using the hypothesis of Hoffmeister and Multer (1968) and Hoffmeister (1974), several subsequent researchers have reached similar conclusions about environmental conditions during development of the Key Largo fossil assemblage. Greenstein and Pandolfi (1997; see also Greenstein, Pandolfi, and Curran 1998) compared life and death assemblages of coral taxa found on living patch reefs and the offshore reef tract with fossil coral assemblages of the Key Largo Limestone. They determined, based on the paucity of acroporids, that the living composition of modern patch reefs was statistically more similar to that of the Key Largo Limestone than to that of the modern offshore reef tract. This similarity led them to conclude that the Key Largo Limestone must have been a series of low-energy patch reefs, supporting the original interpretation of Hoffmeister and Multer (1968). Sixty percent of their modern patch-reef death assemblage, however, was formed by the two major acroporid species, including *A. palmata*

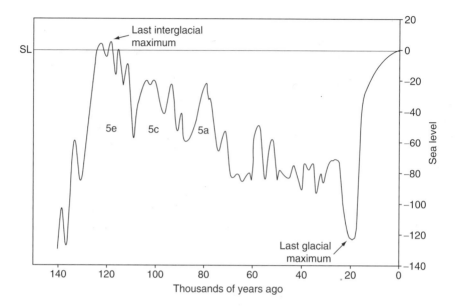

Figure 9.5. Generalized sea-level curve for the last 140 ky. Episodes of reef growth occurred during times of sea-level maximum. Marine Isotope Stages of the major highstands are labeled. Data are from numerous sources, including Chappell (1983), Chappell and Shackleton (1986), and Buddemeier and Kinzie (1998). SL, modern sea level.

(Greenstein and Pandolfi 1997). We agree with Greenstein and Pandolfi that although branching corals are probably overrepresented in the death assemblage, it is a reasonable proxy for the fossil assemblage that will result eventually (see also Kissling 1977a). Pandolfi (2002) concluded that there is a high degree of fidelity between the original coral life assemblages and both their adjacent death assemblages and their fossil counterparts in the Caribbean (see also Kidwell 2001). Greenstein and Pandolfi (1997) specifically highlighted the fact that life assemblages examined in the patch-reef environment were dominated by *M. annularis* and *P. astreoides*, whereas the death assemblage was dominated by *A. cervicornis*. It should be noted that their surveys were performed well after the excision of the acroporids from these reefs as discussed earlier. Their result of abundant *Acropora* spp. (including the presence of *A. palmata*) in the modern patch reef death assemblage should therefore not be surprising as Gilmore and Hall (1976) indicated populations of *A. cervicornis* on patch reefs of the upper Keys, while Shinn et al. (1989) noted that several large patch reefs contained significant populations of *A. palmata* on their seaward sides before their demise of the past two decades.

    If the death assemblages of modern patch reefs of Florida are the appropriate analogues for the fossil record, as postulated by Greenstein and Pandolfi (1997; see also Greenstein and Curran 1997), what happened to all the *Acropora* in the Key Largo Limestone? An explanation to support this hypothesis would be the taphonomic diminution of the acroporids in the transition from the death assemblage to the fossilized reef. However, based on widespread preservation of the acroporids throughout the Caribbean in fossil reefs of the same age and in similar environments, the patch reef model of Greenstein and Pandolfi (1997) is not tenable.

Hodges (1977) supported the low-energy origin for the Key Largo Limestone based on the orientation of head corals in Pleistocene outcrop, where measurements revealed that upright in-place colonies dominate. Recently, Meyer et al. (2003) also concluded that because of high hurricane frequency in the Florida Keys, these in-place colonies are a consequence of their protected setting behind the shelf-margin reef tract. Geister (1980) explained that the almost exclusive occurrence of *M. annularis* in the Key Largo Limestone indicates protected conditions or a growth depth below normal wave base. It has been shown, however, that corals of the *M. annularis* species complex are extremely robust and resistant to the effects of storms, including hurricanes (Stoddart 1963; Woodley et al. 1981; Graus, Macintyre, and Herchenroder 1984; Woodley 1992; Bythell, Hillis-Starr, and Rogers 2000; Jackson and Johnson 2000), and that their distribution and zonation are due to a combination of factors including, but not limited to, competitive interactions with other corals—notably the *Acropora* species (Porter et al. 1981; Rylaarsdam 1983; Tunnicliffe 1983; Huston 1985; Aronson and Precht 2001b). In fact, large, centuries-old colonies of the *M. annularis* species complex appear to be the most resistant corals to catastrophic hurricanes in the Caribbean (Stoddart 1963; Woodley 1992; Aronson and Precht 2001b). Upright *Montastraea* colonies, therefore, do not necessarily indicate conditions of low wave energy.

Graus and Macintyre (1989) demonstrated that the *M. annularis* species complex occurs at a minimum depth of 1 m in fore-reef environments in the absence of *Acropora*. Adey and Burke (1977) also showed that on some reef flats of the eastern Caribbean, colonies of *M. annularis*, *P. astreoides*, and *Diploria* spp. often become as abundant as *A. palmata*. In Barbados, *Montastraea* spp. have replaced *A. palmata* in the shallow fore reef (Lewis 1984). Hunter and Jones (1996) found coral associations dominated by *M. annularis*, *Diploria* spp., and *Siderastrea* spp. in both patch-reef and reef-tract facies of the Pleistocene Ironshore Formation of Grand Cayman. They distinguished between habitats on the presence/absence of rare corals such as *Meandrina meandrites*, with *Meandrina* generally being absent from lagoonal patch reefs. *Meandrina* colonies are an integral part of the coral fauna of the Key Largo Limestone leading again to the interpretation of a reef-tract setting for the Key Largo fossil assemblage.

Also important in eliminating Hypothesis 2 is the occurrence of *M. annularis* in reef-crest positions from locations along the Florida reef tract. Vaughan (1919b) identified *M. annularis* as the predominant coral species in the shallow Tortugas reef community (Fig. 9.6). Also, Shinn et al. (1977), citing modern examples from the Florida Keys, noted that in areas of low *A. palmata* abundance or absence, head corals predominate in the reef-crest facies. Specifically, they stated, "[T]he reefs at Marker G and those at Dry Tortugas have the same coral fauna as the Key Largo formation, but the lack of *A. palmata* shows that it is no longer necessary

Figure 9.6. High-energy, reef-crest community dominated by monospecific stands of massive *M. annularis*. Photo is from the Dry Tortugas. This modern-day coral assemblage is an excellent counterpart to the fossil reef exposures of the Pleistocene Key Largo Limestone.

to call on a patch reef origin. Reefs can form and keep pace with sea level at the platform margin, for reasons not (yet) understood, without the help of *A. palmata*."

Hypothesis 3: Dodd, Hattin, and Liebe (1973) proposed a modification of the Hoffmeister and Multer (1968) patch-reef hypothesis for the exposed Key Largo Limestone. They speculated that the linear nature of the exposed Key Largo Limestone reef system is analogous to an elongate series of modern patch reefs dominated by head corals forming on a bathymetric high on the lagoon floor near Big Pine Key. Antecedent control can explain the linear nature of these reefs, but the rationale used above to falsify the patch-reef origin of Hypothesis 2 is the same here.

Hypothesis 4: Perkins (1977) proposed an explanation based on an analogy with the modern shelf-margin sand-shoal complex, White Bank, and associated inner-shelf patch reefs. The present-day White Bank is a continuous sand shoal that is 40 km long and 1 to 2 km wide. White Bank is actively migrating in a landward direction due primarily to transport by storms (Ball, Shinn, and Stockman 1967; Enos 1977). Adjacent, patch reefs are often engulfed by these migrating sands. Perkins (1977) proposed that similar sand shoals and patch reefs migrated landward during Pleistocene (Q5) sea-level rise on a carbonate ramp and did not require a seaward reef barrier for its inception. At sea-level maximum, the leading shallowest edge of the transgressive unit was preserved as an arcuate patch-reef and skeletal-sand complex.

Perkins could not reconcile the absence of *A. palmata* in the Pleistocene without calling on a patch-reef origin. He commented that the modern patch-reef assemblages on the inner shelf were mostly devoid of *A. palmata* and were more in accord with those of the exposed Q5 coralline facies. Although Perkins' assertion that a hypothetical reef barrier seaward of the present-day Keys did not exist was recently borne out by detailed subsurface investigations (Multer et al. 2002a), we can eliminate the patch-reef origin of Perkins using the same rationale discussed above for Hypothesis 2. Moreover, biotic constituent analysis of sediment types from White Bank has little resemblance to the sediments preserved in the exposed Key Largo Limestone (Stanley 1966; Enos 1977).

Hypothesis 5: Multer et al. (2002a) recently proposed that the exposed MIS 5e reefs of the Key Largo Limestone were shallow-water patch reefs that developed on an antecedent bathymetric high (the Northern Keys High). Based on core and outcrop investigations, they identified well-developed patch reefs with grainstones and packstones forming on the highest elevations of the underlying older Key Largo units. Multer et al. (2002a, p. 254) specifically noted, "In the upper Keys, Q5e patch reefs, grainstones and packstones became well developed on the highest elevations of the Northern Keys High. Some can be seen today exposed in the Windley Key Fossil Reef State Park (formally called the Windley Key Quarry) and along the Key Largo Cross Canal." Multer et al. (2002a,b) argued that the steeply seaward-dipping geometry of this unit and the absence of an Atlantic breakwater dominated by *A. palmata* supports the contention that the seaward edge of the MIS 5e platform had a ramp-type character.

Reconstruction of the entire south Florida platform during MIS 5e reveals that the reefs of the upper and middle Keys separate a broad, flat, shallow carbonate platform dominated by back-reef mudstones and skeletal packstones rich in bryozoans to the west (Perkins 1977), from a gently seaward-dipping, coral studded platform to the east (Figs. 9.7 and 9.8). The location of the Upper Keys High thus formed the locus of an arcuate shelf-break between fore- and back-reef environments at the peak of the last interglacial maximum, negating a ramp-type setting proposed by Multer et al. (2002a,b); see Ahr (1973). The Northern Keys High controlled the position of the reef-crest community at this shelf edge and was inherited from underlying older (Q3 and Q4) Pleistocene reefal limestones (Harrison and Coniglio 1985). This reef-on-reef stratigraphic sequence indicates that subsequent episodes of reef development magnified the underpinnings of the Northern Keys High. Hence, this shelf edge reveals that the exposed reefs of MIS 5e were the breakwater reefs of the Florida platform, contrary to the "patch-reefs developed on a carbonate ramp" interpretation of Multer et al. (2002a,b).

Hypothesis 6: This brings us to the final hypothesis of Shinn et al. (1977), Harrison and Coniglio (1985), and others. Understanding what we do about the role of inimical bank waters, the susceptibility and vitality of the acroporids, and the multihabit distribution of *M. annularis*, we can now reconstruct the reef community and associated environments that formed during the higher sea level of the last major interglacial episode.

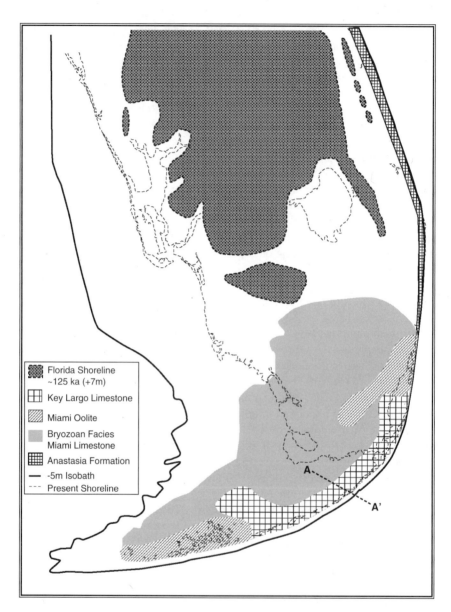

Figure 9.7. Paleogeographic reconstruction of the Florida Peninsula at ~125 ka during Marine Isotope Stage 5e, when sea level was ~7 m higher than today. Modified from numerous sources, including Perkins (1977), Harrison and Coniglio (1985), and Multer et al. (2002a). Note the location of stratigraphic cross section (A-A') shown in Fig. 9.8.

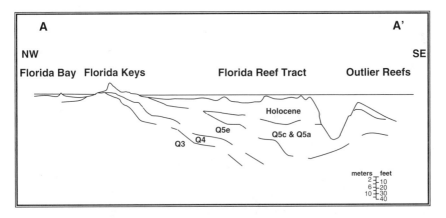

Figure 9.8. Stratigraphic cross section (A-A′) across the Florida Keys and reef tract. Stratigraphic units follow terminology of Perkins (1977) and Multer et al. (2002a). Modified from numerous sources including Perkins (1977), Harrison and Coniglio (1985), Lidz et al. (1997a), Lidz (2000a), and Multer et al. (2002a).

Based on time-space relationships in paleoecology and basic ecological principles we would expect that fossil abundance would be greatest near the center of the geographic range of the species in question and progressively decrease to the margin, as optimal conditions graded to less favorable ones (Brown 1984). Because coral reefs of south Florida are at the latitudinal extreme of reef development in the western Atlantic, the absence of *A. palmata* in the Key Largo Limestone is not surprising, due to contraction and expansion of species ranges in response to changing environmental conditions through time and space. At the apex of the last major interglacial, the entire south Florida Peninsula was flooded, creating a broad, shallow Bahamas Bank-type carbonate platform (Fig. 9.7). This would have allowed unimpeded flow from a much-enlarged Gulf of Mexico over a broad, shallow carbonate bank in the west to the Straits of Florida and open Atlantic in the east. Reef communities forming at the interface between the two would have been bathed by clear, warm, oceanic waters from the Gulf Stream on incoming tides and waters of variable quality, especially in regard to temperature, salinity, and sediment on outgoing tides. With no exposed islands (Keys) to protect the reefs, during the winter months, chilled and sediment-laden bank waters would have periodically poured onto these shelf-margin reefs. Under this scenario the resulting coral community would have been devoid of the shallow, thermophilic reef-crest species *A. palmata* (Harrison and Coniglio 1985; Shinn 1988). At slightly deeper depths on the fore reef, ephemeral coppices of *A. cervicornis* would have filled in sheltered pockets in a *Montastraea–Diploria–Porites* head-coral assemblage.

One line of circumstantial evidence supporting the inimical-conditions model for the Key Largo Limestone comes from the measured growth rates of *M. annularis* in the fossil record compared to their living counterparts. Hudson (1981a,b) showed that under optimum conditions, the modern growth rate of *M. annularis* along the modern Florida reef tract is ~10 mm year^{-1}, whereas in the Key Largo

Limestone it is half that (Hoffmeister and Multer 1964; Shinn et al. 1989). Landon (1975) noted a similar relationship for the extension rates of all forms of the *M. annularis* complex, *Siderastrea siderea*, and *Porites porites*. These reduced growth rates most probably indicate fluctuating and hostile environmental conditions that were unfavorable to all but the hardiest reef-building species during the last major interglacial in Florida.

Stratigraphically, the reconstruction of the exposed Q5 unit of the Key Largo Limestone is straightforward, with the arcuate reef margin of the *Montastraea–Diploria–Porites* community forming in the position of the present-day islands of the Florida Keys. The location of the exposed Pleistocene reef tract is controlled by the underlying Q3 reefal unit (Perkins 1977; Harrison and Coniglio 1985; Multer et al. 2002a). The sediment accommodation space is filled to its maximum at the point of the shallowest reef community on this topographic feature, forming a series of discontinuous, shallow, bank-barrier type reefs stretching for almost 170 km (Fig. 9.7). This interpretation requires fewer assumptions about complex geologic processes; specifically tectonic movement, tilting, or large-scale erosion of the platform are not necessary to explain the geography, topography, and species composition of the Key Largo Limestone. Of greatest significance, it does not require *A. palmata* to have been extant in these high-energy, shallow-water reef communities.

Accepting Hypothesis 6 leads to an interesting question: What would otherwise-healthy Caribbean reefs look like in the absence of the acroporids? To address this question, we have to understand why, under optimum conditions, the acroporids are competitively superior in the first place. In the case of *A. palmata*, extremely rapid growth and asexual propagation favor the buildup of dense stands in shallow water that are resistant to all but the most extreme storms. These dense coral forests grow above and ultimately shade out most other corals, including the slower-growing *M. annularis* species complex (Jackson 1991). At intermediate depths, the more fragile, branching *A. cervicornis* has similar traits, enabling the species to overwhelm other corals and form monospecific thickets (Tunnicliffe 1983). If the acroporids were excised from the reef, subordinate understory corals, including the *M. annularis* species complex and others, would be released from spatial competition. Those remaining species would then respond based on their individual ecologies, with some becoming the new spatial dominants.

Using Hypothesis 6 as our model for the Key Largo Limestone, we can predict the future direction of the reef communities in Florida under a scenario of declining importance of *Acropora* spp. Originally, the coral community would be dominated by early successional brooding species, especially *Agaricia* spp. and *Porites* spp. (Hughes 1985, 1994; Kojis and Quinn 1994; Aronson and Precht 2001c; Edmunds and Carpenter 2001; Knowlton 2001; Cho and Woodley 2002). Many of these brooding species (e.g., *P. astreoides*), however, are relatively short-lived (Kissling 1977b; Hughes 1985). Because of their longevity and their ability to build large, massive, wave-resistant colonies, over time (probably centuries) the *M. annularis* species complex would become the spatial dominants in this depth range (see discussion of Edmunds and Bruno (1996) in Aronson and Precht 2001b; see also Hughes and Tanner 2000).

Testing this prediction in real time requires a modern example of a shallow-water, relatively high-energy Caribbean/western Atlantic reef that is, and always has been, totally devoid of *Acropora* species. The rim and main terrace reefs (2–15 m depth) of Bermuda as described by Logan (1988) fit these requirements. The Bermudian coral fauna is depauperate compared to the rest of the Caribbean–West Indian biogeographic province, with low winter water temperatures invoked as the main cause. These rim/terrace communities are dominated by a *Diploria–Montastraea–Porites* assemblage and have high coral cover (Dodge, Logan, and Antonius 1982; Logan 1988; Smith et al. 2002). Growth, form, composition, percent cover, and spatial orientation of corals from these Bermudian reefs are in many respects similar to those observed in the Key Largo Limestone (Thaddeus Murdoch personal communication).

9.4.2 The Late Pleistocene: Marine Isotope Stages 5c, 5b, and 5a

Toscano and Lundberg (1998, 1999)—see also Lidz et al. (1991), Ludwig et al. (1996), Lidz et al. (1997b), Multer et al. (2002a), Lidz, Reich, and Shinn (2003)—have shown that the reefs that lie beneath the Holocene and modern reef tract are equivalent to MIS 5c and 5a, and they have been dated at approximately 106 to 112 ka and 78 to 86 ka, respectively. Recently, Lidz (2004) documented MIS 5b reefal deposits dating between 87 and 94 ka sandwiched between the MIS 5c and 5a outlier reefs. These outlier-reef deposits exhibit a shelf-margin, wedge geometry, onlapping the seaward-dipping MIS 5e surface (Fig. 9.8), and they form the antecedent underpinning for Holocene reef development in the Florida Keys (Multer et al. 2002a; Lidz, Reich, and Shinn 2003). At the time of the MIS 5a high stand, sea level stood ~9 m below the present. Paleogeographic reconstruction of the MIS 5a Florida Peninsula reveals a broad, exposed, low-relief plateau with reef development confined to the seaward edge of the platform. This fossil reef subcrops along the entire length of the Florida Keys (Lidz, Reich, and Shinn 2003).

The younger MIS 5c, 5b, and 5a reef tracts had the advantage of being sheltered from deleterious bank waters by the emergent Florida platform. These protected shelf-margin reefs were probably responsible for the reappearance of *A. palmata* on the Florida reef tract in the late Pleistocene. Evaluation of the stratigraphic position and subsea elevation of the MIS 5c and 5a reefal units by Toscano and Lundberg (1998) also confirms the tectonic stability of the Florida platform throughout the late Quaternary.

9.4.3 The Latest Pleistocene and Holocene History of Reef-Building

The last glacial maximum (LGM) occurred during the Pleistocene Epoch about 18 ka. At that time, global sea level was ~120 m lower than today (Matthews 1986; Fairbanks 1989; Bloom 1994). CLIMAP (1976) hindcast models indicated that minimum sea-surface temperatures during this time were less than 2 °C cooler than those associated with modern Caribbean seas. Since then debate has arisen as to whether tropical SSTs may have been as much as 4 to 5 °C colder during this period (Guilderson, Fairbanks, and Rubenstone 1994; Beck et al. 1997). However, recent data and global climate models estimate a drop of no more than ~2.5 °C (Crowley 2000). Trend-Staid and Prell (2002) found only a 1.0 to 1.5 °C drop at the LGM in samples from near Barbados. The data, coupled with other data from all ocean basins, led them to the conclusion that SSTs in the western tropics and subtropics were essentially indistinguishable from modern values. This suggests that temperature was not low enough to terminate reef development through the last major glacial–interglacial cycle. The coral faunas that comprised the reef-building episodes of the previous interglacial high stands show amazing similarity in species composition throughout the Pleistocene and Holocene, so we can speculate that these corals and reef communities persisted in downslope refugia during the glacial low sea stands.

Pandolfi (1999) speculated that the major drop in sea level at the LGM resulted in a dramatic loss of habitat, causing the rapid extinction of some widespread Caribbean coral species. The fact that Indo-Pacific corals did not experience similar extinctions under essentially the same sea-level and temperature history argues against these as being the primary causal agents (Budd, Johnson, and Stemann 1994). This is also contrary to evidence from Pleistocene molluscan assemblages which reveal species responding to sea level and climate change by fluid geographic shifts rather than by extinction or speciation (Roy et al. 1996). Thus, quantitative habitat loss associated with geological-scale processes seems an unlikely extinction mechanism, especially in light of the fact that Kleypas (1997) calculated that during the LGM coral reef growth was more extensive than previously thought. Paulay (1990) indicated that reef geomorphology was an important control in the distribution of bivalve species through glacial–interglacial cycles. He noted that although inner-reef lagoons are stranded during glacial low stands, resulting in extirpation of soft-bottom bivalve fauna, coral-dominated fore-reef environments persisted, albeit displaced downslope. As the continental shelves were flooded at 8 to 10 ka the total area available for reef growth increased. Goreau (1969) used the phrase "post-Pleistocene urban renewal of coral reefs" to describe the Holocene reemergence of reefs associated with the flooding of platforms during the last deglaciation.

SSTs in the subtropical western Atlantic show a warming between 14 and 10 ka, a maximum that is warmer than today between 10 and 6 ka, and a late-Holocene cooling to modern values (Balsam 1981; Ruddiman and Mix 1991; Bradley 2000). Evidence from coring submerged reefs off Barbados (Fairbanks 1989) shows a backstepping of *A. palmata* communities that essentially track postglacial sea level from 17.1 to 7.8 ka. Sea-level curves derived from dates of *A. palmata* allow for the calculation of the rate of glacial meltwater discharge into the Atlantic (Fairbanks 1990). Rates of rise of sea level were rapid during this period, punctuated by two periods of extremely rapid rise (meltwater pulses) in the latest Pleistocene. During these rapid bursts, sea-level rise apparently outstripped the accumulation rates of *A. palmata*, leading to brief hiatuses of reef growth (i.e., incipient drowning), followed by a backstepping of the shallow-reef communities to more shoreward positions higher on the shelves. In south Florida, such drowned shelf features have been identified using high-resolution seismic, side-scan sonar and in situ sampling by submersibles in water depths of 50 to 124 m (Locker et al. 1996). Although the rate of rise in sea level during the Holocene (the last 10 ky) has never exceeded the maximum accretion rate of *A. palmata*, reef growth was not continuous but punctuated by a series of reef ridges and troughs (Lighty 1977; Lighty, Macintyre, and Stuckenrath 1978; Lidz and Shinn 1991; Toscano and Lundberg 1998; Precht et al. 2000; Lidz, Reich, and Shinn 2003).

In some areas, numerous distinct relict reef ridges are present (Lighty, Macintyre, and Stuckenrath 1978) while in other areas there is a gradational separation (Toscano and Macintyre 2003). These relict Holocene reefs, comprised primarily of *A. palmata,* are found at water depths between 20 and 7 m south of Miami and between 27 and 7.8 m depth to the north (Toscano and Lundberg 1998; Toscano and Macintyre 2000). In all, very few reefs have been sampled and dated, making it difficult to evaluate the precise chronology of the turn-on and turn-off of this reef system.

Throughout most of the early to middle Holocene, however, oceanic conditions were favorable for the growth and accumulation of *A. palmata* along much of the Florida reef tract. In fact, major *Acropora*-dominated reefs were well-developed off Palm Beach County (Lighty 1977; Lighty, Macintyre, and Stuckenrath 1978), indicating that water quality along the platform margin was more favorable to coral growth than today. A buried topographic feature lying off Fort Pierce may be the northernmost extent of this mid-Holocene reef tract (Hine 1997).

We interpret this northward range expansion of the *Acropora* spp. as a reflection of northward excursions of warm water related to a unique conjunction of factors. First, during this time lower sea level placed the active shelf-margin reef system in closer proximity to the equable waters of the Gulf Stream. Second, the period from about 9 to 5 ka corresponds to the mid-Holocene Warm (Zubakov and Borzenkova 1990; Pielou 1991; Maul 1992; Thompson et al. 1998; Zhou, Baoyin, and Petit-Marie 1998; Kerwin et al. 1999). Terrestrial records indicate that the climate in North America was 2 to 4 °C warmer than at present during this interval (Folland, Karl, and Vinnikov 1990). Climate simulations suggest that North Atlantic SSTs were at least 4 °C warmer than today at 6 ka (Kerwin et al. 1999). Stable oxygen-isotopic values recorded in sediments from the extratropical North Atlantic also indicate warmer-than-present SSTs during the mid-Holocene (Ruddiman and Mix 1991). Evidence from both terrestrial and coastal regions shows that warming during this interval allowed many organisms to increase their ranges northward (Clarke et al. 1967; COHMAP 1988; Delcourt and Delcourt 1991; Pielou 1991; Salvigsen, Forman, and Miller 1992; Dyke, Dale, and McNelly 1996; Strasser 1999; Dahlgren, Weinberg, and Halanych 2000; Jackson and Overpeck 2000). This period of thermal optimum during the Holocene also correlates with the northernmost position and expansion of coral reefs in the Pacific (Taira 1979; Veron 1992). Warmer conditions apparently allowed the northward expansion of acroporid-dominated reefs throughout southeast Florida (Precht and Aronson 2004). As temperatures have cooled since the middle Holocene (~5 ka to present), the conditions for acroporid reef growth have diminished and the northward limit of the reef tract has progressively moved south to its present position. While reefs at their latitudinal extremes have responded rapidly to climate flickers, results from coring studies of Holocene reefs in the insular tropical Caribbean support the notion that tropical oceanic climates have been buffered from climatic variability (Gill, Hubbard, and Dickson 1999).

Dead and senescent Holocene reefs are also present throughout the middle Florida Keys (Lidz, Robbin, and Shinn 1985; Shinn et al. 1989; Shinn, Lidz, and Harris 1994; Lidz and Hallock 2000). Geologic investigation of these reefs has shown that the acroporids were again the major constructors and frame builders until the late Holocene. Reconstruction of these shallow shelf-margin reefs in relation to sea-level rise by Lidz and Shinn (1991) showed that the flooding of Florida Bay around 2 to 3 ka was contemporaneous with the localized demise of *A. palmata* on many of these reefs. Specifically, Lidz, Reich and Shinn (2003) commented that this flooding led to deteriorating water quality by tidal and storm-driven mixing of bay and Gulf waters, causing the coral reef system to decline. Shinn et al. (1981) indicated that the deep-reef spurs (15–20 m) at Looe Key initiated around 6.5 ka, with the shallow-reef spurs developing ~4 ka and ceasing ~800 years ago. In the lee of large islands, however, some reefs maintained active growth and were dominated by the acroporids until the combined events of the last few decades led to their decline. Such reefs can have Holocene deposits >14 m thick (e.g., Grecian Rocks and Key Largo Dry Rocks).

These spatial examples of reefs with differing geologic histories confirm the inimical-waters model of Ginsburg and Shinn (1964, 1994). They also attest to the fact that conditions favorable to expansive reef growth have been diminishing on the Florida platform for the past few millennia.

9.5 The Present Status of the Florida Reef Tract

All of the classic environmental issues faced by coastal communities and coral reefs worldwide (Wilkinson 2002) are concentrated in the narrow strip of land that is the Florida Keys. These issues include development, ship groundings, pollution, and overharvesting. Regional and global perturbations that affect coral reefs, such as coral bleaching and coral disease, are also present in Florida. The decline of living reefs in Florida was first reported by Voss (1973) and was a concern in the Florida Keys as early as the 1950s. Indeed, the perceived degradation of coral reefs from overfishing, coastal development, and increasing visitation in Key Largo were the main reasons for the creation of the world's first coral reef marine protected area, John Pennekamp Coral Reef State Park, in 1960 (Voss 1960). Although various forms of reef protection in Florida have been around since the 1960s, most of these have been set ad hoc and without programs in place to assess their efficacy. Recently a systematic, ecosystem-based management approach has been implemented to protect Florida's reefs (NOAA 2003).

We are aware of only one quantitative reef-monitoring program from the early 1970s for the lower Florida Keys (Kissling 1977a). This study integrated hydro-logical, sedimentological, and ecological elements of nine reef areas located from Looe Key to Sand Key. Kissling's data indicated diverse living coral assemblages, with death assemblages dominated by fragments of *Acropora* spp. Subsequent monitoring studies by Dustan (1977), Dustan and Halas (1987), Jaap, Halas, and Muller (1988), and Porter and Meier (1992) all documented examples of coral decline but only from single sites or at limited spatial scales. Extrapolating these results to the entire Florida reef tract is problematic due to variability at scales larger than those of individual reefs (Murdoch and Aronson 1999). Further, these studies focused on the emergent reef systems in the Keys that represent a habitat type that comprises a small fraction of total area in the region. Additional one-time assessments without temporal follow-ups include work in the early 1980s in John Pennekamp Coral Reef State Park and the Key Largo Coral Reef Marine Sanctuary (Voss 2002), Biscayne National Park (Burns 1985) and Looe Key (Wheaton and Jaap 1988); in 1995 throughout the Keys (Murdoch and Aronson 1999); and in 1996 for 20 reef sites in the upper Keys (Chiappone and Sullivan 1997). At Looe Key, landscape-scale reef censuses were performed in 1983 and 2000 (Miller, Bourque, and Bohnsack 2002).

The US Environmental Protection Agency (EPA) in conjunction with the Florida Marine Research Institute (FMRI) and NOAA started the first Keys-wide status-and-trends program in 1996 (Jaap et al. 2001). This program documented coral cover during the period 1996 to 2000 (and continues to the present), with estimated declines as high as 36% and averaging about 6% to 10% across 40 sta-tions sampled in multiple coral reef habitats. Although it was spatially extensive, the program began long after substantial coral loss had already occurred, partic-ularly the demise of the branching corals *A. palmata* and *A. cervicornis*. Of the 40 stations in the study, only seven contained *A. palmata* in 1996. At these seven sites, the average loss of *A. palmata* was 85% between 1996 and 1999 (Patterson et al. 2002). At first blush these results are alarming; however, they need to be placed in perspective. The average percent cover of *A. palmata* at the seven sites was around ~5% in 1996 and had declined to ~2% by 1999, which is an absolute change in coral cover of only 3%. One of the stations sampled included the reef at Rock Key. There, *A. palmata* cover dropped from <4% to slightly less than 2%. This is the same reef that Enos (1977) described as having no living *A. palmata* in the early 1970s.

It could well be that these small populations of *A. palmata* represent residual, transient, or marginal populations that are highly susceptible to both biotic and physical disturbances including disease, predation, storms, cold fronts, and coral bleaching events. Interestingly, the period 1996 to 2000 of the Jaap and Patterson studies spanned two major coral bleaching events in 1997 to 1998, continued out-breaks of pests and disease, and the passage of a major hurricane directly over many of the sampling stations. At this writing, there has not been a coral-bleaching event in Florida since 1998, and recently published data from the EPA study show a leveling off of coral decline after that last regional disturbance (Wilkinson 2004).

The EPA program and two other Keys-wide sampling programs (Murdoch and Aronson 1999; Miller, Swanson, and Chiappone 2002) are important because they documented patterns of community structure, and underlying processes that drive community structure can sometimes be inferred from patterns that emerge across multiple spatial scales (Cale, Henebry, and Yeakley 1989; Aronson 1994, 2001; Done 1999). The EPA results focused on change over time (Jaap et al. 2001), but understanding change across spatial scales is also important. For example, across deeper reef sites (13–19 m) in the Keys, spatial variability was found to be greater between adjacent reefs than across geographic regions (Murdoch and Aronson 1999). Assessments of marine protected areas across the Florida Keys, which were originally selected to capture much of the shallow spur-and-groove habitat in the FKNMS, revealed that these reefs are as likely to be different from each other as they are from surrounding reference sites or those subjected to fishing (Miller, Swanson, and Chiappone 2002). Both studies suggest that processes operating at multiple spatial scales determine coral community structure. Furthermore, these results are important from a management perspective because it is easier and more cost-effective to manage a system if one knows the constraints of overall system variability and history instead of trying to achieve results that are outside the bounds of the system (Allen and Hoeskstra 1992; Jameson, Tupper, and Ridley 2002). This premise will be discussed in more detail below.

The longest historical record available for the Florida reef tract is derived from maps of community distributions in the Dry Tortugas, meticulously prepared by Alexander Agassiz in 1881 and then redrawn from new surveys in 1976 (Agassiz 1883; Davis 1982). A number of interesting changes in coral community structure occurred over that century, all of which could be attributed to natural system variability. For example, the spatial extent of octocoral-dominated communities increased from ~5 to 17%. Thousands of square meters of A. palmata were alive in 1881, but they were largely reduced to alga-covered rubble by the late 1970s, with only two living patches (~600 m²) in 1976. A. cervicornis cover was not common in the late 1880s, according to Agassiz's map; it dominated many locations by 1976 (Davis 1982), only to suffer >90% morality during the winter cold fronts of 1977 described earlier. These results suggest a dynamic system in the Dry Tortugas, just west of the Florida Keys, outside the bounds of direct human influences such as coastal development.

One peculiar event recorded in the reefs of the Dry Tortugas occurred in 1878 and was referred to as "black water" (Mayer 1902; see also Davis 1982). This event purportedly eliminated almost all the A. cervicornis from the Dry Tortugas, yet historical records show some recovery of the affected populations only a few years later (Agassiz 1883). In 2002 a large mass of black water originally of unknown origin was spotted off the southwest coast of Florida. Preliminary results indicate that a bloom of diatoms commingled with a "red tide" were the main constituents of the black water (South-West Florida Dark-Water Observations Group 2002). This water mass moved toward the Dry Tortugas and lower Florida Keys, with some reports of mortality to sponges and corals (Hu et al. 2003).

Ongoing investigations and surveys will hopefully shed some light on this enigmatic condition and, in retrospect, on the event of 1878.

Evaluating the present state of Florida Keys reefs and predicting their future requires that managers and scientists integrate historical information with an understanding of the agents that drive community change at multiple spatial and temporal scales in the present. In practice, managers address these challenges at local scales and over careers that are shorter than many of the underlying ecological processes that affect the way coral reefs look and function. From a societal perspective, working for short-term gains is the easiest and often most politically expedient course of action, but the recent regional and global declines in coral reefs over the last two decades require bold and unconventional approaches (Buddemeier 2001). The following sections describe factors that affect the present state of reefs in Florida. We will use this information to make predictions about the future trajectory of these reefs in ecologic and geologic time.

Such discussions typically follow a standard approach by listing and distinguishing natural from human-caused stresses to reefs. We will instead identify what we believe to be the principal threats caused by humans and then dismiss most of them as not relevant to the long-term condition of reefs in the Florida Keys. For example, physical impacts such as damage by divers or snorkelers and individual dredge-and-fill projects are clearly important as factors that cause damage locally (Taylor and Saloman 1968) but generally do not have systemwide effects on the condition of reefs. Similarly, harvesting of live rock (coral rubble with living coelobites, endoliths, and epibionts) may have been important historically when live rock was routinely extracted from reefs in the Florida Keys, but such activity is now prohibited. Furthermore, it is well documented that marine collectors extract large numbers of ornamental species, but this reduction is not thought to impact long-term reef condition or growth. Vessel groundings are among the most destructive anthropogenic factors causing significant localized damage to Florida's reefs; however, detailed management strategies addressing these insults have been evaluated and are now being implemented (Precht 2006). Introduction of exotic species is a potential problem (Stachowicz et al. 2002) but is not currently viewed as a factor affecting reef condition. It is possible that many marine diseases are invasive species but this linkage has not been established. We do not mean to imply that these factors are unimportant to the condition of the Florida reef tract. Indeed, it is precisely the fact that these and other impacts were important at the scale of individual reefs along the Florida Keys that caused them to be managed and protected in the first place (Jaap 1994; Davidson 2002; NOAA 2003).

9.5.1 Water Quality

The most publicized environmental issue related to the demise of coral reefs is declining water quality. The reason water quality is so often cited as the cause of coral reef decline in the Keys is that the increased abundance of algae observed on reefs over the last two decades supports an intuitive assumption that nutrient enrichment must be the cause (Szmant 2002). Those who advocate nutrient pollution as the primary factor behind coral reef decline in the Keys have proposed connections from agricultural runoff in the Lake Okeechobee region, through the Everglades, into Florida Bay, and ultimately to the offshore coral reefs of the Keys (Brand 2002; Lapointe, Matzie, and Barile 2002). Although the Everglades system is eutrophic in its northern reaches, including the Lake Okeechobee watershed, path analysis of nutrient concentrations reveals an extremely oligotrophic system to the south, through Everglades National Park and into Florida Bay (Stober et al. 2001). If agricultural nutrients do not impact the southern reaches of Everglades National Park and Florida Bay, or do so minimally, then it is even less likely that effects will be seen on the offshore reefs. At an even larger scale, it has been suggested that the entire Caribbean basin may be enriched due to its semienclosed geography and increased runoff from poor land-use practices, coastal development, and deforestation (Muller-Karger et al. 1989; Hallock, Muller-Karger, and Halas 1993). Regionally, large volumes of Mississippi River water have been found embedded in the Florida Current from Key West to Miami, confirming large-scale connectivity (Lee et al. 2002), and this water may have an influence on reef growth and development (Ogden et al. 1994).

In addition to supposed linkages with agricultural runoff from the Everglades, sewage pollution from local sources is often emphasized as *the* cause of ecosystem decline in the Florida Keys (Lapointe, O'Connell, and Garrett 1990). What, then, explains the unambiguous increases in algae that are seen Caribbean-wide? Recent reviews of the role of nutrient enrichment on coral reef decline (Szmant 2002) and the relationships among nutrients, algae, and competition between algae and corals (McCook 2001; McCook, Jompa, and Diaz-Pulido 2001) suggest that the increased abundance of algae is not due to nutrients but instead is explained by the availability of new substrate for algal colonization following coral mortality (Aronson and Precht 2001b; Williams, Polunin, and Hendrick 2001; McManus and Polsenberg 2004). The role of herbivores is clearly also a factor in explaining increased algal abundance on reefs in Florida, from the demise of *D. antillarum* throughout the Caribbean (Lessios 1988), to observations that an inverse relationship exists between algae and herbivorous fish biomass surveyed on 19 reefs throughout the Caribbean (Williams and Polunin 2001). Distinctions are typically not made between benthic habitats nearshore (especially hard-bottom communities and seagrass beds close to shore) and the main reef tract located 5 to 7 km offshore. While few data exist to suggest that changes in offshore reef communities were caused by local pollution sources in Florida, and although few historical data exist regarding nutrient pollution, algal abundance, or coral cover (Szmant 2002), strenuous debate about the relative importance of top-down (Hughes et al. 1999; Miller et al. 1999; Aronson and Precht 2000) versus bottom-up (Lapointe 1997, 1999) processes continues (see Davidson 1998).

The water-quality debate is additionally fueled in Florida by reports of increased nutrient concentrations (Lapointe, O'Connell, and Garrett 1990), the presence of viral tracers (Paul et al. 1995) and fecal indicator bacteria (Paul et al. 1997) that are detectable in nearshore waters and corals (Lipp et al. 2002), but other studies have not found increased nutrients or these microbial constituents in offshore waters or sediments (Szmant and Forrester 1996), seagrass beds (Fourqurean and Zieman 2002), or algae (Hanisak and Siemon 1999). A three-year study of nearshore waters (Keller and Itkin 2002) throughout the Keys found that total nitrogen and phosphorus concentrations did not differ significantly between developed and undeveloped shorelines, although variability was high. Boyer and Jones (2002) revealed that ambient water quality along the Florida Keys and throughout the FKNMS was most strongly influenced by natural oceanographic parameters. While they noted that differences in various water-quality parameters were significantly different across many of their 150 stations, the absolute differences were small and not likely to be biologically important.

Sewage disposal via septic tanks and cesspits represents a pollution threat to canals and nearshore waters due to the highly porous Key Largo Formation, upon which the Keys have been developed. Rates of contaminated groundwater flow to nearshore waters have been measured from injection wells and septic systems (Lapointe and Clark 1992; Shinn, Reese, and Reich 1994; Paul et al. 1995). Factors that influence groundwater flow include the difference in water level between Florida Bay and the Atlantic Ocean, daily tidal cycles, and meteorology (Reich et al. 2002; but see Shinn, Reich, and Hickey 2002). Additional evidence for reduced water quality stems from patterns of groundwater discharge detected using chemical tracers (Corbett et al. 1999) and the presence of over 20,000 septic systems, some 2000 remaining cesspits, and approximately 5000 shallow-water injection wells located in porous limestones throughout the Keys (Shinn 1996; Kruczynski and McManus 2002). Influences related to storm-water runoff are also sources of pollution (Lapointe and Matzie 1996; Brad Rosov personal communication). Hudson et al. (1994) contended that domestic sewage from the Miami area has been a major factor in reducing coral growth rates in the upper Keys since the 1950s. However, Ginsburg, Gischler, and Kiene (2001) demonstrated that corals from these upper Keys reefs had less dead tissue per colony than those on reefs in the middle and lower Keys.

Chiappone (1996) and Miller, Chiappone, and Swanson (unpublished data) found that along a gradient from onshore to offshore, coral cover was highest and algal cover lowest on nearshore patch reefs with a switch to algal dominance as one moves to reefs offshore throughout the FKNMS. This is contrary to the results one would expect if land-based sources of nutrients and pollutants were responsible. Miller, Weil, and Szmant (2000) found similar relationships in both coral cover and coral recruitment on reefs of the northern Keys.

Risk et al. (2001) indicated that it is possible to discern natural versus anthropogenic nutrients (especially sewage) using geochemical techniques. Ward-Paige, Risk, and Sherwood (2005) measured $\delta^{15}N$ in gorgonian corals and Lapointe, Barile, and Matzie (2004) measured $\delta^{15}N$ in macroalgae, both concluding that anthropogenic nitrogen pollution was responsible for the increased $\delta^{15}N$ values they found. However, Swart, Saied, and Lamb (2005) measured $\delta^{15}N$ in the scleractinian *Montastraea faveolata* and did not find any indication of nitrogen pollution. Furthermore, based on extensive sampling and a thorough literature review, Swart (personal communication) cautions that for several reasons it is not possible to use specific $\delta^{15}N$ values alone as an indicator for sewage, including: (1) variable isotopic signals for sewage that are not always heavy; (2) $\delta^{15}N$ results measured for particulate organic material that did not differ along nearshore to offshore transects throughout the Keys; and (3) discovery that the $\delta^{15}N$ of particulate organic matter from inshore stations is largely derived from seagrasses and macroalgae, not sewage. Even if sewage pollution makes its way to offshore reefs, the amount supplied must be measured relative to substantial natural nutrient sources due to Gulf Stream flow, tidal flushing (Szmant and Forrester 1996), and upwelling (Leichter, Stewart, and Miller 2003). The Gulf Stream and tidal movements must also be considered a diluting influence relative to potential sewage inputs.

One of the largest quantified sources of episodic nutrient additions to offshore reefs in the Keys is frequent cold-water upwelling events that deliver high concentrations of nitrate and phosphate on a regular basis (Lapointe and Smith 1987; Leichter and Miller 1999; Leichter, Stewart, and Miller 2003; Smith et al. 2004; for a similar analysis from the Great Barrier Reef, Australia, see Andrews and Gentien 1982; Sammarco et al. 1999). Leichter, Stewart, and Miller (2003) noted enriched $\delta^{15}N$ at depth related to internal tidal bores. More recently, Leichter (personal communication) measured the $\delta^{15}N$ values of the upwelled nitrate in these waters and found them to be sufficiently high to confound interpretations that sewage or runoff is responsible for increased $\delta^{15}N$ measured in seaweed collected along the outer reef tract (Lapointe, Barile, and Matzie 2004). These upwelling events deliver water with 10- to 40-fold higher nutrient concentrations than published estimates of inputs to nearshore waters from wastewater and stormwater. Based on the physical oceanographic processes responsible for the transport and delivery of these high-concentration nutrient pulses, it is likely that the offshore reefs of the Florida reef tract have historically and periodically been subjected to high concentrations of nonanthropogenic nitrate and phosphate. Periodic advection of coastal waters may also deliver pulses of nutrient-rich water to the offshore reefs (Pitts 2002).

Szmant (1997, 2002) suggested that the level of topographic complexity on a reef indirectly influences the effects of nutrient input and uptake. In the absence of overfishing or mass mortalities of herbivores, topographic complexity determines the availability of shelter for herbivores. The recent mass mortality of the acroporids has greatly reduced the topographic complexity of these reefs, reducing herbivore populations and thereby reducing rates of algal consumption in the face of pulsed or chronic nutrification (Szmant 1997). For coral reef and seagrass habitats, however, it has been shown that intact herbivore populations can compensate for effects of increased nutrient supply by locating and consuming nutrient-enriched algae (Boyer et al. 2004). While the necessary modeling has not been conducted to quantify the relationships among various nutrient sources and sinks in the Keys, blaming sewage from leaking cesspits or runoff from the Everglades Agricultural Area as the primary causes of coral-reef decline in Florida is almost certainly an oversimplification.

Water quality, however, explains some important things about the Florida reef tract. For instance, in the last few thousand years, reef development has been greater seaward of the Keys than seaward of the tidal passes between them. As previously discussed, this inimical-waters model explains the discontinuous nature of reefs along the reef tract. One would therefore expect a similar effect today with coral assemblages being healthier and more robust where the Keys block the flow from Florida Bay. This, in fact, is not what is happening, as essentially all reef communities of the Keys—even those blocked from Florida Bay—are in decline (Murdoch and Aronson 1999). It may well be that all reef areas of the Keys, not just those seaward of the tidal passes, are being exposed to poor-quality water. However, water quality along the Florida reef tract is best seaward of Key Largo (Boyer personal communication); yet coral loss has historically been highest in this area. This is exactly opposite of Lapointe's hypothesis for local nutrient sources as the proximal cause of reef decline in the Keys. It is becoming increasingly apparent that regional- and global-scale causes of reef decline are most important in structuring modern reef communities including those in Florida.

Gardner et al. (2003) used meta-analysis to assess the extent of coral decline across the Caribbean since the 1970s. Their results revealed that reefs from all regions were affected. Interestingly, Florida's reefs showed lower levels of loss than other regions. In general, the same maladies affecting coral reefs in Florida are present throughout the Caribbean, with reefs both near and far from anthropogenic sources of nutrients displaying similar community shifts from coral to macroalgal domination (Aronson and Precht 2001c). For instance, Lapointe, Littler, and Littler (1997) used reefs of the offshore barrier reef tract of Belize in the vicinity of Carrie Bow Cay as an example of a low-nutrient (oligotrophic) reef system and as a counterpoint to the Florida reef tract (Lapointe 1997). Like Florida, these Belizean reefs have recorded significantly reduced coral populations, especially the acroporids, with concomitant increases in macroalgae over the past two decades in the absence of nutrient enrichment (Rützler and Macintyre 1982; Littler et al. 1987; Aronson et al. 1994; McClanahan and Muthiga 1998; McClanahan et al. 1999; Aronson and Precht 2001b; Precht and Aronson 2002). Likewise, McClanahan et al. (2003) showed through experimental manipulation that coral mortality combined with low herbivory is responsible for the high levels of macroalgae reported on remote patch reefs at Glovers Reef Atoll, Belize. In another manipulative field experiment off Key Largo, Miller et al. (1999) found strong effects on increasing frondose macroalgal biomass as a response to exclusion of large herbivorous fishes and only negligible effects from nutrient enrichment.

To determine if there were linkages between deteriorating water quality and reef condition in the Florida Keys, Dustan (1994) performed coral vitality studies off Key Largo. He used reefs from San Salvador Island in the easternmost Bahamas as his "unaltered" control. Nutrient-depleted, oligotrophic pelagic waters bathe these Bahamian reefs and local anthropogenic impacts were deemed minimal. Interestingly, there was no statistical difference in diseased and degraded corals between the Bahamian reefs and the reefs in the Florida Keys. The Bahamian reefs had suffered the same catastrophic losses of acroporid corals in the early 1980s related to the Caribbean pandemic of WBD (Shinn 1989; Aronson and Precht 2001b). More recently, a coral-bleaching event recorded in 1995 at San Salvador resulted in a significant decline of the remaining coral species, with concomitant increases in macroalgae. This community shift occurred without changes in herbivore abundance or nutrient concentrations (Ostrander et al. 2000).

Bryant et al. (1998) profiled the Florida Keys reefs as one of the world's twelve most threatened. However, they stated, "[D]espite years of research, it is difficult to lay blame for damage on specific anthropogenic stresses." Association between nearshore nutrient enrichment and coral mortality followed by a shift toward macroalgal dominance does not by itself prove a causal connection. In fact, evidence linking anthropogenic nutrients to coral reef decline in Florida remains correlative. The most important consequences of increased nutrification may, however, lie in an increase in the severity and duration of coral diseases (Bruno et al. 2004) and in the inability of coral reefs to recover from disturbance events, especially if human activities serve to increase the frequency and/or intensity of the disturbance regime (McCook, Wolanski, and Spagnol 2001).

Lastly, Kinsey (1991), Smith and Buddemeier (1992), and Buddemeier and Kinzie (1998) have pointed out that elevated nutrient levels are commonly regarded as intrinsically damaging to the physiological function of corals. Nevertheless, corals often grow well at high nutrient levels (Atkinson, Carlson, and Crow 1995; Szmant 1997). Furthermore, corals can persist in close juxtaposition to macroalgae that are commonly regarded as superior competitors in high-nutrient situations (see de Ruyter van Steveninck, Van Mulekon, and Breenan 1988; Miller 1998; McCook 2001; Jompa and McCook 2002). Clearly the role of nutrients on coral reefs is not completely understood.

9.5.2 Climate Change

On coral reefs, most predictions of the effects of global climate change have been confined to temperature-induced coral bleaching (Hoegh-Guldberg 1999; Walther et al. 2002), rising sea level (Graus and Macintyre 1998), and changing ocean chemistry (Kleypas, Buddemeier, and Gattuso 2001; Buddemeier, Kleypas, and Aronson 2004). Recent reports have also identified the poleward expansion of reef corals as a response to climatic warming (Precht and Aronson 2004). The effects of climate change are most likely to be seen in community and species responses to the changes in the nature, intensity, and/or frequency of extreme events.

In the last two decades, documented coral bleaching events have increased in frequency and intensity both globally and regionally (Hoegh-Guldberg 1999). Evidence from sclerochronological studies supports the suggestion that widespread coral bleaching is a recent phenomenon (Halley and Hudson, Chapter 6). Stone et al. (1999) proposed that the sudden occurrence of mass bleaching episodes is coincident with intensification of El Niño–Southern Oscillation (ENSO) events, which may be augmented by global warming (Hoegh-Guldberg 1999; Aronson et al. 2002b; Hughes et al. 2003).

In Florida, bleaching events occurred during 1983, 1987, 1989, 1990, 1991, 1997, and 1998 (Causey 2001). Mortality after the 1983 bleaching event, affecting only the outer reefs in the lower Keys, was estimated at 3 to 5% (Causey 2001). The first major Keys-wide bleaching event occurred in 1987, with recovery of most of the affected colonies. The worst years on record for coral bleaching were 1997 and 1998, worldwide and in Florida (Causey et al. 2000; Wilkinson 2000). In the Keys, corals remained bleached or mottled from 1997 into 1998, when they experienced another episode of bleaching due to excessively warm summertime conditions (Edmunds, Gates, and Gleason 2003). These consecutive bleaching events in Florida caused alarm among local scientists and managers. Declines in coral cover were reported during the 1997 to 1998 period (Jaap et al. 2001), but because high-frequency before-during-and-after sampling was not conducted, the relationship between coral mortality and bleaching remains correlative. Two hurricanes also affected the region during this period, further confounding interpretations. Although estimates of bleaching-related coral mortality are not available for Florida, anecdotal evidence suggests that bleaching has killed large numbers of corals (Causey 2001) and must be considered a significant threat to reefs in Florida (Szmant 2002). The projected continuing increases of bleaching events on Florida's coral reefs related to global warming are likely to decrease coral abundance into the future (Hoegh-Guldberg 1999; Walther et al. 2002; Sheppard 2003). Bleaching was not observed in Florida in 1995, yet major bleaching with associated mortality was observed in Belize (McField 1999) and at San Salvador Island in the Bahamas (Ostrander et al. 2000).

Predicted effects of changes in seawater chemistry on reef growth (carbonate production) are based largely on geochemical evidence and modeling. Essentially all of the experiments that have manipulated seawater carbonate chemistry and measured effects on corals, calcifying algae, and coral reef mesocosms indicate that calcification approaches a linear response to decreased carbonate saturation state (Kleypas, Chapter 12). Based on the factors that control $CaCO_3$ precipitation, it is very likely that dissolution will increase on reefs as atmospheric CO_2 concentrations increase, leading to reduced capacity for reef building.

Extreme weather and climate events will likely increase in the future (Easterling et al. 2000), especially in mid- to high-latitude settings. For coral reefs the most profound changes are likely to be seen at the latitudinal extremes of reef building, including the reefs of Florida. Increases in extreme high temperatures, decreases in extreme low temperatures, and increases in tropical storms and hurricanes will likely add stress to a reef system already suffering from numerous natural and anthropogenic perturbations.

9.5.3 Overfishing

Overfishing is often considered to be one of the most significant threats to coral reef ecosystems (Roberts 1995; Jackson et al. 2001). At high intensities, fishing reduces species diversity and can lead to extirpations of not only target species, but other species as well. Most importantly, overfishing interacts with other disturbances such as disease outbreaks in populations of herbivorous sea urchins, reducing the capacity of reefs to recover from natural perturbations such as hurricanes (Hughes 1994; Nyström, Folke, and Moberg 2000; Knowlton 2001; Bellwood et al. 2004).

In the Florida Keys overfishing of finfish and shellfish has resulted from a combination of commercial pressure and substantial recreational effort (Bohnsack, Harper, and McClellan 1994; Ault, Bohnsack, and Meester 1998; Bohnsack 2003). Substantial declines have occurred in historically productive snapper and grouper fisheries (Ault et al. 2001). Densities of herbivorous fish, however, remain high compared to many sites in the Caribbean, mostly because fishermen do not target these species (Aronson et al. 1994; Bohnsack, Harper, and McClellan 1994). This is especially relevant in Florida because herbivorous fish may have ameliorated the effects of the 1983 to 1984 mass mortality of the sea urchin *Diadema antillarum* compared to other sites with depauperate herbivorous fish populations, which suffered substantial increases in macroalgal cover and biomass after the urchin dieoff (Hay 1984; Carpenter 1990a,b; Aronson and Precht 2000, 2001b).

One model of coral-reef ecology holds that the persistence of reef communities turns on positive interactions between corals and herbivorous fish (Hay 1981; Hay and Goertemiller 1983; Hay and Taylor 1985; Lewis 1986; Szmant 1997). Several studies have demonstrated the negative effects of fishing on tropical resources (Hughes 1994; Roberts 1995; Rogers and Beets 2001). Results from marine protected areas also suggest that such connections can be important (McClanahan and Muthiga 1988). Equally important, some authors have argued historical declines in the population densities and body sizes of herbivorous and predatory fish have gone unrecognized as potential agents of change in coral reef habitats (Jackson 2001; Jackson et al. 2001).

Pandolfi et al. (2003) argued that the only reasonable explanation for the long-term degradation of reef systems is overfishing. The loss of large predators, including groupers, snappers, and sharks, has likely had substantial cascading effects on the structure of fish assemblages (Aronson 1990; Ault, Bohnsack, and Meester 1998; Pauly et al. 1998; Steneck 1998), which in turn has led to collapse of whole reef ecosystems at the global scale (Pandolfi et al. 2003).

The idea that overfishing has been a main cause of the collapse of coral assemblages on reefs of Florida and the Caribbean has been the subject of contentious debate (Aronson et al. 2003; Precht et al. 2005b) that continues to the present (Precht and Aronson 2006). It is clear that overfishing can eliminate entire functional groups and lead to trophic cascades on reefs (Pauly et al. 1998), but there is no evidence that reef decline in Florida or elsewhere in the Caribbean have been compromised by these actions.

Coral reef ecosystems throughout the Caribbean and western Atlantic including Florida have been overfished for decades and perhaps centuries (see Wing and Wing 2001). It may well be that reefs with their full complement of trophic levels and resources will be more resilient to disturbances than those that are over-exploited (Knowlton 2001), but it remains premature to assume that coral assemblages will automatically recover to historic levels if fish populations are restored.

Effective management strategies such as the creation of marine reserves should help restore and protect fishery resources, possibly resulting in increased reef resilience. In the FKNMS, the recent establishment of some 23 no-take marine reserves is a first step.

9.5.4 Diseases

Major disease events in recent years have changed the way reefs look over relatively short time scales. For example, in the early 1980s a disease epidemic killed almost all the *D. antillarum* that once thrived in the Caribbean region (Lessios, Robertson, and Cubit 1984), including the Florida Keys (Forcucci 1994). *D. antillarum* was an important grazer, and after the mass mortality substantial and long-term increases in macroalgal populations generally resulted, especially on reefs subjected to intense fishing pressure (Hay 1984; Carpenter 1990a,b; Hughes 1994).

Urchin recovery has begun at a few locations in the Caribbean (Aronson and Precht 2000; Edmunds and Carpenter 2001; Cho and Woodley 2002) but not in the Florida Keys, where a second dieoff was observed in the early 1990s (Forcucci 1994). A recent Keys-wide survey counted few urchins but discovered evidence of significant adult populations in the Dry Tortugas (Chiappone et al. 2001). Although the cause of the urchin dieoff remains unknown (but see Bauer and Agerter 1987), it represents one of two ecologically critical disease outbreaks that swept through the Caribbean and Florida in the last 25 years.

The second epidemic, WBD (discussed previously), still affects acroporid corals in Florida (Precht and Aronson 1997; Porter et al. 2001; Santavy et al. 2001). The loss of the acroporids due to WBD significantly reduced the three-dimensional structure of the reefs and effectively eliminated two of the major reef-building corals from the region (Sheppard 1993; Aronson and Precht 2001c; Precht et al. 2002). Coring studies in Belize and Jamaica have revealed that the regional mass mortality of acroporid corals from WBD was novel on a centennial to millennial scale (Aronson et al. 2002a; Wapnick, Precht, and Aronson 2004). The combined effects of the two lethal, essentially concurrent, Caribbean-wide epizootics resulted in what was apparently an unprecedented, regional shift from coral to macroalgal dominance (Knowlton 2001).

Estimates of the prevalence of WBD in remnant *A. cervicornis* and *A. palmata* populations in the FKNMS were obtained over the period 1999 to 2001 and were 3.2% and 5.6%, respectively (Chiappone, Miller, and Swanson 2002). In addition to WBD, another lethal disease known as white pox, which is specific to *A. palmata*, has been identified throughout the Florida reef tract (Holden 1996). The common bacterium *Serratia marcescens* has been implicated (Patterson et al. 2002). Because *S. marcescens* can be a human fecal enterobacterium, Patterson et al. (2002) inferred that its pathway into the marine environment is likely via human waste. However, further evidence is needed to substantiate this species as more than an incidental pathogen. Others have noted that *S. marcescens* is commonly found throughout the marine realm, including in the gut tracts of fish (Bruckner 2002). As enterobacteria are ubiquitous in the marine environment, it should not

278 William F. Precht and Steven L. Miller

be surprising that some should be opportunistic pathogens of fish and marine invertebrates, including corals that are already stressed.

Beginning in the 1970s and continuing through today, coral diseases or disease-like syndromes have appeared in other coral species throughout the Caribbean (Antonius 1977; Edmunds 1991; A. Bruckner, Bruckner, and Williams 1997; Nagelkerken et al. 1997; Santavy et al. 1999; Garzon-Ferreira et al. 2001), Bermuda (Garrett and Ducklow 1975), and the Florida Keys (Dustan and Halas 1987; Kuta and Richardson 1996; Richardson et al. 1998; Porter et al. 2001; Santavy et al. 2001; Patterson et al. 2002). Other environmental stressors such as pollution, increased nutrients, increased iron supply, African dust, and temperature may be associated with coral disease outbreaks, yet no firm connections have been established (Shinn 1996, 2001; Epstein et al. 1998; Hayes and Goreau 1998; Harvell et al. 1999; Hayes et al. 2001; Jackson et al. 2001; Richardson and Aronson 2002; Bruno et al. 2004). Antonius (1977) described a white syndrome in acroporid corals, which he termed shut-down reaction (SDR). He noted that even a slight break in coral tissue in "stressed" colonies triggered SDR. The symptoms of SDR are similar to WBD but highly accelerated. It may be that stress due to increased temperature contributes to white syndromes in the acroporids. The long-term impacts of coral diseases on populations remain difficult to assess, but disease is clearly a major factor in the continued decline of corals and reef function in Florida.

9.6 The Future

Future climatic and environmental changes are emerging as major research priorities in the new millennium. The National Research Council (1988) recommended that studies documenting population, community, and ecosystem responses to rapid environmental changes of the Quaternary could provide the necessary insight into the rates and directions of biotic changes in the future. It has been argued that Quaternary reefs from the Caribbean display stability and persistence in coral assemblage structure through time and, therefore, constitute the most important database for the study of modern reefs (Jackson 1991, 1992; Pandolfi and Jackson 1997; Pandolfi 1999, 2002). While we agree that these studies demonstrate predictable patterns of reef community membership in tropical areas during glacial–interglacial cycles, we disagree with the notion that persistent, tropical coral assemblages represent biologically integrated faunas (e.g., Jackson 1994). Rather, persistence was likely due to individual species tracking suitable habitats, favorable environmental conditions, and fluctuating sea levels in time and space (see Brett, Miller, and Baird 1990; Holland 2000).

Much can be learned about coral ecology from Pleistocene reefs of the Caribbean: the persistence of reef zonation schemes through time, the continuity of environments during sea level fluctuations, and the fidelity of community and species composition during successive high stands in the tropics. It may well be, however, that reefs living under suboptimal conditions in more thermally reactive areas, including reefs in Florida tell a different story and may be more useful for predictive studies. We believe that had Jackson, Pandolfi, and colleagues looked in the other direction, at the latitudinal extremes of subtropical reef systems, they would have come to a very different conclusion about the stability of reef ecosystems. For instance, the reef community at Rottnest Island off Western Australia (32°S) has some 25 species of zooxanthellate corals, most occurring at the southern limits of their ranges, with *Acropora* spp. being rare (Marsh 1992). However, during the last major interglacial (MIS 5e) major reefs were formed by both staghorn and tabular *Acropora* spp. (Szabo 1979; Playford 1988). In a similar analysis, Veron (1992) showed that on the world's highest-latitude coral reef in Tateyama, Japan (33.5°N), differences in species composition between the mid-Holocene and the modern indicate that even a brief period of regional or global warming can produce significant changes in coral assemblages. Conversely, in the insular tropics the mid-Holocene temperature maximum had no effect on species composition or reef-building (Gill, Hubbard, and Dickson 1999).

When conditions have been favorable in Florida, as they were during the middle Holocene, acroporids have dominated the shallow-reef community. When conditions have deteriorated, as in the Pleistocene, head corals have dominated and persisted. Species replacements and range expansions in the past, especially of the *Acropora* spp., emphasize the resilience of reef ecosystems and the individualistic responses of coral species to rapid environmental change in the absence of major human modification of the seascape.

Recent changes to Florida's coral reefs could persist for decades or longer, and it is unclear how future global climate change will interact with disease and other stresses (Brown 1997; Harvell et al. 1999; Kleypas et al. 1999). The virulence of some coral diseases may be positively correlated with increases in temperature, which could doom reefs in the face of global warming (Rosenberg and Ben-Haim 2002). The combined effects of numerous anthropogenic impacts could alter ecosystem resilience and prevent recovery of the affected reefs (Knowlton 1992; Nyström, Folke, and Moberg 2000; McCook, Wolanski, and Spagnol 2001).

What is required for recovery in the Florida Keys? Of course, the recent demise of the *Acropora* in the Caribbean has not yet been tied to extinction. In ecological time, the prolific growth of the acroporids (see Shinn 1966, 1976) could essentially repopulate all of the habitats occupied in Florida prior to the WBD epidemic. Populations of *Acropora* spp. persist throughout their geographic range, including Florida, and substantial recovery could occur on a decadal scale (Woodley 1992; see Idjadi et al. 2006). If the acroporids do recover, we would expect prolific and expansive acroporid thickets in the lee of large islands or seaward of the mainland of the Florida Peninsula, in areas protected from inimical waters.

Recent observations in the Keys show that short-term, sporadic, and spatially limited acroporid thickets can develop from remnant populations. Demise of these reoccurring smaller populations remains highly probable, but in the face of global warming a northward expansion of the acroporids and the reef tract is also possible, mimicking the conditions of the mid-Holocene (Precht and Aronson 2004). The recent occurrence of *A. cervicornis* thickets off Fort Lauderdale presents an interesting case. Are these populations a harbinger of impending global change, or do they merely represent the temporary range expansion of remnant stands, which are likely to be killed by the next cold front or hurricane?

Factors related to biology may limit recovery of the acroporids. Dependence on asexual propagation in the acroporids comes at the expense of larval production (Hughes 1989; Knowlton, Lang, and Keller 1990). *A. palmata* and *A. cervicornis* are broadcast spawners, releasing their eggs into the water column for fertilization and development. Because these corals are now rare, they may be experiencing an Allee effect; colonies may be too far apart for high fertilization success (Knowlton 1992). Low levels of sexual recruitment by *A. cervicornis* have slowed its recovery at most Caribbean localities. Although rapid growth and fragmentation are also strategies employed by *A. palmata*, this species shows slightly higher rates of sexual recruitment to disturbed areas than *A. cervicornis* (Highsmith 1982; Rosesmyth 1984; Jordán-Dahlgren 1992). Currently, brooding coral species, which retain their eggs after internal fertilization and release their offspring as planula larvae, are recruiting more successfully than broadcast spawners in the Caribbean (Fig. 9.9). All Caribbean representatives of the families Agariciidae and Poritidae are brooders, and *Agaricia agaricites, P. astreoides*, and *P. porites* are among the first coral species to appear on disturbed reef surfaces, including *Acropora* rubble fields (Bak and Engel 1979; Rylaarsdam 1983; Rogers et al. 1984; Hughes 1989; Edmunds et al. 1998; Aronson and Precht 2001c; Knowlton 2001).

What else could prevent recovery? Disease pathogens are still present (Santavy et al. 2001; Aronson and Precht 2001c; Patterson et al. 2002), cold-weather fronts still occur, bleaching events are increasing in frequency and intensity, hurricanes

Figure 9.9. Present-day coral assemblage in many shallow fore-reef settings. Brooding species dominate, especially *Porites astreoides, P. porites*, and *Agaricia agaricites* (forma *danai*).

can set back reef development, and predators are continuing to take their toll on remnant populations (Knowlton, Lang, and Keller 1990; Bruckner, Bruckner, and Williams 1997; Miller 2001), especially when other stressors are also present (Hughes 1989, 1994; Knowlton 1992; Hughes and Connell 1999). Suitable substrate may also limit larval recruitment where algal populations remain high (Hughes and Tanner 2000). If one or both of the *Acropora* species do not recover, we predict continued high abundance of fleshy macroalgae and an increasing role for weedy, fecund, brooding corals such as *Porites* and *Agaricia*. At longer time scales of decades to centuries, we predict that slower-growing, eurytopic, head-coral species, especially the *Montastraea* species, will prevail.

In retrospect, we are lucky to have reefs in Florida at all. Disease outbreaks; coral bleaching; extreme weather events, including hurricanes and cold fronts; and the average annual position of the Gulf Stream all affect the nature and presence of reefs in Florida. A major challenge to scientists and managers working in the Keys is to understand the combined effects of natural system variability and human-caused damage. Can we use our knowledge of reefs through time in Florida and the rest of the Caribbean to establish guidelines, methods, and strategies for protection and stewardship?

9.7 Management Implications at the Interface of Geology and Ecology in South Florida

Wood (1999) stated, "[A]ncient reefs are too often studied solely as geological phenomena with little regard to the detailed biological interactions within their constructional communities." Her premise was that if one does not understand the biology of a reef, one does not understand the reef at all. Conversely, Aronson and Precht (2001b) commented ". . . that paleobiologists pay close attention to ecology, but ecologists largely ignore the fossil record." They noted that if the important questions for the future are to be answered, we must combine the two disciplines as Gould (1981) suggested.

Paleoecological studies of Pleistocene and Holocene reefs: (1) make it possible to investigate ecological processes that occur over temporal scales longer than most long-term (decadal-scale) ecological studies; (2) provide a baseline for the condition of reefs prior to human influence; and (3) enable us to evaluate the response of corals and coral reefs to environmental changes of magnitudes that are beyond historical values but within the range of projected global change. Understanding natural change in reef community structure preserved in fossil sequences can provide a major source of information for predicting future change. Specifically, the lessons learned from paleoecological studies can form the basis for the restoration and management of modern ecosystems (Foster, Schoonmaker, and Pickett 1990). If present approaches to conservation are not sufficient to preserve or restore existing or damaged reef systems (Buddemeier 2001), then can we do anything that will make a difference?

Coral reefs in the Florida Keys that are typically described as the ones in need of salvation represent only a fraction—about 2%—of the total coral habitat in the

Florida Keys (FMRI 1998). These are the shallow spur-and-groove locations that until recently were dominated by *A. palmata* and *A. cervicornis*, and they are the reefs that kept pace with rising sea level through the Holocene; they are the named reefs on nautical charts (Fig. 9.2). What about the other 98% of coral habitat in the Keys? There are vast stretches of shallow and deep hard-bottom habitats characterized by low hard-coral cover, variable but often-high algal cover, and variably abundant gorgonians and sponges (Enos 1977; Chiappone and Sullivan 1994). Many of these habitats are probably little changed over the last 20 years (and possibly much longer), but monitoring programs do not typically include such sites in their studies (Hughes 1992).

It has been argued that the coral reef scientific and management community has failed the world's coral reefs (Risk 1999; Bellwood et al. 2004). In Florida, this implied failure is based on the hypothesis that diminished water quality is the single most important cause of environmental degradation in the Everglades, in Florida Bay, and on the reefs of the Florida Keys, and that the woes that beset these systems are local, human-induced, and reversible. It is easy to come away with this view after reading the literature (Lapointe 1997; Bacchus 2002; Brand 2002; Lapointe, Matzie, and Barile 2002; Lipp et al. 2002; Patterson et al. 2002; Porter et al. 2002) and being exposed to the idea as presented in the popular media (e.g., Lapointe 1989; Torrance 1991; Dustan 1997; Barnett 2003), but evidence linking nutrient loading to reef degradation remains elusive (Szmant 2002).

We contend that the focus on nutrient pollution as the primary factor responsible for the decline of coral reefs in the Florida Keys is misplaced. Unfortunately, politicians, managers, and the public are receptive to such arguments because poor sewage treatment and runoff from the Everglades Agricultural Area are things that make intuitive sense as causal agents and also have strong emotional appeal. As a result, the paradigm of nutrient pollution and coral decline has dominated debate in Florida for over 15 years. Nobody is going to argue that sewage pollution or untreated agricultural runoff is a good thing. However, intuition and emotional appeal are nonscientific approaches that seldom leave room for alternative or complementary views. One of our primary motivations for writing this chapter is that we believe it is time to move to a more balanced dynamic between the multiple stakeholders throughout south Florida and the scientists who help inform them.

Winter cold fronts, hurricanes, global warming combined with increasing ENSO intensity (causing coral bleaching), and coral and urchin diseases are stressors in the Florida Keys that have well-known cause-and-effect relationships at multiple spatial and temporal scales. These punctuated extreme events have strongly influenced the trajectory of these coral communities, yet authors advocating the primacy of diminished water quality in reef decline have essentially overlooked them. As Pennisi (1997) commented, "[The] lack of a clear smoking gun even for reefs as well studied as Florida's have been due partly to the tendency of reef scientists to focus their research on a single parameter—say, water quality." Nutrient pollution is clearly a threat, and it is possible that gradual deterioration of water quality at the regional scale has predisposed the reef biota to the sudden and radical shifts observed in the past few decades (Dubinsky and

Stambler 1996; Vitousek et al. 1997). Management steps are already under review to clean up nearshore waters in the Florida Keys (i.e., the Florida Keys Water Quality Improvement Program; Kruczynski and McManus 2002), and systemwide water-quality and habitat-restoration efforts are ongoing under the Comprehensive Everglades Restoration Plan (Causey 2002). In 1998 the Governor of Florida issued Executive Order 98-309, directing local and state agencies to coordinate with Monroe County to eliminate cesspits, failing septic systems, and other substandard sewage systems. Additionally, in 2001 the Florida Keys Water Quality Act (Public Law 106-554) was approved by the U.S. Congress and authorized the U.S Army Corps of Engineers to provide technical and financial assistance for wastewater treatment and stormwater management projects to improve water quality of the FKNMS. When implemented, engineering solutions to improve the quality of nearshore waters certainly will not hurt the offshore reefs, and they may provide some benefit; however, it is absurd to think that the Florida reef tract will recover within our lifetime if we do not address the larger-scale causes of reef decline.

Saving coral reefs rests on our ability to forge a comprehensive, sophisticated, and accurate understanding of the mechanisms responsible for coral reef decline and regeneration. Therefore, it is imperative that managers and policymakers directly address the range of factors that have devastated coral populations and hinder their recovery. As Knowlton (2001) noted, it would be prohibitively expensive to repair reefs crippled by the negative synergistic effects of different types of stressors, including global warming and emergent diseases, yet these stressors need to be addressed at multiple scales if reefs as we know them are to survive.

9.8 Summary and Conclusions

By studying reef communities through the Quaternary and their changes over time, it is possible to understand more fully the history of development, structure, and function of modern reef systems. Conversely, modern reefs form the foundation for understanding the multiscale processes behind the preserved sedimentary architecture and biofacies of the fossil reef sequences. The principle of uniformitarianism forms the basis for understanding changes in reefs through time and space. Studies of Caribbean Quaternary reefs emphasize the persistence in composition and zonation over broad spatial and temporal scales. It may well be, however, that Quaternary reefs living under nonoptimal conditions, including those in Florida, may prove more useful for predicting the future trajectory of reef communities.

In this chapter, we have attempted to synthesize the Quaternary history of reef development in Florida. Using the maxim "the past is the key to the future," we have tried to tease out those factors most responsible for reef decline in Florida, both historically and recently. Living coral reefs of Florida have been in a state of flux for the past 25 years. Among these changes has been the near-elimination of the dominant coral species *Acropora palmata* and *A. cervicornis*, with concomitant increases in macroalgae. Winter cold fronts, hurricanes, global warming (causing coral bleaching), and coral disease are stressors in the Florida Keys that have well-known cause-and-effect relationships. These rapid, rare, and extreme events have strongly influenced the composition of reef communities in the Florida Keys.

Whether these recent changes are natural cyclic events or the result of human activities has been a topic of strenuous debate. To address this issue, we asked the question, "Did episodes of reef degradation occur in the past, before the era of human interference, or is the current state of coral reefs unique to our time?" Because coral reefs are both geologic and biologic entities, it is possible to observe the effects of various disturbances in ecological time, detect historical changes in the paleoecological record, and deduce the multiscale processes behind those patterns. For Florida at least, the present reef community assemblage, highlighted by diminishing *Acropora* populations, is *not* unique in space or in time. This is contrary to results from other areas of the Caribbean.

Being at the northern limit of reef growth in the western Atlantic, Florida is subjected to a host of conditions unfavorable to prolific reef development. In fact, we are lucky to have reefs at all in Florida. Disease epidemics, coral bleaching events, extreme weather conditions including hurricanes and cold fronts, as well as the average annual position of the Gulf Stream affect the ecological history of reefs in Florida. Throughout the Quaternary, Florida's reefs have been subjected to numerous large-scale disturbances that have reorganized the coral community structure. Although in many cases the causes of these disturbance events have been different, the resulting community shift away from dominance by *Acropora* spp. has been similar. When conditions have been favorable on time scales of millennia, as in the early to middle Holocene, acroporids have dominated the shallow-reef community. When conditions have deteriorated, as in the Pleistocene, head corals have dominated and persisted. Quaternary reefs from Florida emphasize the resilience of reef ecosystems and the individualistic responses of coral species to rapid environmental change in the absence of major human modification of the seascape. These ecological shifts in coral community composition are preserved in the fossil and subfossil record of these reefs and allow us to use the past as a key to predicting the future of reefs in a world now besieged by numerous disturbances and the influence of man. As Gene Shinn has repeatedly emphasized, the story written in the geologic history of coral reefs is our most reliable guide to their uncertain future.

Acknowledgements. Many of the ideas expressed in this chapter were built on foundations provided by previous workers. In that regard, special thanks go to Gene Shinn and Barbara Lidz of the USGS for freely sharing their views on the history of the Florida reef tract. We would also like to thank Hemmo Bosscher, John Bruno, Billy Causey, Mark Chiappone, Pete Edmunds, Bill Fitt, Peter Glynn, Jack Kindinger, Les Kaufman, Joanie Kleypas, Judy Lang, Jim Leichter, Diego Lirman, Ian Macintyre, Thad Murdoch, Conrad Neumann, Harry Roberts, Caroline Rogers, and Dione Swanson for discussions over the years that led to consolidation of many of the ideas presented herein. Leslie Duncan helped with sorting out the references. Thad Murdoch drafted Fig. 9.2, and Beth Zimmer and

Adam Gelber prepared Figs. 9.4, 9.7, and 9.8. Thad Murdoch, Martha Robbart, Barbara Lidz, Billy Causey, and Gene Shinn critiqued versions of the manuscript and Ian Macintyre, Bernhard Riegl, and Maggie Toscano provided critical reviews of the final draft. We thank Brian Keller of the FKNMS for his thoughtful analysis and editorial advice. Sincere thanks to our friend and colleague Rich Aronson for his patience and dedication in pulling this volume together and for his insight and comments throughout the entire process that greatly enhanced this chapter.

This chapter is dedicated to the memory of the late Donald R. Moore (1921–1997), Professor of Marine Geology at the University of Miami. Don was one of the early pioneers working on some of the problems discussed in this chapter. All who knew him will miss his sage advice and kind nature.

References

Adey, W.H., and R.B. Burke. 1977. Holocene bioherms of Lesser Antilles: Geologic control of development. In *Reefs and Related Carbonates—Ecology and Sedimentology*, eds. S.H. Frost, M.P. Weiss, and J.B. Saunders, 67–82. Studies in Geology No. 4. Tulsa: American Association of Petroleum Geologists.

Agassiz, A. 1883. The Tortugas and Florida reefs. *Mem. Am. Acad. Arts Sci.* 11: 107–134.

Agassiz, A. 1896. The elevated reef of Florida. *Bull. Mus. Comp. Zool. Harvard Coll.* 28:1–62.

Ahr, W.M. 1973. The carbonate ramp: An alternative to the shelf model. *Trans. Gulf Coast Assoc. Geol. Soc.* 23:221–225.

Allen, T.F.H., and T.W. Hoeskstra. 1992. *Toward a Unified Ecology*. New York: Columbia University Press.

Andréfouët, S., P.J. Mumby, M. McField, C. Hu, and F.E. Muller-Karger. 2002. Revisiting coral reef connectivity. *Coral Reefs* 21:43–48.

Andrews, J.C., and P. Gentien. 1982. Upwelling as a source of nutrients for the Great Barrier Reef ecosystems: A solution to Darwin's question? *Mar. Ecol. Prog. Ser.* 8:257–269.

Antonius, A. 1977. Coral mortality in reefs: A problem for science and management. *Proc. Third Int. Coral Reef Symp., Miami* 2:617–623.

Antonius, A. 1981. The "band" diseases in coral reefs. *Proc. Fourth Int. Coral Reef Symp., Manila* 2:6–14.

Antonius, A. 1994a. Endangered elkhorn coral, *Acropora palmata* (Lamarck). In *Rare and Endangered Biota of Florida Series, Vol. IV. Invertebrates*, eds. M. Deyrup and R. Franz, 33–34. Gainesville: University of Florida Press.

Antonius, A. 1994b. Endangered staghorn corals, *Acropora cervicornis* (Lamarck), *Acropora prolifera* (Lamarck). In *Rare and Endangered Biota of Florida Series, Vol. IV. Invertebrates*, eds. M. Deyrup and R. Franz, 34–36. Gainesville: University of Florida Press.

Aronson, R.B. 1990. Onshore–offshore patterns of human fishing activity. *Palaios* 5:88–93.

Aronson, R.B. 1994. Scale-independent biological interactions in the marine environment. *Oceanogr. Mar. Biol. Ann. Rev.* 32:435–460.

Aronson, R.B. 2001. The limits of detectability: Short-term events and short-distance variation in the community structure of coral reefs. *Bull. Mar. Sci.* 69:331–332.

Aronson, R.B., J.F. Bruno, W.F. Precht, P.W. Glynn, C.D. Harvell, L. Kaufman, C.S. Rogers, E.A. Shinn, and J.F. Valentine. 2003. Causes of reef degradation. *Science* 302:1502.

Aronson, R.B., P.J. Edmunds, W.F. Precht, D.W. Swanson, and D.R. Levitan. 1994. Large-scale, long-term monitoring of Caribbean coral reefs: Simple, quick, inexpensive techniques. *Atoll Res. Bull.* 421:1–19.

Aronson, R.B., I.G. Macintyre, W.F. Precht, C.M. Wapnick, and T.J.T. Murdoch. 2002a. The expanding scale of species turnover events on coral reefs in Belize. *Ecol. Monogr.* 72:233–249.

Aronson, R.B., and W.F. Precht. 1995. Landscape patterns of reef coral diversity: A test of the intermediate disturbance hypothesis. *J. Exp. Mar. Biol. Ecol.* 192:1–14.

Aronson, R.B., and W.F. Precht. 1997. Stasis, biological disturbance, and community structure of a Holocene coral reef. *Paleobiology* 23:326–346.

Aronson, R.B., and W.F. Precht. 2000. Herbivory and algal dynamics on the coral reef at Discovery Bay, Jamaica. *Limnol. Oceanogr.* 45:251–255.

Aronson, R.B., and W.F. Precht. 2001a. Applied paleoecology and the crisis on Caribbean coral reefs. *Palaios* 16:195–196.

Aronson, R.B., and W.F. Precht. 2001b. Evolutionary paleoecology of Caribbean coral reefs. In *Evolutionary Paleoecology: The Ecological Context of Macroevolutionary Change*, eds. W.D. Allmon and D.J. Bottjer, 171–233. New York: Columbia University Press.

Aronson, R.B., and W.F. Precht. 2001c. White-band disease and the changing face of Caribbean coral reefs. *Hydrobiologia* 460:25–38.

Aronson, R.B., W.F. Precht, I.G. Macintyre, and T.J.T. Murdoch. 2000. Coral bleach-out in Belize. *Nature* 405:36.

Aronson, R.B., W.F. Precht, M.A. Toscano, and K.H. Koltes. 2002b. The 1998 bleaching event and its aftermath on a coral reef in Belize. *Mar. Biol.* 141:435–447.

Atkinson, M.J., B. Carlson, and G.L. Crow. 1995. Coral growth in high-nutrient, low-pH seawater: A case study of corals cultured at the Waikiki Aquarium, Honolulu, Hawaii. *Coral Reefs* 14:215–223.

Ault, J.S., J.A. Bohnsack, and G.A. Meester. 1998. A retrospective (1979–1996) multispecies assessment of coral reef fish stocks in the Florida Keys. *Fish. Bull.* 96:395–414.

Ault, J.S., S.G. Smith, J. Luo, G.A. Meester, J.A. Bohnsack, and S.L. Miller. 2001. Baseline multispecies coral reef fish stock assessments for the Dry Tortugas. Final Report to the National Park Service. Miami: Rosenstiel School of Marine and Atmospheric Science, University of Miami.

Bacchus, S. 2002. The "ostrich" component of the multiple stressor model: Undermining south Florida. In *The Everglades, Florida Bay, and Coral Reefs of the Florida Keys—An Ecosystem Sourcebook*, eds. J.W. Porter and K.G. Porter, 677–748. Boca Raton: CRC Press.

Bak, R.P.M. 1983. Aspects of community organization in Caribbean stony corals. In *Coral Reefs, Seagrass Beds and Mangroves: Their Interaction in the Coastal Zones of the Caribbean*, eds. J.C. Ogden and E.H. Gladfelter. UNESCO. Reports in Marine Science 23. Montevideo: UNESCO.

Bak, R.P.M., and M.S. Engel. 1979. Distribution, abundance and survival of juvenile hermatypic corals (Scleractinia) and the importance of life history strategies in the parent coral community. *Mar. Biol.* 54:341–352.

Bak, R., and G. van Eys. 1975. Predation of the sea urchin *Diadema antillarum* Philippi on living coral. *Oecologia* 20:111–115.

Ball, M.M., E.A. Shinn, and R.W. Stockman. 1967. The geologic effects of Hurricane Donna in south Florida. *J. Geol.* 75:583–597.

Balsam, W. 1981. Late Quaternary sedimentation in the western North Atlantic: Stratigraphy and paleoceanography. *Palaeogeogr. Palaeoclimatol. Palaeoecol.* 35:215–240.

Barnett, C. 2003. 'Distress syndrome'. *Florida Trend* June: 74–79.

Bauer, J.C., and C.J. Agerter. 1987. Isolation of bacteria pathogenic for the sea urchin *Diadema antillarum* (Echinodermata: Echinoidea). *Bull. Mar. Sci.* 40:161–165.

Baums, I.B., M.W. Miller, and A.M. Szmant. 2003. Ecology of a corallivorous gastropod, *Coralliophila abbreviata*, on two scleractinian hosts. I. Population structure of snails and corals. *Mar. Biol.* 142:1083–1091.

Beck, J.W., J. Recy, F. Taylor, R.L. Edwards, and G. Cabioch. 1997. Abrupt changes in early Holocene tropical sea surface temperature derived from coral records. *Nature* 385:705–707.

Bell, P.R.F., and T. Tomascik. 1994. The demise of the fringing coral reefs of Barbados and of regions in the Great Barrier Reef (GBR) lagoon—Impacts of eutrophication. In *Proceedings of the Colloquium on Global Aspects of Coral Reefs: Health, Hazards and History*, compiler R.N. Ginsburg, 319–325. Miami: Rosenstiel School of Marine and Atmospheric Science, University of Miami.

Bellwood, D.R., T.P. Hughes, C. Folke, and M. Nyström. 2004. Confronting the coral reef crisis. *Nature* 429:827–833.

Blair, S.M., T.L. McIntosh, and B.J. Mostkoff. 1994. Impacts of Hurricane Andrew on the offshore reef systems of central and northern Dade County, Florida. *Bull. Mar. Sci.* 54:961–973.

Blanchon, P. 1997. Architectural variation in submerged shelf-edge reefs: The hurricane control hypothesis. *Proc. Eighth Int. Coral Reef Symp., Panama* 1:547–554.

Blanchon, P., B. Jones, and W. Kalbfleisch. 1997. Anatomy of a fringing reef around Grand Cayman: Storm rubble, not coral framework. *J. Sediment. Res.* 67:1–16.

Bloom, A.L. 1994. The coral record of late glacial sea level rise. In *Proceedings of the Colloquium on Global Aspects of Coral Reefs: Health, Hazards and History*, compiler R.N. Ginsburg, 1–6. Miami: Rosenstiel School of Marine and Atmospheric Science, University of Miami.

Bohnsack, J.A. 2003. Shifting baselines, marine reserves, and Leopold's biotic ethic. *Gulf Carib. Res.* 14:1–7.

Bohnsack, J.A., D.E. Harper, and D.B. McClellan. 1994. Fisheries trends from Monroe County, Florida. *Bull. Mar. Sci.* 54:982–1018.

Bosscher, H. 1992. *Growth Potential of Coral Reefs and Carbonate Platforms.* Amsterdam: Proefschrift Vrije Universiteit.

Boyer, J.N., and R.D. Jones. 2002. A view from the bridge: External and internal forces affecting the ambient water quality of the Florida Keys National Marine Sanctuary (FKNMS). In *The Everglades, Florida Bay, and Coral Reefs of the Florida Keys—An Ecosystem Sourcebook*, eds. J.W. Porter and K.G. Porter, 609–628. Boca Raton: CRC Press.

Boyer, K.E., P. Fong, A.R. Armitage, and R.A. Cohen. 2004. Elevated nutrient content of tropical macroalgae increases rates of herbivory in coral, seagrass, and mangrove habitats. *Coral Reefs* 23:530–538.

Bradley, R.S. 2000. Past global changes and their significance for the future. *Quat. Sci. Rev.* 19:391–402.

Brand, L.E. 2002. The transport of terrestrial nutrients to south Florida coastal waters. In *The Everglades, Florida Bay, and Coral Reefs of the Florida Keys—An Ecosystem Sourcebook*, eds. J.W. Porter and K.G. Porter, 361–414. Boca Raton: CRC Press.

Brett, C.E., K.B. Miller, and G.C. Baird. 1990. A temporal hierarchy of paleoecologic processes within a middle Devonian epeiric sea. *Paleontol. Soc. Spec. Publ.* 5: 178–209.

Bright, T.J., G.P. Kraemer, G.A. Minnery, and S.T. Viada. 1984. Hermatypes of the Flower Garden Banks, northwestern Gulf of Mexico: A comparison to other western Atlantic reefs. *Bull. Mar. Sci.* 34:461–476.

Broecker, W.S., and D.L. Thurber. 1964. Uranium-series dating of corals and oolites from Bahaman and Florida Keys limestone. *Science* 149:58–60.

Brown, B.E. 1997. Disturbances to reefs in recent times. In *Life and Death of Coral Reefs*, ed. C. Birkeland, 354–379. New York: Chapman and Hall.

Brown, J.H. 1984. On the relationship between abundance and distribution of species. *Am. Nat.* 124:255–279.

Bruckner, A.W. 2002. *Priorities for Effective Management of Coral Diseases*. NOAA Tech. Mem. NMFS-OPR-22.

Bruckner, A.W., R.J. Bruckner, and E.H. Williams, Jr. 1997. Spread of a black-band disease epizootic through the reef system in St. Ann's Bay, Jamaica. *Bull. Mar. Sci.* 61:919–928.

Bruckner, R.J., A.W. Bruckner, and E.H. Williams, Jr. 1997. Life history strategies of *Coralliophila abbreviata* Lamarck (Gastropoda: Coralliophilidae) on the southwest coast of Puerto Rico. *Proc. Eighth Int. Coral Reef Symp., Panama* 1: 627–632.

Bruno, J.F., L.E. Petes, C.D. Harvell, and A. Hettinger. 2004. Nutrient enrichment can increase the severity of coral diseases. *Ecol. Lett.* 6:1056–1061.

Bryant, D., L. Burke, J. McManus, and M. Spalding. 1998. *Reefs at Risk—A Map-Based Indicator to Threats to the World's Coral Reefs*. Washington, DC: World Resources Institute.

Budd, A.F., K.G. Johnson, and T.A. Stemann. 1994. Plio-Pleistocene extinctions and the origin of the modern Caribbean reef-coral fauna. In *Proceedings of the Colloquium on Global Aspects of Coral Reefs: Health, Hazards and History, 1993*, compiler R.N. Ginsburg, 7–13. Miami: Rosenstiel School of Marine and Atmospheric Science, University of Miami.

Buddemeier, R.W. 2001. Is it time to give up? *Bull. Mar. Sci.* 69:317–326.

Buddemeier, R.W., and R.A. Kinzie III. 1998. Reef science: Asking all the wrong questions in all the wrong places? *Reef Encounter* 23:29–34.

Buddemeier, R.W., J.A. Kleypas, and R.B. Aronson. 2004. *Coral Reefs and Global Climate Change: Potential Contributions of Climate Change to Stresses on Coral Reef Ecosystems*. Arlington, VA: Pew Center on Global Climate Change.

Burns, T.P. 1985. Hard-coral distribution and cold-water disturbances in south Florida. *Coral Reefs* 4:117–124.

Bythell, J.C., E.H. Gladfelter, and M. Bythell. 1993. Chronic and catastrophic natural mortality of three common Caribbean reef corals. *Coral Reefs* 12:143–152.

Bythell, J.C., Z.M. Hillis-Starr, and C.S. Rogers. 2000. Local variability but landscape stability in coral reef communities following repeated hurricane impacts. *Mar. Ecol. Prog. Ser.* 204:93–100.

Bythell, J.C., and C. Sheppard. 1993. Mass mortality of Caribbean shallow corals. *Mar. Pollut. Bull.* 26:296–297.

Cale, W.G., G.M. Henebry, and J.A. Yeakley. 1989. Inferring process from pattern in natural communities. *BioScience* 39:600–605.

Carpenter, R.C. 1990a. Mass mortality of *Diadema antillarum*. 1. Long-term effects on sea urchin population dynamics and coral reef algal communities. *Mar. Biol.* 104:67–77.

Carpenter, R.C. 1990b. Mass mortality of *D. antillarum*. 2. Effects on population densities and grazing intensity of parrotfishes and surgeonfishes. *Mar. Biol.* 104:79–86.

Causey, B.D. 2001. Lessons learned from the intensification of coral bleaching from 1980–2000 in the Florida Keys, USA. In *Coral Bleaching and Marine Protected Areas. Proceedings of the Workshop on Mitigating Coral Bleaching Impact through MPA Design*, eds. R.V. Salm and S.L. Coles, 60–66. Honolulu: Asia Pacific Coastal Marine Program Report #0102.

Causey, B.D. 2002. The role of the Florida Keys National Marine Sanctuary in the south Florida ecosystem restoration initiative. In *The Everglades, Florida Bay, and Coral Reefs of the Florida Keys—An Ecosystem Sourcebook*, eds. J.W. Porter and K.G. Porter, 883–894. Boca Raton: CRC Press.

Causey, B., J. Delaney, E. Diaz, D. Dodge, J.R. Garcia, J. Higgins, W. Jaap, C.A. Matos, G.P. Schmahl, C. Rogers, M.W. Miller, and D.D. Turgeon. 2000. Status of coral reefs in the US Caribbean and Gulf of Mexico: Florida, Texas, Puerto Rico, U.S. Virgin Islands and Navassa. In *Status of Coral Reefs of the World: 2000*, ed. C. Wilkinson, 239–259. Cape Ferguson: Australian Institute of Marine Science.

Causey, B., G. Garrett, B. Haskell, W. Jaap, and A. Szmant. 1998. The 1997–1998 mass bleaching event around the world—Florida (USA). In *Status of Coral Reefs of the World: 1998*, ed. C. Wilkinson, 34–35. Cape Ferguson: Australian Institute of Marine Science.

Chappell, J. 1983. Sea level changes and coral reef growth. In *Perspectives on Coral Reefs*, ed. D.J. Barnes, 46–55. Manuka: Brian Clouston.

Chappell, J., and N.J. Shackleton. 1986. Oxygen isotopes and sea level. *Nature* 324:137–140.

Chiappone, M. 1996. *Coral Watch program summary. A report on volunteer and scientific efforts to document the status of reefs in the Florida Keys National Marine Sanctuary.* Summerland Key: The Nature Conservancy.

Chiappone, M., S.L. Miller, and D.W. Swanson. 2002. Status of *Acropora* corals in the Florida Keys: Habitat utilization, coverage, colony density, and juvenile recruitment. In *Proceedings of the Caribbean Acropora Workshop: Potential Application of the U.S. Endangered Species Act as a Conservation Strategy*, ed. A. Bruckner, 125–135. NOAA–OPR-24.

Chiappone, M., S.L. Miller, D.W. Swanson, J.S. Ault, and S.G. Smith. 2001. Comparatively high densities of the long-spined sea urchin in the Dry Tortugas, Florida. *Coral Reefs* 20:137–138.

Chiappone, M., and K.M. Sullivan. 1994. Ecological structure and dynamics of nearshore hard-bottom communities in the Florida Keys. *Bull. Mar. Sci.* 54:747–756.

Chiappone, M., and K.M. Sullivan. 1997. Rapid assessment of reefs in the Florida Keys: Results from a synoptic survey. *Proc. Eighth Int. Coral Reef Symp., Panama* 2:1509–1514.

Cho, L.L., and J.D. Woodley. 2002. Recovery of reefs at Discovery Bay, Jamaica and the role of *D. antillarum. Proc. Ninth Int. Coral Reef Symp., Bali* 1:331–338.

Clarke, A.H., D.J. Stanley, J.C. Medcof, and R.E. Drinnan. 1967. Ancient oyster and bay scallop shells from Sable Island. *Nature* 215:1146–1148.

CLIMAP Project Members. 1976. The surface of the ice-age earth. *Science* 191:1131–1144.

COHMAP Members. 1988. Climatic changes of the last 18,000 years: Observations and model simulations. *Science* 241:1043–1052.

Coniglio, M., and R.S. Harrison. 1983. Facies and diagenesis of late Pleistocene carbonates from Big Pine Key, Florida. *Bull. Can. Pet. Geol.* 31:135–147.

Connell, J.H. 1978. Diversity in tropical rain forests and coral reefs. *Science* 199:1302–1310.

Connell, J.H. 1997. Disturbance and recovery of coral assemblages. *Coral Reefs* 16:S101–S113.

Cook, C.B., E.M. Mueller, M.D. Ferrier, and E. Annis. 2002. The influence of nearshore waters on corals of the Florida Reef Tract. In *The Everglades, Florida Bay, and Coral Reefs of the Florida Keys—An Ecosystem Sourcebook*, eds. J.W. Porter and K.G. Porter, 771–788. Boca Raton: CRC Press.

Corbett, D.R., J. Chanton, W. Burnett, K. Dillon, C. Rutkowski, and J.W. Fourqurean. 1999. Patterns of groundwater discharge into Florida Bay. *Limnol. Oceanogr.* 44:1045–1055.

Cortés, J. 1994. A reef under siltation stress: A decade of degradation. In *Proceedings of the Colloquium on Global Aspects of Coral Reefs: Health, Hazards and History*, compiler R.N. Ginsburg, 240–246. Miami: Rosenstiel School of Marine and Atmospheric Science, University of Miami.

Cortés, J., M.M. Murillos, H.M. Guzman, and J. Acuna. 1984. Perdida de zooxantelas y muerte de corales y otros organismos arrecifales en el Atlantico y Pacifico de Costa Rica. *Rev. Biol. Trop.* 32:227–231.

Cronin, T.M., B.J. Szabo, T.A. Ager, J.E. Hazel, and J.P. Owens. 1981. Quaternary climates and sea levels of the U.S. Atlantic Coastal Plain. *Science* 211:233–240.

Crowley, T.J. 2000. CLIMAP SSTs re-visited. *Clim. Dynam.* 16:241–255.

Dahlgren, T.G., J.R. Weinberg, and K.M. Halanych. 2000. Phylogeography of the ocean quahog (*Artica islandica*): Influences of paleoclimate on genetic diversity and species range. *Mar. Biol.* 137:487–495.

Dana, J.D. 1843. On the temperature limiting the distribution of corals. *Am. J. Sci.* 45:130–131.

Davidson, M.G. 2002. Protecting coral reefs: Principal national and international legal instruments. *Harv. Eniviron. Law Rev.* 26:499–546.

Davidson, O.G. 1998. Once more to the Keys. In *The Enchanted Braid—Coming to Terms with Nature on the Coral Reefs*, 207–224. New York: John Wiley & Sons.

Davis, G.E. 1982. A century of natural change in coral distribution at the Dry Tortugas: A comparison of reef maps from 1881 and 1976. *Bull. Mar. Sci.* 32:608–623.

Davis, R.A., A.C. Hine, and E.A. Shinn. 1992. Holocene coastal development on the Florida Peninsula. In *Quaternary Coasts of the United States: Marine and Lacustrine Systems*, eds. C. Fletcher and J. Wehmiller, 193–212. SEPM Special Publication 48. Tulsa: Society for Sedimentary Geology.

Delcourt, H.R., and P.A. Delcourt. 1991. *Quaternary Ecology—A Paleoecological Perspective*. London: Chapman & Hall.

de Ruyter van Steveninck, E.D., L.L. Van Mulekon, and A.M. Breenan. 1988. Growth inhibition of *Lobophora variegata* (Lamouroux) Womersley by scleractinian corals. *J. Exp. Mar. Biol. Ecol.* 115:169–178.

Dodd, R., D.E. Hattin, and R.M. Liebe. 1973. Possible living analog of the Pleistocene Key Largo reefs of Florida. *Geol. Soc. Am. Bull.* 84:3995–4000.

Dodge, R.E., A. Logan, and A. Antonius. 1982. Quantitative reef assessment studies in Bermuda: A comparison of methods and preliminary results. *Bull. Mar. Sci.* 32:745–760.

Done, T.J. 1999. Coral community adaptability to environmental change at the scales of regions, reefs and reef zones. *Am. Zool.* 39:66–79.

Dubinsky, Z., and N. Stambler. 1996. Marine pollution and coral reefs. *Global Change Biol.* 2:511–526.

Dustan, P. 1977. Vitality of coral populations off Key Largo, Florida: Recruitment and mortality. *Environ. Geol.* 2:51–58.

Dustan, P. 1994. Developing methods for assessing coral reef vitality: A tale of two scales. In *Proceedings of the Colloquium on Global Aspects of Coral Reefs: Health, Hazards and History*, compiler R.N. Ginsburg, 38–44. Miami: Rosenstiel School of Marine and Atmospheric Science, University of Miami.

Dustan, P. 1997. What is killing our coral reefs? *Calypso Log* October.

Dustan, P., and J.C. Halas. 1987. Changes in the reef-coral community of Carysfort Reef, Key Largo, Florida: 1974–1982. *Coral Reefs* 6:91–106.

Dyke, A.S., J.E. Dale, and R.N. McNelly. 1996. Marine molluscs as indicators of environmental change in glaciated North America and Greenland during the last 18,000 years. *Geogr. Phys. Quat.* 50:125–184.

Eakin, C.M., J.W. McManus, M.D. Spalding, and S.C. Jameson. 1997. Coral reef status around the world: Where are we and where do we go from here? *Proc. Eighth Int. Coral Reef Symp., Panama* 1:277–282.

Easterling, D.R., G.A. Meehl, C. Parmesan, S.A. Changnon, T.R. Karl, and L.O. Mearns. 2000. Climate extremes: Observations, modeling, and impacts. *Science* 289:2068–2074.

Edmunds, P.J. 1991. Extent and effect of black-band disease on a Caribbean reef. *Coral Reefs* 10:161–165.

Edmunds, P.J., R.B. Aronson, D.W. Swanson, D.R. Levitan, and W.F. Precht. 1998. Photographic versus visual census techniques for the quantification of juvenile corals. *Bull. Mar. Sci.* 62:437–446.

Edmunds, P.J., and J.F. Bruno. 1996. The importance of sampling scale in ecology: Kilometer-wide variation in coral reef communities. *Mar. Ecol. Prog. Ser.* 143:165–171.

Edmunds, P.J., and R.C. Carpenter. 2001. Recovery of *Diadema antillarum* reduces macroalgal cover and increases abundances of juvenile corals on a Caribbean reef. *Proc. Natl. Acad. Sci. USA* 98:5067–5071.

Edmunds, P.J., R.D. Gates, and D.F. Gleason. 2003. The tissue composition of *Montastraea franksi* during a natural bleaching event in the Florida Keys. *Coral Reefs* 22:54–62.

Enos, P. 1977. Holocene sediment accumulations of the south Florida shelf margin. In *Quaternary Sedimentation in South Florida, Memoir 147*, eds. P. Enos and R.D. Perkins, 1–130. Boulder: Geological Society of America.

Epstein, P.R., B. Sherman, E. Spanger-Siegfried, A. Langston, S. Prasad, and B. McKay. 1998. Marine ecosystems—Emerging diseases as indicators of change. *Health Ecological and Economic Dimensions of Global Change Program*. Cambridge: Harvard University.

Evans, C.C. 1987. The relationship between the topography and internal structure of an ooid shoal sand complex: The upper Pleistocene Miami Limestone. *Miami Geol. Soc. Mem.* 3:18–41.

Fairbanks, R.G. 1989. A 17,000-year glacio-eustatic sea level record: Influence of glacial melting rates on the Younger Dryas event and deep-ocean circulation. *Nature* 342:637–642.

Fairbanks, R.G. 1990. The age and origin of the "Younger Dryas Climate Event" in Greenland ice cores. *Paleoceanography* 5:937–948.

Fitt, W.K., B.E. Brown, M.E. Warner, and R.P. Dunne. 2001. Coral bleaching: Interpretation of thermal tolerance limits and thermal thresholds in tropical corals. *Coral Reefs* 20:51–65.

Fitt, W.K., H.J. Spero, J. Halas, M.W. White, and J.W. Porter. 1993. Recovery of the coral *Montastrea annularis* in the Florida Keys after the 1987 Caribbean "bleaching event." *Coral Reefs* 12:57–64.

FMRI. 1998. *Benthic Habitats of the Florida Keys.* Tech. Report TR-3. St. Petersburg: Florida Marine Research Institute.

Folland, C.K., T.R. Karl, and K.Ya. Vinnikov. 1990. Observed climate variations and change. In *Climate Change, The IPCC Scientific Assessment*, eds. J.T. Houghton, G.J. Jenkins, and J.J. Ephraums, 195–238. Cambridge: Cambridge University Press.

Fong, P., and D. Lirman. 1995. Hurricanes cause population expansion of the branching coral *Acropora palmata* (Scleractinia): Wound healing and growth patterns of asexual recruits. *Mar. Ecol.* 16:317–335.

Forcucci, D. 1994. Population density, recruitment and 1991 mortality event of *Diadema antillarum* in the Florida Keys. *Bull. Mar. Sci.* 54:917–928.

Foster, D.R., P.K. Schoonmaker, and S.T.A. Pickett. 1990. Insights from paleoecology to community ecology. *Trends Ecol. Evol.* 5:119–122.

Fourqurean, J.W., and J.C. Zieman. 2002. Seagrass nutrient content reveals regional patterns of relative availability of nitrogen and phosphorous in the Florida Keys, FL, USA. *Biogeochemistry* 61:229–245.

Friedman, G.M. 1977. The Bahamas and southern Florida: A model for carbonate deposition. In *Field Guide to Some Carbonate Rock Environments—Florida Keys and Western Bahamas*, ed. H.G. Multer, 384–391. Dubuque: Kendall/Hunt.

Fruijtier, C., T. Elliott, and W. Schlager. 2000. Mass-spectrometric ^{234}U-^{230}Th ages from the Key Largo Formation, Florida Keys, United States: Constraints on diagenetic age disturbance. *Geol. Soc. Am. Bull.* 112:267–277.

Gardner, T.A., I.M. Côté, J.A. Gill, A. Grant, and A.R. Watkinson. 2003. Long-term region-wide declines in Caribbean corals. *Science* 301:958–960.

Gardner, T.A., I.M. Côté, J.A. Gill, A. Grant, and A.R. Watkinson. 2005. Hurricanes and Caribbean coral reefs: Impacts, recovery patterns, and role in long-term decline. *Ecology* 86:174–184.

Garrett, P., and P. Ducklow. 1975. Coral disease in Bermuda. *Nature* 253:349–350.

Garzon-Ferreira, J., D.L. Gil-Agudelo, L.M. Barrios, and S. Zea. 2001. Stony coral diseases observed in southwestern Caribbean reefs. *Hydrobiologia* 460:65–69.

Geister, J. 1977. The influence of wave exposure on the ecological zonation of Caribbean coral reefs. *Proc. Third Int. Coral Reef Symp., Miami* 1:23–29.

Geister, J. 1980. Calm-water reefs and rough-water reefs of the Caribbean Pleistocene. *Acta Palaeontol. Polon.* 25:541–556.

Geister, J. 1983. Holozäne westindische Korallenriffe: Geomorphologie, Ökologie und Fazies. *Facies* 9:173–284.

Gentry, R.C. 1984. Hurricanes in south Florida. In *Environments of South Florida Present and Past II*, ed. P.J. Gleason, 510–519. Coral Gables: Miami Geological Society.

Gill, I., D. Hubbard, and J.A.D. Dickson. 1999. Corals, reefs, and six millennia of Holocene climate. 11th Bathurst Meeting, Cambridge, *J. Conf. Abs.* 4:921.

Gilmore, M.D., and B.R. Hall. 1976. Life history, growth habits, and constructional roles of *Acropora cervicornis* in the patch reef environment. *J. Sediment. Petrol.* 40:519–522.

Ginsburg, R.N. (compiler). 1994. *Proceedings of the Colloquium on Global Aspects of Coral Reefs: Health, Hazards and History.* Miami: Rosenstiel School of Marine and Atmospheric Science, University of Miami.

Ginsburg, R.N., E. Gischler, and W.E. Kiene. 2001. Partial mortality of massive reef-build-ing corals: An index of patch reef condition, Florida reef tract. *Bull. Mar. Sci.* 69:1149–1173.

Ginsburg, R.N., and N.P. James. 1974. Spectrum of Holocene reef-building communities in the western Atlantic. In *Principles of Benthic Community Analysis, Sedimenta IV*, eds. A.M. Ziegler, K.R. Walker, E.J. Anderson, E.G. Kauffman, R.N. Ginsburg, and N.P. James, 7.1–7.22. Miami: Rosenstiel School of Marine and Atmospheric Science, University of Miami.

Ginsburg, R.N., and E.A. Shinn. 1964. Distribution of the reef-building community in Florida and the Bahamas. *Am. Assoc. Petrol. Geol. Bull.* 48:527.

Ginsburg, R.N., and E.A. Shinn. 1994. Preferential distribution of reefs in the Florida Reef Tract: The past is the key to the present. In *Proceedings of the Colloquium on Global Aspects of Coral Reefs: Health, Hazards and History*, compiler R.N. Ginsburg, 21–26. Miami: Rosenstiel School of Marine and Atmospheric Science, University of Miami.

Gittings, S.R., K.J.P. Deslarzes, D.K. Hagman, and G.S. Boland. 1992. Reef coral popula-tions and growth at the Flower Garden Banks, northwest Gulf of Mexico. *Proc. Seventh Int. Coral Reef Symp., Guam* 1:90–96.

Gladfelter, W.B. 1982. White band disease in *Acropora palmata*: Implications for the structure and growth of shallow reefs. *Bull. Mar. Sci.* 32:639–643.

Gleason, P.J. 1984. Saving the wild places—a necessity for growth. In *Environments of South Florida Past and Present II*, ed. P.J. Gleason, viii–xxiii. Coral Glabes: Miami Geological Society.

Glynn, P.W. 1973. Aspects of the ecology of coral reefs in the western Atlantic region. In *Biology and Geology of Coral Reefs*, eds. O.A. Jones and R. Endean, 271–324. New York: Academic Press.

Glynn, P.W. 1983. Extensive bleaching and death of reef corals on the Pacific coast of Panama. *Environ. Conserv.* 10:149–154.

Glynn, P.W. 1991. Coral reef bleaching in the 1980's and possible connections with global warming. *Trends Ecol. Evol.* 6:175–179.

Glynn, P.W. 1993. Coral reef bleaching: Ecological perspectives. *Coral Reefs* 12:1–17.

Glynn, P.W. 1996. Coral reef bleaching: Facts, hypotheses and implications. *Global Change Biol.* 2:495–509.

Glynn, P.W., L.R. Almodóvar, and J.G. González. 1964. Effects of Hurricane Edith on marine life in La Parguera, Puerto Rico. *Caribb. J. Sci.* 4:335–345.

Glynn, P.W., and L. D'Croz. 1990. Experimental evidence for high temperature stress as the cause of El Niño-coincident coral mortality. *Coral Reefs* 8:181–191.

Goldberg, W.M. 1973. The ecology of the coral–octocoral communities off the southeast Florida coast: Geomorphology, species composition, and zonation. *Bull. Mar. Sci.* 23:465–487.

Goreau, T.F. 1959. The ecology of Jamaican coral reefs. I. Species composition and zona-tion. *Ecology* 40:67–90.

Goreau, T.F. 1969. Post Pleistocene urban renewal in coral reefs. *Micronesica* 5:323–326.

Goreau, T.F., and N.I. Goreau. 1973. The ecology of Jamaican coral reefs. II. Geomorphology, zonation and sedimentary phases. *Bull. Mar. Sci.* 23:399–464.

Gould, S.J. 1981. Palaeontology plus ecology as palaeobiology. In *Theoretical Ecology: Principles and Applications*, ed. R.M. May, 295–317. Sunderland: Sinauer Associates.

Graus, R.R., and I.G. Macintyre. 1989. The zonation of Caribbean coral reefs as controlled by wave and light energy input, bathymetric setting and reef morphology: Computer simulation experiments. *Coral Reefs* 8:9–18.

Graus, R.R., and I.G. Macintyre. 1998. Global warming and the future of Caribbean coral reefs. *Carbonates and Evaporites* 13:43–47.

Graus, R.R., I.G. Macintyre, and B.E. Herchenroder. 1984. Computer simulation of the reef zonation at Discovery Bay, Jamaica: Hurricane disruption and long-term physical oceanographic controls. *Coral Reefs* 3:59–68.

Greenstein, B.J., and H.A. Curran. 1997. How much ecological information is preserved in fossil reefs and how reliable is it? *Proc. Eighth Int. Coral Reef Symp., Panama* 1:417–422.

Greenstein, B.J., H.A. Curran, and J.M. Pandolfi. 1998. Shifting ecological baselines and the demise of *Acropora cervicornis* in the Western Atlantic and Caribbean Province: A Pleistocene perspective. *Coral Reefs* 17:249–261.

Greenstein, B.J., L.A. Harris, and H.A. Curran. 1998. Comparison of recent coral life and death assemblages to Pleistocene reef communities: Implications for rapid faunal replacement on recent reefs. *Carbonates and Evaporites* 13:23–31.

Greenstein, B.J., and J.M. Pandolfi. 1997. Preservation of community structure in modern coral life and death assemblages of the Florida Keys: Implications for the Quaternary fossil record of coral reefs. *Bull. Mar. Sci.* 61:431–452.

Greenstein, B.J., J.M. Pandolfi, and H.A. Curran. 1998. The completeness of the Pleistocene fossil record: Implications for stratigraphic adequacy. In *The Adequacy of the Fossil Record*. eds. S.K. Donovan and C.R.C. Paul, 75–109. London: Wiley.

Guilderson, T.P., R.G. Fairbanks, and J.L. Rubenstone. 1994. Tropical temperature variations since 20,000 years ago: Modulating interhemispheric climate change. *Science* 263:663–665.

Halley, R.B., and C.C. Evans. 1983. *The Miami Limestone—A Guide to Selected Outcrops and their Interpretation*. Miami Geological Society Publications.

Halley, R.B., H.L. Vacher, and E.A. Shinn. 1997. Geology and hydrogeology of the Florida Keys. In *Geology and Hydrogeology of Carbonate Islands,* eds. H.L. Vacher and T.M. Quinn, 217–248. New York: Elsevier.

Hallock, P., F.E. Muller-Karger, and J.C. Halas. 1993. Coral reef decline. *Natl. Geogr. Res. Expl.* 9(3):358–378.

Hanisak, M.D., and L.W. Siemon. 1999. Macroalgal tissue nutrients as indicators of nitrogen and phosphorus status in the Florida Keys. *J. Phycol.* 14:28.

Harrison, R.S., and M. Coniglio. 1985. Origin of the Pleistocene Key Largo Limestone, Florida Keys. *Bull. Can. Soc. Petrol. Geol.* 33:350–358.

Harrison, R.S., L.D. Cooper, and M. Coniglio. 1984. Late Pleistocene carbonates of the Florida Keys. In *Carbonates in Subsurface and Outcrop*, 291–306. 1984 CSPG Core Conference, Alberta: Canadian Society of Petroleum Geologists.

Harvell, C.D., K. Kim, J.M. Burkholder, R.R. Colwell, P.R. Epstein, D.J. Grimes, E.E. Hofmann, E.K. Lipp, A.D.M.E. Osterhaus, R.M. Overstreet, J.W. Porter, G.W. Smith, and G.R. Vasta. 1999. Emerging marine diseases—Climate links and anthropogenic factors. *Science* 285:1505–1510.

Hay, M.E. 1981. Herbivory, algal distribution, and the maintenance of between-habitat diversity on a tropical fringing reef. *Am. Nat.* 118:520–540.

Hay, M.E. 1984. Patterns of fish and urchin grazing on Caribbean coral reefs: Are previous results typical? *Ecology* 65:446–454.

Hay, M.E., and T. Goertemiller. 1983. Between-habitat differences in herbivore impact on Caribbean coral reefs. In *The Ecology of Deep and Shallow Coral Reefs*, ed. M.L. Reaka, 97–102. Washington, DC: NOAA Symposia Series for Undersea Research.

Hay, M.E., and P.R. Taylor. 1985. Competition between herbivorous fishes and urchins on Caribbean reefs. *Oecologia* 65:591–598.

Hayes, M.L., J. Bonaventura, T.P. Mitchell, J.M. Prospero, E.A. Shinn, F. Van Dolah, and R.T. Barber. 2001. How are climate and marine biological outbreaks functionally linked? *Hydrobiologia* 460:213–220.

Hayes, R.L., and N.I. Goreau. 1998. The significance of emerging diseases in the tropical coral reef ecosystem. *Rev. Biol. Trop.* 46:173–185.

Highsmith, R.C. 1982. Reproduction by fragmentation in corals. *Mar. Ecol. Prog. Ser.* 7:207–226.

Highsmith, R.C., A.C. Riggs, and C.M. D'Antonio. 1980. Survival of hurricane-generated coral fragments and a disturbance model of reef calcification/growth rates. *Oecologia* 46:322–329.

Hine, A.C. 1997. Structural and paleoceanographic evolution of the margins of the Florida platform. In *The Geology of Florida*, eds. A.F. Randazzo and D.S. Jones, 169–194. Gainesville: University Press of Florida.

Hodges, L.T. 1977. Coral size and orientation relationships of the Key Largo Limestone of Florida. *Proc Third Int. Coral Reef Symp., Miami* 2:347–352.

Hoegh-Guldberg, O. 1999. Climate change, coral bleaching and the future of the world's coral reefs. *Mar. Freshwater Res.* 50:839–866.

Hoffmeister, J.E. 1974. *Land from the Sea—A Geologic Story of South Florida*. Miami: University of Miami Press.

Hoffmeister, J.E., J.I. Jones, J.D. Milliman, D.R. Moore, and H.G. Multer. 1964. *Living and Fossil Reef Types of South Florida*. Field Trip No. 3. Miami: Geological Society of America.

Hoffmeister, J.E., and H.G. Multer. 1964. Growth rate estimates of a Pleistocene coral reef of Florida. *Geol. Soc. Am. Bull.* 75:353–358.

Hoffmeister, J.E., and H.G. Multer. 1968. Geology and origin of the Florida Keys. *Geol. Soc. Am. Bull.* 79:1487–1502.

Holden, C. 1996. Coral disease hot spot in Florida Keys. *Science* 274:2017.

Holland, S.M. 2000. The quality of the fossil record: A sequence stratigraphic approach. *Paleobiology* 26:148–168.

Hu, C., K.E. Hackett, M.K. Callahan, S. Andréfouët, J.L. Wheaton, J.W. Porter, and F.E. Muller-Karger. 2003. The 2002 ocean color anomaly in the Florida Bight: A cause of local coral reef decline? *Geophys. Res. Lett.* 30:1151.

Hubbard, D.K. 1988. Controls of modern and fossil reef development: Common ground for biological and geological research. *Proc. Sixth Int. Coral Reef Symp., Townsville* 1:243–252.

Hubbard, D.K., E.H. Gladfelter, and J.C. Bythell. 1994. Comparison of biological and geological perspectives of coral-reef community structure at Buck Island, U.S. Virgin Islands. In *Proceedings of the Colloquium on Global Aspects of Coral Reefs: Health, Hazards and History*, compiler R.N. Ginsburg, 201–207. Miami: Rosenstiel School of Marine and Atmospheric Science, University of Miami.

Hubbard, D.K., K.M. Parsons, J.C. Bythell, and N.D. Walker. 1991. The effects of Hurricane Hugo on the reefs and associated environments of St. Croix, U.S. Virgin Islands. *J. Coast. Res. Spec. Iss.* 8:33–48.

Hudson, J.H. 1981a. Growth rates in *Montastraea annularis*: A record of environmental change in Key Largo Coral Reef Marine Sanctuary, Florida. *Bull. Mar. Sci.* 31:444–459.

Hudson, J.H. 1981b. Response of *Montastraea annularis* to environmental change in the Florida Keys. *Proc. Fourth Int. Coral Reef Symp., Manila* 2:233–40.

Hudson, J.H., K.J. Hanson, R.B. Halley, and J.L. Kindinger. 1994. Environmental implications of growth rate changes in *Montastraea annularis*: Biscayne National Park, Florida. *Bull. Mar. Sci.* 54:647–669.

Hughes, T.P. 1985. Life histories and population dynamics of early successional corals. *Proc. Fifth Int. Coral Reef Congr., Tahiti* 4:101–106.

Hughes, T.P. 1989. Community structure and diversity of coral reefs: The role of history. *Ecology* 70:275–279.

Hughes, T.P. 1992. Monitoring of coral reefs: A bandwagon? *Reef Encounter* 11:9–11.

Hughes, T.P. 1994. Catastrophes, phase shifts and large-scale degradation of a Caribbean coral reef. *Science* 265:1547–1551.

Hughes, T.P., A.H. Baird, D.R. Bellwood, M. Card, S.R. Connolly, C. Folke, R. Grosberg, O. Hoegh-Guldberg, J.B.C. Jackson, J. Kleypas, J.M. Lough, P. Marshall, M. Nyström, S.R. Palumbi, J.M. Pandolfi, B. Rosen, and J. Roughgarden. 2003. Climate change, human impacts, and the resilience of coral reefs. *Science* 301:929–933.

Hughes, T.P., and J.H. Connell. 1999. Multiple stressors on coral reefs: A long-term perspective. *Limnol. Oceanogr.* 44:932–940.

Hughes, T.P., A.M. Szmant, R. Steneck, R. Carpenter, and S. Miller. 1999. Algal blooms on coral reefs: What are the causes? *Limnol. Oceanogr.* 44:1583–1586.

Hughes, T.P., and J.E. Tanner. 2000. Recruitment failure, life histories, and long-term decline of Caribbean corals. *Ecology* 81:2250–2263.

Hunter, I.G., and B. Jones. 1996. Coral associations of the Pleistocene Ironshore Formation, Grand Cayman. *Coral Reefs* 15:249–267.

Huston, M.A. 1985. Patterns of species diversity on coral reefs. *Ann. Rev. Ecol. Syst.* 16:149–177.

Hutton, J. 1785. The system of the earth, its duration, and stability. Abstracted from a dissertation read to the Royal Society of Edinburgh in 1785. In *Philosophy of Geohistory: 1785–1970*, ed. C.C. Albritton, Jr., 24–52. Stroudsburg: Dowden, Hutchinson & Ross.

Idjadi, J.A., S.C. Lee, J.F. Bruno, W.F. Precht, L. Allen-Requa, and P.J. Edmunds. 2006. Rapid phase-shift reversal on a Jamaican coral reef. *Coral Reefs*.

Jaap, W.C. 1979. Observations on zooxanthellae expulsion at Middle Sambo Reef, Florida Keys. *Bull. Mar. Sci.* 29:414–422.

Jaap, W.C. 1984. The ecology of the south Florida coral reefs: A community profile. FWS OBS-82/08 and MMS 84-0038.

Jaap, W.C. 1985. An epidemic zooxanthellae expulsion during 1983 in the lower Florida Keys coral reefs: Hyperthermic etiology. *Proc. Fifth Int. Coral Reef Congr., Tahiti* 6:142–148.

Jaap, W.C. 1994. Coral and coral reef management. In *Rare and Endangered Biota of Florida Series, Vol. IV. Invertebrates*, eds. M. Deyrup and R. Franz, 26–29. Gainesville: University Press of Florida.

Jaap, W.C. 1998. Boom-bust cycles in *Acropora*. *Reef Encounter* 23:12–13.

Jaap, W.C., J.C. Halas, and R.G. Muller. 1988. Community dynamics of stony corals (Milleporina and Scleractinia) at Key Largo National Marine Sanctuary, Florida, during 1981–1986. *Proc. Sixth Int. Coral Reef Symp., Townsville* 2:237–243.

Jaap, W.C., and P. Hallock. 1990. Coral reefs. In *Ecosystems of Florida*, eds. R.L. Meyers and J.J. Ewel, 574–616. Orlando: University of Central Florida Press.

Jaap, W.C., J.W. Porter, J. Wheaton, K. Hackett, M. Lybolt, M.K. Callahan, C. Tsokos, and G. Yanev. 2001. EPA/FKNMS Coral Reef Monitoring Project: Updated executive summary, 1996–2000. Report to Steering Committee, August 2001.

Jaap, W.C., and F.J. Sargent. 1994. The status of the remnant population of *Acropora palmata* (Lamarck, 1816) at Dry Tortugas National Park, Florida, with a discussion of possible causes of changes since 1881. In *Proceedings of the Colloquium on Global Aspects of Coral Reefs: Health, Hazards and History*, compiler R.N. Ginsburg, 101–105. Miami: Rosenstiel School of Marine and Atmospheric Science, University of Miami.

Jackson, J.B.C. 1991. Adaptation and diversity of reef corals. *BioScience* 41:475–482.

Jackson, J.B.C. 1992. Pleistocene perspectives of coral reef community structure. *Am. Zool.* 32:719–731.

Jackson, J.B.C. 1994. Community unity? *Science* 264:1412–1413.

Jackson, J.B.C. 2001. What was natural in the coastal oceans? *Proc. Natl. Acad. Sci.* USA 98:5411–5418.

Jackson, J.B.C., A.F. Budd, and J.M. Pandolfi. 1996. The shifting balance of natural communities? In *Evolutionary Paleoecology*, eds. D. Jablonski, D.H. Erwin, and J.H. Lipps, 89–112. Chicago: University of Chicago Press.

Jackson, J.B.C., and K.G. Johnson. 2000. Life in the last few million years. *Paleobiology* 26:221–235.

Jackson, J.B.C., M.X. Kirby, W.H. Berger, K.A. Bjorndal, L.W. Botsford, B.J. Bourque, R.H. Bradbury, R. Cooke, J. Erlandson, J.A. Estes, T.P. Hughes, S. Kidwell, C.B. Lange, H.S. Lenihan, J.M. Pandolfi, C.H. Peterson, R.S. Steneck, M.J. Tegner, and R.R. Warner. 2001. Historical overfishing and the recent collapse of coastal ecosystems. *Science* 293:629–638.

Jackson, S.T., and J.T. Overpeck. 2000. Responses of plant populations and communities to environmental changes of the late Quaternary. *Paleobiology* 26:194–220.

James, N.P., C.W. Stern, and R.S. Harrison. 1977. *Field Guide Book to Modern and Pleistocene Reef Carbonates—Barbados*. Miami: Third International Coral Reef Symposium.

Jameson, S.C., M.H. Tupper, and J.M. Ridley. 2002. The three screen doors: Can marine "protected" areas be effective? *Mar. Pollut. Bull.* 44:1177–1183.

Jokiel, P., and S. Coles. 1990. Response of Hawaiian and other Indo-Pacific corals to elevated temperatures. *Coral Reefs* 8:155–162.

Jompa, J., and L.J. McCook. 2002. The effects of nutrients and herbivory on competition between a hard coral (*Porites cylindrica*) and a brown alga (*Lobophora variegata*). *Limnol. Oceanogr.* 47:527–534.

Jones, J.A. 1977. Morphology and development of southeastern Florida patch reefs. *Proc. Third Int. Coral Reef Symp., Miami* 2:231–235.

Jordán-Dahlgren, E. 1992. Recolonization patterns of *Acropora palmata* in a marginal environment. *Bull. Mar. Sci.* 51:104–117.

Kaufman, L. 1977. The three spot damselfish: Effects on benthic biota of Caribbean coral reefs. *Proc. Third Int. Coral Reef Symp., Miami* 1:559–564.

Keller, B.D., and A. Itkin. 2002. Shoreline nutrients and chlorophyll a in the Florida Keys, 1994–1997: A preliminary analysis. In *The Everglades, Florida Bay, and Coral Reefs of the Florida Keys—An Ecosystem Sourcebook*, eds. J.W. Porter and K.G. Porter, 649–658. Boca Raton: CRC Press.

Kerwin, M., J.T. Overpeck, R.S. Webb, A. DeVernal, D.H. Rind, and R.J. Healy. 1999. The role of oceanic forcing in mid-Holocene Northern Hemisphere climatic change. *Paleoceanography* 14:200–210.

Kidwell, S.M. 2001. Preservation of species abundance in marine death assemblages. *Science* 294:1091–1094.

Kindinger, J.L. 1986. Geomorphology and tidal-belt depositional model of lower Florida Keys. *Am. Assoc. Petrol. Geol. Bull.* 70:607.

Kinsey, D.W. 1991. The coral reef: An owner-built, high-density, fully-serviced, self-sufficient housing estate in the desert—or is it? *Symbiosis* 10:1–22.

Kinzie, R.A., III. 1973. The zonation of West Indian gorgonians. *Bull. Mar. Sci.* 23: 93–155.

Kissling, D.L. 1977a. Coral reefs in the Lower Florida Keys: A preliminary report. In *Field Guide to Some Carbonate Rock Environments—Florida Keys and Western Bahamas*, ed. H.G. Multer, 209–215. Dubuque: Kendall/Hunt.

Kissling, D.L. 1977b. Population structure characteristics for some Paleozoic and modern colonial corals. In *Second International Symposium on Corals and Fossil Coral Reefs, Paris, September 1975*, 497–506. Paris: Memoires Du B.R.G.M. No. 89

Kleypas, J. 1997. Modeled estimates of global reef habitat and carbonate production since the last glacial maximum. *Paleoceanography* 12:533–545.

Kleypas, J.A., R.W. Buddemeier, D. Archer, J.-P. Gattuso, C. Langdon, and B.N. Opdyke. 1999. Geochemical consequences of increased atmospheric carbon dioxide on coral reefs. *Science* 284:118–120.

Kleypas, J.A., R.W. Buddemeier, and J.-P. Gattuso. 2001. The future of coral reefs in an age of global change. *Geol. Rundsch.* 90:426–437.

Kleypas, J.A., J. McManus, and L.B. Menez. 1999. Environmental limits to coral reef development: Where do we draw the line? *Am. Zool.* 39:146–159.

Knowlton, N. 1992. Thresholds and multiple stable states in coral reef community dynamics. *Am. Zool.* 32:674–682.

Knowlton, N. 2001. The future of coral reefs. *Proc. Natl. Acad. Sci. USA* 98:5419–5425.

Knowlton, N., J.C. Lang, and B.D. Keller. 1988. Fates of staghorn coral isolates on hurricane-damaged reefs in Jamaica: The role of predators. *Proc. Sixth Int. Coral Reef Symp., Townsville* 2:83–88.

Knowlton, N., J.C. Lang, and B.D. Keller. 1990. Case study of natural population collapse: Post-hurricane predation on Jamaican staghorn corals. *Smithsonian Contrib. Mar. Sci.* 31:1–25.

Knowlton, N., J.L. Maté, H.M. Guzmàn, R. Rowan, and J. Jara. 1997. Direct evidence for reproductive isolation among the three species of the *Montastraea annularis* complex in Central America (Panamá and Honduras). *Mar. Biol.* 127:705–711.

Knowlton, N., E. Weil, L.A. Weigt, and H.M. Guzmán. 1992. Sibling species in *Montastraea annularis*, coral bleaching, and the coral climate record. *Science* 255:330–333.

Kojis, B.L., and N.J. Quinn. 1994. Biological limits to Caribbean reef recovery: A comparison with western South Pacific reefs. In *Proceedings of the Colloquium on Global Aspects of Coral Reefs: Health, Hazards and History*, compiler R.N. Ginsburg, 353–359. Miami: Rosenstiel School of Marine and Atmospheric Science, University of Miami.

Kramer, P.R. (compiler). 2002. Report from the status and trends working group. In *Proceedings of the Caribbean Acropora Workshop: Potential Application of the U.S. Endangered Species Act as a Conservation Strategy*, ed. A. Bruckner, 28–37. NOAA–OPR-24.

Kramer, P.A., and P.R. Kramer. 2000. *Ecological Status of the Mesoamerican Barrier Reef System: Impacts of Hurricane Mitch and 1998 Coral Bleaching*. Final Report to the World Bank. Miami: Rosenstiel School of Marine and Atmospheric Science, University of Miami.

Kruczynski, W.L., and F. McManus. 2002. Water quality concerns in the Florida Keys: Sources, effects, and solutions. In *The Everglades, Florida Bay, and Coral Reefs of the Florida Keys—An Ecosystem Sourcebook*, eds. J.W. Porter and K.G. Porter, 827–881. Boca Raton: CRC Press.

Kuta, K.G., and L.L. Richardson. 1996. Abundance and distribution of black band disease on coral reefs in the northern Florida Keys. *Coral Reefs* 15:219–223.

Landon, S.M. 1975. *Environmental Controls on Growth Rates in Hermatypic Corals from the Lower Florida Keys.* MS Thesis. Binghamton: SUNY.

Lang, J.C., H.R. Lasker, E.H. Gladfelter, P. Hallock, W.C. Jaap, F.J. Losada, and R.G. Muller. 1992. Spatial and temporal variability during periods of "recovery" after mass bleaching on western Atlantic coral reefs. *Am. Zool.* 32:696–706.

Lapointe, B.E. 1989. Caribbean coral reefs: Are they becoming algal reefs? *Sea Front.* 35:82–91.

Lapointe, B.E. 1997. Nutrient thresholds for bottom up control of marcroalgal blooms on coral reefs in Jamaica and southeast Florida. *Limnol. Oceanogr.* 42:1119–1131.

Lapointe, B.E. 1999. Simultaneous top-down and bottom-up forces control macroalgal blooms on coral reefs. *Limnol. Oceanogr.* 44:1586–1592.

Lapointe, B.E., P.J. Barile, and W.R. Matzie. 2004. Anthropogenic nutrient enrichment of seagrass and coral reef communities in the Lower Florida Keys: Discrimination of local versus regional nitrogen sources. *J. Exp. Mar. Biol. Ecol.* 308:23–58.

Lapointe, B.E., and M.W. Clark. 1992. Nutrient inputs from the watershed and coastal eutrophication in the Florida Keys. *Estuaries* 15:465.

Lapointe, B.E., M.M. Littler, and D.S. Littler. 1997. Macroalgal overgrowth of fringing coral reefs at Discovery Bay, Jamaica: Bottom-up versus top-down control. *Proc. Eighth Int. Coral Reef Symp., Panama* 1:927–932.

Lapointe, B.E., and W.R. Matzie. 1996. Effects of stormwater discharges on euthrophication processes in nearshore waters of the Florida Keys. *Estuaries* 19:422–435.

Lapointe, B.E., W.R. Matzie, and P.J. Barile. 2002. Biotic phase-shifts in Florida Bay and fore reef communities of the Florida Keys: Linkages with historical freshwater flows and nitrogen loading from Everglades runoff. In *The Everglades, Florida Bay, and Coral Reefs of the Florida Keys—An Ecosystem Sourcebook*, eds. J.W. Porter and K.G. Porter, 629–648. Boca Raton: CRC Press.

Lapointe, B.E., J.E. O'Connell, and G.S. Garrett. 1990. Nutrient couplings between on-site sewage disposal systems, groundwaters, and nearshore surface waters of the Florida Keys. *Biogeochemistry* 10:289–307.

Lapointe, B.E., and N.P. Smith. 1987. A preliminary investigation of upwelling as a source of nutrients to Looe Key National Marine Sanctuary. *NOAA Technical Memorandum NOS MEMD 9.* Washington, DC: NOAA.

Lee, T.N., E. Williams, E. Johns, D. Wilson, and N.P. Smith. 2002. Transport processes linking south Florida ecosystems. In *The Everglades, Florida Bay, and Coral Reefs of the Florida Keys—An Ecosystem Sourcebook*, eds. J.W. Porter and K.G. Porter, 309–342. Boca Raton: CRC Press.

Leichter, J.J., and S.L. Miller. 1999. Predicting high frequency upwelling: Spatial and temporal patterns of temperature anomalies on a Florida coral reef. *Cont. Shelf Res.* 19:911–928.

Leichter, J.J., H.L. Stewart, and S.L. Miller. 2003. Episodic nutrient transport to Florida coral reefs. *Limnol. Oceanogr.* 48:1394–1407.

Lessios, H.A. 1988. Mass mortality of *Diadema antillarum* in the Caribbean: What have we learned? *Ann. Rev. Ecol. Syst.* 19:371–393.

Lessios, H.A., D.R. Robertson, and J.D. Cubit. 1984. Spread of *Diadema* mass mortality through the Caribbean. *Science* 226:335–337.

Lewis, J.B. 1984. The *Acropora* inheritance: A reinterpretation of the development of fringing reefs in Barbados, West Indies. *Coral Reefs* 3:117–122.

Lewis, S.M. 1986. The role of herbivorous fishes in the organization of a Caribbean reef community. *Ecol. Monogr.* 56:183–200.

Lidz, B.H. 1997. *Fragile Coral Reefs of the Florida Keys: Preserving the Largest Reef Ecosystem in the Continental U.S.* Open-File Report 97-453. St. Petersburg: US Geological Survey.

Lidz, B.H. 2000a. *Bedrock Beneath Reefs: The Importance of Geology in Understanding Biological Decline in a Modern Ecosystem.* Open-File Report 00-046. St. Petersburg: US Geological Survey.

Lidz, B.H. 2000b. *Reefs, Corals, and Carbonate Sands: Guides to Reef Ecosystem Health and Environments.* Open-File Report 00-164. St. Petersburg: US Geological Survey.

Lidz, B.H. 2004. Coral reef complexes at an atypical windward platform margin: Late Quaternary, southeast Florida. *Geol. Soc. Am. Bull.* 116:974–988.

Lidz, B.H., and P. Hallock. 2000. Sedimentary petrology of a declining reef ecosystem, Florida Reef Tract (USA). *J. Coast. Res.* 16:675–697.

Lidz, B.H., A.C. Hine, E.A. Shinn, and J.L. Kindinger. 1991. Multiple outer-reef tracts along the south Florida bank margin: Outlier reefs, a new windward-margin model. *Geology* 19:115–118.

Lidz, B.H., C.D. Reich, and E.A. Shinn. 2003. Regional Quaternary submarine geomorphology in the Florida Keys. *Geol. Soc. Am. Bull.* 115:845–866.

Lidz, B.H., D.M. Robbin, and E.A. Shinn. 1985. Holocene carbonate sedimentary petrology and facies accumulation, Looe Key National Marine Sanctuary, Florida. *Bull. Mar. Sci.* 36:672–700.

Lidz, B.H., and E.A. Shinn. 1991. Paleoshorelines, reefs, and a rising sea: South Florida, USA. *J. Coast. Res.* 7:203–229.

Lidz, B.H., E.A. Shinn, M.E. Hansen, R.B. Halley, M.W. Harris, S.D. Locker, and A.C. Hine. 1997a. Sedimentary and biological environments, depth to Pleistocene bedrock, and Holocene sediment and reef thickness, Key Largo, south Florida. *US Geological Survey Miscellaneous Investigations Series*, Map I-2505.

Lidz, B.H., E.A. Shinn, A.C. Hine, and S.D. Locker. 1997b. Contrasts within an outlier-reef system: Evidence for differential Quaternary evolution, south Florida windward margin, USA. *J. Coast. Res.* 13:711–731.

Lighty, R.G. 1977. Relict shelf-edge Holocene coral reef, southeast coast of Florida. *Proc. Third Int. Coral Reef Symp., Miami* 2:215–221.

Lighty, R.G. 1981. Fleshy-algal domination of a modern Bahamian barrier reef: Example of an alternate climax reef community. *Proc. Fourth Int. Coral Reef Symp., Manila* 1:722.

Lighty, R.G., I.G. Macintyre, and A.C. Neumann. 1980. Demise of a Holocene barrier-reef complex, northern Bahamas. *Geol. Soc. Am. Abstr. Prog.* 12:471.

Lighty, R.G., I.G. Macintyre, and R. Stuckenrath. 1978. Submerged early Holocene barrier reef south-east Florida shelf. *Nature* 276:59–60.

Lighty, R.G., I.G. Macintyre, and R. Stuckenrath. 1982. *Acropora palmata* reef framework: A reliable indication of sea level in the western Atlantic for the past 10,000 years. *Coral Reefs* 1:125–130.

Lipp, E.K., J.L. Jarrell, D.W. Griffin, J. Jacukiewicz, J. Lukasik, and J.B. Rose. 2002. Preliminary evidence for human fecal contamination in corals of the Florida Keys, USA. *Mar. Pollut. Bull.* 44:666–670.

Lirman, D. 2000. Fragmentation in the branching coral *Acropora palmata* (Lamarck): Growth, survivorship, and reproduction of colonies and fragments. *J. Exp. Mar. Biol. Ecol.* 251:41–57.

Lirman, D. 2001. Competition between macroalgae and corals: Effects of herbivore exclusion and increased algal biomass on coral survivorship and growth. *Coral Reefs* 19:392–399.

Lirman, D., and P. Fong. 1997. Patterns of damage to the branching coral *Acropora palmata* following Hurricane Andrew: Damage and survivorship of hurricane-generated asexual recruits. *J. Coast. Res.* 13:67–72.

Littler, M.M., P.R. Taylor, D.S. Littler, R.H. Sims, and J.N. Norris. 1987. Dominant macrophyte standing stocks, productivity and community structure on a Belizean barrier reef. *Atoll Res. Bull.* 302:1–24.

Locker, S.D., A.C. Hine, L.P. Tedesco, and E.A. Shinn. 1996. Magnitude and timing of episodic sea-level rise during the last deglaciation. *Geology* 24:827–830.

Logan, A. 1988. *Holocene Reefs of Bermuda, Sedimenta XI.* Miami: Rosenstiel School of Marine and Atmospheric Science, University of Miami.

Lough, J.M. 2000. 1997–98: Unprecedented thermal stress to coral reefs? *Geophys. Res. Lett.* 23:3901–3904.

Ludwig, K.R., D.R. Muhs, K.R. Simmons, R.B. Halley, and E.A. Shinn. 1996. Sea-level records at ~80ka from tectonically stable platforms: Florida and Bermuda. *Geology* 24:211–214.

Lugo, A.E., C. Rogers, and S. Nixon. 2000. Hurricanes, coral reefs and rainforests: Resistance, ruin and recovery in the Caribbean. *Ambio* 29:106–114.

Macintyre, I.G. 1988. Modern coral reefs of western Atlantic: New geological perspective. *Am. Assoc. Petrol. Geol. Bull.* 72:1360–1369.

Macintyre, I.G. 2006. Demise, regeneration, and survival of some western Atlantic reefs during the Holocene transgression. In *Geological Approaches to Coral Reef Ecology*, ed. R.B. Aronson, 181–200. New York: Springer-Verlag.

Macintyre, I.G., and P.G. Glynn. 1976. Evolution of a modern Caribbean fringing reef, Galeta Point, Panama. *Am. Assoc. Petrol. Geol. Bull.* 60:1054–1072.

Macintyre, I.G., and O.H. Pilkey. 1969. Tropical reef corals tolerance of low temperatures on the North Carolina continental shelf. *Science* 166:374–375.

Marsh, L.M. 1992. The occurrence and growth of *Acropora* in extra-tropical waters off Perth, Western Australia. *Proc. Seventh Int. Coral Reef Symp., Guam* 2:1233–1238.

Marszalek, D.S., G. Babashoff, M.R. Noel, and D.R. Worley. 1977. Reef distribution in south Florida. *Proc. Third Int. Coral Reef Symp., Miami* 2:223–229.

Matthews, R.K. 1986. Quaternary sea-level change. In *Sea-Level Change—Studies in Geophysics*, 88–103. Washington, DC: National Research Council.

Maul, G.A. 1992. Temperature and sea level change. In *Global Warming: Physics and Facts*, eds. B.G. Levi, D. Hafemeister, and R.A. Schribner, 78–112. New York: American Institute of Physics.

Mayer, A. 1902. The Tortugas as a station for research in biology. *Science* 17:190–192.

Mayer, A.G. 1914. The effects of temperature upon tropical marine animals. *Papers Tortugas Laboratory, Carnegie Institute of Washington* 6:1–14.

Mayer, A.G. 1915. The lower temperature at which reef corals lose their ability to capture food. *Yearb. Carnegie Inst. Washington Publ.* 183:1–24.

Mayer, A.G. 1918. Toxic effects due to high temperature. *Papers Tortugas Laboratory, Carnegie Institute of Washington* 252:172–178.

McClanahan, T.R., R.B. Aronson, W.F. Precht, and N.A. Muthiga. 1999. Fleshy algae dominate remote coral reefs of Belize. *Coral Reefs* 18:61–62.

McClanahan, T.R., and N.A. Muthiga. 1988. Changes in Kenyan coral reef community structure and function due to exploitation. *Hydrobiologia* 166:269–276.

McClanahan, T.R., and N.A. Muthiga. 1998. An ecological shift among patch reefs of Glovers Reef Atoll, Belize over 25 years. *Environ. Conserv.* 25:122–130.

McClanahan, T.R., E. Sala, P.A. Stickels, B.A. Cokos, A.C. Baker, C.J. Stager, and S.H. Jones IV. 2003. Interaction between nutrients and herbivory in controlling algal communities and coral condition on Glover's Reef, Belize. *Mar. Ecol. Prog. Ser.* 261:135–147.

McCook, L.J. 2001. Competition between corals and algal turfs along a gradient of terrestrial influence in the nearshore central Great Barrier Reef. *Coral Reefs* 19:419–425.

McCook, L.J., J. Jompa, and G. Diaz-Pulido. 2001. Competition between corals and algae on coral reefs: A review of evidence and mechanisms. *Coral Reefs* 19:400–417.

McCook, L.J., E. Wolanski, and S. Spagnol. 2001. Modeling and visualizing interactions between natural disturbances and eutrophication as causes of coral reef degradation. In *Oceanographic Processes of Coral Reefs: Physical and Biological Links in the Great Barrier Reef*, ed. E. Wolanski, 113–125. Boca Raton: CRC Press.

McField, M.D. 1999. Coral response during and after mass bleaching in Belize. *Bull. Mar. Sci.* 64:155–172.

McManus, J.W., and J.F. Polsenberg. 2004. Coral-algal phase-shifts on coral reefs: Ecological and environmental aspects. *Prog. Oceanogr.* 60:263–279.

Mesolella, K.J. 1967. Zonation of uplifted Pleistocene coral reefs on Barbados, West Indies. *Science* 156:638–640.

Meyer, D.L., J.M. Bries, B.J. Greenstein, and A.O. Debrot. 2003. Preservation of in situ reef framework in regions of low hurricane frequency: Pleistocene of Curaçao and Bonaire, southern Caribbean. *Lethaia* 36:273–285.

Miller, M.W. 1998. Coral/seaweed competition and the control of reef community structure within and between latitudes. *Oceanogr. Mar. Biol. Ann. Rev.* 36:65–96.

Miller, M.W. 2001. Corallivorus snail removal: Evaluation of impact on *Acropora palmata*. *Coral Reefs* 19:293–295.

Miller, M.W. (compiler). 2002. *Acropora* corals in Florida: Status, trends, conservation, and prospects for recovery. In *Proceedings of the Caribbean Acropora Workshop: Potential Application of the U.S. Endangered Species Act as a Conservation Strategy*, ed. A. Bruckner, 59–70. NOAA–OPR-24.

Miller, M.W., I.B. Baums, D.E. Williams, and A.M. Szmant. 2002. Status of candidate coral, *Acropora palmata*, and its snail predator in the upper Florida Keys National Marine Sanctuary. *NOAA Technical Memorandum* NMFS-SEFSC-479.

Miller, M.W., A.S. Bourque, and J.A. Bohnsack. 2002. An analysis of the loss of acroporid corals at Looe Key, Florida, USA: 1983–2000. *Coral Reefs* 21:179–182.

Miller, M.W., M.E. Hay, S.L. Miller, D. Malone, E.E. Sotka, and A.M. Szmant. 1999. Effects of nutrients versus herbivores on reef algae: A new method for manipulating nutrients on coral reefs. *Limnol. Oceanogr.* 44:1847–1861.

Miller, M.W., E. Weil, and A.M. Szmant. 2000. Coral recruitment and juvenile mortality as structuring factors for reef benthic communities in Biscayne National Park, USA. *Coral Reefs* 19:115–123.

Miller, S.L., and M.P. Crosby. 1999. The extent and condition of U.S. coral reefs. In *NOAA's State of the Coast Report*, 1–34. Silver Spring: NOAA.

Miller, S.L., and D.W. Swanson. 1999. Rapid assessment methods for monitoring marine protected areas in the Florida Keys National Marine Sanctuary: Program design and effects of Hurricane Georges on reefs in the middle and lower Keys. In *Abstracts with Program, International Conference on Scientific Aspects of Coral Reef Assessment, Monitoring, and Restoration*, 139–140. Ft. Lauderdale: NCRI.

Miller, S.L., D.W. Swanson, and M. Chiappone. 2002. Multiple spatial scale assessment of coral reef and hard-bottom community structure in the Florida Keys National Marine Sanctuary. *Proc. Ninth Int. Coral Reef Symp. Bali* 1:69–74.

Mitterer, R.M. 1974. Pleistocene stratigraphy in southern Florida based on amino acid diagenesis in fossil *Mercenaria. Geology* 2:425–428.

Muhs, D.R., B.J. Szabo, L. McCartan, P.B. Maat, C.A. Bush, and R.B. Halley. 1992. Uranium-series estimates of corals from Quaternary marine sediments of southern Florida. In *The Plio-Pleistocene Stratigraphy and Paleontology of Southern Florida*, eds. T.M. Scott and W.D. Allmon. Florida Geological Survey Special Publication 36:41–49.

Muller-Karger, F.E., C.R. McClain, T.R. Fisher, W.E. Esaias, and R. Varela. 1989. Pigment distribution in the Caribbean Sea: Observations from space. *Prog. Oceanogr.* 23:23–64.

Multer, H.G. 1977. Pleistocene back reef environments. In *Field Guide to Some Carbonate Rock Environments—Florida Keys and Western Bahamas*, ed. H.G. Multer, 258–271. Dubuque: Kendall/Hunt.

Multer, H.G., E. Gischler, J. Lundberg, K.R. Simmons, and E.A. Shinn. 2002a. Key Largo Limestone revisited: Pleistocene shelf-edge facies, Florida Keys, USA. *Facies* 46:229–272.

Multer, H.G., E. Gischler, J. Lundberg, K.R. Simmons, and E.A. Shinn. 2002b. Key Largo Limestone revisited: Pleistocene shelf-edge facies, Florida Keys, USA. *Geol. Soc. Am. Abstr. Progr.* 34:388.

Multer, H.G. and J.E. Hoffmeister. 1977. Petrology and significance of the Key Largo (pleistocene) Limestone. In *Field Guide to Some Carbonate Rock Environments—Florida Keys and Western Bahamas*, ed. H.G. Multer, 261–264. Dubuque: Kendall/Hunt.

Murdoch, T.J.T., and R.B. Aronson. 1999. Scale-dependent spatial variability of coral assemblages along the Florida reef tract. *Coral Reefs* 18:341–351.

Nagelkerken, I.K., C.D. Harvell, C. Heberer, K. Kim, C. Petrovic, L. Pors, and P. Yoshioka. 1997. Wide-spread disease in Caribbean sea fans: II. Patterns of infection and tissue loss. *Mar. Ecol. Prog. Ser.* 160:255–263.

National Research Council. 1988. *Toward an Understanding of Global Change.* Washington, DC: National Academy Press.

Neumann, A.C., and P.J. Hearty. 1996. Rapid sea-level changes at the close of the last interglacial (stage 5e) recorded in Bahamian Island geology. *Geology* 24:775–778.

Neumann, A.C., and I.G. Macintyre. 1985. Reef response to sea level rise: Keep-up, catch-up or give-up. *Proc. Fifth Int. Coral Reef Congr., Tahiti* 3:105–110.

Neumann, C.J., B.R. Jarvinen, C.J. McAdie, and J.D. Elms. 1993. *Tropical Cyclones of the North Atlantic Ocean, 1871–1992.* Asheville: NOAA National Climatic Data Center.

Newell, N.D., J. Imbrie, E.G. Purdy, and D.L. Thurber. 1959. Organism communities and bottom facies, Great Bahamas Bank. *Am. Mus. Nat. Hist. Bull.* 117:181–228.

NOAA (National Oceanic and Atmospheric Administration). 2003. The Florida Keys: A progression in protection. www.noaanews.noaa.gov/magazine/stories/mag92.htm.

Nyström, M., C. Folke, and F. Moberg. 2000. Coral reef disturbance and resilience in a human-dominated environment. *Trends Ecol. Evol.* 15:413–417.

Ogden, J.C., J.W. Porter, N.P. Smith, A.M. Szmant, W.C. Jaap, and D. Forcucci. 1994. A long-term interdisciplinary study of the Florida Keys seascape. *Bull. Mar. Sci.* 54:1059–1071.

Osmund, J.K., J.R. Carpenter, and H.L. Windom. 1964. Th^{230}/U^{234} age of Pleistocene corals and oolites of Florida. *J. Geophys. Res.* 70:1843–1847.

Ostrander, G.K., K.M. Armstrong, E.T. Knobbe, D. Gerace, and E.P. Scully. 2000. Rapid transition in the structure of a coral reef community: The effects of coral bleaching and physical disturbance. *Proc. Natl. Acad. Sci. USA* 97:5297–5302.

Pandolfi, J.M. 1999. Response of Pleistocene coral reefs to environmental change over long temporal scales. *Am. Zool.* 39:113–130.

Pandolfi, J.M. 2001. Pleistocene persistence and the recent decline in Caribbean coral communities. *PaleoBios* 21:100.

Pandolfi, J.M. 2002. Coral community dynamics at multiple scales. *Coral Reefs* 21:13–23.

Pandolfi, J.M., R.H. Bradbury, E. Sala, T.P. Hughes, K.A. Bjorndal, R.G. Cooke, D. McArdle, L. McClenachan, M.J.H. Newman, G. Paredes, R.R. Warner, and J.B.C. Jackson. 2003. Global trajectories of the long-term decline of coral reef ecosystems. *Science* 301:955–958.

Pandolfi, J.M., and J.B.C. Jackson. 1997. The maintenance of diversity on coral reefs: Examples from the fossil record. *Proc. Eighth Int. Coral Reef Symp., Panama* 1:397–404.

Pandolfi, J., and J.B.C. Jackson. 2001. Community structure of Pleistocene coral reefs of Curaçao, Netherlands Antilles. *Ecol. Monogr.* 71:49–67.

Pasley, D. 1972. *Field Guide—Field Trip #1, Windley Key Quarry (Key Largo Limestone).* Miami: American Quaternary Association Second National Conference.

Patterson, K.L., J.W. Porter, K.B. Ritchie, S.W. Polson, E. Mueller, E.C. Peters, D.L. Santavy, and G.W. Smith. 2002. The etiology of white pox, a lethal disease of the Caribbean elkhorn coral, *Acropora palmata. Proc. Natl. Acad. Sci. USA* 99:8725–8730.

Paul, J.H., J.B. Rose, J.K. Brown, E.A. Shinn, S. Miller, and S.R. Farrah. 1995. Viral tracer studies indicate contamination of marine waters by sewage disposal practices in Key Largo, FL. *Appl. Environ. Microbiol.* 61:2230–2234.

Paul, J.H., J.B. Rose, S.C. Jiang, X. Zhou, P. Cochran, C.A. Kellog, J.B. Kang, D.W. Griffin, S.R. Farrah, and J. Lulasik. 1997. Evidence for groundwater and surface marine water contamination by waste disposal wells in the Florida Keys. *Water Res.* 32:1448–1454.

Paulay, G. 1990. Effects of late Cenozoic sea-level fluctuations on the bivalve faunas of tropical oceanic islands. *Paleobiology* 16:415–434.

Pauly, D.W., J. Christiansen, J. Dahsgaard, R. Froese, and F.C. Torres, Jr. 1998. Fishing down marine food webs. *Science* 279:860–863.

Pennisi, E. 1997. Brighter prospects for the world's coral reefs? *Science* 277:491–493.

Perkins, R.D. 1977. Depositional framework of Pleistocene rocks in south Florida. In *Quaternary Sedimentation in South Florida, Memoir 147*, eds. P. Enos and R.D. Perkins, 131–198. Boulder: Geological Society of America.

Perkins, R.D., and P. Enos. 1968. Hurricane Betsy in the Florida-Bahama area—Geologic effects and comparison with Hurricane Donna. *J. Geol.* 76:710–717.

Peters, E.C. 1997. Diseases of coral reef organisms. In *Life and Death of Coral Reefs*, ed. C. Birkeland, 114–139. New York: Chapman and Hall.

Peters, E.C., and H.B. McCarty. 1996. Carbonate crisis? *Geotimes* 41:20–23.

Peters, E.C., P.P. Yevich, and J.J. Oprandy. 1983. Possible causal agent of 'white band disease' in Caribbean acroporid corals. *J. Invertebr. Pathol.* 41:394–396.

Pielou, E.C. 1991. *After the Ice Age.* Chicago: University of Chicago Press.

Pitts, P.A. 2002. The role of advection in transporting nutrients to the Florida reef tract. *Proc. Ninth Int. Coral Reef Symp., Bali* 2:1219–1224.

Playford, P.E. 1988. Guidebook of the geology of Rottnest Island. Geological Society of Australia.

Podestá, G.P., and P.W. Glynn. 1997. Sea surface temperature variability in Panamá and Gálapagos: Extreme temperatures causing coral bleaching. *J. Geophys. Res.* 102:15749–15759.

Podestá, G.P., and P.W. Glynn. 2001. The 1997–98 El Niño event in Panamá and Galápagos: An update of thermal stress indices relative to coral bleaching. *Bull. Mar. Sci.* 69:43–59.

Porter, J.W., J.F. Battey, and J.G. Smith. 1982. Perturbation and change in coral reef communities. *Proc. Natl. Acad. Sci. USA* 79:1678–1681.

Porter, J.W., P. Dustan, W.C. Jaap, K.L. Patterson, V. Kosmynin, O.W. Meier, M.E. Patterson, and M. Parsons. 2001. Patterns of spread of coral diseases in the Florida Keys. *Hydrobiologia* 460:1–24.

Porter, J.W., W.K. Fitt, H.J. Spero, C.S. Rogers, and M.W. White. 1989. Bleaching in reef corals: Physiological and stable isotopic responses. *Proc. Natl. Acad. Sci. USA* 86:9342–9346.

Porter, J.W., V. Kosmynin, K.L. Patterson, K.G. Porter, et al. 2002. Detection of coral reef change by the Florida Keys Coral Reef Monitoring Project. In *The Everglades, Florida Bay, and Coral Reefs of the Florida Keys; An Ecosystem Sourcebook*, eds. J.W. Porter and K.G. Porter, 749–769. Boca Raton: CRC Press.

Porter, J.W., S.K. Lewis, and K.G. Porter. 1999. The effect of multiple stressors on the Florida Keys coral reef ecosystem: A landscape hypothesis and a physiological test. *Limnol. Oceanogr.* 44:941–949.

Porter, J.W., and O.W. Meier. 1992. Quantification of loss and change in Floridian reef coral populations. *Am. Zool.* 32:625–640.

Porter, J.W., and J.I. Tougas. 2001. Reef ecosystems: Threats to their biodiversity. In *Encyclopedia of Biodiversity*, Vol. 5, 73–95. New York: Academic Press.

Porter, J.W., J.D. Woodley, G.J. Smith, J.E. Neigel, J.F. Battey, and D.F. Dallmeyer. 1981. Population trends among Jamaican reef corals. *Nature* 294:249–250.

Precht, W.F. 1993. Stratigraphic evidence from reef studies for a double-high sea stand during the last interglacial maximum. *Am. Assoc. Petrol. Geol. Bull.* 77:1473.

Precht, W.F. (ed.). 2006. *Reef Restoration Handbook—The Rehabilitation of an Ecosystem Under Siege.* Boca Raton: CRC Press.

Precht, W.F., and R.B. Aronson. 1997. White band disease in the Florida Keys—A continuing concern. *Reef Encounter* 22:14–16.

Precht, W.F., and R.B. Aronson. 2002. The demise of *Acropora* in the Caribbean: A tale of two reef systems. In *Proceedings of the Caribbean Acropora Workshop: Potential Application of the U.S. Endangered Species Act as a Conservation Strategy*, ed. A. Bruckner, 147. NOAA–OPR-24.

Precht, W.F., and R.B. Aronson. 2004. Climate flickers and range shifts of reef corals. *Front. Ecol. Environ.* 6:307–313.

Precht, W.F., and R.B. Aronson. 2006. Death and resurrection of Caribbean reefs: A palaeoecological perspective. In *Coral Reef Conservation*, eds. I. Côté and J. Reynolds, 40–77. Cambridge: Cambridge University Press.

Precht, W.F., R.B. Aronson, P.J. Edmunds, and D.R. Levitan. 1993. Hurricane Andrew's effect on the Florida reef tract. *Am. Assoc. Petrol. Geol. Bull.* 77:1473.

Precht, W.F., R.B. Aronson, S.L. Miller, B.D. Keller, and B. Causey. 2005a. The folly of coral restoration programs following natural disturbances in the Florida Keys National Marine Sanctuary. *Ecol. Restor.* 23:24–28.

Precht, W.F., A.W. Bruckner, R.B. Aronson, and R.J. Bruckner. 2002. Endangered acroporid corals of the Caribbean. *Coral Reefs* 21:41–42.

Precht, W.F., and W.H. Hoyt. 1991. Reef facies distribution patterns, Pleistocene (125 Ka) Falmouth Formation, Rio Bueno, Jamaica, WI. *Am. Assoc. Petrol. Geol. Bull.* 75:656–657.

Precht, W.F., I.G. Macintyre, R.E. Dodge, K. Banks, and L. Fisher. 2000. Backstepping of Holocene reefs along Florida's east coast. *Abstr. Prog., Ninth Int. Coral Reef Symp., Bali* 321.

Precht, W.F., S.L. Miller, R.B. Aronson, J.F. Bruno, and L. Kaufman. 2005b. Reassessing U.S. coral reefs. *Science* 308:1741.

Randazzo, A.F., and R.B. Halley. 1997. Geology of the Florida Keys. In *The Geology of Florida*, eds. A.F. Randazzo and D.S. Jones, 251–259. Gainesville: University Press of Florida.

Raymond, W.F. 1972. A geologic investigation of the offshore sands and reefs of Broward County, Florida. MS Thesis. Tallahassee: Florida State University.

Reich, C.D., E.A. Shinn, T.D. Hickey, and A.B. Tihansky. 2002. Tidal and meteorological influences on shallow marine groundwater flow in the upper Florida Keys. In *The Everglades, Florida Bay, and Coral Reefs of the Florida Keys—An Ecosystem Sourcebook*, eds. J.W. Porter and K.G. Porter, 827–881. Boca Raton: CRC Press.

Richardson, L.L. 1998. Coral diseases: What is really known? *Trends Ecol. Evol.* 13:438–443.

Richardson, L.L., and R.B. Aronson. 2002. Infectious diseases of reef corals. *Proc. Ninth Int. Coral Reef Symp., Bali* 2:1225–1230.

Richardson, L.L., W.M. Goldberg, K.G. Kuta, R.B. Aronson, G.W. Smith, K.B. Richie, J.C. Halas, J.S. Feingold, and S.L. Miller. 1998. Florida's mystery coral-killer identified. *Nature* 393:557–558.

Risk, M.J. 1999. Paradise lost: How marine science failed the world's coral reefs. *Mar. Freshwater Res.* 50:831–837.

Risk, M.J., J.M. Heikoop, E.N. Edinger, and M.V. Erdmann. 2001. The assessment 'toolbox': Community-based reef evaluation methods coupled with geochemical techniques to identify sources of stress. *Bull. Mar. Sci.* 69:443–458.

Ritchie, K.B., and G.W. Smith. 1998. Type II white-band disease. *Rev. Biol. Trop.* 46:199–203.

Roberts, C.M. 1995. Effects of fishing on the ecosystem structure of coral reefs. *Conserv. Biol.* 9:988–995.

Roberts, C.M. 1997. Connectivity and management of Caribbean coral reefs. *Science* 278:1454–1457.

Roberts, H.H., L.J. Rouse, Jr., and N.D. Walker. 1983. Evolution of cold-water stress conditions in high-latitude reef systems: Florida Reef Tract and the Bahama Banks. *Caribb. J. Sci.* 19:55–60.

Roberts, H.H., L.J. Rouse, Jr., N.D. Walker, and J.H. Hudson. 1982. Cold-water stress in Florida Bay and northern Bahamas: A product of winter cold-air outbreaks. *J. Sediment. Petrol.* 52:145–155.

Roberts, H.H., P.A. Wilson, and A. Lugo-Fernandez. 1992. Biologic and geologic responses to physical processes: Examples from modern reef systems of the Caribbean-Atlantic region. *Cont. Shelf Res.* 12:809–834.

Robinson, A. 1973. Natural vs. visitor-related damage to shallow water corals: Recommendations for visitor management and the design of underwater nature trails in the Virgin Islands. *National Park Service Report*, 1–23.

Rogers, C.S. 1983. Sublethal and lethal effects of sediments applied to common Caribbean reef corals in the field. *Mar. Pollut. Bull.* 14:378–382.

Rogers, C.S. 1985. Degradation of Caribbean and western Atlantic coral reefs and decline of associated fisheries. *Proc. Fifth Int. Coral Reef Congr., Tahiti* 6:491–496.

Rogers, C.S. 1990. Responses of coral reefs and reef organisms to sedimentation. *Mar. Ecol. Prog. Ser.* 62:185–202.

Rogers, C.S. 1993a. Hurricanes and coral reefs: The intermediate disturbance hypothesis revisited. *Coral Reefs* 12:127–137.

Rogers, C.S. 1993b. A matter of scale: Damage from Hurricane Hugo (1989) to U.S. Virgin Islands reefs at the colony, community, and whole reef level. *Proc. Seventh Int. Coral Reef Symp., Guam* 1:127–133.

Rogers, C.S., and J. Beets. 2001. Degradation of marine ecosystems and decline of fishery resources in marine protected areas in the US Virgin Islands. *Environ. Conserv.* 28:312–322.

Rogers, C.S., H.C. Fitz III, M. Gilnack, J. Beets, and J. Hardin. 1984. Scleractinian coral recruitment patterns at Salt River Submarine Canyon, St. Croix, U.S. Virgin Islands. *Coral Reefs* 3:69–76.

Rogers, C., W. Gladfelter, D. Hubbard, E. Gladfelter, J. Bythell, R. Dunsmore, C. Loomis, B. Devine, Z. Hillis-Starr, and B. Phillips. 2002. *Acropora* in the US Virgin Islands: A wake or an awakening? In *Proceedings of the Caribbean Acropora Workshop: Potential Application of the U.S. Endangered Species Act as a Conservation Strategy*, ed. A. Bruckner, 95–118. NOAA–OPR-24.

Rosenberg, E., and Y. Ben-Haim. 2002. Microbial diseases of corals and global warming. *Environ. Microbiol.* 4:318–326.

Rosesmyth, M.C. 1984. Growth and survival of sexually produced *Acropora* recruits: A post-hurricane study at Discovery Bay. In *Advances in Reef Science*, eds. P.W. Glynn, P.K. Swart, and A.M. Szmant-Froelich, 105–106. Miami: Rosenstiel School of Marine and Atmospheric Science, University of Miami.

Roy, K., J.W. Valentine, D. Jablonski, and S.M. Kidwell. 1996. Scales of climatic variability and time averaging in Pleistocene biotas: Implications for ecology and evolution. *Trends Ecol. Evol.* 11:458–463.

Ruddiman, W.F., and A.C. Mix. 1991. The north and equatorial Atlantic at 9000 and 6000 yr B.P. In *Global Climates Since the Last Glacial Maximum*, eds. H.E. Wright, Jr., J.E. Kutzbach, T. Webb III, W.F. Ruddiman, F.A. Street-Perrott, and P.J. Bartlein, 94–124. Minneapolis: University of Minnesota Press.

Rützler, K., and I.G. Macintyre. 1982. The habitat distribution and community structure of the barrier reef complex at Carrie Bow Cay, Belize. In *The Atlantic Barrier Reef Ecosystem at Carrie Bow Cay, Belize, I. Structure and Communities*, eds. K. Rützler and I.G. Macintyre, 9–45. Washington, DC: Smithsonian Institution Press.

Rylaarsdam, K.W. 1983. Life histories and abundance patterns of colonial corals on Jamaican reefs. *Mar. Ecol. Prog. Ser.* 13:249–260.

Salvigsen, O., S.L. Forman. and G.H. Miller. 1992. Thermophilous molluscs on Svalbard during the Holocene and their paleoclimatic implications. *Polar Res.* 11:110.

Sammarco, P.W., M.J. Risk, H.P. Schwarcz, and J.M. Heikoop. 1999. Cross-continental shelf trends in $\delta^{15}N$ on the Great Barrier Reef: Further consideration of the reef nutrient paradox. *Mar. Ecol. Prog. Ser.* 180:131–138.

Sanford, S. 1909. The topography and geology of southern Florida. *Florida Geological Survey Second Annual Report*, 175–231.

Santavy, D.L., E. Mueller, E.C. Peters, L. MacLaughlin, J.W. Porter, K.L. Patterson, and J. Campbell. 2001. Quantitative assessment of coral diseases in the Florida Keys: Strategy and methodology. *Hydrobiologia* 460:39–52.

Santavy, D.L., and E.C. Peters. 1997. Microbial pests: Coral diseases in the western Atlantic. *Proc. Eighth Int. Coral Reef Symp., Panama* 1:607–612.

Santavy, D.L., E.C. Peters, C. Quirolo, J.W. Porter, and C.N. Bianchi. 1999. Yellow-blotch disease outbreak on reefs of the San Blas Islands, Panama. *Coral Reefs* 19:97.

Sheppard, C. 1993. Coral reef environmental science: Dichotomies, not the Cassandras, are false. *Reef Encounter* 14:12–13.

Sheppard, C.R.C. 2003. Predicted recurrences of mass coral mortality in the Indian Ocean. *Nature* 425:294–297.

Shinn, E.A. 1963. Spur and groove formation on the Florida reef tract. *J. Sediment. Petrol.* 33:291–303.

Shinn, E.A. 1966. Coral growth rate, an environmental indicator. *J. Paleontol.* 40:233–240.

Shinn, E.A. 1976. Coral reef recovery in Florida and the Persian Gulf. *Environ. Geol.* 1:241–254.

Shinn, E.A. 1980. Geologic history of Grecian Rocks, Key Largo Coral Reef Marine Sanctuary. *Bull. Mar. Sci.* 30:646–656.

Shinn, E.A. 1984. Geologic history, sediment and geomorphic variations within the Florida reef tract. In *Advances in Reef Science*, 113–114. Miami: Rosenstiel School of Marine and Atmospheric Science, University of Miami.

Shinn, E.A. 1988. The geology of the Florida Keys. *Oceanus* 31:46–53.

Shinn, E.A. 1989. What is really killing the corals? *Sea Front.* 35:72–81.

Shinn, E.A. 1996. No rocks, no water, no ecosystem. *Geotimes* 41:16–19.

Shinn, E.A. 2001. African dust causes widespread environmental distress. St. Petersburg: US Geological Survey Open-File Report 01-246.

Shinn, E.A., J.H. Hudson, R.B. Halley, and B.H. Lidz. 1977. Topographic control and accumulation rate of some Holocene coral reefs, south Florida and Dry Tortugas. *Proc. Third Int. Coral Reef Symp., Miami* 2:1–7.

Shinn, E.A., J.H. Hudson, D.M. Robbin, and B.H. Lidz. 1981. Spurs and grooves revisited: Construction versus erosion, Looe Key Reef, Florida. *Proc. Fourth Int. Coral Reef Symp., Manila* 1:475–483.

Shinn, E.A., B.H. Lidz, and M.W. Harris. 1994. Factors controlling distribution of Florida Keys reefs. *Bull. Mar. Sci.* 54:1084.

Shinn, E.A., B.H. Lidz, J.L. Kindinger, J.H. Hudson, and R.B. Halley. 1989. *Reefs of Florida and the Dry Tortugas. Field Trip Guidebook T176.* Washington, DC: American Geophysical Union.

Shinn, E.A., R.S. Reese, and C.D. Reich. 1994. Fate and Pathways of Injection-Well Effluent in the Florida Keys. St. Petersburg: US Geological Survey Open-File Report 94–276.

Shinn, E.A., C.D. Reich, and T.D. Hickey. 2002. Seepage meters and Bernoulli's revenge. *Estuaries* 25:126–132.

Shinn, E.A., C.D. Reich, T.D. Hickey, and B.H. Lidz. 2003. Staghorn tempestites in the Florida Keys. *Coral Reefs* 22:91–97.

Shinn, E.A., G.W. Smith, J.M. Prospero, P. Betzer, M.L. Hayes, V. Garrison, and R.T. Barber. 2000. African dust and the demise of Caribbean coral reefs. *Geophys. Res. Lett.* 27:3029–3032.

Smith, J.E., C.M. Smith, P.S. Vroom, K.L. Beach, and S. Miller. 2004. Nutrient and growth dynamics of *Halimeda tuna* on Conch Reef, Florida Keys: Possible influence of internal tides on nutrient status and physiology. *Limnol. Oceanogr.* 49: 1923–1936.

Smith, N.P. 1994. Long-term Gulf-to-Atlantic transport trough tidal channels in the Florida Keys. *Bull. Mar. Sci.* 54:602–609.

Smith, N.P., and P.A. Pitts. 2002. Regional-scale and long-term transport processes in the Florida Keys. In *The Everglades, Florida Bay, and Coral Reefs of the Florida Keys— An Ecosystem Sourcebook*, eds. J.W. Porter and K.G. Porter, 343–360. Boca Raton: CRC Press.

Smith, S.R., G. Webster, S. De Putron, T. Murdoch, S. McKenna, D. Hellin, L. Grayston, and A.M. Stanley. 2002. Coral population dynamics on Bermuda's reefs: A key to understanding reef development and persistence at high latitude. *ISRS European Meeting Abstracts Volume,* 91. Cambridge: International Society for Reef Studies.

Smith, S.V., and R.W. Buddemeier. 1992. Global change and coral reef ecosystems. *Ann. Rev. Ecol. Syst.* 23:89–118.

South-West Florida Dark-Water Observations Group. 2002. Satellite images track "black water" event off Florida coast. *Eos* 83: 281, 285.

Stachowicz, J., H. Fried, R.W. Osman, and R.B. Whitlatch. 2002. Biodiversity, invasion resistance, and marine ecosystem function: Reconciling pattern and process. *Ecology* 83:2575–2590.

Stanley, S.M. 1966. Paleoecology and diagenesis of Key Largo Limestone, Florida. *Am. Assoc. Petrol. Geol. Bull.* 50:1927–1947.

Stemann, T.A., and K.G. Johnson. 1992. Coral assemblages, biofacies, and ecological zones in the mid-Holocene reef deposits of the Enriquillo Valley, Dominican Republic. *Lethaia* 25:231–241.

Steneck, R.S. 1994. Is herbivore loss more damaging to reefs than hurricanes? Case studies from two Caribbean reef systems. In *Proceedings of the Colloquium on Global Aspects of Coral Reefs: Health, Hazards and History*, compiler R.N. Ginsburg, 220–226. Miami: Rosenstiel School of Marine and Atmospheric Science, University of Miami.

Steneck, R.S. 1998. Human influences on coastal ecosystems: Does overfishing create trophic cascades? *Trends Ecol. Evol.* 13:429–430.

Stober, Q.J., K. Thornton, R. Jones, J. Richards, C. Ivey, R. Welch, M. Madden, J. Trexler, E. Gaiser, D. Scheidt, and S. Rathbun. 2001. South Florida Ecosystem Assessment— Phase I/II—Everglades stressor interactions: Hydropatterns, euthrophication, habitat alteration, and mercury contamination. Monitoring for adaptive management: Implications for ecosystem restoration. United States Environmental Protection Agency, EPA 904-R-01-002.

Stoddart, D.R. 1963. Effects of Hurricane Hattie on the British Honduras reefs and cays, October 30–31, 1961. *Atoll Res. Bull.* 95:1–142.

Stoddart, R., and T.P. Scoffin. 1979. Microatolls: Review of form, origin and terminology. *Atoll Res. Bull.* 224:1–17.

Stokstad, E. 2001. Humans to blame for coral loss. *Science* 293:593.

Stone, L., A. Huppert, B. Rajagopalan, H. Bhasin, and Y. Loya. 1999. Mass coral reef bleaching: A recent outcome of increased El-Niño activity? *Ecol. Lett.* 2:325–330.

Strasser, M. 1999. *Mya arenaria*—An ancient invader of the North Sea coast. *Helgol. Wiss. Meeresunters.* 52:309-324.

Swart, P.K., A. Saied, and K. Lamb. 2005. Temporal and spatial variation in the δ^{15}N and δ^{13}C of coral tissue and zooxanthellae in *Montastraea faveolata* collected from the Florida reef tract. *Limnol. Oceangr.* 50:1049–1058.

Szabo, B.J. 1979. Uranium-series age of coral reef growth on Rottnest Island, Western Australia. *Mar. Geol.* 29:M11–M15.

Szmant, A.M. 1997. Nutrient effects on coral reefs: A hypothesis on the importance of topographic and trophic complexity to reef nutrient dynamics. *Proc. Eighth Int. Coral Reef Symp., Panama* 2:1527–1532.

Szmant, A.M. 2002. Nutrient enrichment on coral reefs: Is it a major cause of coral reef decline? *Estuaries* 25:743–766.

Szmant, A.M., and A. Forrester. 1996. Water column and sediment nitrogen and phosphorus distribution patterns in the Florida Keys, USA. *Coral Reefs* 15:21–41.

Taira, K. 1979. Holocene migrations of the warm-water front and sea-level fluctuations in the northwestern Pacific. *Palaeogeogr. Palaeoclimatol. Palaeoecol.* 28:197–204.

Taylor, J.F., and C.H. Saloman. 1968. Some effects of hydraulic dredging and coastal development in Boca Ciega Bay, Florida. *US Fish Wildl. Serv.* 67:213–241.

Textoris, S.D., J.M. Fuhr, and D.F. Merriam. 1989. Patch reefs in the Pleistocene of south Florida and their implications. *Geol. Soc. Am. Abstr. Progr.* 21:41.

Thompson, L.G., M.E. Davis, E. Mosley-Thompson, T.A. Sowers, K.A. Henderson, V.S. Zagorodnov, P.-N. Lin, V.N. Mikhalenko, R.K. Campen, J.F. Bolzan, J. Cole-Dai, and B. Francou. 1998. A 25,000-year tropical climate history from Bolivian ice cores. *Science* 282:1858–1864.

Tomascik, T., and F. Sander. 1987. Effects of eutrophication on reef-building corals. II. Structure of scleractinian coral communities on fringing reefs, Barbados, West Indies. *Mar. Biol.* 94:53–75.

Torrance, D.C. 1991. Deep ecology: Rescuing Florida's coral reefs. *Nature Conservancy* 4:9–17.

Toscano, M.A., and J. Lundberg. 1998. Early Holocene sea-level record from submerged fossil reefs on the southeast Florida margin. *Geology* 26:255–258.

Toscano, M.A., and J. Lundberg. 1999. Submerged Late Pleistocene reefs on the tectonically-stable S.E. Florida margin: High-precision geochronology, stratigraphy, resolution of Substage 5a sea-level elevation, and orbital forcing. *Quat. Sci. Rev.* 18:753–767.

Toscano, M.A., and I.G. Macintyre. 2000. Response of southeast Florida and Bahamas Holocene relict reefs to deglacial sea-level rise. *Abstr. Progr., Ninth Int. Coral Reef Symp., Bali* 59.

Toscano, M.A., and I.G. Macintyre. 2003. Corrected western Atlantic sea-level curve for the last 11,000 years based on calibrated ^{14}C dates from *Acropora palmata* and mangrove intertidal peat. *Coral Reefs* 22:257–270.

Treml, E., M. Colgan, and M. Keevican. 1997. Hurricane disturbance and coral reef development: A geographic information system (GIS) analysis of 501 years of hurricane data from the Lesser Antilles. *Proc. Eighth Int. Coral Reef Symp., Panama* 1:541–546.

Trend-Staid, M., and W.L. Prell. 2002. Sea surface temperature at the Last Glacial Maximum: A reconstruction using the modern analog technique. *Paleoceanography* 17:1065–1083.

Tunnicliffe, V. 1981. Breakage and propagation of the stony coral *Acropora cervicornis*. *Proc. Natl. Acad. Sci. USA* 78:2427–2431.

Tunnicliffe, V. 1983. Caribbean staghorn coral populations—Pre-Hurricane Allen conditions in Discovery Bay, Jamaica. *Bull. Mar. Sci.* 33:132–151.

Vargas-Ángel, B., J.D. Thomas, and S.M. Hoke. 2003. High-latitude *Acropora cervicornis* thickets off Fort Lauderdale, Florida, USA. *Coral Reefs* 22:465–474.

Vaughan, T.W. 1914. Investigations of the geology and geologic processes of the reef tracts and adjacent areas of the Bahamas and Florida. *Carnegie Inst. Washington Yearb.* 12:1–183.

Vaughan, T.W. 1918. The temperature of the Florida coral reef tract. *Carnegie Inst. Washington Publ.* 213:321–339.

Vaughan, T.W. 1919a. Fossil corals from Central America, Cuba, and Puerto Rico, with an account of the American Tertiary, Pleistocene and Recent coral reefs. *US Natl. Mus. Bull.* 103:189–524.

Vaughan, T.W. 1919b. Corals and the formation of coral reefs. *Smithsonian Inst. Ann. Rep.—1917,* 189–238.

Veron, J.E.N. 1992. Environmental control of Holocene changes to the world's most northern hermatypic coral outcrop. *Pac. Sci.* 46:405–425.

Vitousek, P.M., J. Aber, R.W. Howarth, G.E. Likens, P.A. Matson, D.W. Schindler, W.H. Schlesinger, and D.G. Tilman. 1997. Human alteration of the global nitrogen cycle: Sources and consequences. *Ecol. Appl.* 7:737–750.

Vollmer, S.V., and S.R. Palumbi. 2002. Hybridization and the evolution of reef coral diversity. *Science* 296:2023–2025.

Voss, G.L. 1960. First underseas park. *Sea Front.* 6:87–94.

Voss, G.L. 1973. Sickness and health in Florida's coral reefs. *Nat. Hist.* 82:40–47.

Voss, G.L. 2002. An environmental assessment of the John Pennekamp Coral Reef State Park and the Key Largo Coral Reef Marine Sanctuary (Unpublished 1983 Report). In *NOAA Technical Memorandum NOS NCCOS CCMA 161*, eds. N. Voss, A.Y. Cantillo, and M.J. Bello. Joint NOAA/UMiami report. NOAA LISD Current References 2002-6. University of Miami RSMAS TR 2002-03.

Walker, N.D., H.H. Roberts, L.J. Rouse, Jr., and O.K. Huh. 1982. Thermal history of reef-associated environments during a record cold-air outbreak event. *Coral Reefs* 1:83–87.

Walker, N.D., L.J. Rouse, Jr., and O.K. Huh. 1987. Response of subtropical shallow-water environments to cold-air outbreak events: Satellite radiometry and heat flux modeling. *Cont. Shelf Res.* 7:735–757.

Walther, G.-R., E. Post, P. Convey, A. Menzel, C. Parmesan, T.J.C. Beebee, J.-M. Fromentin, O. Hoegh-Guldberg, and F. Bairlein. 2002. Ecological responses to recent climate change. *Nature* 416:389–395.

Wapnick, C., W.F. Precht, and R.B. Aronson. 2004. R.B. Millennial-Scale dynamics of staghorn coral in Discovery Bay, Jamaica. *Ecol. Lett.* 7:354–361.

Ward, F. 1990. Florida's coral reefs are imperiled. *Natl. Geogr. Mag.* 178:115–132.

Ward-Paige, C.A., M.J. Risk, and O.A. Sherwood. 2005. Reconstruction of nitrogen sources on coral reefs: $\delta^{15}N$ and $\delta^{13}C$ in gorgonians from Florida Reef Tract. *Mar. Ecol. Prog. Ser.* 296:155–163.

Warzeski, E.R. 1977. Data relating to water and climatic conditions in the Florida Keys. In *Field Guide to Some Carbonate Rock Environments Florida Keys and Western Bahamas*, ed. H.G. Multer, 317–323. Dubuque: Kendall/Hunt.

Weiss, M.P., and D.A. Goddard. 1977. Man's impact on coastal reefs—An example from Venezuela. In *Reefs and Related Carbonates—Ecology and Sedimentology*, eds. S.H. Frost, M.P. Weiss, and J.B. Saunders, 111–124. Tulsa: American Association of Petroleum Geologists.

Wellington, G.M., P.W. Glynn, A.E. Strong, S.A. Navarette, E. Wieters, and D. Hubbard. 2001. Crisis on coral reefs linked to climate change. *Eos* 82:1, 5.

Wells, S., and N. Hanna. 1992. *The Greenpeace Book of Coral Reefs*. New York: Stirling.

Wheaton, J.L., and W.C. Jaap. 1988. Corals and other prominent benthic cnidaria of Looe Key National Marine Sanctuary. *Fla. Mar. Res. Publ.* 43:1–25.

White, B., H.L. Curran, and M.A. Wilson. 1998. Bahamian coral reefs yield evidence of a brief sea-level lowstand during the last interglacial. *Carbonates Evaporites* 13:10–22.

Wilkinson, C.R. 1993. Coral reefs of the world are facing widespread devastation: Can we prevent this through sustainable management practices? *Proc. Seventh Int. Coral Reef Symp., Guam* 1:11–21.

Wilkinson, C. (ed.). 2000. *Status of Coral Reefs of the World: 2000.* Cape Ferguson and Dampier: Australian Institute of Marine Science.

Wilkinson, C. (ed.). 2002. *Status of Coral Reefs of the World: 2002.* Cape Ferguson and Dampier: Australian Institute of Marine Science.

Wilkinson, C. (ed.). 2004. *Status of Coral Reefs of the World: 2004.* Cape Ferguson and Dampier: Australian Institute of Marine Science.

Williams, E.H., Jr., and L. Bunkley-Williams. 1990. The world-wide coral reef bleaching cycle and related sources of coral mortality. *Atoll Res. Bull.* 335:1–71.

Williams, E.H., Jr., and L. Williams. 2000. Marine major ecological disturbances of the Caribbean. *Infect. Dis. Rev.* 2:110–127.

Williams, I.D., and N.V.C. Polunin. 2001. Large-scale associations between macroalgal cover and grazer biomass on mid-depth reefs in the Caribbean. *Coral Reefs* 19:358–366.

Williams, I.D., N.V.C. Polunin, and V.J. Hendrick. 2001. Limits to grazing by herbivorous fishes and the impact of low coral cover on macroalgal abundance on a coral reef in Belize. *Mar. Ecol. Prog. Ser.* 222:187–196.

Wing, S.R., and E.S. Wing. 2001. Prehistoric fisheries in the Caribbean. *Coral Reefs* 20:1–8.

Wood, R. 1999. *Reef Evolution.* New York: Oxford University Press.

Woodley, J.D. 1992. The incidence of hurricanes on the north coast of Jamaica since 1870: Are the classic reef descriptions atypical? *Hydrobiologia* 247:133–138.

Woodley, J.D., E.A. Chornesky, P.A. Clifford, J.B.C. Jackson, L.S. Kaufman, N. Knowlton, J.C. Lang, M.P. Pearson, J.W. Porter, M.C. Rooney, K.W. Rylaarsdam, V.J. Tunnicliffe, C.M. Wahle, J.L. Wulff, A.S.G. Curtis, M.D. Dallmeyer, B.P. Jupp, M.A.R. Koehl, J. Neigel, and E.M. Sides. 1981. Hurricane Allen's impact on Jamaican coral reefs. *Science* 214:749–755.

Zhou, Z., Y. Baoyin, and N. Petit-Marie. 1998. Paleoenvironments in China during the Last Glacial Maximum and the Holocene Optimum. *Episodes* 21:152–158.

Zubakov, V.A., and I.I. Borzenkova. 1990. *Global Paleoclimate of the Late Cenozoic.* Developments in Palaeontology and Stratigraphy 12. Amsterdam: Elsevier.

Part IV. Coral Reefs and Global Change

10. Extreme Climatic Events and Coral Reefs: How Much Short-Term Threat from Global Change?

Bernhard Riegl

10.1 Introduction

Throughout much of the literature dealing with climate and coral reefs it is almost an accepted fact that observable global change is happening and that we have the opportunity to study its effects (Lough 1994, 2000; Kleypas, McManus, and Menez 1999a; Kleypas et al. 1999b; Houghton et al. 2001; Hughes et al. 2003; Pandolfi et al. 2003). Coral reefs appear to be sensitive indicators. During earth history, crises or changes in coral reefs have often preceded crises or changes in other biota (e.g., Copper 1994; Veron 1995; Chadwick-Furman 1996; Stanley 2002), and it therefore seems appropriate to pay special attention to the fate of coral reefs.

It appears that major, even catastrophic, impacts are now being recorded with increasing frequency from more and more reef areas (e.g., Wilkinson 2000, 2004). This could be a function of more money being spent on reef research, more people observing, and a generally better grasp of what is going on. This tight net of observations could therefore simply show things that hitherto passed unnoticed. Or, and this is more likely, we are now carefully monitoring ecosystems in an undisputedly changing atmosphere (Houghton et al. 2001) and seeing in detail the effects of a changing world (Wellington et al. 2001; Pandolfi et al. 2003). Evidence points to these presently observed climatic changes being mainly anthropogenically driven (Houghton et al. 2001).

Global climatic change is nothing new. It has been happening throughout earth history and is driven by a multitude of factors (Ruddiman 2001). The geological record faithfully preserves many of these changes and our understanding is increasing. However, it is still far from trivial to decipher the relatively rapid, yearly or decadal climatic oscillations and to interpret their place within the larger-scale trends and cycles of which they are part.

Whether anthropogenically modified or not, global change has a marked effect on coral reefs. Climate and aragonite saturation state of the ocean are strongly influenced by the concentration of greenhouse gases in the atmosphere. With further increases, oceanic aragonite saturation state would change in a way that could seriously reduce coral calcification (Kleypas, McManus, and Menez 1999a; Kleypas et al. 1999b; Guinotte, Buddemeier, and Kleypas 2003; Kleypas, Chapter 12), if the system is not buffered by the dissolution of resident carbonates and increased input through rivers (the latter via increased dissolution of terrestrial carbonate reservoirs by acid rain). Increased prelocated inferior CO_2 in seawater also appears to have negative effects on the vitality of certain coral species (Renegar and Riegl 2005). In the atmosphere, increased greenhouse gas concentrations influence the planet's heat balance and therefore have a strong influence on climate and sea level, leading to thermal expansion of the oceans and melting of the ice caps (Houghton et al. 2001). The functioning of coral reefs and their associated carbonate accretionary processes are strongly dependent on local climate, which in turn is forced by large-scale teleconnections: long-distance effects of changes in atmospheric dynamics in one region on those in another, faraway region. Therefore, the implications of any climate change, whether global or regional, are serious for coral reefs.

Extreme climatic events on small, intermediate, and large scales have received considerable attention due to their potentially destructive influence not only on coral reefs, but also on the entire coastal human socioeconomic system. Among the most notable and best publicized events are tropical cyclones (hurricanes; e.g., Gardner et al. 2005) and temperature anomalies, such as those associated with El Niño–Southern Oscillations (ENSO; e.g., Glynn 1990; Pittock 1999; Glynn and Colley 2001).

This chapter provides a brief overview of some extreme climatic events and their influence on coral-associated systems: coral reefs, coral carpets, and non-framebuilding communities (sensu Riegl and Piller 2000). It focuses on tropical cyclones and temperature anomalies, and their spatial and temporal patterns and associated perturbations such as emergent diseases (inasmuch as present information allows a linkage to climatic factors), and it examines whether anticipated future changes could lead to an increased threat to coral reefs. It builds on case studies in the Pacific, Indian, and Atlantic Oceans and uses these to highlight what may be relevant to all coral reefs.

10.2 Study Areas

This chapter uses case studies from the Arabian Gulf, South Africa, the Bahamas, Turks and Caicos, Cayman Islands, and St Croix, which represent a cross section of habitats, and biogeographical and climatological zones. The study sites are situated within and outside the world's regions influenced most strongly by ENSO sea-surface temperature (SST) variability and the cyclone basins. Most have recently been impacted in one way or the other. Site selection was opportunistic and reflects the author's personal experience with these reef systems (Riegl et al. 1995; Riegl and Piller 2000, 2001, 2003; Riegl 2001, 2002, 2003). Although these sites and studies might not be ideal to illustrate the phenomena discussed, they nonetheless provide representative samples.

10.3 Global Climate Change

The Intergovernmental Panel on Climate Change published the following opinion regarding ongoing climate change: ". . . an increasing body of observations gives a collective picture of a warming world and other changes in the climate system . . . temperatures have risen during the past four decades in the lowest 8 kilometers of the atmosphere . . . snow cover and ice extent have decreased . . . global average sea level has risen and ocean heat content has increased . . . emissions of greenhouse gases and aerosols due to human activities continue to alter the atmosphere in ways that are expected to affect the climate . . ." (Houghton et al. 2001, pp. 3–5). Both earth's surface temperatures over the last 140 years and the average global SST anomalies in this century have markedly increased. Given this background, observed impacts on coral reef ecosystems are likely to reflect present changes, and future impacts are to be expected (see also Kleypas, McManus, and Menez 1999a; Kleypas, Chapter 12).

10.3.1 El Niño–Southern Oscillation

The El Niño–Southern Oscillation (ENSO; for comments on the different use of the terms El Niño and ENSO in the coral literature, see Massel (1999) and Glynn and Colley (2001)) is an extreme expression of a circulating cell of air masses with sinking and rising motions occurring at opposite times over northern Australia/ Indonesia and the south-central Pacific. It is a fluctuation on the scale of a few years in the ocean-atmospheric system involving large changes in the Walker circulation (the large-scale west-to-east atmospheric circulation over the tropical Pacific, spanning Indonesia to South America) and Hadley cells (the north–south component of the trade-wind circulation) throughout the tropical Pacific Ocean region (Philander 1990). The trade winds pile up warmer surface water in the western equatorial Pacific. During El Niño years, the trade winds weaken, the Walker circulation changes, and pressure differences across the tropical Pacific (high in the West, low in the East; expressed in the Southern Oscillation Index, SOI) allow the warm water to reach the eastern Pacific. Upon reaching the Americas, the warm water travels north and south along the coast, bringing unusually warm air and precipitation. In the western Pacific, Australia and Indonesia experience dry conditions. Warm water and low cloud cover (the Indonesian and

Northwest Australian thunderstorm belt moves into the Pacific) thus produce coral bleaching conditions first only in the West and then across virtually the entire Pacific. The oscillation and timing of the event seems to be determined by the time taken by the eastward-moving Kelvin waves to reach the Americas and then return to the Australian/Indonesian margin as Rossby waves that establish "normal" conditions (Massel 1999; Sugimoto, Kimura, and Tadakoro 2001). In years following the El Niño event, the "normal" state can be overshot and abnormally cool SSTs can be produced in the eastern Pacific. This phenomenon has been variously called La Niña (Ruddiman 2001; Sugimoto, Kimura, and Tadakoro 2001), anti-El Niño, non-El Niño, or El Viejo.

Records from the tropical Pacific and the Andes indicate a possible ENSO age of about 5000 to 7000 years, at least in the form that we know it today (Tudhope et al. 2001; Moy et al. 2002; Wellington and Glynn, Chapter 11). A 12,000-year-long record in Andean lake sediments in Ecuador (Moy et al. 2002) suggests an increase in ENSO frequency throughout the Holocene from 7000 ybp, to a peak around 1200 ybp and a subsequent decline. Some authors trace ENSO back to the Pleistocene (Beaufort et al. 2002; Koutavas et al. 2002) and even earlier. The historical record of 115 events in 465 years (records since 1525 in ships' logs and missionaries' reports) yields an average of four years between successive events; however, the actual recurrence time varies (Torrence and Webster 1999; Ruddiman 2001). Nine very severe El Niños occurred during these 465 years, averaging one every 50 years. The latest severe events in 1983 and 1998 were unusually close-spaced (Ruddiman 2001), and both caused spectacular and well-documented mortality on coral reefs (Glynn 1990; Glynn and Colley 2001). Had it not been for the Mt. Pinatubo eruption, which led to a brief cooling of the world's climate, it is likely that the 1990 to 1991 ENSO would have triggered an SST anomaly at least as strong as the 1998 event (Wigley 2000). No consensus seems to exist at present on whether the spacing of recurrences has increased in the recent past. Whereas some authors claim that increased frequency of El Niño events is likely under higher greenhouse gas concentrations (Timmermann et al. 1999) or that an increase in ENSO frequency is already observable (Huckleberry and Billman 2003), others state that the high variability in the past and the uncertainty inherent in climate models do not yet allow confident prediction (Enfield 2001; Ruddiman 2001). Projections in Houghton et al. (2001) show little change in the amplitude of El Niño events over the next 100 years. Some archeological evidence points to an increase in ENSO frequency over the last 50 years (Huckleberry and Billman 2003); however, it is unclear whether this is just a short-term effect.

The ENSO strongly interacts with climate in remote basins via a series of teleconnections. In the Atlantic, the most robust response to ENSO is a significant positive SST anomaly in the tropical North Atlantic, which may be via a weakening of northeasterly trade winds and consequent reduction of evaporative cooling (Curtis and Hastenrath 1995; Elliott, Jewson, and Sutton 2001). This creates bleaching conditions for corals: high SST, low wind, and low cloud cover. ENSO events also influence areas outside the Pacific Ocean and play a key role in the Asian and Indian monsoon systems. ENSO influences the Indian Ocean via the Indian Ocean

Dipole/Zonal Mode (IOZM; Saji et al. 1999; Webster et al. 1999; Loschnigg and Webster 2000; Loschnigg et al. 2003). The influence of ENSO on peripheral Indian Ocean basins such as the Arabian Gulf and the Red Sea, which have strong local weather patterns, is not well understood; however, it is likely that again some connection exists via the Indian IOZM (Purkis and Riegl 2005).

10.3.2 ENSO Effects on Coral Reefs

Much has been written about the effects of ENSO on coral reefs, dating back to the 1983 ENSO when Glynn (1984) first made the link between ENSO-associated SST heating and coral bleaching. Subsequently, especially after the 1997 to 1998 super-ENSO, impacts have been recorded in great detail in many areas within the reef belt (e.g., Wilkinson 2000; and numerous other studies). Effects of temperature variation on reefs caused by ENSO are best understood in the Eastern Pacific (Wellington and Glynn, Chapter 11), where the impacts of several ENSO events were documented. These studies showed that while it was previously believed that Eastern Pacific coral communities were primarily influenced by extreme cold temperature, they are apparently even more susceptible to extreme warm temperatures (Wellington and Glynn, Chapter 11).

Areas with wide, shallow banks in relatively high latitudes, such as the Bahamas, Florida, and the Arabian Gulf, have been known for a long time to be controlled by extreme temperature lows, but an increasing body of literature shows their susceptibility to hot anomalies. The Bahamas and surrounding areas in the northern tropical Atlantic are strongly influenced by ENSO-teleconnected heating when the northeasterly trade winds are reduced (Curtis and Hastenrath 1995; Elliott, Jewson, and Sutton 2001). During these doldrum-like conditions, corals are more likely to bleach due to high SST and increased radiation, both PAR (photosynthetically active radiation) and ultraviolet (Gleason and Wellington 1993; Lesser 1997; Glynn 2000; Dunne and Brown 2001; Mumby et al. 2002a,b). Heating is most intense over the shallow bank areas (of which the Great Bahama Bank alone has over 110,000 km^2). Due to increased evapotranspiration, water can become hyperpycnal and drain as density flow off the platforms, causing the bleaching of shelf-edge reefs at intermediate depth (Lang, Wicklund, and Dill 1988; Smith 2001; Riegl and Piller 2003). Oceanographic details of such phenomena are outlined by Hickey et al. (2000) and Smith (2001).

Due to their proximity to North America, the Bahamas also experience extreme cooling events associated with cold fronts (Roberts, Wilson, and Lugo-Fernandez 1992), similar to effects felt in the Florida Keys (Walker et al. 1982; Burns 1985). Coral mortality is due to unusually low and high temperatures and resultant hyperpycnal flows (Wilson and Roberts 1992, 1995). Whether hot or cold, the generation of lethal hyperpycnal flows involves evaporative heat and moisture loss to the air (either to hot air or to cold, dry continental air), and, during cold fronts, loading of the water with suspended fine sediment due to wind-driven wave action. Several heat-related, but no cold-related, major coral kills have been reported from this area over the past decade.

Despite ENSO's large-scale impacts, however, it is often local ocean dynamics that lead to differential patterns of mortality and recovery. This was demonstrated a year after the 1997 to 1998 ENSO in the Bahamas (Riegl and Piller 2003). Where reefs were blocked from heated bank waters and had cooler water protecting the corals, such as in southern Cat Island (as opposed to Eleuthera Island), corals were overall healthier (70% healthy versus 44%), less diseased (12% versus 24%), and less bleached (10% versus 26%). In the Cayman Islands, which have no wide shelf areas at all and consequently reefs are bathed in essentially oceanic waters, in the same season only 0.1% of corals bleached and 4.2% were diseased. In the Turks and Caicos Islands, where most reefs are also protected from bank waters by islands, no bleaching was observed and 2% (Turks Bank) to 10.6% (Mouchoir Bank) were diseased (Riegl and Piller 2003). The sites in more open, oceanic settings showed better recovery from bleaching events than sites influenced by wide bank areas, as predicted by Glynn (1996).

The importance of regional oceanographic setting in mediating the effects of strong ENSO-linked temperature excursions is also well demonstrated in South Africa, a region where ENSO has important and marked influence on the terrestrial climate (Houghton et al. 2001). While the 1997 to 1998 ENSO led to significant bleaching and coral death in East Africa and Madagascar, corals in South Africa barely bleached (1%), and also no temperature-related epizootics were observed (0.5% of corals diseased; Jordan and Samways 2001). In South Africa, an absence of major disturbances before 1999 had even allowed an increase in total macrobenthic cover between 1992 (57.4%) and 1999 (72.6%; Riegl et al. 1995; Jordan and Samways 2001). From 1999 to 2001, years without major bleaching in the Indian Ocean, bleaching in South Africa reached 10% locally, affecting mainly the scleractinian genus *Montipora* (Celliers and Schleyer 2002; Floros 2002; Riegl 2003; Floros, Samways, and Armstrong 2004). In this out-of-ENSO phase bleaching, we likely find an expression of the IOZM, which lags behind the ENSO. Corals in South Africa were protected by favorable oceanographic conditions. Each year, cold spikes decrease maximum temperatures by about 1 °C (Riegl 2003). In summer 1997 to 1998, the cold spike did not occur and 1% bleaching was observed. In 1998 to 1999, when bleaching occurred on nearby Indian Ocean reefs (Madagascar, Tanzania, Kenya, Comores, Reunion; Wilkinson 2000), the cold spike did occur and corals did not bleach. Cold spikes are linked to small-scale, localized upwelling events caused by colder water upwelling at the inshore edge of the Agulhas current, due to bottom Ekman veering. An alternative and/or extension to this hypothesis is that the water flow imparted by the Agulhas Current onto the reef corals alone could have helped prevent bleaching. Nakamura and van Woesik (2001) observed that corals in high-flow regimes survived bleaching better than corals in low flow. They also suggested that high flow may help to avoid bleaching or retard the onset of bleaching by increasing passive diffusion of hydroxyl radicals (Yamasaki 2000). Several studies in the Eastern Pacific showing protection of reefs by local ocean dynamics, including local upwelling and an open ocean setting, can be found in Glynn and Colley (2001).

Severe impacts linked to ENSO and the IOZM were also experienced in the Arabian Gulf. There, coral reefs persist in an area with annual temperature fluctuations of up to 20 °C (Kinsman 1964). Although the Arabian Gulf is a good example of how corals can adapt to extreme temperature variability, it also highlights the dire effects of relatively small temperature excursions away from the long-term averages, modulated by global climate change. In addition to its high-latitude position and small, shallow nature, local climate in the Arabian Gulf is dynamic and variable. Both extreme cold and warm excursions are influenced by the position of the Subtropical Jetstream; when in a northern position it allows hot air from the Arabian Peninsula to reach the Gulf, but when in a southern position pushes cold air from the Anatolian Plateau and Iran, known as the Shamal, into the Gulf basin (Nasrallah, Nieplova, and Ramadan 2004). Shinn (1976), Downing (1985), Coles and Fadlallah (1991), Sheppard, Price, and Roberts (1992), and Fadlallah, Allen, and Estudillo (1995) attributed coral die-back in the southern Gulf to Shamal-related cold events, which were long believed to be the primary factor shaping coral community and framework processes.

In 1996, 1998, and 2002, previously unrecorded, sustained positive temperature excursions raised average summer SST by >2 °C for almost three months (George and John 1999, 2004; Riegl 2002, 2003; Sheppard and Loughland 2002). In all three years, the SST anomaly started around April and lasted until about September. Maximum in situ measured temperatures in the study area were 35 to 37 °C, which is about 2 to 4 °C above the usually recorded maxima. The large-scale climatic factors associated with these positive anomalies need further investigation, but Purkis and Riegl (2005) observed in spectral analysis of combined Comprehensive Ocean Atmosphere Data Set (COADS) and National Center for Environmental Prediction (NCEP) data sets a roughly 60-month recurrence of peak temperature events and hypothesized a possible connection with the IOZM (Loschnigg et al. 2003). Heat waves in the Gulf region have indeed increased over the last decades (Nasrallah, Nieplova, and Ramadan 2004), which is in step with the regional climate change scenarios suggesting greater-than-average warming of atmospheric temperatures in the Arabian region (Houghton et al. 2001). This is also supported by the findings in Casey and Cornillon (2001). Arabian Gulf corals can apparently expect little respite (Sheppard 2003).

The impacts of these heat events on the coral communities showed a clear evolution. In 1996 coral bleaching and mortality were restricted largely to *Acropora* and reduced its live cover to essentially zero, but in 1998 mortality affected all remaining corals without clear taxonomic pattern. In 2002 all corals bleached again, but this time with the exception of those *Acropora* that had resettled the area (Fig. 10.1). Some *Acropora* did not even bleach, which might suggest rapid acclimatization of survivors (Riegl 2003; Purkis and Riegl 2005).

With the death of *Acropora*, which is characterized by dense, interlocking colonies and rapid growth, the primary framework-producing component was lost in 1996 over much of the southeastern Gulf. Wide areas are today characterized by expansive fields of *Acropora* skeletons. The dead corals were first covered by fine

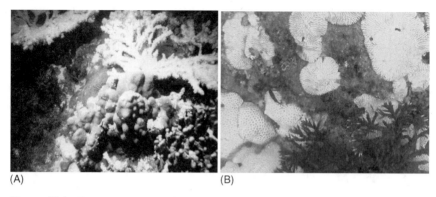
(A)   (B)

Figure 10.1. Temperature anomalies and corals. (**A**) Arabian Gulf (Jebel Ali) one year after the 1996 event, which killed almost all *Acropora*. *Acropora* in the image are dead, *Porites harrisoni*, *Platygyra lamellina*, and *Cyphastrea microphthalma* are normally colored. (**B**) Arabian Gulf (Sir Abu Nuair) 2002 bleaching event; in contrast to 1996, this time all corals except *Acropora* bleached. The *Acropora* were either survivors or propagules of survivors of the 1996 and 1998 events. Therefore, some hope for rapid acclimatization of some corals may indeed exist.

algal turf, and intense grazing by echinoids, particularly *Echinometra mathaei*, caused erosion of surficial structures. Within one to three years, the branches became heavily encrusted by layers of oysters (*Chama* sp. and *Spondylus* sp.) and coralline algae, and were extensively bored by clionid sponges. Breakdown of the tabular colonies proceeded and within five to seven years the frameworks began to degrade to rubble (Fig. 10.2). Recruits generally settled anywhere, but the recruits that had grown on coral branches were subsequently lost due to breakage of their substratum (Riegl 2001; George and John 2004). Thus, framebuilding had to begin anew on the original substratum. The temperature anomalies had not only decimated the corals, but also the reefs themselves were rapidly being eroded. Without significant coral regeneration, frameworks will likely be increasingly eroded, to the point of total disappearance.

Similar to the situation in the Bahamas, extreme heat events of longer-than-usual duration appear to be novel phenomena in the Arabian Gulf, or at least there is no mention of them in the literature despite ample mention of cold events (Downing 1985; Coles and Fadlallah 1991). The hypothesis of climatological novelty can be tested using synthetic long-term data sets, such as the COADS, NCEP, or HadISST1 data sets (Sheppard and Loughland 2002; Sheppard 2003; Riegl 2003; Riegl and Piller 2003), and such a finding would be consistent with global warming trends observed by Lough (2000) and Houghton et al. (2001). It would also be consistent with reports from the Eastern Pacific (Wellington and Glynn, Chapter 12). Corals seem to be destined to feel the heat.

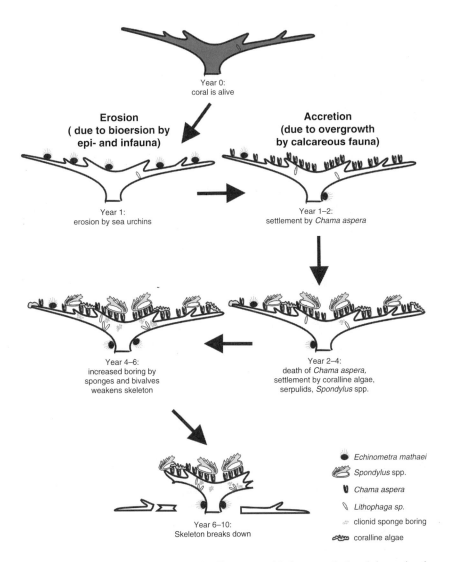

Figure 10.2. Taphonomic processes leading to rapid framework breakdown in the Arabian Gulf.

10.4 Disturbance by Tropical Cyclones and Associated Storm Turbulence

Storm swells generated by tropical cyclones are well known to be a major factor shaping the geomorphology and community composition of coral reefs. Coral assemblages can suffer catastrophic damage (Harmelin-Vivien and Laboute 1986; Rogers, McLain, and Tobias 1991; Done 1992; Rogers 1993b; Scoffin

1993; Bourrouilh-Le Jan 1998; Lugo, Rogers, and Nixon 2000). The combination of abrasion and breakage can modify the morphology of reefs (Scoffin 1993; Bourrouilh-Le Jan 1998) or can create rubble reefs (Blanchon and Jones 1997; Hubbard et al. 2005) and ramparts (Scoffin 1993). Hurricanes occur on a regular basis in seven "basins" that more or less coincide with the subtropical reef belts (Fig. 10.3). The Indian and Pacific Oceans have both a northern and a southern cyclone basin, whereas in the Atlantic, no tropical cyclones are found in the southern hemisphere. This means that of all Atlantic reefs, only those of Brazil are outside the influence of hurricanes (Maida and Ferreira 1997; Castro and Pires 2001); however, hurricanes are rare in the southern Caribbean (Bries, Debrot, and Meyer 2004). Damage from cyclones is determined by intensity (maximum surface wind speeds or minimum surface pressure), frequency, and geographical distribution, so it is of interest to investigate whether anticipated global change will affect these factors and, therefore, the persistence of certain taxa on coral reefs.

In order to understand how global warming and/or atmospheric CO_2 enrichment can affect cyclone activity, it is useful to understand the six prerequisite conditions for the formation of tropical cyclones (Gray 1979, 1988; Landsea 2000a,b).

1. SST above 26 °C.
2. Unstable atmosphere, cooling rapidly enough with height to allow for convection or thunderstorm activity, which allows release of heat stored in ocean water for tropical cyclone development.
3. Moist layers near the midtroposphere (5 km altitude); dry layers do not allow enough thunderstorm activity.

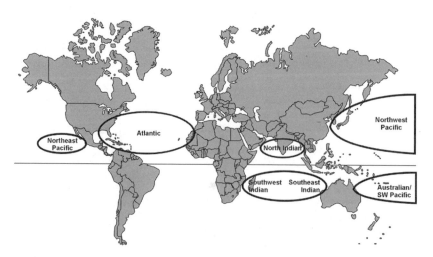

Figure 10.3. The seven basins within which tropical cyclones and hurricanes regularly form. Note how an expansion into higher latitudes could affect reef areas around Arabia (North Indian Basin), in southern Africa (Southwest Indian Basin), West Australia (Southeast Indian Basin), and East Australia (Southwest Pacific Basin). Also the southern Caribbean/Brazil region is presently rarely affected by hurricanes.

4. Minimum distance of 500 km from the equator, since in lower latitudes the Coriolis effect does not impart enough spin.
5. Preexisting disturbance at low level with sufficient rotation and convergence to trigger a cyclone.
6. Low values of vertical wind shear (<30 km h^{-1}) between surface and upper troposphere (12 km altitude).

Can global change interfere here? There is an upper limit to the intensity that a storm can achieve in a given SST and atmospheric environment. This theoretical limit (very few storms ever reach maximum intensity) increases with the amount of greenhouse gas in the atmosphere (Emanuel 1987; Knutson, Tueleya, and Kurihara 1998; Knutson and Tuleya 2001), but the extent to which this happens is not known (Emanuel 1997). It appears, however, that hurricanes have indeed become more destructive over the past decades (Emanuel 2005). Since the frequency of hurricanes depends on Gray's six conditions and the existence of a trigger, it is unclear whether any or exactly which changes in frequency would occur. With increased concentration of greenhouse gases, the strength of large-scale tropical circulation should increase (Henderson-Sellers et al. 1998) and therewith vertical wind shear, which would hinder the formation of cyclones. However, this more vigorous large-scale circulation could also provide stronger cyclone triggers, such as stronger Easterly Waves (atmospheric disturbances moving from West Africa into the subtropical Atlantic frequently turning into hurricanes), counteracting any action of increased wind shear.

Changing circulation patterns could also change the geographical distribution of tropical cyclones, with corresponding increases or decreases in the rate at which tropical cyclones move out of their genesis region and into higher latitudes. It is unlikely that global warming would substantially expand or contract the area of the cyclone basins, since cyclones are formed in areas where the atmosphere is slowly ascending (warm SST, little vertical wind shear, and easy convection due to cooling by thunderstorms). About as much atmosphere is descending as ascending; therefore, the total area experiencing ascent is unlikely to change (Landsea 1997, 2000a,b; Henderson-Sellers et al. 1998). Some studies, however, predict increases in hurricane frequency (Krishnamurti et al. 1998). Houghton et al. (2001) conclude that although the regional frequencies of tropical cyclones could change, their regions of formation might not. They also point to evidence that peak intensity could increase by 5 to 10% and that precipitation rates could increase by 20 to 30%, which would create additional stresses for coral reefs due to increased runoff from land.

The intensity of the Atlantic hurricane season, which has a marked effect on the development of Caribbean coral reefs (Blanchon and Jones 1997; Hubbard et al. 2005), is teleconnected to such diverse processes as iceberg rafting in the high polar Atlantic and rainfall in the West African Sahel zone. Increased southward iceberg rafting introduces cold, low-salinity water into the North Atlantic, which decreases the intensity of the thermohaline circulation and thus the formation of North Atlantic Deep Water (NADW). A relatively colder North Atlantic means a relatively warmer South Atlantic. This interferes with the West African monsoon and decreases rainfall in the Sahel, creating a high-pressure system on the West African coast, which in turn reduces the frequency of Easterly Waves needed to spawn hurricanes. The opposite scenario (warm North Atlantic and cold South Atlantic) means reduced vertical wind shear in the main hurricane development region, favoring the formation and intensification of hurricanes (Landsea 2000a,b). There is a marked, decadal-scale variability in years with active or inactive thermohaline circulation in the North Atlantic and, therefore, intense or weak hurricane seasons. It is unclear whether the addition of greenhouse gases has had any influence on this cyclicity (Landsea 2000a,b).

Another factor influencing tropical cyclone frequency is ENSO and the stratospheric Quasi-Biennial Oscillation (QBO), which is an east–west oscillation of stratospheric winds near the equator that has a strong effect on Atlantic Ocean (more activity in west phase), Southwest Indian Ocean (more activity in east phase) and Northwest Pacific Ocean regions (more activity in west phase; Gray and Sheaffer 1991; Landsea 2000a,b). Local effects, such as sea level pressure, SST, and trade-wind and monsoon circulations, also exert strong controls (Landsea 2000a,b). ENSO has a great influence on global atmospheric circulation patterns, affecting tropical cyclone frequencies by altering the lower tropospheric source of vorticity and changing the vertical shear profile (the difference in speed and direction of winds; Landsea 2000a,b). However, these changes are not uniform throughout the tropics. Near Australia (90°E to 165°E) tropical cyclone frequency is decreased (increased) during El Niño (La Niña), which is compensated for by an increase (decrease) in the South Pacific east of 165°E. The center of cyclone genesis shifts due to a weakening of the Australian monsoonal trough. The Northwest Pacific sees changes in the location of tropical cyclone genesis, without changes in frequency. West of 160°E there are reduced numbers of cyclones forming, whereas they increase in number east of 160°E (Landsea 2000a,b). Near Hawaii, more tropical cyclones are spawned during El Niño years (Landsea 2000a,b). Atlantic hurricanes are influenced by increases (decreases) of vertical wind shear due to changes in the climatological Westerlies in the upper troposphere. The Northwest Pacific, Southwest Indian, and North Indian cyclone basins do not seem to show ENSO-forced variations (Landsea 2000a,b).

Since some uncertainty remains as to how exactly ENSO frequency will change over the next 100 years in a warmer climate, some uncertainty about the teleconnected effects on tropical cyclones remains (Houghton et al. 2001). Goldenberg et al. (2001) observed a 2.5-fold increase of hurricane activity in the North Atlantic in a recent 6-year period compared to the previous 24 years, which they traced to increased North Atlantic SST and decreased vertical wind shear. They conclude that the present high level of hurricane activity is likely to persist for the next 10 to 40 years. If recent hurricane activity in the Atlantic is any guide to the future, corals and any other dwellers of the hurricane belt will experience more and more destructive hurricanes (Goldenberg et al. 2001; Emanuel 2005).

10.4.1 Tropical Cyclones and Coral Health

Much has been written about this topic (e.g., Lugo, Rogers, and Nixon 2000), and some local examples are presented here. St. Croix is one of the best-described case studies of hurricane effects, due to repeated, direct hits in the recent past. The island suffered severely from a category 5 hurricane in 1928. In 1989, the coral reefs of St. Croix were severely damaged by Hurricane Hugo (Hubbard et al. 1991). St. Croix was directly hit by two hurricanes, Luis and Marilyn, in 1995 (Rogers, Garrison, and Grober-Dunsmore 1997; Treml, Colgan, and Keevican 1997), by a hurricane and a tropical storm in 1916, and by two tropical storms in 1933. Over the period 1886 to 1998, St. Croix was hit by 12 hurricanes and 6 tropical storms, which gives a recurrence period of approximately 9 years for hurricanes and approximately 6 years for hurricanes and tropical storms. Minimum recurrence time for two consecutive hurricanes was 2–3 years in three instances (1876–1878, 1891–1893, 1995–1998), and 6 years in two instances (1910–1916, 1989–1995) and for tropical storms and hurricanes 1 year in one instance (1898–1899). The longest recurrence time gap between two hurricanes was 33 years (1956–1989), and between a hurricane and a tropical storm 30 years (1959–1989) (Neumann et al. 1999). This was the longest period on record allowing unhindered coral growth in shallow water.

The Cayman Islands are also frequently impacted by intense storms and hurricanes from the east or southeast. Thirty-eight hurricanes passed within 80 km of Grand Cayman over the 264 years before 1996 (Blanchon and Jones 1997), which gives a roughly 7-year recurrence time. Storms passing within 10 km of the island have an average 20-year recurrence, ranging from 1 to 55 years (Blanchon et al. 1997), which is the effective disturbance-free window for coral growth in shallow water. Therefore, hurricane-induced wave stress on the shallow biota is a powerful factor affecting potential reef framework development (Blanchon and Jones 1997; Blanchon, Jones, and Kalbfleisch 1997; Riegl 2001; Fig. 10.4). The hurricane history on Little Cayman is similar but not identical. For example, Little Cayman was not directly in the path of Hurricane Gilbert in 1988, which caused much destruction on Grand Cayman. Major devastation was caused in 1980 by Hurricane Allen, which also caused severe damage in Jamaica (Woodley et al. 1981), and lately in 2004, when Hurricanes Charlie and Ivan barely missed Grand Cayman and caused extensive damage due to a combination of high winds and storm surge.

In view of the frequency of recent hurricane impacts, it is not surprising that in St. Croix and the Caymans (and indeed in many other places throughout the Caribbean) coral cover was significantly reduced. The previously dense zone of *Acropora palmata* (Rigby and Roberts 1976; Logan 1994) was transformed into a "stump-and-boulder zone" (Fig. 10.4) with isolated colonies remaining alive in the Caymans and a much-reduced area of active framework building in St. Croix; these changes are thoroughly described in the literature (Edmunds and Witman 1991; Hubbard et al. 1991, 2005; Bythell, Bythell, and Gladfelter 1993; Aronson, Sebens, and Ebersole 1994; Blanchon, Jones, and Kalbfleisch 1997). Even where

Figure 10.4. While hurricanes destroy corals ((**A**) undamaged *Acropora palmata* at Andros, Bahamas) and lead to a loss of apparent rugosity ((**B**) *Acropora palmata* stumps after passage of a hurricane at Grand Cayman), the generated rubble itself can form reef-like structures that, in some areas, maintain the wave-breaking function of the fringing reef ((**C**) rubble fringing reef at Ensenada Honda, Vieques). Rubble beds themselves provide shelter to many coral reef organisms ((**D**) *Acropora palmata* rubble at St. Croix).

frameworks are intact, the corals are frequently in poor shape due to the prevalence of diseases, reported in St. Croix since the work of Antonius and Weiner (1982), Gladfelter (1982), and Rogers (1993a; see also Smith, Rogers, and Bouchon 1997). Like in the aftermath of bleaching events, it appears that the weakened condition of the corals after a hurricane leads to disease outbreaks. Thus, regeneration has been slowed, if not halted entirely, by the high incidence of diseases. This seems to vindicate the suggestions of Aronson, Precht, and Macintyre (1998) and Aronson and Precht (2001) that coral disease outbreaks alone or in the aftermath of extreme climatic events could indeed lead to the disappearance of dominant frame-builders such as *Acropora*. Whether such phase shifts are unprecedented or even permanent remains unclear in the light of observations by Shinn et al. (2003) and Hubbard et al. (2005), who show repeated hiatuses in the historical occurrence of *A. cervicornis* and *A. palmata* in Florida and St. Croix, respectively.

In an interesting contrast to the Caribbean, where diseases increased after coral mortality events caused by hurricanes or bleaching, diseases decreased in the Arabian Gulf after heat-related mortality. White-band disease (WBD), black-band disease

(BBD), and yellow-band disease (YBD) were present previous to bleaching and mortality in 1996 and favored *Acropora*, with the highest frequencies in dense areas of coral (Riegl 2002). There was also seasonality prior to the 1996 mass mortality (~15% in summer, ~8% in winter). Seasonality disappeared after the bleaching event, leaving infection at reduced winter levels (~7%, with only massive species infected).

Where corals die, their populations obviously need to be replenished by recruitment. However, several studies in the Caribbean (Lang 2003; Moulding 2005; Quinn and Kojis 2005) show the most common recruits to be of the genera *Porites* and *Agaricia*, with a clear absence of *Acropora*. As a result of heavy mortality in adult colonies, there is now a lack of recruitment and it remains to be seen whether *Acropora* will be able to make it through this Allee effect (Case 2000).

In the Arabian Gulf, *Acropora* recruitment did not decline as strongly, and apparently corals are able to complete gametogenesis and produce viable larvae despite SST anomalies. This is now reported for low- and high-temperature stress (Coles and Fadlallah 1991; Fadlallah 1996; Riegl 2003). If this was a general phenomenon, it would be a mechanism for the maintenance of corals in regions with high disturbance frequency, such as the Arabian Gulf. Where sexual reproduction fails, asexual reproduction can be of increasing importance. In the Caribbean, *Acropora palmata*, for example, is able to regenerate from surviving tissue pieces and "re-sheet" dead branches (Kramer and Ginsburg personal communication). In the Indo-Pacific, *Acropora* thickets maintain live branch tips inside the framework from where regeneration begins if all tissues at the colony surface are dead (Riegl and Piller 2001). In the face of recent mortality events, asexual regeneration and propagation remain important at the level of the individual colony but are likely not significant for regeneration at the scale of entire reefs.

10.5 Discussion

A review of some key climatic factors shows that coral reefs around the world are linked by an intricate web of climatic teleconnections. Pacific corals suffered significant mortality during the positive SST anomaly of 1998, brought on by an exceptionally strong ENSO. By increasing the strength of upper-tropospheric winds that eventually reached western Africa, the 1998 ENSO also interfered with tropical waves shed off West Africa. By creating higher wind shear, disturbances that could have been incipient hurricanes were dissipated. Although this may have led to less hurricane damage on Caribbean reefs, the doldrum-like conditions associated with the ENSO-related weakening of the trade winds caused bleaching conditions on many Caribbean reefs. Via its connected, but out-of-phase IOZM, the same ENSO caused bleaching years later in the westernmost Indian Ocean. Thus, the same climatic event had worldwide effects on coral reefs. Only some reefs protected by local oceanographic conditions, like some in South Africa or the Eastern Pacific, did not suffer.

Having established that anthropogenically driven climate change causes coral mortality, the question remains whether it could also interfere with reef frame-building processes. Will, for example, hurricanes change in a way that could compromise reef building? Climate models are still contradictory with regard to the effects of added greenhouse gases and hurricane frequency, but if the past decade is any indication of what is yet to come, increased impacts have to be expected. A great threat is a possible extension of the areas into which tropical cyclones will travel after their formation in the usual zones. Recent studies suggest that global warming could result in a northward shift of storm frequency in the Northern Hemisphere (McCabe, Clark, and Serreze 2001). Whether that will mean major changes for coral reefs is debatable, because most Atlantic and Pacific coral reefs are already situated inside the tropical cyclone belt. Many high-latitude reefs feel hurricane impact from swells, even if they are not regularly hit by tropical cyclones (e.g., Dollar and Tribble 1993; Riegl 2001). Suddenly being directly hit by tropical cyclones could mean changes for higher-latitude reef areas in the Indian Ocean, including southern Africa, Australia and the Red Sea, and the reefs of the southern Caribbean/tropical South Atlantic, which at present are largely outside the cyclone belts. The severity of such impacts was demonstrated by Bries, Debrot, and Meyer (2004) in Curaçao.

Because it is unclear whether global climate change will change the frequency of El Niño events in the next 100 years, it is also unclear whether such catastrophic SST anomalies as recorded in 1982 to 1983 (Glynn 1990) and 1998 will recur. According to Sheppard (2003), more similar bleaching events can be expected and all available evidence certainly supports this finding. In combination with drier conditions (less cloud cover, resulting in higher UV radiation) this could mean greater stress conditions for reefs via increased bleaching. Although warmer temperatures could mean an expansion of the area within which coral reef growth is theoretically possible (i.e., a poleward shift of the 18 °C isotherm), the real picture is likely more complicated due to changes in ocean surface-water chemistry, which will be most serious first at higher latitudes (Kleypas et al. 1999b; Guinotte et al. 2003). Gattuso, Allemand, and Frankignoulle (1999), Kleypas, McManus, and Menez (1999a), Kleypas et al. (1999b), and Guinotte, Buddemeier, and Kleypas (2003) draw attention to concomitant changes in ocean aragonite saturation state. At first sight a warmer ocean may appear good for reefs, but a more detailed look at climate and ocean chemistry seems to reveal the opposite (see Kleypas, Chapter 12). Not only will increased CO_2 likely lead to less coral calcification (Reynaud et al. 2003) and thus make them more susceptible to breakage and abrasion during storms, but it will also negatively influence coral survival (Renegar and Riegl 2005).

Climate change and teleconnections notwithstanding, comparison of post-disturbance ecology on Indo-Pacific and Atlantic reefs shows one striking difference. Although regeneration in the Indo-Pacific has been strong, at least locally, it appears to have slowed significantly in the Caribbean (Lugo, Rogers, and Nixon 2000; Lang 2003; Gardner et al. 2005). Gardner et al. (2005) demonstrate in a meta-analysis of 286 coral reef studies in the Caribbean that eight years after disturbance, no recovery was recorded anywhere. A key factor is the diseases that have decimated the major framebuilders. Detailed studies by Perry (1999, 2001),

Pandolfi (1999), Hubbard (2002), Meyer et al. (2003), and others show that Caribbean reefs coped with repeated hurricane disturbance throughout the Pleistocene. Investigations into reef frameworks (Macintyre, Burke, and Stuckenrath 1981; Macintyre et al. 1985; Hubbard, Burke, and Gill 1986; Hubbard, Miller, and Scaturo 1990; Hubbard 1997; Hubbard, Burke, and Gill 1998; Hubbard et al. 2005) reveal that shallow frameworks in particular are composed mostly of broken and tossed *A. palmata*. This clearly shows that catastrophic breakage events have occurred before (Perry 2001). The thickness and repetitiveness of stacked cycles of *A. palmata* suggest rapid recolonization, hints of which were indeed observed in the Caymans. On the other hand, coral diseases, and in particular white-band disease, recently have exacted such a high toll among survivers and recolonizers that new accumulation of *A. palmata* framework has so far been suppressed in most areas.

Woodley (1992) may have been correct in suggesting that early descriptions of Jamaican reefs, pointing to abundant and lush growth by *Acropora*, were atypical because they were made after a relatively long period without major disturbances. However, the near-absence of regeneration of the major framebuilders (particularly *Acropora*) at present is unusual and not evident in the geological record (Aronson and Precht 1997, 2001; Perry 2001). On the other hand, Shinn et al. (2003) and Hubbard et al. (2005) observed lengthy absences respectively of *A. cervicornis* in tempestites in the Florida Keys and of *A. palmata* in reef frameworks in St. Croix, suggesting that the current severe restriction of these species may, at least locally, not be entirely without precedent. Aronson and Precht (1997) and Aronson, Precht, and Macintyre (1998) argue that coral diseases, by removing the dominant framebuilders, exert an extrinsic control over reef building. The same authors and Greenstein, Harris, and Curran (1998) suggest that this is a novel event, unprecedented in the younger geological record of the region. Aronson and Precht (2001) argue that hurricanes without doubt are important local factors in reef destruction but that they alone cannot explain the recent pattern of coral mortality across the Caribbean.

Emergent biological factors may ultimately control future reef building and the maintenance of coral-dominated systems, and they may also be linked to climatic change (Harvell et al. 2002). Harvell et al. (1999) implicated a 25-year trend of warming winter temperatures along the U.S. East Coast in the northward movement of oyster diseases. It is not unreasonable to assume that global warming will be advantageous for certain pathogens (Richardson and Aronson 2002). Kushmaro et al. (1997) and Toren et al. (1998), for example, have shown temperature dependence in the virulence of *Vibrio* strains that can lead to coral bleaching.

In a world with increased coral mortality, the maintenance of reefal carbonate accretion could be in peril, at least in extreme settings such as high latitudes or areas of nutrient loading. Results presented in the literature suggest that coral mass mortality usually also leads to intense bioerosion. Both in the Arabian Gulf and in the eastern Pacific, bioerosion was so strong after coral mass mortality that virtually the entire framework was removed by a combination of bioerosion and physical forces acting on the weakened skeletons (Colgan 1990; Eakin 1996,

2001; Reaka-Kudla, Feingold, and Glynn 1996; Cortes 1997; Glynn 1997; Perry 1999; Riegl 2001; Purkis and Riegl 2005). Recruits had to settle on the bedrock, and reef-building and framework development were reset to zero. If this happens repeatedly in the future, reef development in these areas will be switched off permanently.

The present paper has not taken sea-level rise into account which, largely due to thermal expansion, is expected to be in the range of 0.11 to 0.77 m by 2100 (Houghton et al. 2001). Such a rise will likely not lead to the drowning of most shallow reefs, but it could create problems for corals in deeper habitats (Graus and Macintyre 1998), where they are at the edge of their photoadaptation capacity. There are vast areas of deeper reef habitat (McManus 1997; Riegl and Piller 2000), which could be affected.

10.6 Summary and Conclusions

Extreme climatic events, although spectacular in the damage they cause, do not appear directly to endanger the existence of coral-associated systems (e.g., Rogers 1993b; Bythell, Hillis-Starr, and Rogers 2000; Lugo, Rogers, and Nixon 2000), but they could facilitate important and dramatic changes. Extreme climatic events recur irregularly, but their frequencies generally follow cycles. As far as ENSO is concerned, these cycles are increasingly well understood, but whether the recurrence and/or severity of ENSO will change in the short term is still being debated. Atlantic hurricane frequency seems to undergo 15- to 20-year cycles, and an active hurricane period began in the mid-1990s. Tropical cyclone frequency may or may not increase in a greenhouse world, but peak intensities and moisture content will likely increase, making the storms more devastating. A more vigorous large-scale tropical circulation could provide more cyclone triggers and could change how cyclones move into both higher and lower latitudes. Tropical cyclone basins could shift and thus more reef areas could be exposed. This would cause extensive damage until the newly impacted systems acclimatize.

Due to climatic teleconnections, effects of changes in large-scale climatic patterns (ENSO frequency, surface warming, etc.) will be felt worldwide and on virtually every reef. Only a very few reefs, with favorable oceanographic conditions, will be less impacted. Most coral reef systems are able to withstand or recover from the physical forces associated with extreme climatic events. The greater threats come from changes in ocean chemistry and outbreaks of emergent diseases, at least some of which will thrive in a warmer ocean. Reef survival may not be so much a function of how well reefs can adapt to more extreme physical impacts, but rather how well they are able to withstand the changed chemistry and (micro)biology of anthropogenically altered oceans.

Acknowledgements. Thanks to R. Aronson for inviting me to write this paper. Support was provided through NOAA/NCCOS grant NA16OA1443 to the National Coral Reef Institute. This is NCRI publication # 73.

References

Antonius, A., and A. Weiner. 1982. Coral reefs under fire. *Mar. Ecol.* 3:255–277.

Aronson, R.B., and W.F. Precht. 1997. Stasis, biological disturbance, and community structure of a Holocene coral reef. *Paleobiology* 23:326–346.

Aronson, R.B., and W.F. Precht. 2001. White-band disease and the changing face of Caribbean coral reefs. *Hydrobiologia* 460:25–38.

Aronson, R.B., W.F. Precht, and I.G. Macintyre. 1998. Extrinsic control of species replacement on a Holocene reef in Belize: The role of coral diseases. *Coral Reefs* 17:223–230.

Aronson, R.B., K.B. Sebens, and J.P. Ebersole. 1994. Hurricane Hugo's impact on Salt River submarine canyon, St. Croix, U.S. Virgin Islands. In *Proceedings of the Colloquium on Global Aspects of Coral Reefs: Health, Hazards and History, 1993*, ed. R.N. Ginsburg, 189–195. Miami: Rosenstiel School of Marine and Atmospheric Science, University of Miami.

Beaufort L., T. de Garidel-Thoron, A. Mix, and N.G. Pisias. 2002. ENSO-like forcing on oceanic primary production during the late Pleistocene. *Science* 293:2440–2444.

Blanchon, P., and B. Jones. 1997. Hurricane control on shelf-edge reef architecture around Grand Cayman. *Sedimentology* 44:479–506.

Blanchon, P., B. Jones, and W. Kalbfleisch. 1997. Anatomy of a fringing reef around Grand Cayman: Storm rubble, not coral framework. *J. Sediment. Res.* 67:1–16.

Bourrouilh-Le Jan, F.G. 1998. The role of high-energy events (hurricanes and/or tsunamis) in the sedimentation, diagenesis and karst initiation of tropical shallow water carbonate platforms and atolls. *Sediment. Geol.* 118:3–36.

Bries, J., A.O. Debrot, and D.L. Meyer. 2004. Damage to the leeward reefs of Curaçao and Bonaire, Netherlands Antilles, from a rare storm event: Hurricane Lenny, November 1999. *Coral Reefs* 23:297–307.

Burns, T.P. 1985. Hard coral distribution and cold-water disturbances in South Florida: Variation with depth and location. *Coral Reefs* 4:117–124.

Bythell, J.C., M. Bythell, and E.H. Gladfelter. 1993. Initial results of a long-term coral reef monitoring program: Impact of Hurricane Hugo at Buck Island Reef National Monument, St. Croix, U.S. Virgin Islands. *J. Exp. Mar. Biol. Ecol.* 172:171–183.

Bythell, J.C., Z.M. Hillis-Starr, and C.S. Rogers. 2000. Local variability but landscape stability in coral reef communities following repeated hurricane impacts. *Mar. Ecol. Prog. Ser.* 204:93–100.

Case, T.D. 2000. *An Illustrated Guide to Theoretical Ecology*. Oxford: Oxford University Press.

Casey, K.S., and P. Cornillon. 2001. Global and regional sea surface trends. *J. Climate* 14:3801–3818.

Castro, C.B., and D.O. Pires. 2001. Brazilian coral reefs: What we already know and what is still missing. *Bull. Mar. Sci.* 69:357–371.

Celliers, L., and M.H. Schleyer. 2002. Coral bleaching on high-latitude marginal reefs at Sodwana Bay, South Africa. *Mar. Pollut. Bull.* 44:1380–1387.

Chadwick-Furman, N.E. 1996. Reef coral diversity and global change. *Global Change Biol.* 2:559–568.

Coles, S.L., and Y.H. Fadlallah. 1991. Reef coral survival and mortality at low temperatures in the Arabian Gulf: New species-specific lower temperature limits. *Coral Reefs* 9:231–237.

Colgan, M.W. 1990. El Nino and the history of eastern Pacific reef building. In *Global Ecological Consequences of the 1982–83 El Nino-Southern Oscillation*, ed. P.W. Glynn, 183–229. Elsevier Oceanographic Series 52.

Copper, P. 1994. Ancient reef ecosystem expansion and collapse. *Coral Reefs* 13:3–12.

Cortes, J. 1997. Biology and geology of eastern Pacific coral reefs. *Proc. Eighth Int. Coral Reef Symp., Panama* 1:57–64.

Curtis, S., and S. Hastenrath. 1995. Forcing of anomalous sea surface temperature evolution in the tropical Atlantic during El Niño warm events. *J. Geophys. Res.* 100:15835–15847.

Dollar, S.J., and G.W. Tribble. 1993. Recurrent storm disturbance and recovery: A long-term study of coral communities in Hawaii. *Coral Reefs* 12:223–233.

Done, T.J. 1992. Effects of tropical cyclone waves on ecological and geomorphological structures on the Great Barrier Reef. *Cont. Shelf Res.* 12:859–872.

Downing, N. 1985. Coral reef communities in an extreme environment: The northwestern Arabian Gulf. *Proc. Fifth Int. Coral Reef Congr., Tahiti* 6:343–348.

Dunne, R.P., and B.E. Brown. 2001. The influence of solar radiation on bleaching of shallow water reef corals in the Andaman Sea, 1993–1998. *Coral Reefs* 20:201–210.

Eakin, C.M. 1996. Where have all the carbonates gone? A model comparison of calcium carbonate budgets before and after the 1982–83 El Niño at Uva Island in the eastern Pacific. *Coral Reefs* 15:109–119.

Eakin, C.M. 2001. A tale of two ENSO events: Carbonate budgets and the influence of two warming disturbances and intervening variability, Uva Island, Panama. *Bull. Mar. Sci.* 69:171–186.

Edmunds, P.J., and J.D. Witman. 1991. Effects of Hurricane Hugo on the primary framework of reefs along the south shore of St. John, U.S. Virgin Islands. *Mar. Ecol. Prog. Ser.* 78:201–204.

Elliott, J.R., S.P. Jewson, and R.T. Sutton. 2001. The impact of the 1997/98 El Niño event on the Atlantic ocean. *J. Climate* 14:1069–1077.

Emanuel, K.A. 1987. The dependence of hurricane intensity on climate. *Nature* 326:483–485.

Emanuel, K.A. 1997. Climate variations and hurricane activity: Some theoretical issues. In *Hurricanes—Climate and Socioeconomic Impacts*, eds. H.F. Diaz and R.S. Pulwarty, 55–65. Berlin: Springer.

Emanuel, K. 2005. Increasing destructiveness of tropical cyclones over the past 30 years. *Nature* 436:686–688.

Enfield, D.B. 2001. Evolution and historical perspective of the 1997–1998 El Niño-Southern Oscillation event. *Bull. Mar. Sci.* 69:7–25.

Fadlallah, Y.H. 1996. Synchronous spawning of *Acropora clathrata* coral colonies from the western Arabian Gulf (Saudi Arabia). *Bull. Mar. Sci.* 59:209–216.

Fadlallah, Y.H., K.W. Allen, and R.A. Estudillo. 1995. Mortality of shallow reef corals in the western Arabian Gulf following aerial exposure in winter. *Coral Reefs* 14:99–107.

Floros, C.D. 2002. Bleaching and polychaete loading of coral at Sodwana Bay. MSc thesis, University of Natal at Pietermaritzburg.

Floros, C.D., M. Samways, and B. Armstrong. 2004. Taxonomic patterns of bleaching within a South African coral assemblage. *Biodivers. Conserv.* 13:1175–1194.

Gardner, T.A., I.M. Cote, J.A. Gill, A. Grant, and A.R. Watkinson. 2005. Hurricanes and Caribbean coral reefs: Impacts, recovery patterns, and role in long-term decline. *Ecology* 86:174–184.

Gattuso, J.-P., D. Allemand, and M. Frankignoulle. 1999. Photosynthesis and alcification at cellular, organismal, and community levels in coral reefs: A review of interactions and control by carbonate chemistry. *Am. Zool.* 39:160–183.

George, J.D., and D.M. John. 1999. High sea temperatures along the coast of Abu Dhabi (UAE), Arabian Gulf—Their impact upon corals and macroalgae. *Reef Encounter* 25:21–23.

George, J.D., and D.M. John. 2000. The effects of the recent prolonged high seawater temperatures on the coral reefs of Abu Dhabi (UAE). *Int. Symp. on Extent of Coral Bleaching* 28–29.

George, J.D., and D.M. John. 2004. The coral reefs of Abu Dhabi, United Arab Emirates: Past, present and future. In *Marine Atlas of Abu Dhabi*, eds R.A. Loughland, F.S. Al Muhairi, S.S. Fadel, A.M. Almehdi, and P. Hellyer, 142–156. Emirates Heritage Club. Milan, Italy: Centro Poligrafico Milano SpA.

Gladfelter, W.B. 1982. White band disease in *Acropora palmata*: Implications for the structure and function of shallow reefs. *Bull. Mar. Sci.* 32:639–643.

Gleason, D.F., and G.M. Wellington. 1993. Ultraviolet radiation and coral bleaching. *Nature* 365:836–838.

Glynn, P.W. 1984. Widespread coral mortality and the 1982/83 El Nino warming event. *Environ. Conserv.* 11:133–146.

Glynn, P.W. (ed.). 1990. *Global Ecological Consequences of the 1982–83 El Niño-Southern Oscillation.* Amsterdam: Elsevier Oceanographic Series 52.

Glynn, P.W. 1993. Coral bleaching: Ecological perspectives. *Coral Reefs* 12:1–18.

Glynn, P.W. 1996. Coral reef bleaching: Facts, hypotheses and implications. *Global Change Biol.* 2:495–509.

Glynn, P.W. 1997. Bioerosion and coral reef growth. In *Life and Death of Coral Reefs,* ed. C. Birkeland, 68–95. New York: Chapman and Hall.

Glynn, P.W. 2000. El Niño-Southern Oscillation mass mortalities of reef corals: A model of high temperature marine extinctions? In *Carbonate Platform Systems: Components and Interactions*, eds. E. Insalaco, P.W. Skelton, and P.J. Palmer, 117–133. London: Geol. Soc. Spec. Pub.

Glynn, P.W., and S.B. Colley. (eds.). 2001. A collection of studies on the effects of the 1997–98 El Niño-Southern Oscillation event on corals and coral reefs in the eastern tropical Pacific. *Bull. Mar. Sci.* 69:1–288.

Goldenberg, S.B., C.W. Landsea, A.M. Mestaz-Nuñez, and W.M. Gray. 2001. The recent increase in hurricane activity: Causes and implications. *Science* 293:474–479.

Graus, R.R., and I.G. Macintyre. 1998. Global warming and the future of Caribbean reef building. *Carbonates and Evaporites* 13:43–47.

Gray, W.M. 1979. Hurricanes: Their formation, structure and likely role in the tropical circulation. Supplement to *Meteorology over the Tropical Oceans,* ed. D.B. Shaw, 155–218. Bracknell: Royal Meteorological Society.

Gray, W.M. 1988. Environmental influences on tropical cyclones. *Aust. Meteorol. Mag.* 36:127–139.

Gray, W.M., and J.D. Sheaffer. 1991. El Niño and QBO influences on tropical cyclone activity. In *Teleconnections Linking Worldwide Climate Anomalies: Scientific Basis and Societal Impact*, eds. M.H. Glantz, R.W. Katz, and N. Nicholls, 257–284. Cambridge: Cambridge University Press.

Greenstein, B.J., L.A. Harris, and H.A. Curran. 1998. Comparison of recent life and death assemblages to Pleistocene reef communities: Implications for rapid faunal replacement on recent reefs. *Carbonates and Evaporites* 13:23–31.

Guinotte, J.M., R.W. Buddemeier, and J.A. Kleypas. 2003. Future coral reef habitat marginality: Temporal and spatial effects of climate change in the Pacific basin. *Coral Reefs* 22:551–558.

Harmelin-Vivien, M.L., and P. Laboute. 1986. Catastrophic impact of hurricanes on atoll outer reef slopes in the Tuamotu (French Polynesia). *Coral Reefs* 5:55–62.

Harvell, C.D., K. Kim, J.M. Burkholder, R.R. Colwell, P.R. Epstein, D.J. Grimes, E.E. Hofmann, E.K. Lipp, A.D.M.E. Osterhaus, R.M. Overstreet, J.W. Porter, G.W. Smith, and G.R. Vasta. 1999. Emerging marine diseases—Climate links and anthropogenic factors. *Science* 285:1505–1510.

Harvell, C.D., C.E. Mitchell, J.R. Ward, S. Altizer, A.P. Dobson, R.S. Ostfeld, and M.D. Samuel. 2002. Climate warming and disease risks for terrestrial and marine biota. *Science* 296: 2158–2162.

Henderson-Sellers, A., H. Zhang, G. Berz, K. Emanuel, W. Gray, C. Landsea, G. Holland, J. Lighthill, S.-L. Shieh, P. Webster, and K. McGuffie. 1998. Tropical cyclones and global climate change: A post IPCC assessment. *Bull. Am. Meteorol. Soc.* 79:19–38.

Hickey, B.M., P. MacCready, E. Elliott, and N.B. Kachel. 2000. Dense saline plumes in Exuma Sound, Bahamas. *J. Geophy. Res.* 105C5:11471–11488.

Houghton, J.T., Y. Ding, D.J. Griggs, M. Noguer, P.J. van der Linden, X. Dai, K. Maskell, and C.A. Johnson. 2001. *Climate Change 2001: The Scientific Basis.* Cambridge: Cambridge University Press.

Hubbard, D.K. 1997. Reefs as dynamic systems. In *Life and Death of Coral Reefs,* ed. C. Birkeland, 43–67. New York: Chapman and Hall.

Hubbard, D.K. 2002. The role of framework in modern reefs and its application to ancient systems. In *The History and Sedimentology of Ancient Reef Systems,* ed. G.D. Stanley, Jr., 351–386. New York: Kluwer Academic/Plenum Publishers.

Hubbard, D.K., R.P. Burke, and I.P. Gill. 1986. Styles of reef accretion along a steep, shelf-edge reef, St. Croix, U.S. Virgin Islands. *J. Sediment. Petrol.* 56:848–861.

Hubbard, D.K., R.B. Burke, and I.P. Gill. 1998. Where's the reef: The role of framework in the Holocene. *Carbonates and Evaporites* 13:3–9.

Hubbard, D.K., A.I. Miller, and D. Scaturo. 1990. Production and cycling of calcium carbonate in a shelf-edge reef system (St. Croix, U.S. Virgin Islands): Applications to the nature of reef systems in the fossil record. *J. Sediment. Petrol.* 60:335–360.

Hubbard, D.K., K.M. Parsons, J.C. Bythell, and N.D. Walker. 1991. The effects of Hurricane Hugo on the reefs and associated environments of St. Croix, U.S. Virgin Islands—A preliminary assessment. *J. Coast. Res. Spec. Iss.* 8:33–48.

Hubbard, D.K., H. Zankel, I. van Heerden, and I.P. Gill. 2005. Holocene reef development along the northeastern St. Croix shelf, Buck Island, U.S. Virgin Islands. *J. Sediment. Res.* 75:97–113.

Huckleberry, G.A., and B.R. Billman. 2003. Geoarchaeological insights gained from surficial geologic mapping, middle Moche Valley, Peru. *Geoarchaeology* 18:505–521.

Hughes, T.P., A.H. Baird, D.R. Bellwood, M. Card, S.R. Connolly, C. Folke, R. Grosberg, O. Hoegh-Guldberg, J.B.C. Jackson, J. Kleypas, J.M. Lough, P. Marshall, M. Nystrom, S.R. Palumbi, J.M. Pandolfi, B. Rosen, and J. Roughgarden. 2003. Climate change, human impacts, and the resilience of coral reefs. *Science* 301:929–993.

Jordan, I.E., and M.J. Samways. 2001. Recent changes in the coral assemblages composition of a South African coral reef, with recommendations for long-term monitoring. *Biodivers. Conserv.* 10:1027–1037.

Kinsman, D.J.J. 1964. Reef coral tolerance of high temperatures and salinities. *Nature* 202:1280–1282.

Kleypas, J.A., R.W. Buddemeier, D. Archer, J.-P. Gattuso, C. Langdon, and B.N. Opdyke. 1999b. Geochemical consequences of increased atmospheric carbon dioxide on coral reefs. *Science* 284:118–120.

Kleypas, J.A., J.W. McManus, and L.A.B. Menez. 1999a. Environmental limits to coral reef development: Where do we draw the line? *Am. Zool.* 39:146–159.

Knutson, T.R., and R.E. Tuleya. 2001. Impact of CO_2-induced warming on hurricane intensities as simulated in a hurricane model with ocean coupling. *J. Climate* 14:2458–2468.

Knutson, T.R., R.E. Tuleya, and Y. Kurihara. 1998. Simulated increase of hurricane intensities in a CO_2-warmed climate. *Science* 279:1018–1020.

Koutavas, A., J. Lynch-Stieglitz, T.M. Marchitto, Jr., and J.P. Sachs. 2002. El Niño-like pattern in ice age tropical Pacific sea surface temperature. *Science* 297:226–230.

Krishnamurti, T.N., R. Correa-Torres, M. Latif, and G. Daughenbaugh. 1998. The impact of current and possibly future SSt anomalies on the frequency of Atlantic hurricanes. *Tellus* 50A:186–210.

Kushmaro, A., E. Rosenberg, M. Fine, and Y. Loya. 1997. Bleaching of the coral *Oculina patagonica* by *Vibrio* AK-1. *Mar. Ecol. Prog. Ser.* 147:159–165.

Landsea, C.W. 1997. Comments on "Will greenhouse gas-induced warming over the next 50 years lead to higher frequency and greater intensity of hurricanes?". *Tellus* 49A:622–623.

Landsea, C.W. 2000a. Climate variability of tropical cyclones: Past, present and future. In *Storms 2000,* eds. R.A. Pileke, Sr. and R.A. Pileke, Jr., 220–241. New York: Routledge.

Landsea, C.W. 2000b. ElNino–Southern Oscillation and the seasonal predictability of tropical cyclones. In *El Niño and the Southern Oscillation,* eds. H.F. Diaz and V. Markgraf, 149–182. Cambridge: Cambridge University Press.

Lang, J.C. (ed.). 2003. Status of the coral reefs in the Western Atlantic: Results of initial surveys, Atlantic and Gulf Rapid Reef Assessment (AGRRA) program. *Atoll Res. Bull.* 496, 630.

Lang, J.C., R.I. Wicklund, and R.F. Dill. 1988. Depth- and habitat-related bleaching of zooxanthellate reef organisms near Lee Stocking Island, Exuma Cays, Bahamas, *Proc. Sixth Int. Coral Reef Symp., Australia* 1:269–274.

Lesser, M.J. 1997. Oxidative stress causes coral bleaching during exposure to elevated temperatures. *Coral Reefs* 16:187–192.

Logan, A. 1994. Reefs and lagoons of Cayman Brac and Little Cayman. In *The Cayman Islands—Natural History and Biogeography,* eds. M.A. Brunt and J.E. Davie, 105–124. Dordrecht: Kluwer Academic.

Loschnigg, J., G.A. Meehl, P.J. Webster, J.M. Arblaster, and J.B. Compo. 2003. The Asian monsoon, the tropospheric biennial oscillation, and the Indian Ocean zonal mode in the NCAR CSM. *J. Climate* 16:1617–1642.

Loschnigg, J., and P.J. Webster. 2000. A coupled ocean–atmosphere system of SST modulation for the Indian Ocean. *J. Climate* 13:3342–3360.

Lough, J.M. 1994. Climate variation and El Niño-Southern Oscillation events on the Great Barrier Reef: 1958 to 1987. *Coral Reefs* 13:181–195.

Lough, J.M. 2000. 1997–8: Unprecedented thermal stress to coral reefs? *Geophys. Res. Lett.* 27:3901–3904.

Lugo, A.E., C. Rogers, and S. Nixon. 2000. Hurricanes, coral reefs and rainforests: Resistance, ruin and recovery in the Caribbean. *Ambio* 29:106–114.

338 Bernhard Riegl

Macintyre, I.G., R.B. Burke, and R. Stuckenrath. 1981. Core holes in the outer fore-reef off Carrie Bow Cay, Belize: A key to the Holocene history of the Belizean barrier reef complex. *Proc. Fourth Int. Coral Reef Symp., Manila* 1:567–574.

Macintyre, I.G., H.G. Multer, H.L. Zankl, D.K. Hubbard, M.P. Weiss, and R. Stuckenrath. 1985. Growth and depositional facies of a windward reef complex (Nonsuch Bay, Antigua, W.I.). *Proc. Fifth Int. Coral Reef Congr., Tahiti* 6:605–610.

Maida, M., and B.P. Ferreira. 1997. Coral reefs of Brazil: An overview. *Proc. Eighth Int. Coral Reef Symp., Panama* 1:263–274.

Massel, S.R. 1999. *Fluid Mechanics for Marine Ecologists*. Berlin: Springer-Verlag.

McCabe, G.J., M.P. Clark, and M.C. Serreze. 2001. Trends in northern hemisphere surface cyclone frequency and intensity. *J. Climate* 14:2763–2768.

McManus, J.W. 1997. Tropical marine fisheries and the future of coral reefs: A brief review with emphasis on Southeast Asia. *Coral Reefs* 16:S121–S128.

Meyer, D.L., J.M. Bries, B.J. Greenstein, and A.O. Debrot. 2003. Preservation of in situ reef framework in regions of low hurricane frequency: Pleistocene of Curaçao and Bonaire, southern Caribbean. *Lethaia* 36:273–285.

Moulding, A.L. 2005. Coral recruitment patterns in the Florida Keys. *Rev. Biol. Trop.* 53:75–82.

Moy, C.C., G.O. Seltzer, D.T. Rodbell, and D.M. Anderson. 2002. Variability of El Niño/Southern Oscillation activity at millennial timescales during the Holocene epoch. *Nature* 420:162–165.

Mumby, P.J., J. Chisholm, A. Edwards, C. Clark, E. Roark, S. Andrefouet, and J. Jaubert. 2002a. Unprecedented bleaching-induced mortality in Porites spp. at Rangiroa Atoll, French Polynesia. *Mar. Biol.* 139:183–89.

Mumby P.J., J.R.M. Chisholm, A.J. Edwards, S. Andrefouet, and J. Jaubert. 2002b. Cloudy weather may have saved Society Island reef corals during the 1998 ENSO event. *Mar. Ecol. Prog. Ser.* 222:209–216.

Nakamura, T., and R. van Woesik. 2001. Water-flow rates and passive diffusion partially explain differential survival of corals during the 1998 bleaching event. *Mar. Ecol. Prog. Ser.* 212:301–304.

Nasrallah, H.A., E. Nieplova, and E. Ramadan. 2004. Warm season extreme temperature events in Kuwait. *J. Arid Environ.* 56: 357–371.

Neumann, C.J., B.R. Jarvinen, C.J. McAdie, and G.R. Hammer. 1999. *Tropical Cyclones of the North Atlantic Ocean, 1871–1998*. Historical Climatology Series 6-2, National Climate Data Center, Asheville, NC.

Pandolfi, J.M. 1999. Responses of Pleistocene coral reefs to environmental change over long temporal scales. *Am. Zool.* 39:113–130.

Pandolfi, J.M., R.H. Bradbury, E. Sala, T.P. Hughes, K.A. Bjorndal, R.G. Cooke, D. McArdle, L. McClenachan, M.J.H. Newman, G. Paredes, R.R. Warner, J.B.C. Jackson. 2003. Global trajectories of the long-term decline of coral reef ecosystems. *Science* 301: 955–958.

Perry, C.T. 1999. Reef framework preservation in four contrasting modern reef environments, Discovery Bay, Jamaica. *J. Coast. Res.* 15:796–812.

Perry, C.T. 2001. Storm-induced coral rubble deposition: Pleistocene records of natural reef disturbance and community response. *Coral Reefs* 20:171–183.

Philander, S.G. 1990. *El Niño, La Niña, and the Southern Oscillation*. San Diego: Academic Press.

Pittock, A.B. 1999. Coral reefs and environmental change: Adaptation to what? *Am. Zool.* 39:10–29.

Purkis, S.J., and B. Riegl. 2005. Spatial and temporal dynamics of Arabian Gulf coral assemblages quantified from remote-sensing and *in situ* monitoring data. *Mar. Ecol. Prog. Ser.* 287:99–113.

Quinn, N.J., and B.L. Kojis. 2005. Patterns of sexual recruitment for acroporid coral populations on the West Fore Reef at Discovery Bay, Jamaica. *Rev. Biol. Trop.* 53:83–90.

Reaka-Kudla, M.L., J.S. Feingold, and P.W. Glynn. 1996. Experimental studies of rapid bioerosion of coral reefs in the Galapagos islands. *Coral Reefs* 15:101–107.

Renegar, D.A., and B.M. Riegl. 2005. Effects of nutrient enrichment and elevated CO_2 partial pressure on growth rate of Atlantic scleractinian coral *Acropora cervicornis. Mar. Ecol. Prog. Ser.* 293:69–76.

Reynaud, S., N. Leclercq, S. Romaine-Lioud, C. Ferrier-Pagès, J. Jaubert, and J.P. Gattuso. 2003. Interacting effects of CO_2 partial pressure and temperature on photosynthesis and calcification in a scleractinian coral. *Global Change Biol.* 9:1660–1668.

Reynolds, R.W., and T.M. Smith. 1994. Improved global sea surface temperature analyses using optimum interpolation. *J. Climate* 7:929–948.

Richardson, L.L., and R.B. Aronson. 2002. Infectious diseases of reef corals. *Proc. Ninth Int. Coral Reef Symp., Bali* 2:1225–1230.

Riegl, B. 2001. Inhibition of reef framework by frequent disturbance: Examples from the Arabian Gulf, South Africa, and the Cayman Islands. *Palaeogeogr. Palaeoclimatol. Palaeoecol.* 175:79–101.

Riegl, B. 2002. Effects of the 1996 and 1998 positive sea-surface temperature anomalies on corals, coral diseases and fish in the Arabian Gulf (Dubai, UAE). *Mar. Biol.* 140:29–40.

Riegl, B. 2003. Global climate change and coral reefs: Different effects in two high latitude areas (Arabian Gulf, South Africa). *Coral Reefs* 22:433–446.

Riegl, B., and W.E. Piller. 2000. Reefs and coral carpets in the northern Red Sea as models for organism–environment feedback in coral communities and its reflection in growth fabrics. In *Carbonate Platform Systems: Components and Interactions,* eds. E. Insalaco, P.W. Skelton, and T.J. Palmer, 71–88. London: Geol. Soc. London. Spec. Pub. 178.

Riegl, B., and W.E. Piller. 2001. Cryptic tissues inside *Acropora* frameworks: A mechanism to enhance tissue survival in hard times while also enhancing framework density. *Coral Reefs* 5:67–68.

Riegl, B., and W.E. Piller. 2003. Possible refugia for reefs in times of environmental stress. *Int. J. Earth Sci. (Geol. Rundsch.)* 92:520–531.

Riegl, B., M.H. Schleyer, P.J. Cook, and G.M. Branch. 1995. Structure of Africa's southernmost coral communities. *Bull. Mar. Sci.* 56:676–691.

Rigby, J.K., and H.H. Roberts. 1976. Geology, reefs and marine communities of Grand Cayman Island, British West Indies. *Brigham Young Univ. Geol. Stud. Spec. Pub.* 4.

Roberts, H.H., P.A. Wilson, and A. Lugo-Fernandez. 1992. Biologic and geologic responses to physical processes: Examples from modern reef systems of the Caribbean-Atlantic region. *Cont. Shelf Res.* 12:809–834.

Rogers, C.S. 1993a. Hurricanes and anchors: Preliminary results from the National Park Service regional reef assessment program. In *Global Aspects of Coral reefs: Health, Hazards and History,* ed. R.N. Ginsburg, 214–219. Miami: University of Miami, Rosenstiel School of Marine and Atmospheric Science.

Rogers, C.S. 1993b. Hurricanes and coral reefs: The intermediate disturbance hypothesis revisited. *Coral Reefs* 12: 127–137.

Rogers, C.S., V. Garrison, and R. Grober-Dunsmore. 1997. A fishy story about hurricanes and herbivory: Seven years of research on a reef in St. John, U.S. Virgin Islands. *Proc. Eighth Int. Coral Reef Symp.* 1:555–560.

Rogers, C.S., L.N. McLain, and C.R. Tobias. 1991. Effects of Hurricane Hugo (1989) on a coral reef in St. John, USVI. *Mar. Ecol. Prog. Ser.* 78:189–199.

Ruddiman, W.F. 2001. *Earth's Climate. Past and Future.* New York: W.H. Freeman.

Saji, N.H., B.N. Goswami, P.N. Vinayachandran, and T. Yamagata. 1999. A dipole mode in the tropical Indian Ocean. *Nature* 401:360–361.

Salm, R.V. 1993. Coral reefs of the Sultanate of Oman. *Atoll Res. Bull.* 380:1–83.

Schleyer, M.H. 1999. A synthesis of KwaZulu/Natal coral research. *Oceanogr. Res. Inst. Durban Spec. Pub.* 5:1–36.

Scoffin, T.P. 1993. The geological effects of hurricanes on coral reefs and the interpretation of storm deposits. *Coral Reefs* 12:203–221.

Sheppard, C.R.C. 2003. Predicted recurrence of coral mass mortality in the Indian Ocean. *Nature* 425:294–297.

Sheppard, C.R.C., and R. Loughland. 2002. Coral mortality and recovery in response to increasing temperature in the southern Arabian Gulf. *Aquat. Ecosyst. Health Manage.* 5:395–402.

Sheppard, C.R.C., A.R.G. Price, and C.M. Roberts. 1992. *Marine Ecology of the Arabian Region: Patterns and Processes in Extreme Tropical Environments.* London: Academic Press.

Shinn, E.A. 1976. Coral reef recovery in Florida and the Persian Gulf. *Environ. Geol.* 1:241–254.

Shinn, E.A., C.D. Reich, D.A. Hickey, and B.A. Lidz. 2003. Staghorn tempestites in the Florida Keys. *Coral Reefs* 22:91–97.

Smith, A.H., C.S. Rogers, and C. Bouchon. 1997. Status of western Atlantic coral reefs in the Lesser Antilles. *Proc. Eighth Int. Coral Reef Symp., Panama* 1:351–356.

Smith, N.P. 2001. Weather and hydrographic conditions associated with coral bleaching: Lee Stocking Island, Bahamas. *Coral Reefs* 20:415–422.

Stanley, G.D., Jr. 2002. *The History and Sedimentology of Ancient Reef Systems.* New York: Kluwer Academic/Plenum Publishers.

Sugimoto, T., S. Kimura, and K. Tadakoro. 2001. Impact of El Niño events and climate regime shift on living resources in the western North Pacific. *Progr. Oceanogr.* 49:113–127.

Timmermann, A., J. Oberhuber, A. Bacher, M. Esch, M. Latif, and E. Roeckner. 1999. Increased El-Niño frequency in a climate model forced by future greenhouse warming. *Nature* 398:694–696.

Toren, A., L. Landau, A. Kushmaro, Y. Loya, and E. Rosenberg. 1998. Effect of temperature on adhesion of *Vibrio* strain AK-1 to *Oculina patagonica* and on coral bleaching. *Appl. Environ. Microbiol.* 64:1379–1384.

Torrence, C., and P.J. Webster. 1999 Interdecadal changes in the ENSO-monsoon system. *J. Climate* 12:2679–2690.

Treml, E., M. Colgan, and M. Keevican. 1997. Hurricane disturbance and coral reef development: A geographic information system (GIS) analysis of 501 years of hurricane data from the Lesser Antilles. *Proc. Eighth Int. Coral Reef Symp., Panama* 1:541–546.

Tudhope, A.W., C.P. Chilcott, M.T. McCulloch, E.R. Cook, J. Chappell, R.M. Ellam, D.W. Lea, J.M. Lough, and G.B. Shimmield. 2001. Variability in the El Niño-Southern Oscillation through a glacial-interglacial cycle. *Science* 291:1511–1517.

Veron, J.E.N. 1995. *Corals in Space and Time*. Sydney: UNSW Press.

Walker, N.D., H.H. Roberts, L.J. Rouse, and O.K. Huh. 1982. Thermal history of reef-associated environments during a record cold-air outbreak event. *Coral Reefs* 1:83–87.

Webster, P.J., A.M. Moore, J.P. Loschnigg, and R.R. Leben. 1999. Coupled ocean-atmosphere dynamics in the Indian Ocean during 1997–98. *Nature* 401:356–360.

Wellington, G.M., P.W. Glynn, A.E. Strong, S.A. Nauarrete, E. Wieters, and D. Hubbard. 2001. Crisis on coral reefs linked to climate change. *EOS* 82(1):1–6.

Wigley, T.M.L. 2000. ENSO, volcanoes and record breaking temperature. *Geophys. Res. Lett.* 27:4101–4104.

Wilkinson, C. (ed.). 2000. *Status of Coral Reefs of the World: 2000*. Townsville: Aust. Inst. Mar. Sci.

Wilkinson, C. (ed.). 2004. *Status of Coral Reefs of the World: 2004*, Vols. 1 and 2. Townsville: Aust. Inst. Mar. Sci.

Wilson, P.A., and H.H. Roberts. 1992. Carbonate-periplatform sedimentation by density flows: A mechanism for rapid off-bank and vertical transport of shallow water fines. *Geology* 20:713–716.

Wilson, P.A., and H.H. Roberts. 1995. Density cascading: Off-shelf sediment transport, evidence and implications, Bahama Banks. *J. Sediment. Res.* A65:45–56.

Woodley, J.D. 1992. The incidence of hurricanes on the north coast of Jamaica since 1870: Are the classic reef descriptions atypical? *Hydrobiologia* 247:133–138.

Woodley, J.D., E.A. Chornesky, P.A. Clifford, J.B.C. Jackson, L.S. Kaufman, J.C. Lang, M.P. Pearson, J.W. Porter, M.C. Rooney, K.W. Rylaarsdam, V.J. Tunnicliffe, C.M. Wahle, J.L. Wulff, A.S.G. Curtis, M.D. Dallmeyer, B.P. Jupp, M.A.R. Koehl, J. Neigel, and E.M. Sides. 1981. Hurricane Allen's impact on Jamaican coral reefs. *Science* 214:749–755.

Yamasaki, H. 2000. Nitrite-dependent nitric oxide production pathway: Implications for involvement of active nitrogen species in photoinhibition *in vivo*. *Philos. Trans. R. Soc. London Ser. B* 355:1477–1488.

11. Responses of Coral Reefs to El Niño–Southern Oscillation Sea-Warming Events

Gerard M. Wellington and Peter W. Glynn

11.1 Introduction

Prior to Chesher's (1969) stunning prediction that the corallivorous sea star, *Acanthaster planci*, could spell the initial phases of extinction of reef-building corals in the Pacific Ocean, coral reefs were regarded as highly stable and resilient ecosystems in near equilibrium with their physical and biotic environmental controls (e.g., Odum and Odum 1955; Wells 1957; Margalef 1968). Widespread outbreaks of *A. planci* and resulting coral mortality over the Indo-Pacific region have caused considerable alarm, and although the causes of these outbreaks are still unknown, some workers link them to favorable conditions in the water column promoting larval survival followed by heavy settlement and high recruitment (Birkeland and Lucas 1990). The effects of this predator are still a significant management problem in many areas (Wilkinson 1990; Wilkinson and Macintyre 1992; DeVantier and Done, Chapter 4), but concerns about *A. planci* disturbances have been increasingly supplanted by concerns about other global-scale disturbances that are linked to significant coral reef decline, including the deterioration of water quality from runoff (Ginsburg 1994), overfishing (Jackson 1997; Pandolfi et al. 2003), coral diseases (Harvell et al. 1999), and coral bleaching (Glynn 1993; Brown 1997; Wilkinson 2000; Hughes et al. 2003). This essay focuses on coral reef degradation due to coral bleaching, which is considered by many to be the largest threat to coral reef ecosystems (Hoegh-Guldberg 1999; Wilkinson 2002).

Beginning with the very strong 1982 to 1983 El Niño event, and continuing at high frequency through the 1980s, workers became aware of unprecedented coral reef bleaching: the sudden loss of endosymbiotic dinoflagellates that often leads to widespread coral mortality (Glynn 1984, 1990a; Brown 1987; Guzmán et al. 1987; Carriquiry et al. 1988; Williams and Bunkley-Williams 1988, 1990). Similar to the alarm precipitated by *Acanthaster* disturbances of the previous decade, the global coral bleaching and mortality events stimulated widespread concern in the coral reef research community, resulting in workshops, symposium addresses, and special journal publications focusing on this issue (e.g., Ogden and Wicklund 1988; Brown 1990; D'Elia et al. 1991; Buddemeier 1992). Several field and laboratory studies have demonstrated that the majority of these recent coral bleaching and mortality events are caused by elevated seawater temperatures that exceed the stress-response thresholds of temperature- and light-sensitive coral–algal symbioses (Glynn and D'Croz 1990; Jokiel and Coles 1990; Glynn et al. 1992, 2001; Gleason and Wellington 1993; Goreau and Hayes 1994; Rowan et al. 1997; Hoegh-Guldberg 1999; Lough 2000; Berkelmans 2002). A multitude of stressors often associated with sea warming disturbances, either during the event or with some time lag, often exacerbate the already weakened state of corals. These may include increased rainfall, flooding and runoff, reduced cloud cover and wind speed (favoring increased light penetration), sedimentation, pollution, violent storms, intensified upwelling, and sea level lowering during La Niña events. Additionally, several recent studies have demonstrated increasing incidences of epizootics that are often correlated with elevated thermal conditions (Kushmaro et al. 1996; Harvell et al. 1999; Porter 2001). Finally, taking into account predicted sea-level rise and changes in the carbonate mineral saturation state, effects that could interact to reduce calcification rates and reef building (Smith and Buddemeier 1992; Kleypas et al. 1999, 2001; Pittock 1999), it will become necessary to deal with a suite of disturbances affecting coral reefs during global warming.

New and stronger evidence indicates that most of the global warming observed over the last 50 years is attributable to anthropogenic forcings, namely, greenhouse gas and aerosol emissions (compare IPCC 1992, 1996, and 2001). The observed mean sea-surface temperature (estimated subsurface bulk temperature; i.e., the upper few meters depth) has increased from 0.4 to 0.8 °C since the late 19th century. The sea-surface temperature data obtained from 1961 to 1990 show a 0.23 to 0.27 °C warming during this 30-year period. The two warmest years globally were 1998 and 2005, which were the highest observed over an instrumental record beginning in 1880 (Shein 2006). Positive anomalies above the 1961 to 1990 mean were +0.50 °C (1998) and +0.53 °C (2005), the latter in the absence of a strong El Niño signal. The globally averaged surface temperature is projected to increase by 1.4 to 5.8 °C over the period 1990 to 2100 (full range of 35 IPCC emission scenarios based on several climate models). The projected rate of warming is greater than the observed changes during the 20th century and, based on paleoclimate data, is very likely to be without precedent during at least the last 10,000 years (IPCC 2001).

In this chapter we address sea-warming disturbances associated with El Niño–Southern Oscillation (ENSO) events, with a focus on the eastern tropical Pacific and the effects of ENSO disturbances on coral bleaching, mortality, and recovery. While no generally accepted definition of El Niño exists (see Trenberth 1997), historically it refers to local sea warming off coastal Peru. The ENSO phenomenon is more wide-ranging, involving the migration of the south Pacific warm pool across the International Dateline, and causing atmospheric and oceanic perturbations along the entire eastern Pacific as well as the western Atlantic and Indian Oceans. Here we use El Niño and ENSO for local (eastern tropical Pacific) and global effects, respectively. The eastern Pacific region is particularly suited for an assessment of ENSO effects on coral reefs because it experiences a strong warming signal and because a long-term database of the condition of pre- and post-disturbance coral communities is available. Other coral reef areas are also discussed where they provide insight into how the varied effects of sea warming disturbances influence coral reef ecosystems.

11.2 The Nature of ENSO Events

In the marine realm, one of the first and most notable effects of ENSO is a sudden warming of the central to eastern tropical Pacific Ocean. In the eastern Pacific, elevated sea level, thermocline depression, local flooding or drought conditions, and increased storm activity often accompany sea warming. The frequency of ENSO events over the past century is once about every four years, but their onset is erratic, and their intensity, duration, and spatial extent are highly variable (Philander 1990; Enfield 1992; Allan et al. 1996). The atmospheric manifestation of ENSO, the Southern Oscillation, is synchronous with the marine phase and involves linkages between remote climatic perturbations—so-called teleconnections—that can influence all of the world's oceans and continental regions (Riegl, Chapter 10). For example, El Niño warming in the eastern Pacific is usually correlated with: (1) high SSTs in the western equatorial and southwest Pacific (Great Barrier Reef region) about one year before and following an event (Rasmusson and Carpenter 1982; Lough 1994); (2) high SSTs contemporaneously and basin-wide in the Indian Ocean (Baquero-Bernal et al. 2002); and (3) high SSTs in the northwestern tropical Atlantic and Caribbean Sea 4 to 5 months following the eastern Pacific mature phase (Enfield and Mayer 1997).

In terms of ENSO activity, the 1980s and 1990s were unusual in several respects. The two strongest ENSO events in recorded history—1982 to 1983 and 1997 to 1998—occurred only 15 years apart, and during this same time one of the most protracted El Niño warming periods, lasting from 1990 to 1995, was recorded (Trenberth and Hoar 1996). Some workers, however, have concluded that this was not an extended El Niño event, but rather a prolonged warm phase of a Pacific decadal mode of variability (Lau and Weng 1999). The 1982 to 1983 event was touted as the "event of the century" (Cane 1983, 1986; Hansen 1990) and the 1997 to 1998 event was considered comparable or even stronger than the earlier event (McPhaden 1999; Enfield 2001). The warming intervals were comparable during the two events, with each about 14 months in duration. A comparison of SST metrics in the eastern Pacific, in Niño 1 + 2 and Niño 3 zones, shows both SSTs and SSTAs (SST anomalies) rose faster in mid-1997 than in mid-1982 (Fig. 11.1). These data are from an optimally interpolated (OI) analysis of ship, drifter, and satellite measurements produced by the NOAA National Centers for Environmental Prediction (NCEP; Reynolds and Smith 1994). Niño 3 SSTAs also were ~0.5 °C higher in 1997 to 1998 than in 1982 to 1983.

Several regional- to local-scale differences between the 1982 to 1983 and 1997 to 1998 events have been recognized, including diverse physical and biotic characteristics and responses. For example, coral bleaching and mortality associated

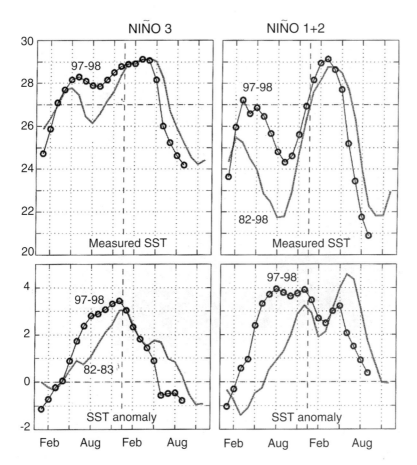

Figure 11.1. Comparisons of SSTs and SSTAs in Niño zones 3 (90°–150°W, 5°N–5°S) and 1+2 (80°–90°W, 0°–10°S) during the 1982 to 1983 and 1997 to 1998 ENSO events. These Niño regions were selected to coincide with historical ship tracks where reliable data were available (Rasmusson and Carpenter 1982). Niño 3 is centered offshore and far to the west along the equator, and Niño 1+2 refers to the region near the South American coast. Horizontal dashed lines are plotted at 27 °C (upper) and 0 °C (lower) for reference. After Enfield (2001).

with ENSO-elevated SSTs in 1982 to 1983 were largely confined to the eastern
Pacific, the central and northern sectors of the Great Barrier Reef, the Java Sea,
southern Japan, the western Indian Ocean, and the Caribbean Sea and adjacent
waters (Glynn 1984; Brown 1987; Coffroth et al. 1990), and included 68 reported
bleaching events. In contrast, during the 1997 to 1998 ENSO, frequent and widespread coral bleaching and mortality were reported in all major coral reef regions,
totaling 2,070 events (Figs. 11.2A,B, 11.3). Notable newly impacted areas were
Belize, Brazil, northeastern Australia, the Philippines, Southeast Asia, south India,

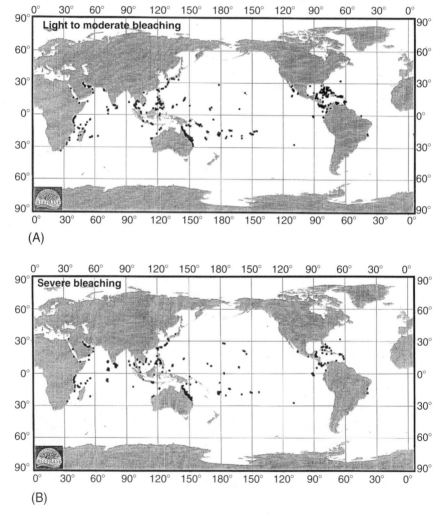

Figure 11.2. Global distribution of all known coral bleaching events (1963–2002). (**A**) Light
(<10%) to moderate (10–50%) coral bleaching (*n* = 1629 events). (**B**) Severe (>50%) coral
bleaching (*n* = 955 events). Map produced on 15 November 2002 from ReefBase (2002).

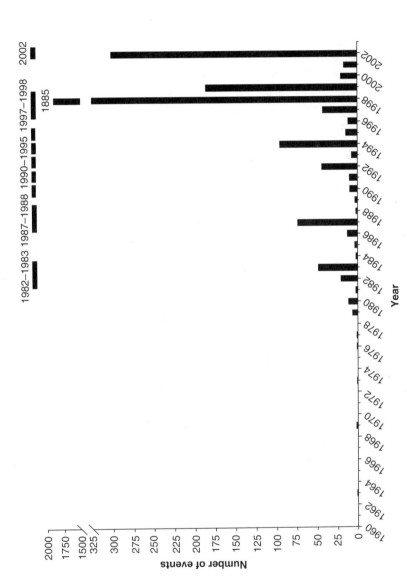

Figure 11.3. The total number of coral reef bleaching observations for each year globally, 1963 to 2002. The 1998 bleaching records ($n = 1885$) are severely truncated. From ReefBase (2002) data set, 10 August 2002. The duration of ENSO events during the 1980s and 1990s is noted above.

the Laccadive and Maldive Islands, continental East Africa, and the Red Sea. The only region where no bleaching was reported is off western tropical Africa, perhaps due to the absence of observers. The latest data available for 2002 indicate the occurrence of 452 coral bleaching events (ReefBase 2002). This was the greatest bleaching year ever reported for the Great Barrier Reef region, and was related to a large thermal anomaly observed by NOAA. The other reports in 2002 may be due to local thermal events more than to any major regional-scale pattern. Some proportion of the large increase in extent and number of reports between the earlier and latter events is certainly due to a greater awareness and documentation in recent years and the use of the Internet to aid transmission and collection of reports.

Given the general similarity of the 1982 to 1983 and 1997 to 1998 El Niño events in the eastern Pacific, we still note some marked local-scale differences between the two disturbances in both occurrence and timing of temperature stressors, and the severity and distribution of coral mortality. An example of one area where corals experienced severe bleaching and mortality in 1982 to 1983 (with overall mean mortality ~90%), and no adverse effects in 1997 to 1998 is the Pearl Islands in the Gulf of Panamá (Table 11.1). Seasonal upwelling typically occurs in the Gulf from late December to about the end of April. Upwelling was largely suppressed in 1983, and coral bleaching/mortality coincided with high maximum SSTs (>30 °C) and high stressful temperature duration (degree days, DD > 400; a "degree-days" index is the combined effect of SST anomalies and their duration; Podestá and Glynn 1997). Upwelling occurred immediately preceding and during the height of El Niño warming in 1998 (Glynn et al. 2001), with maximum SSTs <30 °C and degree days duration <400. Another difference was noted in the Galápagos Islands, which experienced similarly stressful high sea temperatures during 1982 to 1983 and 1997 to 1998. Critically high temperatures persisted in the Galápagos Islands in 1997 to 1998, resulting in coral bleaching and mortality (~24% overall), but of significantly lower magnitude than in 1982 to 1983 (~97%) (Feingold 2001; Podestá and Glynn 2001). The more recent, relatively low coral mortality could be attributed in part to marked semidiurnal temperature fluctuations, ranging from 2 to 4 °C, which occurred at sites where bleaching and mortality were not severe. A third difference probably was related to a respite from elevated sea temperatures in 1997 to 1998. In 1982 to 1983, coral bleaching and mortality occurred more or less continuously over the El Niño period, from 4 to 10 months depending on locality. In 1997 to 1998, two distinct bleaching bouts were detected at three equatorial eastern Pacific sites, each corresponding to local sea-warming pulses. The initial period of elevated temperatures and bleaching occurred over varying time intervals from August to November 1997 in Colombia (Vargas-Ángel et al. 2001), Panamá (Glynn et al. 2001), and Costa Rica (Jiménez et al. 2001). These were followed by 4- to 6-month intervals of lower temperatures, then a return of stressful temperatures from March to August 1998, depending upon locality. These high-temperature pulses were recorded in situ in coral communities, and also were visible as transitory filaments online in NOAA near-real-time weekly animated satellite imagery. In summary, several

Table 11.1. Mean (± SE) coral mortality at various eastern Pacific sites during the 1982 to 1983 and 1997 to 1998 El Niño warming events

Locality	Latitude	Percent mortality		Authority
		1982-83	1997-98	
Ecuadorean Coast	1.0–2.0°S	~80[2]	8.1 ±0.6	Glynn et al. (2001)
Galápagos Islands	1.5°S–1.5°N	97.0 ±1.4	24.3 ±1.0	Glynn et al. (2001)
Gorgona Island	3N	>50[3]	<1	Vargas-Ángel et al. (2001)
Malpelo Island	4N	?	<1	Vargas-Ángel et al. (2001)
Cocos Island	5N	96.8	20.4 ± 22.0	Guzmán and Cortés (1992), and unpublished data
Utría-Tebada	6.0–6.5°N	?	<1	Vargas-Ángel et al. (2001)
Gulf of Chiriquí	7.0–8.0°N	76.3 ±6.5	13.2 ±0.5	Glynn et al. (2001)
Gulf of Panamá	7.5–9.0°N	89.9[4] ±4.6	0	Glynn et al. (2001)
Caño Island	9°N	63.4[4] ±7.3	2.7	Guzmán and Cortés (1992)
Costa Rican coast[1]	10.5–11.0°N	?	5.7	Jiménez et al. (2001)
Huatulco	16°N	?	98.9[6]	Reyes-Bonilla et al. (2002)
Banderas Bay	20°N	?	62.1 ±7.8[7]	Carriquiry et al. (2001)
Baja California	23–25°N	10.2[5] ±4.0	18.2 ±1.8	Reyes-Bonilla (2001)

[1]Three localities were surveyed: Murciélagos Islands, Culebra Bay, and Golfo Dulce.
[2]Based on surveys conducted in 1986, which revealed wholly dead pocilloporid reefs similar to those in the Galápagos Islands that experienced 100% mortality in 1983 (Glynn et al. 2001). Numerous massive corals, contributing relatively little to live coral cover, survived the warming event with varying amounts of partial colony mortality.
[3]According to Prahl (1983, 1985), about 85% of the corals on a Gorgona Island reef were bleached and a few months later most of these were dead and covered with macroalgae.
[4]The percent coral mortalities at two localities in Figure 2 in Glynn et al. (1988) are incorrect and should be emended to 89.9% (from 85%) for the Gulf of Panamá and to 63.4% (from 51%) for Caño Island.
[5]This figure was reported for the 1987 El Niño bleaching event; apparently corals in the southern sector of Baja California were not affected in 1982-83, however, no data are available for this period (Reyes-Bonilla 1993).
[6]This figure is from two reefs surveyed, namely, Tijera and Mazunte reefs.
[7]This figure is for sampling sites at the north end of Banderas Bay. The 62.1% mean mortality reported in Carriquiry et al. (2001) is for sampling sites at the north end of Banderas Bay. The 96.0% mean mortality is for all 9 sites sampled across the bay (A. L. Cupul-Magaña, personal communication).

factors may have contributed to the reduced coral mortality in 1997 to 1998 in the equatorial eastern Pacific: (1) timing of the largest SST anomalies, which occurred in the Galápagos Islands during the cool season; (2) unabated upwelling in the Gulf of Panamá; (3) temperature fluctuations in the Galápagos Islands; (4) a 4- to 6-month respite from stressful temperatures at several sites; and (5) the presence of host/symbiont combinations more resistant to high temperatures in corals that survived the 1982 to 1983 bleaching event. Evidence for the latter effect is presented by Glynn et al. (2001), who proposed a bleaching model to explain the variable coral bleaching responses observed during El Niño events in Panamá and the Galápagos Islands.

Perhaps the greatest intraregional difference between events was the relatively low coral mortality rates in 1997 to 1998 compared to 1982 to 1983 (Glynn and Colley 2001; Glynn 2002). In general, every surveyed site in the equatorial eastern Pacific, over a latitudinal spread of 11° (2°S to 9°N), experienced notably lower coral mortalities in 1997 to 1998 compared with 1982 to 1983 (Table 11.1). At higher eastern Pacific latitudes, off the west coast of Mexico (16 to 20°N), coral mortalities were high at Banderas Bay during the 1997 to 1998 El Niño event and along the Huatulco reef tract in 1998. At Banderas Bay, overall mean coral mortality ranged from 62% to 96%, and the high coral cover in 1991 indicated no prior ENSO-induced changes, at least in the short term (Carriquiry and Reyes-Bonilla 1997). The high mean coral mortality of 98.9% reported at Huatulco occurred during a La Niña cool phase, but the lowest available temperatures, monthly mean values of 25 and 26 °C in February and May, respectively, did not appear to reach critically low levels (Reyes-Bonilla et al. 2002). (La Niña refers to large-scale changes in atmospheric and oceanographic conditions opposite to El Niño, e.g., the westerly location of the equatorial warm pool, and in the eastern tropical Pacific shoaling of the thermocline, anomalously cool sea surface temperatures, nutrient-rich surface waters, and high primary and secondary productivity.) Coral cover was also high on the Huatulco reef tract in 1996, but relatively large sections of dead and eroded pocilloporid reef frames at some sites and massive corals that recruited in 1989 suggest that a thermal disturbance might have occurred during the 1987 El Niño event (Glynn and Leyte Morales 1997). Coral mortality off southern Baja California was unknown in 1982 to 1983, but it was relatively low after the 1987 (10.2%) and 1997 to 1998 (18.2%) ENSO events (Table 11.1). The relatively low-level coral bleaching and mortality at Baja California is attributed to the lower and less prolonged stressful temperatures there, and the seasonal timing of positive temperature anomalies, which tend to occur during the low-temperature periods of winter and spring (Reyes-Bonilla et al. 2002).

Violent storms affecting reef areas in the eastern tropical Pacific were markedly different during 1997 to 1998 compared to 1982 to 1983. Four cyclones traversed the west coast of Mexico in 1997 and 1998. While the greatest damage on shallow pocilloporid reefs amounted to ~50% to 60% mortality, with some dislodgment of reef frame blocks, most reefs were not seriously affected (Lirman et al. 2001; S. S. González and H. Reyes-Bonilla personal communication).

No storm damage to corals was reported off Mexico or any Pacific Central American region in 1982 to 1983, but large swells and contrary seas caused considerable damage to coral reefs in the Galápagos Islands. Storm-generated seas uprooted branching and massive corals at Floreana and Pinta Islands in the Galápagos, depositing large amounts of coral and reef-associated animals on the shoreline (Robinson 1985). In 1982 to 1983, storm activity was greater in the South Pacific. Between December 1982 and April 1983, six hurricanes passed through French Polynesia, causing 50 to 100% destruction of corals on some deep (40 to >100 m) atoll reef slopes (Laboute 1985; Harmelin-Vivien and Laboute 1986).

11.3 ENSO-Related Disturbances to Coral Reefs

ENSO disturbances to coral reefs may be classified into short-term and longer-term temporal scales. Elevated sea water temperatures, often in combination with high irradiance levels (including UV radiation), are short-term disturbances that can cause coral responses of bleaching, tissue loss, and partial or whole colony mortality. Longer-term or delayed ENSO-related disturbances are: (1) violent storms; (2) high rainfall and flooding with increased coastal runoff and sedimentation; (3) periods of sudden and marked sea-temperature decline, resulting from thermocline shoaling and upwelling; (4) elevated nutrient concentrations; (5) dinoflagellate blooms; (6) disease epizootics; and (7) subaerial exposure of corals due to sea-level fluctuations. Detailed information on these sorts of disturbances is available in the reviews of Williams and Bunkley-Williams (1990), Glynn (1993, 2000), Brown (1997), and Hoegh-Guldberg (1999). Conditions (3), (4), (5), and (7), often associated with post-ENSO activities (La Niña events), can affect corals several months to more than a year following a period of elevated temperature.

The loss of zooxanthellae and/or decline in chlorophyll concentration in surviving tissues retards calcification and skeletal growth (Goreau and Macfarlane 1990; Glynn 1993; Hoegh-Guldberg 1999), interferes with reproduction (Szmant and Gassman 1990; Glynn et al. 2000; Hirose and Hidaka 2000; Omori et al. 2001), and lowers a coral's capacity to repair tissue damage and resist epizootics (Meesters and Bak 1993; Mascarelli and Bunkley-Williams 1999; Harvell et al. 2001). Lipid production also declines rapidly in bleached corals, negatively impacting the obligate crustacean symbionts of branching corals that are dependent on their hosts for trophic sustenance (Glynn et al. 1985). In some areas and for some coral taxa, e.g., *Oculina* in the eastern Mediterranean, bacteria have been shown to cause coral bleaching and may therefore precede a bleaching response induced solely by high-temperature stress (Kushmaro et al. 1996). However, it is likely that an increase in temperature may weaken a coral's resistance and increase the virulence of pathogenic bacteria (Kushmaro et al. 1996; Harvell et al. 2001). Finally, long-term consequences of bleaching that may continue for many years include: (8) changing patterns of predation on corals; (9) lowered rates of coral recruitment; (10) bioerosion and loss of stable coral frameworks; and (11) local to regional-scale coral extinctions.

11.4 History of ENSO Events

After the annual cycle, the Pacific El Niño-Southern Oscillation (ENSO) and its climatic impacts around the world constitute the strongest, most spatially coherent climate signal that exists in both the ocean and the atmosphere.

—Enfield and Mestas-Nuñez (1999).

Prior to the early 1970s little was known about the phenomenon now referred to as ENSO. The earliest accounts of these events, described by Murphy (1926) and others, came from the coastal shores of Ecuador and northern Perú during the mid- to late 1800s and were characterized by prolonged periods of high SSTs, accompanied by heavy precipitation and the cessation of coastal winds. During the 1920s Walker and Bliss (1930) described the occurrence of periodic "seesaw" shifts in atmospheric pressure systems resulting in the reversal of ocean currents and atmospheric pressures across the Pacific and Indian Oceans. However, it was not until Bjerknes (1969, 1972) that it became clear that the El Niño phenomenon was related to the Southern Oscillation through teleconnections linking the interaction of both atmospheric and ocean processes. Finally, Wyrtki (1973) demonstrated that the El Niño phenomenon could be explained by Pacific-wide changes in sea level responding to shifts in the trade winds. Even though much is now known about the dynamics of ENSO events, our ability to predict the occurrence and intensity of this phenomenon remains elusive.

   In this review we describe several recent studies that have shed light on the history of ENSO events ranging from recent to millennial time scales. How long have ENSOs and ENSO-like phenomena existed? And, how might their frequency and strength change under a future global warming scenario? Several recent studies provide compelling evidence linking anthropogenic factors as contributing significantly to the severity of both the 1982 to 1983 and 1997 to 1998 ENSO events.

   One of the first multicentury, coral-based climate reconstructions was conducted by Dunbar et al. (1994) in the Galápagos Islands, located at the epicenter of ENSO activity. A 10-m-wide, 5-m-high coral colony was sampled at a site located on the equator, at the "center of action" for recording thermal anomalies in the eastern Pacific. In this landmark study, a 367-year growth chronology and 347-year annual $\delta^{18}O$ analysis produced a climate record extending from 1586 to 1953 AD. In addition to the coral proxy records, written historical archives extend the record from 1586 to recent times (Quinn et al. 1987; Quinn and Neal 1995). This isotopic record made it possible to examine eastern Pacific SST variability over a time span not previously available and, more importantly, at such a pivotal site for recording ENSO events. An evolutionary spectral plot of the isotopic data reveals distinct step-like shifts in the frequency of ENSOs (Fig. 11.4). From 1750 to 1750 AD an ENSO event occurred every 6 to 4.6 years. By 1850 it shifted to 3.4 years. The comprehensive approach by Quinn and Neal (1995) involved the reconstruction of past El Niño events through an investigation of historical records. Their definition of an El Niño event is as follows: "the appearance of anomalously warm water along the coast of Ecuador and Perú as far south as

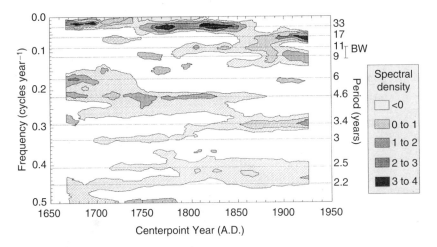

Figure 11.4. Evolutionary spectral density plot based on $\delta^{18}O$ coral showing progression of dominant oscillatory modes in the Urvina Bay (Galápagos Islands) annual record. The shaded areas indicate relative concentrations of variance. Results are consistent with historical records showing a progressive increase in frequency of ENSO from 6 to 2.3 years between events. Reproduced from Dunbar et al. (1994) with permission from the American Geophysical Union.

Lima (12° South) during which a normalized sea surface temperature (SST) anomaly exceeded one standard deviation for at least four consecutive months at three or more coastal stations." Under this criterion, Quinn and Neal were able to identify the El Niño events of 1957, 1965, 1972 to 1973, and 1976, based on 1956 to 1981 data. They classified the powerful 1982 to 1983 event as very strong, based on SST anomalies that were three times higher than their minimal criteria. In their essay they outlined 12 characteristics that could be used to identify very strong events, including SST anomalies 6 to 12 °C above normal in peak months, a rise in sea level resulting in coastal flooding, mass mortality of sea birds, and a drastic reduction in fisheries production. The level of confidence in these records is clearly higher after 1800. A simple analysis of Quinn and Neal's records from 1525 to 1987 AD, categorized by the number of strong, severe, and very severe El Niño events (Fig. 11.5), matches extremely well with shifts in the recurrence interval of El Niños reported in Dunbar et al. (1994). The fact that these two independent records are highly correlated provides confidence that this historical record may be largely correct. However, the accuracy of the historically based El Niño records (Quinn et al. 1987) has been called into question. Ortlieb and Machare (1993) argued that the observations of Dunbar et al. (1994) and Quinn and Neal (1995) did not necessarily match those from coastal Ecuador and northern Perú in terms of intensity and timing. This, however, could be related to errors in the historical record and/or the fact that the very strong Galápagos record reflected a more oceanic signal (Quinn and Neal 1995). Quinn and Neal's (1995) reconstruction suggests that the very strong 1982 to 1983 and 1997 to 1998

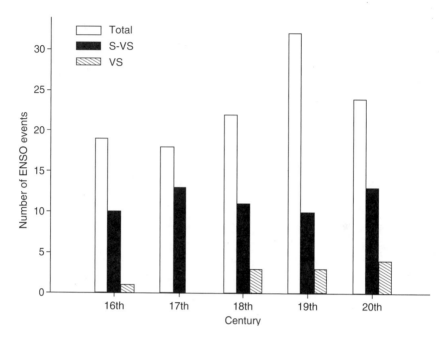

Figure 11.5. Historical occurrences of severe and very severe ENSO events from the 16th through the 20th century, indicating an increased frequency in total ENSO events during the 19th and 20th centuries. Data from Quinn and Neal (1995).

El Niño events may have been unique over the past four centuries in terms of their strength and overwhelming deleterious effects on coral growth and survivorship. These combined events led to the destruction of many of the modest coral reef formations present in the eastern Pacific. Corals living in warm-water areas without seasonal upwelling (e.g., the Gulf of Chiriquí, western Panamá) generally fared better than those experiencing upwelling in the Gulf of Panamá. Surprisingly, however, in 1998 SSTs in the Gulf of Panamá did not surpass levels sufficient for severe bleaching to occur, due in part to the timing and early fluctuations in SST anomalies in 1998 (Glynn et al. 2001).

A recent study in the Galápagos Islands conducted by Riedinger et al. (2002) produced an ~6100-year-long ^{14}C sediment record from a hypersaline lake, providing evidence of the frequency and intensity of El Niño events since the mid-Holocene. Data from their study indicate that El Niño activity was quite high from the present to ~3100 ybp. However, between about 4600 and 7130 ybp El Niño activity, particularly strong to very strong El Niño events, was much reduced in frequency. Their data indicate that from the present to ~3100 ybp, there were 80 strong to very strong events, while from 4000 to 6100 ybp only 23 events of comparable magnitude occurred. Support for these findings includes studies by Sandweiss et al. (1996) who also provide evidence for a prolonged warming period along the Peruvian coast as far as 10°S during the mid-Holocene period from 5000 to 800 ybp (see Fig. 11.6). Their conclusions were based on the presence of shallow, warm-water fossil bivalve assemblages dated to this interval.

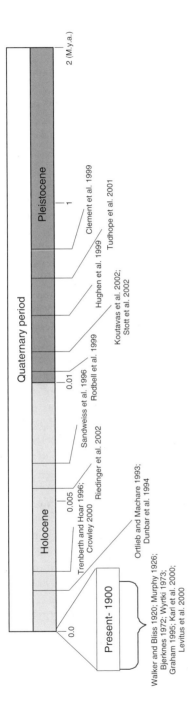

Figure 11.6. The age (or oldest age) of samples in studies of ENSO events from the recent to the middle Pleistocene. <u>Present–1900 AD</u>: developed understanding of ENSO frequency and dynamics. <u>1600 AD</u>: isotopic record and historical El Niño reconstructions in good agreement, showing low temperatures during the early 1600s and early 1800s, and relatively warmer conditions during the 1700s; a slight cooling observed between 1880 and 1940. <u>1000 AD</u>: late 20th century warming that closely agrees with the response predicted from greenhouse gas forcing; prolonged 1990–1995 ENSO unusual with expected occurrence once in 1100 years. <u>~6100 ¹⁴C ybp</u>: laminated sediments from a hypersaline crater lake in the Galápagos Islands indicate that at least 435 moderate to very strong El Niño events have occurred since ~6100 ¹⁴C ybp; the frequency and intensity of events increased at about 3000 ¹⁴C ybp. <u>5000–8000 ybp</u>: stable, warm tropical waters in northern Peru, based on geoarchaeological evidence, suggest that El Niño events did not occur for some millennia preceding 5000 ybp. <u>7000–15,000 ybp</u>: ¹⁴C-dated debris flows suggest that El Niño periodicity ≥15 years to 7000 ybp then increased to 2–8.5 years by 5000 ybp, which is modern El Niño periodicity. <u>20–70 ka</u>: based on δ¹⁸O and Mg/Ca composition of plank-tonic foraminifera, the last glacial maximum cooling between 21 and 23 ka implies a persistent El Niño-like pattern in the tropical Pacific; more frequent and perhaps more severe El Niño events are linked to stadial (glacial) conditions during the last 70 kyr. <u>130 ka</u>: δ¹⁸O and Sr/Ca records from a fossil Indonesian coral indicate robust ENSO activity during the last interglacial period when global climate was slightly warmer than present. <u>150 ka</u>: a model study shows that the mean global climate response of precessional (Milankovitch) forcing is due to an interaction between an altered seasonal cycle and the ENSO.

In support of the Galápagos coral-based study by Dunbar and co-workers (1994), Rodbell et al. (1999) examined dated inorganic laminae derived from storm deposits in an alluvial alpine lake in Ecuador. These data from a 9.2-m core yielded climate information spanning from less than 200 ybp to as far back as 15 ka. They found that El Niño events were longer in periodicity and lower in amplitude. Rodbell et al. (1999) were able to show a strong match between the timing of clastic laminae and historical records of moderate to severe El Niño events occurring from 1800 to 1976. Overall they identified the occurrence of 26 moderate El Niño events between 1800 and 1976 AD. From about 15 to 7 ka, the periodicity of the clastic sediments was ≥15 years followed by a progressive increase toward lower frequencies that became established around 5000 ybp. This result was consistent with the studies of Sandweiss et al. (1996) and Riedinger et al. (2002). Using a simple numerical ocean–atmospheric model, Clement et al. (1999) found that during the mid-Holocene the occurrence of ENSO events was likely to have been less intense in both amplitude and frequency compared to the present.

Focusing on past ENSO behavior during the Last Glacial Maximum (LGM), Koutavas et al. (2002) reconstructed a ^{14}C-dated Mg/Ca SST record based on δ^{18}O values of foraminifera preserved in the marine sediments at the Galápagos Islands. Their results showed that over the past 30 kyr the intensity of the Cold Tongue varied in step with precession-induced seasonal changes. During the LGM, temperatures cooled by 1.2 °C and Koutavas et al. (2002) suggested that such a shift would have resulted in the relaxation of temperature gradients, which in turn would have reduced gradients in both the Hadley and Walker circulation systems. This would have led to a southerly shift in the position of the Intertropical Convergence Zone (ITCZ)—in essence chronic, quasi-continuous ENSO conditions. In support of this putative climate shift, Koutavas et al. (2002) cited evidence of wetter climate conditions in the Bolivian Altiplano and drier conditions in northern South America. Additional studies by Rodbell et al. (1999), Tudhope et al. (2001), and others provide evidence in support of their results.

Stott et al. (2002) recently offered evidence for a "super" ENSO during the Late Pleistocene. Using δ^{18}O and Mg/Ca ratios from planktonic foraminifera as a proxy for temperature and salinity these workers differentiated the calcium δ^{18}O record, revealing a dominant salinity signal that varied with Dansgaard/Oeschger cycles over Greenland. Salinities were found to be higher during periods of latitudinal cooling and lower during interstadials. These variations in salinity shifts are analogous to Pacific ocean–atmospheric cycles. ENSO events correlated with stadials at high latitudes while La Niña conditions correlated with interstadials. According to Stott et al. (2002), at times of cooling at high latitudes, the tropical Pacific experienced either more frequent or more persistent ENSO events.

Millennial-scale shifts in atmospheric convection away from the western tropical Pacific may explain many paleo-observations, including atmospheric CO_2, N_2O, and CH_4 during stadials and patterns of extratropical ocean variability that have tropical source functions positively correlated with ENSO.

The fossil coral records reported by Tudhope et al. (2001) extend back to 130,000 years, through the last glacial–interglacial cycle. These investigators used $\delta^{18}O$ and $\delta^{13}C$ proxies to examine paleo-SST and to establish the time series of the records. The fossil corals analyzed grew when sea level was 70 to 100 m lower than present, ca. 40 and 130 ka, respectively, during the penultimate interglacial period. Interannual variations were interpreted as reflective of a paleo-ENSO system. Spectral analysis further provided data on ENSO periodicity during these periods. One of the conclusions of Tudhope et al. (2001) was that the variance in the ENSO band is greater in the modern coral record compared with fossil corals. However, they did find some periods in the last interglacial and late Holocene that matched current ENSO frequencies. The key point is that ENSO events have been a prominent climatic feature during interglacial periods, extending perhaps from the last interglacial up through the Holocene. Tudhope et al. (2001) suggested that SSTs were similar to present conditions and 2 to 3 °C cooler during four low sea-level stands between 130 ka and the last glacial period.

Tudhope and co-workers found that ENSO was weak and suggested that at 6.5 ka the $\delta^{18}O$ values indicated similar or slightly cooler SSTs. Overall, however, the evidence from several proxy records in the eastern Pacific, including Sandweiss et al. (1996), Rodbell et al. (1999), Riedinger et al. (2002) as well as others, support their arguments. Lastly, Tudhope and co-workers suggested that orbital precession-glacial dampening (Crowley and North 1991) may have controlled the frequency and strength of ENSO events, and that the pacing and strength of ENSOs are higher now than at any time over the past 150,000 years. However, Hughen et al. (1999) reconstructed a fossil record from a 124,000-year-old coral in Indonesia that suggests that during the last interglacial period temperatures were slightly warmer than present but distinct from the recent records since the mid-1970s, when global SSTs shifted to an average warmer state.

From the recent paleoclimate studies presented above, it is clear that ENSO events have been one of the dominant features shaping global climate change at least as far back as the late Pleistocene, and perhaps earlier (Fig. 11.6). The major question at present is how much warmer will it become in the near future and how fast will this change occur?

11.5 What Factors Are Causing Recent Warming?

Certainly the well-documented shift to warmer subsurface sea temperatures associated with El Niño (Guilderson and Schrag 1998) appears to have had a major impact on the heat budget across the Pacific and is thought to be responsible for the increase in frequency and intensity of ENSO events since 1976. Evidence supporting this abrupt shift is based on bomb ^{14}C that showed enrichment in surface water via shoaling of the undercurrent in the equatorial eastern Pacific. In 1960 ^{14}C values ranged seasonally from 20 to 30%. By the mid-1970s, the isotopic values increased from 50 to 100% indicating a significant shoaling of the Equatorial Undercurrent and hence warmer surface waters. Although the phenomenon was described from eastern Pacific surface waters, heat levels have increased in all the major oceans (Fig. 11.7; Levitus et al. 2000).

One of the first studies to assess global temperature trends was conducted by Graham (1995). Using the most recent portion of the observed ocean SST data set (1970–1992), he was able to reproduce the results of the atmospheric models using ocean surface temperatures alone. The conclusion from the modeling effort

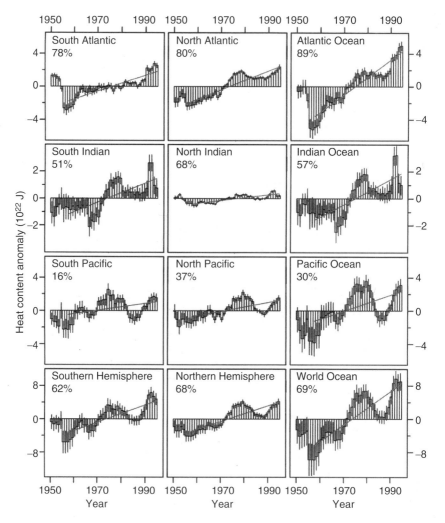

Figure 11.7. Time series of 5-year running composites of heat content (10^{22} J) in the upper 3000 m for each major ocean basin. Vertical lines represent ±1 SE of the 5-year mean estimate of heat content. The linear trend is estimated for each time series for the period 1955 to 1996, which corresponds to the period of best data coverage. The trend is plotted as a black line. The percent variance accounted for by this trend is given in the upper left corner of each panel. Reproduced from Levitus et al. (2000) with permission from the American Association for the Advancement of Science.

indicated that simulated temperatures are caused by enhancement of the tropical hydrologic cycle resulting from increasing SST, in essence suggesting that model results could be due to natural climate variability, but are most likely attributable to increasing levels of atmospheric CO_2.

Further evidence of forced warming was reported in Trenberth and Hoar (1996), who noted that ENSO-level warming occurred continuously in the Niño 4 region from 1990 to 1995, marking the longest ENSO event over the past 113 years. This manifestation was mainly driven by the static position of the Southern Oscillation, which remained negative from early 1990 to June 1995. Statistical analyses based on modern ENSO frequencies predict that this prolonged negative SOI would only be expected to occur every 1100 years and an ENSO as prolonged as the 1990 to 1995 one would occur only once every 1500 to 3000 years.

By identifying and quantifying all potential sources giving rise to recent warming over the past 1000 years, Crowley (2000) presented results from an energy balance model showing that 41 to 64% of pre-anthropogenic (pre-1850) variation could be explained by natural variation in solar radiation and volcanic activity. In addition, the contribution of solar forcing was evaluated and compared with greenhouse gases from the middle of the last century. Those results indicated that the climate effects due to greenhouse gases were four times larger than the effect due to solar variability and that temperature changes since 1850 could not be explained by natural variability alone. Using a linear upwelling/diffusion energy balance model (EBM) to calculate mean annual changes in forcing, the results indicated that during pre-1850 there was a 0.05 °C decrease in the 17th and 18th centuries reflecting a CO_2 decrease of ~6 parts per million in an ice-core record. Combining solar and volcanic variability, model responses indicated only a 0.15 to 0.20 °C increase in temperature between 1905 and 1955, representing one quarter of the total observed 20th-century warming.

Crowley (2000) provided two independent lines of evidence indicating that Northern Hemisphere temperatures were unusually high over the past 1000 years. First, the warming over the past century is unprecedented in the past 1000 years. Second, the same climate model explaining much of the variability in temperature over the interval 1000 to 1850 can only account for about 25% of the 20th-century temperature increase attributable to natural variability. Crowley (2000) concluded by suggesting that recent increases are due to an already present warming trend from greenhouse gas forcing that is likely to accelerate in the near future.

Recent empirical observations provide strong evidence that global temperatures have risen from +0.3 °C to a value closer to +0.6 °C, based on a series of unprecedented warming events that broke consecutive temperature records in the 1990s. To assess the significance of these results, Karl et al. (2000) employed an autoregressive moving-average (ARMA) model to separate the timing of change points in temperatures over the past 500 to 1000 years. The difference in the rates of change over the two periods, 1912 to 1941 and 1976 to 1998, was highly significant. Statistical methods employed ARMA to calculate the probability of 16 consecutive months of record-breaking mean monthly temperatures. The resulting probabilities varied from 0.009 to 0.04, using a rate of warming between 2.5 and 3.0 °C per century.

In attempts to predict future climate change based on the results of rigorous model/observation analyses, incorporating a coupled ocean–atmospheric model, Stott et al. (2000) were able to show that model analyses could successfully simulate global mean land temperature variations over large scales, indicating that external forces (i.e., increased greenhouse gas effects) have had a strong influence on observed temperature change. Their models were also able to explain more than 80% of the observed variation in decadal temperature changes during the 20th century and indicated that temperatures would continue to rise to 2100 at a rate comparable to current levels (~0.2 K per decade).

To assess the effects of natural forcing attributable to changes in stratospheric aerosols, specifically those due to volcanic eruptions and fluctuations in solar irradiance, Stott et al. (2000) evaluated three climate ensembles based on the DADCM3 dynamical climate model. The ensemble comprised four simulations that varied only in their initial conditions, ALL, ANTHRO, and NATURAL forcing. The ALL ensembles revealed temperature changes since 1860, which were consistent with observations of changes over the past 30 years. The NATURAL model showed no warming over the past 30 years, while the ANTHRO model showed a 0.2 °C warming relative to the period 1881 to 1920. By comparing both anthropogenic and natural forcing, their model successfully predicted significant large-scale responses to observed temperatures. The global mean model (ALL) trends were consistent with observations. The NATURAL ensemble was conclusively rejected, indicating that anthropogenic forcing is, in fact, the most likely cause of recent warming.

Perhaps one of the most important and telling pieces of evidence in support of significant climate change, with respect to global warming, is the recent data compiled by Levitus et al. (2000). These workers documented dramatic increases in oceanic heat content over a 40-year period (the mid-1950s to the mid-1990s), corresponding to a warming rate of 0.3 °C between 300 and 1000 m depth in each of the major ocean basins (Fig. 11.7). From 0 to 300 m depth, the measured increase was 0.31 °C. This warming began in the 1950s in the Pacific and Atlantic Oceans, with the Indian Ocean showing a warming trend since the mid-1960s. Maximum heat content was observed during the 1997 to 1998 ENSO event. These data suggest the possibility that enhancement of the Pacific Decadal Oscillation (PDO) contributed significantly to the warming signal.

11.6   ENSO Markers on Coral Reefs

11.6.1 Geomorphological and Ecological Markers

Several lines of geomorphological and ecological evidence have been proposed to help identify ENSO effects on coral reefs, on decadal to centennial time scales. None of these features alone can serve as unique signals of ephemeral ENSO warming events (DeVries 1987), but in combination, and with additional markers from geochemical analyses (see below), they can offer reasonably convincing evidence of past ENSO disturbances.

Geomorphological evidence. The death of entire coral reef frameworks in the Galápagos Islands (Glynn 1994), at Cocos Island (Guzmán and Cortés 1992), and along coastal Ecuador (Glynn 2003) can be attributed to the sudden and prolonged sea warming associated with the 1982 to 1983 El Niño event. Many of these frameworks have suffered from significant erosion and collapse since 1982 to 1983. Several reef structures in the Galápagos Islands have disappeared entirely, whereas some at Cocos Island and along coastal Ecuador are still standing but in varying stages of disintegration. Those frameworks that are still intact could serve as indicators of El Niño for two or possibly more decades. Colgan (1990) offered evidence from a tectonically uplifted and well-preserved coral assemblage that the 1941 El Niño event caused widespread mortality of pocilloporid frameworks at Urvina Bay, Galápagos Islands. The destabilization and disintegration of in situ reef structures may also result in off-reef submarine and/or strand-line clastic deposits that could be dated and possibly related to El Niño activity (Scott et al. 1988; Glynn 2000). For example, an approximately 1-m-high berm of coral debris on the south end of Pinta Island (Galápagos Islands) most likely represents the remnants of an incipient pocilloporid patch reef that was present just offshore before the 1982 to 1983 El Niño event (Fig. 11.8). Coral tempestites were formed in the Galápagos in 1983 due to storms, swell reversals, and high sea-level stands, which caused coral breakage, dislodgment, and transport onto the shoreline (Robinson 1985).

Figure 11.8. Coral rubble berm at the south end of Pinta Island, Galápagos Islands (17 May 2002). This feature, about 1 m high and 5 to 8 m wide, is composed of coral debris originating from a pocilloporid patch reef killed during the 1982 to 1983 El Niño disturbance. Courtesy of T. Smith.

From preliminary analyses of carbonate sediment production rates, and reef and offshore sediment stratigraphy, Scott et al. (1988) argued that intense bioerosion following massive El Niño-induced coral death would accelerate the deposition of coral sediments. These workers reported increases in coral sediment grain sizes after the 1982 to 1983 El Niño, indicative of intense lithophagine bioerosion. Since the size and texture of bioeroded sediments are often uniquely produced by different taxa, it is possible that the composition of postdisturbance sediments can be related to the relative abundances of bioeroder taxa (Glynn 2000). Rapid rates of bioerosion have been quantified on several reefs impacted by El Niño warming disturbances (Glynn 1988; Eakin 1996, 2001; Reaka-Kudla et al. 1996). In an analysis of erosion on an entire 2.5-ha Panamanian reef, Eakin (1996) reported overall net losses of 4800 kg $CaCO_3$/year after the 1982 to 1983 El Niño event, compared with predisturbance net depositional rates of 8600 kg $CaCO_3$/year. The reef continues in a highly variable, habitat-specific erosional mode, with net losses of 3000 to 18,000 kg $CaCO_3$/year in the aftermath of the 1997 to 1998 El Niño event. Much of this variability is due to changes in (a) the community composition of calcifying organisms, (b) topographic complexity, and (c) echinoid population densities. Coral mortalities due to La Niña-forced low tidal exposures (1989 and 1993) actually had a larger effect on these erosional rates than coral mortalities due to elevated SSTs during the 1997 to 1998 El Niño event (Eakin 2001).

The sudden death of long-lived massive corals that have grown continuously for several centuries can also indicate El Niño activity. For example, numerous old colonies of *Porites lobata* and *Pavona clavus* in the Galápagos Islands, with maximum estimated ages ranging from 347 to 423 years, died during the 1982 to 1983 El Niño event (Glynn 1990b). Also, *P. clavus* demonstrated continuous skeletal growth at Urvina Bay (Galápagos Islands) for 367 years before the colony was uplifted and killed in 1954 (Dunbar et al. 1994). These growth records suggest that coral mortality resulting from the 1982 to 1983 ENSO warming event in the Galápagos Islands was a unique event over the prior four centuries.

Ecological evidence. Colgan (1990) observed several markers within branching and massive coral skeletons that he attributed to the 1941 El Niño disturbance. These features were present on corals at the Urvina Bay uplift, which, as of 2002, were still in a high state of preservation. After death, branching coral (*Pocillopora* spp.) frameworks formed towers through the interaction of damselfish that defended and reduced bioersion on the coral summits, and echinoids (*Eucidaris*) that grazed and caused undercutting of the tower bases. This differential erosion eventually led to the collapse and disintegration of the towers. In massive coral colonies (*Porites lobata* and *Pavona* spp.), algae and epifauna colonized the dead upper surfaces until surviving coral tissues on the sides of the colonies eventually spread laterally by regrowth and began to cover the scars on the summits. Relating species-specific growth rates to the amount of regrowth that occurred on these massive corals, Colgan (1990) calculated that this could have occurred between the strong 1941 El Niño event and the tectonic uplift and death of these corals in 1954.

Similar scarring and coral regrowth have produced lobes on the massive coral *Gardineroseris planulata* in Panamá (Fig. 11.9) and on *Porites lobata* (Fig. 11.10)

Figure 11.9. Skeletal regeneration lobes on *Gardineroseris planulata* that formed following partial colony mortality during the 1982 to 1983 ENSO event at Uva Island, Gulf of Chiriquí, Panamá (6 m depth, 15 May 1999). The summits of the lobes, covered with turf algae, bleached and suffered partial mortality again during the 1997 to 1998 ENSO event. The heights of the lobes, excluding losses due to erosion, ranged from 10 to 20 cm.

at Clipperton Atoll. In Panamá, the bottoms and tops of these lobes respectively mark the 1982 to 1983 and 1997 to 1998 El Niño disturbances at the Uva Island reef in the Gulf of Chiriquí (Glynn 2000). At Clipperton, the regenerated lobes on *P. lobata* have been correlated with the 1987 ENSO event, which at that location was more severe than the ENSO event of 1982 to 1983 (based on the magnitude and intensity of SST anomalies; Glynn et al. 1996). Regenerating lobes also form on massive *Porites* spp. colonies that have been partially consumed by the corallivorous sea star *Acanthaster planci* (Done 1987; DeVantier and Done, Chapter 4). Stable oxygen isotope ($^{18}O/^{16}O$) thermometry of skeletal growth immediately preceding and following the partial mortality event should help to distinguish between thermally induced bleaching and death from predation, uplift, or other factors.

Another example of ecological processes that may offer evidence of past El Niño activity relates to predator–prey interactions (Glynn 1985). Before the 1982–83 El Niño, several ~100- to 200-year-old colonies of *G. planulata* on the Uva Island reef were protected from *A. planci* attack by a barrier of pocilloporid corals and their obligate crustacean guards. Almost all of the *G. planulata*

Figure 11.10. Skeletal regeneration lobes on *Porites lobata* that formed following partial colony mortality during the 1987 ENSO event at Clipperton Atoll (10 m depth, 23 April 1994). The live summits bear numerous pufferfish bite marks. The heights of the dated lobes since 1987, excluding losses from bioerosion, ranged from 12 to 15 cm.

colonies survived the elevated temperatures of 1982 to 1983, but the branching pocilloporid corals did not. This resulted in the death of the crustaceans, which allowed *A. planci* access to the more temperature-resistant massive corals within the refuge. Sclerochronologic analysis of cores from the largest *Gardineroseris* colony showed no signs of prior growth discontinuities, which suggested that the biotic barrier had protected this massive coral for up to 200 years. This approach is no longer relevant due to the recent severe reduction in the abundance of *A. planci*. Although a pocilloporid barrier has not re-formed (as of March 2005), *A. planci* predation is no longer an important factor on the Uva Island reef (Fong and Glynn 1998).

11.6.2 Geochemical Markers

The carbonate skeletons of corals are capable of providing accurate, high-resolution climatic data sets stored over a vast range of temporal scales from diurnal to millennial. They are recorders of natural events such as solar activity, volcanic eruptions, human-induced forcing of greenhouse gases, and other environmental changes. In addition, tree rings, varved sediments, ice cores, and other such markers are capable of storing past information on climate change.

Studies of growth patterns in massive corals indicate that two skeletal bands are typically formed annually: one wide, low-density band that usually forms in the cool season, and a narrow, denser band that forms during warmer periods (Fig. 11.11). It should be noted that this pattern varies geographically. Periodic

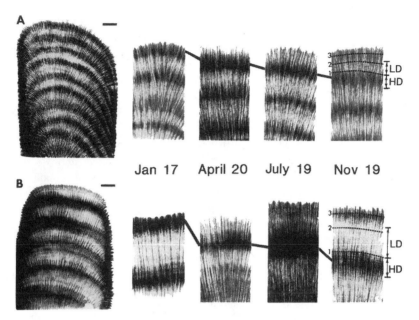

Figure 11.11. X-radiographic positives showing the changes in skeletal density from 17 January to 19 November 1979 for the corals (**A**) *Pavona gigantea* and (**B**) *Pavona clavus* at Contadora Island, Pearl Islands, Gulf of Panamá. The time series in samples to the right were collected from the same colonies over the time periods indicated. Solid lines connect the high-density bands formed in the preceding wet, nonupwelling season (June–December). Scale bars are 1 cm. Linear dimensions of low-density (LD) and high-density (HD) bands are shown on the far right. Dotted lines (1, 2, 3) on the November 19 samples indicate position of Alizarin Red stain lines. Corals were stained on 25 January, 24 May, and 25 September 1979. From Wellington and Glynn (1983).

extremes in either warm or cool temperatures can result in the formation of narrow stress bands. In the case of corals, moderate to severe ENSOs can result in interrupted skeletal growth that can provide a history of ENSO disturbances or extreme cooling events that periodically occur on reefs at higher latitudes during severe winters (Hudson et al. 1976; Hudson 1981; Halley and Hudson, Chapter 6).

One of the first studies to observe and suggest the annual nature of high- and low-density banding in coral was conducted by Ma (1933). Following this initial observation, Knutson et al. (1972) demonstrated that coral bands were formed on an annual basis. This important study paved the way to investigate not only the factors controlling rates of growth but also their capacity to capture a history of climate change based on reasonably well-understood biochemical proxies such as $\delta^{18}O$ and Sr/Ca ratios. Traditionally, $\delta^{18}O$ has been the primary method used to reconstruct proxy sea surface temperatures. This proxy signal, however, is constrained by potential variation in sea surface salinities that can alter the relationship between $\delta^{18}O$ and sea surface temperatures. This problem is largely circumvented where variations in $\delta^{18}O_{seawater}$ are minimal.

In attempts to reconstruct a paleo-SST relationship using Sr/Ca ratios, de Villiers et al. (1995) called into question the validity of a 4 to 6 °C cooling in the tropics, which was reported to have occurred during the LGM. Based on observed variations in the Sr/Ca content of sea surface waters, latitudinal transects across both the Pacific and Atlantic Oceans revealed uncertainties of 2 to 3 °C with regard to the Sr/Ca thermometer. One of the major problems observed, however, was that differences in coral growth rates appear to have an effect on the slope of the Sr/Ca relationship, with slower-growing corals showing higher Sr/Ca levels. de Villiers et al. (1995) concluded that biological factors, such as variations in metabolic rate, could play a significant role in influencing the actual Sr/Ca signal.

In contrast to de Villiers' study, an instrumental calibration study conducted in the laboratory by Schrag (1999) has shown that Sr/Ca ratios in corals have a high potential for providing very accurate information on SSTs, to within ± 0.2 °C. In comparing these results with the field observations of de Villiers et al. (1995) and a recent laboratory study by Meibom et al. (2003), it is obvious that the accuracy of the Sr/Ca tracer technique is inconsistent. The latter workers found that the temperature signal could vary by as much as 10%, resulting from metabolic changes synchronized with the lunar cycle.

Cohen et al. (2001, 2002) evaluated the relationship between SST and skeletal Sr/Ca in a temperate coral living over a broad temperature range. *Astrangia poculata* occurs naturally with or without algal symbionts. In comparing the relationship between SST and Sr/Ca values, it was found that 65% of the Sr/Ca variability in the symbiotic state was related to metabolic activity independent of SST, as earlier shown by de Villiers et al. (1995). Based on the derived Sr/Ca–SST relationship they found that the slopes of the temperature curves were virtually identical between symbiotic and nonsymbiotic colonies. Moreover, Cohen and co-workers showed that variations in Sr/Ca in the asymbiotic form are primarily controlled by variation in SST, whereas the Sr/Ca ratio in the symbiotic coral is under strong biological control. More work on this problem requires robust protocols that can provide accurate proxy SSTs under a variety of biotic and abiotic conditions. Despite the recent recognition of these potential pitfalls in achieving accurate SST proxies, several studies have produced remarkably high correlations between SST and Sr/Ca ratios (Linsley et al. 2000; Fig. 11.12).

11.7 Spatial and Temporal Scales of ENSO-Related Coral Bleaching

Given the high variability of stressful levels of temperature and irradiance distributions on coral reefs and the varying sensitivities of zooxanthellate corals to these conditions, it is not surprising that numerous microscale spatial patterns of bleaching and mortality have been described. Severe coral bleaching often coincides with elevated temperatures in shallow reef habitats, direct exposure to downwelling solar irradiance, and reduced water circulation or mixing. Back-reef corals that are often exposed to greater extremes in temperature than fore-reef corals may not bleach while bleaching is pronounced in other reef habitats.

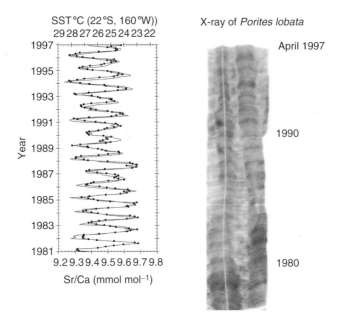

Figure 11.12. Comparison of Rarotonga coral Sr/Ca with IGOSS satellite SST showing a high correlation between SSTs and Sr/Ca (mmol/mol) proxy values. Reproduced from Schrag and Linsley (2002) with permission from the American Association for the Advancement of Science.

Corals in deeper reef zones, exposed to vigorous circulation, subdued irradiance, and mixing with deeper, cooler waters, often are less severely impacted during mass bleaching events. Recent studies have demonstrated that patterns of bleaching in some reef-coral species, both across colonies at different depths and within a single colony, are controlled by diverse communities of symbiotic zooxanthellae with varying tolerance limits for elevated temperature and irradiance stressors (Rowan et al. 1997; Baker 2001; Rowan 2004). These sorts of environmental influences affecting bleaching susceptibility/resistance are addressed extensively in the literature (e.g., Goreau and Macfarlane 1990; Jokiel and Coles 1990; Williams and Bunkley-Williams 1990; Glynn 1993; Brown 1997; Hoegh-Guldberg 1999; Goreau et al. 2000; Feingold 2001; Glynn et al. 2001; Baker et al. 2004; Riegl, Chapter 10).

   In terms of meso- (reefs) to macro-scale (reef regions) patterns of bleaching, whole-reef ecosystems and entire biogeographic regions underwent unprecedented mass bleaching in the 1980s and 1990s. Except for the equatorial west African region, some level of coral reef bleaching has now been observed in all of the world's coral reef provinces, from 27°S (Easter Island) to ~34°N (Japan; H. Yamano personal communication; Fig. 11.2A,B). It should be recognized, however, that intraregional variability in coral bleaching has been observed. For example, Reyes-Bonilla and co-workers (2002) have suggested that Mexican Pacific coral reefs have been relatively little affected by ENSO events because temperature anomalies are less pronounced in that area and coral recruitment has been normal to high, at least since 1997. Compared with the first global compilations of

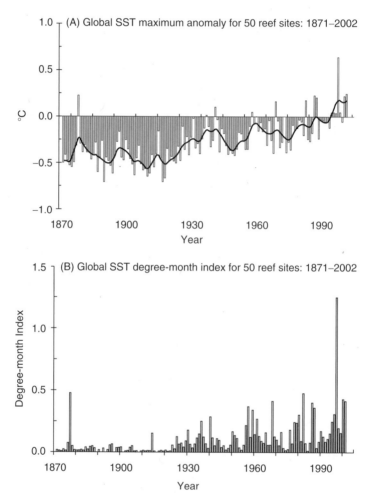

Figure 11.13. (**A**) Mean SST maximum anomaly and (**B**) mean degree-month index at 50 coral reef sites that bleached in 1997 to 1998 (source data 1871–2002, Had1SST, 2000–2002, IGOSS-NMC product). Black curve in (**A**) is a 10-year Gaussian filter. Base period for calculating the anomalies is 1982–1999. The degree-month index combines both the magnitude and duration of warm season SST anomalies by summing positive anomalies for the months the mean maximum monthly SST was exceeded for each year. Courtesy J. M. Lough; see Lough (2000) for additional information on analyses.

bleaching events in the 1980s (Glynn 1984; Brown 1987; Coffroth et al. 1990; Williams and Bunkley-Williams 1990), it is clear that through the 1990s this class of disturbance spread to include nearly every coral reef assemblage worldwide.

It is cautioned, however, that the more recent bleaching events reported by ReefBase include numerous discrete coral reefs whereas the earlier records were often for larger reef regions. To illustrate the high correlation between thermal stress and coral bleaching worldwide during 1997 to 1998, Lough (2000, personal communication) analyzed the occurrence of bleaching at 50 reef sites in the Indian Ocean/ Middle East, Southeast Asia, Pacific Ocean, and the Caribbean/Atlantic. She found that the level of thermal stress at the vast majority of these sites was unmatched during the period 1871 to 1999 (Fig. 11.13A,B).

During the temporal development of a bleaching episode, Hoegh-Guldberg (1999) suggested that coral reefs are generally affected first in the southwestern Pacific and Indian Ocean, then around Southeast Asia, and finally in the Caribbean basin. He further proposed that coral reef bleaching begins in the Southern Hemisphere and spreads into the Northern Hemisphere with the evolution of ENSO events. Coral reef bleaching in the eastern Pacific typically occurs during the warm season about one year following bleaching in the southwestern Pacific (Podestá and Glynn 2001). The timing may vary, however, as bleaching occurred in the eastern Pacific (Panamá) in August to September 1997, preceded by five months of bleaching on the Great Barrier Reef. A second bout of bleaching occurred in Panamá from March to June 1998. In addition, severe bleaching events at Easter Island and Fiji in the Southern Hemisphere occurred during February and March 2000, about two years following the 1997 to 1998 ENSO (Wellington et al. 2001). Clearly, local conditions such as the duration of temperature anomalies, water-column transparency, and cloud cover influence the timing of these events.

11.8 Recovery

Even though hundreds of severe coral bleaching events have been documented during the 1980s and 1990s, relatively few studies are available on the extent of coral community recovery. From quantitative long-term studies, Connell (1997) listed eight examples of changes in coral cover following bleaching events, which were likely caused in large part by elevated seawater temperature or by a combination of elevated temperature and irradiance. This represents only about 10% of the 77 cases available with data spanning at least 4 years. The majority of the changes in coral cover were tentatively assigned to predation, storm damage, sedimentation, reduction in herbivore abundances, and disease epizootics. Sufficient observations were available to judge whether recovery had occurred or not in only five of the eight ENSO-related examples. All western Pacific reefs but one demonstrated significant recovery (>50% increase as percent of loss) over periods of 7 to 13 years. These included coral reefs in the Thousand Islands region of Indonesia (Brown and Suharsono 1990) and at Phuket Island, Thailand (Brown et al. 1993). Some sites at Phuket Island showed no significant declines in coral cover over an 11-year period of study (Chansang and Phongsuwan 1993). The only site showing no recovery was the Uva Island reef (Panamá) in the eastern Pacific (Glynn 1990a).

Here we augment and update the record of changes in coral cover for several equatorial eastern Pacific coral reefs subjected to multiple El Niño/La Niña disturbances over the past two decades. Recovery of coral reefs from the Galápagos Islands to the Gulf of California, Mexico, monitored over the longest periods (15–31 years), is evaluated in terms of live coral cover. Statistical testing of pre- and post-disturbance mean differences in coral cover is noted where possible.

Shallow (1–5 m depth) Galápagos coral reefs that experienced high coral mortality (97% overall, Glynn et al. 1988) in 1982 to 1983 have not demonstrated significant recovery over a period of nearly 20 years, from just after the El Niño disturbance to the most recent sampling (1983–2002; Table 11.2). Indeed, nearly all of the coral reef buildups described in Glynn and Wellington (1983) have been reduced to sediment piles due to intense bioerosion (Glynn 1994; Reaka-Kudla et al. 1996). Although the relatively deep (14–20 m) coral community east of Onslow Island experienced severe to moderate bleaching in 1982 to 1983 and 1997 to 1998, mortality was negligible, with dense aggregations of corals persisting for 25+ years. Declines in coral cover of 16.1% to 56.1% reported by Feingold (2001) over a 2-year period were probably due to the redistribution of the mobile fungiid (*Diaseris*) and siderastreid (*Psammocora*) corals comprising this assemblage. The small areas of pocilloporid buildups at Española and Santa Fe Islands, now consisting of fragmented and dispersed rubble, also did not show signs of recovery when sampled in 2002.

Off the Pacific coast of Colombia, coral reefs at Gorgona Island experienced high mortality (>50%) in 1983 (Table 11.1). By 1998, coral cover was significantly higher on La Azufrada reef than in 1979, demonstrating recovery since the 1982 to 1983 El Niño event (Table 11.2). Panamanian reefs experienced marked declines in coral cover—76% to 90% overall—following the 1982 to 1983 El Niño event (Glynn et al. 1988). Two of the three Panamanian reefs examined here have recovered to pre-El Niño live cover levels. The Saboga Island reef, in the upwelling Gulf of Panamá, recovered fully over only an eight-year period (1984–1992). Coral cover on this reef has remained relatively constant and high during the past 10 years (1992–2002). In the nonupwelling Gulf of Chiriquí, coral cover in the shallow and deep reef zones at the Uva Island reef has fluctuated since 1983, generally showing a downward trend. Reef-flat corals in particular experienced high mortalities in 1989 and 1993, a result of extreme midday low-water exposures. At Caño Island, off the southwestern coast of Costa Rica, coral cover in all shallow to deep reef zones combined (0–14 m) declined significantly in 1985, by nearly 70%, and has remained at a low level over a 15-year period (1985–1999). Dinoflagellate blooms in 1985 caused high mortalities of pocilloporid corals, especially in shallow reef zones (Guzmán and Cortés 2001). The northernmost site examined, the Cabo Pulmo pocilloporid coral reef in the lower part of the Gulf of California, Mexico, suffered nearly a 53% decline in live cover between 1986 and 2002. Over this period, the majority of the corals died from the

Table 11.2. Long-term changes in coral cover in the equatorial eastern Pacific following coral bleaching/mortality caused by elevated sea water temperature associated with El Niño events of the 1980s and 1990s. NA, not applicable

Locality and coral reef site	Depth m	% live coral cover year	% live coral cover mean ± SD	% change	Recovery	No. years observed	t-test	Source
Galápagos Islands Onslow Is. reef	0.5–5	1975 2002	37.1 ±20.8 0.3 ±0.5	−99.2	no	27	p<0.001	Glynn, unpublished data
Onslow Is. east	14–20	1975 2000	80–100[1] 80–100	negligible	yes	25	NA	Glynn and Wellington (1983) Robinson (1985) Feingold (2001)
Española Is. Xarifa coral community	1–3	1975 2002	37.0 ±12.1 0.1 ±0.2	−99.7	no	27	p<0.001	Glynn, unpublished data
Santa Fe Is., NE anchorage	2–3	1976 2002	48.0 ±23.2 0	−100	no	26	NA	Glynn, unpublished data
Colombia Gorgona Is. La Azufrada reef	0.3–15	1979 1998	36.5 ±29.6 66.2 ±20.6	+44.9	yes	19	0.001 <p<0.01	Glynn et al. (1982) Vargas-Ángel et al. (2001)
Panamá Gulf of Panamá Saboga Is. reef	2–3	1971 1984 1992– 2002	45–55[2] 0 47–52	−100 +100	no yes	13 10	NA NA	Glynn, unpublished data Richmond, unpublished data
Gulf of Chiriquí Uva Is. reef	3–5[3]	1974 2002	34.7 ±21.4 12.1 ±9.2	−65.1	no	28	p<0.01	Glynn, unpublished data

(Continued)

Table 11.2. Long-term changes in coral cover in the equatorial eastern Pacific following coral bleaching/mortality caused by elevated sea water temperature associated with El Niño events of the 1980s and 1990s. NA, not applicable (*Continued*)

Locality and coral reef site	Depth m	% live coral cover year	% live coral cover mean ± SD	% change	Recovery	No. years observed	t-test	Source
Uva Is. reef	1 [4]	1974	39.2 ±34.0	−93.6	no	26	$p<0.001$	Glynn (1976)
		2000	2.5 ±5.1					Eakin (2001)
Secas Is. reef	3–7	1974–75	10.6 ±7.6	−28.3	yes	27–28	$p>0.6$	Glynn, unpublished data
		2002	7.6 ±12.1					
Costa Rica								
Caño Is. [5]	0.5–6	1980	17.8 ±7.2	−51.7	NA	4	$0.10>p$ >0.05	Guzmán et al. (1987)
		1984	8.6 ±5.1					
	0–14	1984	32	−68.8	no	15	–	Guzmán and Cortés (2001)
		1999	10					
Mexico								
Gulf of California								
Cabo Pulmo	0–12	1986	45.1 ±14.2	−52.6	no	16	$p<0.001$	Reyes-Bonilla and Calderón
		2002	21.4 ±15.3					Aguilera, unpublished data

[1] Estimates of range of percent coral cover within densest population aggregations of *Diaseris distorta* and *Psammocora stellata*.
[2] Range of percent cover, no replicate sampling conducted; Richmond (personal communication) recorded 100% mortality following the 1982-83 El Niño event; full recovery observed in 1992, then live cover fluctuated between 47 and 52% to 2002.
[3] Windward reef slope and reef base zones.
[4] Reef flat zone.
[5] Sampling conducted on reefs located on the north and east sides of island.

elevated temperature stress that accompanied the 1997 to 1998 ENSO, but coral mortality continued to 2002 from various causes.

In summary, about one-third of the eastern Pacific areas examined have demonstrated varying levels of recovery over periods of from 10 to 20 years. Following El Niño perturbations, coral community decline often continues due to strong upwelling, algal blooms, extreme low tidal exposures, and bioerosion. Considering present limited data, it appears that the potential for reef recovery is higher in the western Pacific than in the eastern Pacific. Possibly this is related to interregional differences in: (1) coral species diversity; (2) connectivity among reefs vis-à-vis source populations, larval dispersal, and recruitment; (3) ENSO intensity and return intervals; and (4) bioerosion.

Finally, although live coral cover is useful as an index of coral reef community recovery, one must be cognizant of the importance of other community attributes such as species composition, relative abundances, contributors to framework construction, integrity of coral frameworks, and the abundances and activities of other community members such as noncoral epibenthos (algae, sponges, zoanthids, *inter alia*), herbivores, corallivores, symbionts, and bioeroders. With the hundreds of coral reefs that have suffered high coral mortalities during recent ENSO disturbances, additional information should soon become available on the recovery process, including the mechanisms and time scales required.

11.9 The Future of Coral Reefs

Bleaching is a complex phenomenon that is generally defined as a breakdown of the symbioses between animal host and algal symbionts, and the resultant loss of algal cells or their pigments, in response to a variety of stressors. These have been characterized as "physiological bleaching," "algal-stress bleaching," and "animal-stress bleaching" by Fitt et al. (2001). While many factors have been implicated in coral bleaching responses (see for example Brown 1997), high and/or prolonged thermal and irradiance anomalies such as those associated with increasing SST and ENSO events are undoubtedly foremost in terms of their role in bleaching.

Since the unusually strong 1982 to 1983 ENSO there has been a progressive increase in sea-surface warming in both the Northern and Southern Hemispheres (Levitus et al. 2000). As a consequence of this warming, virtually every shallow ($\leq$10–15 m) coral reef area worldwide has incurred negative effects in terms of reduced coral cover, with losses ranging from moderate to severe (especially in the eastern Pacific). Only the western African region remains unknown in this regard. As of 2002, nearly every coral reef area in the tropics and extratropics has been affected, including Easter Island (Chile), Baja California (Mexico), the Line Islands in the central Pacific, and the Houtman-Abrolhos Islands (western Australia). In the Line Islands, one of the last to have incurred bleaching for the first time, water temperatures had not previously exceeded coral bleaching thresholds (see Goreau and Hayes 1994; Strong et al. 1997).

Although good physiological models now exist describing the proximate mechanisms responsible for temperature-related coral bleaching (Warner et al. 1996, 1999; Jones et al. 1998, 2000; Gates and Edmunds 1999; Fitt et al. 2001), additional studies are needed to distinguish between the mechanisms accounting for differential responses to elevated SST and UVR. Such studies should include phenotypic responses leading to acclimation, assessment of the efficacy of natural selection, and the interactions of these processes with respect to both the coral host and its algal symbiont composition.

What is the future of coral reefs? Here we address several factors that may have a significant negative impact on reef corals. These major effects include significant increases in downwelling of potentially detrimental levels of UVR (ultraviolet radiation), including both UV-A and UV-B wavelengths. The penetration of potentially harmful UV-B wavelengths, particularly in the upper 10 m, can occur during a reduction in the vertical mixing of the water column. The second factor is sea-surface warming, which can result in coral bleaching. These two physical factors may act synergistically to damage 2 coral symbionts' photosynthetic systems PS1 and PS2 (Jones et al. 1998, 2000).

Factors that might be expected to mitigate the damaging effects of elevated temperature and light include genetic strains of corals/symbiotic algae that are able to tolerate local conditions or rapidly adapt to increased exposure to elevated temperatures and UVB radiation. We know that some coral taxa (e.g., *Porites*) are able to withstand a broad range of conditions while others are physiologically limited to a very narrow range of conditions (Hoegh-Guldberg and Salvat 1995; Marshall and Baird 2000) and are usually the first to manifest signs of bleaching (e.g., *Acropora*). Even closely related species may show marked differences in their tolerances to environmental disturbances. The high mortality and low recovery potential of *Agaricia tenuifolia* compared with *Agaricia agaricites* during and following the 1997 to 1998 ENSO event in Belize suggest that the former species has a lesser ability to produce heat shock proteins for protection against stressful elevated temperatures (Robbart et al. 2004).

What are some of the factors that correlate with an ability to resist bleaching? In general, any physical or environmental variables that serve to reduce stresses to the symbiosis between coral and zooxanthellae will tend to mitigate bleaching. In particular, the reduction of temperature and irradiance stress, increased bleaching tolerance or thresholds, or limits to the cellular damage that can occur in the bleaching response help ensure the resistance of corals and the resilience of the community (reviewed in West and Salm 2003). One common observation often made during bleaching events is that coral colonies exposed to high water flow—currents, wave action, or vertical mixing in the water column—are generally less affected by high temperatures (Jokiel and Coles 1990).

Nakamura and van Woesik (2001) described the physical mechanisms relating to the causes of bleaching in regard to water motion. Using the relationship between diffusion, shear stress, and water velocity passing over a colony, they were able to explain the mechanism responsible for observed differences in bleaching between and among individual colonies. In general, smaller colonies in high water motion environments would be expected to experience lower levels of bleaching than large colonies under the same conditions. Through empirical studies these workers determined that corals subjected to a "high-flow treatment" at 50–70 cm/s showed no bleaching compared to corals in a "low-flow treatment" at <3 cm/s, which suffered high levels of bleaching at the same temperature.

Another important factor comes from the general observation that during bleaching events corals living at deeper depths (> 15 m) generally exhibit reduced levels of bleaching compared to colonies at shallow and intermediate depths. The likely explanation is that UVR attenuates rapidly with depth, particularly the shorter and more potent UV wavelengths between 300 and 320 nm (Gleason and Wellington 1993, 1995; Wellington and Fitt 2003). Hence, depth could provide an important refuge for shade-loving, deep-water corals.

The history of previous exposure to stresses may also play a role in predicting future ability to acclimate to increasing temperatures. For example, Coles and Jokiel (1978) demonstrated that corals exposed to 32.5 °C, after being previously held at 20, 24, 26, and 28 °C, exhibited variable survival rates of 47, 30, 61, and 74%, respectively. It is not clear, however, whether these "adaptive" changes represented differential survivorship based on genetic differences or physiological adaptation.

Arising from this initial work, Buddemeier and Fautin (1993) proposed that corals could adapt to environmental extremes by changing their algal partners involved in the symbiosis. The "adaptive bleaching hypothesis" (ABH) posits the formation of new symbiotic consortiums with different zooxanthellae more suited to current conditions experienced by the host coral. The fundamental tenets of the ABH assume that different types of zooxanthellae respond differently to environmental conditions, particularly temperature, and that bleached corals can acquire zooxanthellae from the environment. Recently, Kinzie et al. (2001) conducted simple tests of the ABH assumptions. They found that (1) bleached adult hosts could acquire algae secondarily from the environment in a dose-dependent manner and (2) genetically different strains of zooxanthellae exhibited variations in growth rate at different temperatures.

Is coral bleaching really adaptive? To address this question, Baker (2001) performed an experiment with several common reef-building species that contained different "high light" and "low light" symbionts at different depths. Corals were reciprocally transplanted between shallow (2–4 m) and deep (20–23 m) sites on a Caribbean reef in Panamá. Baker observed that corals transplanted upward were significantly more bleached than those transplanted downward. However, corals transplanted downward showed significantly higher rates of mortality after one year when compared to upward transplants, despite the fact that they did not initially bleach.

These findings have stimulated interest and continuing investigations. Most work on the systematics of zooxanthellae and their potentially flexible symbioses has, until recently, involved Caribbean species. During the past few years, symbiotic associations in the Indo-Pacific have been assessed in greater detail (LaJeunesse et al. 2003; Fabricius et al. 2004), and the identification of the thermally stress-tolerant clade D zooxanthellae has allowed for further investigations of the ABH. It is now recognized that considerable flexibility of symbioses can and does occur in many, but not all, coral species. Additionally, coral species may tend to harbor en hospite symbionts adapted to ambient environmental conditions, and differential bleaching susceptibilities do exist among zooxanthellae clades, types, and species. Across numerous Indo-Pacific locations, bleaching events resulted in mortality of coral species harboring intolerant zooxanthellae, while species harboring stress-tolerant types not only survived bleaching events, but also increased their populations on reefs formerly dominated by less tolerant symbiosomes (Baker et al. 2004; Fabricius et al. 2004; Rowan 2004). Some workers share a cautiously optimistic view of the future of coral reefs and their ability to adapt to global temperature change (Baker et al. 2004; Rowan 2004) whereas others (Jokiel and Coles 1990; Hoegh-Guldberg 1999; Hughes et al. 2003; McWilliams et al. 2005) are less confident considering the present high rate of environmental change.

11.10 Summary

Despite many seemingly discordant studies indicating variability in the specifics of ENSO-related effects to coral reefs, we can offer several assertions based on a preponderance of data presented here. First, sea surface temperatures are increasing, and these increases are likely to be a signal of global warming. Second, this warming trend is contributing to increasing frequencies and intensities of ENSO events. Third, both warming sea temperatures and ENSO-related climate change have a pronounced deleterious effect on corals and coral reefs, primarily through temperature and ENSO-related coral bleaching events that directly affect the structure and function of reef communities. When added to the numerous other biotic and abiotic stressors currently affecting coral reefs globally, there is considerable uncertainty over the short-term and even long-term persistence of these diverse and critically important biological and geological structures.

Acknowledgments. Thanks are due J. K. Oliver for providing data on bleaching events and permission to reproduce Figures 2A and 2B, and C. M. Eakin, A. Hazra, J.A. Kleypas, J. Luo, J.L. Maté, H. Reyes-Bonilla, and B. Riegl for data and critiques that helped to improve this review. Special thanks are due R.B. Aronson for his various editorial skills. The US National Science Foundation ESH Program (GMW), the Paleoclimate Program of NOAA (GMW), the National Geographic Society (GMW), the National Center for Caribbean Coral Reef Research (PWG), and the Biological Oceanography Program of the U.S. National Science Foundation (PWG) supported much of this work. We acknowledge the generous assistance of the administrative staff of the Charles Darwin Research Station, the Galápagos National Park Service, and the Smithsonian Tropical Research Institute, Panamá. Assistance from E. Borneman and D. Holstein in the final manuscript preparation is also gratefully acknowledged.

References

Allan, R., J. Lindesay, and D. Parker. 1996. El Niño Southern Oscillation and climatic variability. Australia, CSIRO.

Baker, A.C. 2001. Reef corals bleach to survive change. *Nature* 411:765–766.

Baker, A.C., C.J. Starger, T.R. McClanahan, and P.W. Glynn. 2004. Corals' adaptive response to climate change. *Nature* 430:741.

Baquero-Bernal, A., M. Latif, and S. Legutke. 2002. On the dipolelike variability of sea surface temperature in the tropical Indian Ocean. *J. Climate* 15:1358–1368.

Berkelmans, R. 2002. Time-integrated thermal bleaching thresholds of reefs and their variation on the Great Barrier Reef. *Mar. Ecol. Prog. Ser.* 229:73–82.

Birkeland, C., and J.S. Lucas. 1990. *Acanthaster planci*: *Major Management Problem of Coral Reefs*. Boca Raton: CRC Press.

Bjerknes, J. 1969. Atmospheric teleconnections from the equatorial Pacific. *Mon. Weather Rev.* 97:163–172.

Bjerknes, J. 1972. Large-scale atmospheric response to the 1964–65 Pacific equatorial warming. *J. Phys. Oceanogr.* 2:212–217.

Brown, B.E. 1987. Worldwide death of corals—Natural cyclical events or man-made pollution? *Mar. Pollut. Bull.* 18:9–13.

Brown, B.E. (ed.). 1990. Coral bleaching. *Coral Reefs* (special issue) 8:153–232.

Brown, B.E. 1997. Coral bleaching: Causes and consequences. *Proc. Eighth Int. Coral Reef Symp., Panamá* 1:65–74.

Brown, B.E., M.D. Le Tissier, R.P. Dunne, and T.P. Scoffin. 1993. Natural and anthropogenic disturbances on intertidal reefs of S.E. Phuket, Thailand 1979–1992. In *Proceedings of the Colloquium on Global Aspects of Coral Reefs: Health, Hazards and History,* compiler R.N. Ginsburg, 278–285. Rosenstiel School of Marine and Atmospheric Science, University of Miami, Florida.

Brown, B.E., and Suharsono. 1990. Damage and recovery of coral reefs affected by El Niño related seawater warming in the Thousand Islands, Indonesia. *Coral Reefs* 8:163–170.

Buddemeier, R.W. 1992. Corals, climate and conservation. *Proc. Seventh Int. Coral Reef Symp., Guam* 1:3–10.

Buddemeier, R.W., and D.G. Fautin. 1993. Coral bleaching as an adaptive mechanism. *BioScience* 43:320–326.

Cane, M.A. 1983. Oceanographic events during El Niño. *Science* 222:1189–1195.

Cane, M.A. 1986. El Niño. *Ann. Rev. Planet. Sci.* 4:43–70.

Carriquiry, J.D., A.L. Cupul-Magaña, F. Rodríguez-Zaragoza, and P. Medina-Rosas. 2001. Coral bleaching and mortality in the Mexican Pacific during the 1997-98 El Niño and prediction from a remote sensing approach. *Bull. Mar. Sci.* 69:237–249.

Carriquiry, J.D., and H. Reyes-Bonilla. 1997. Community structure and geographic distribution of the coral reefs of Nayarit, Mexican Pacific. *Cienc. Mar.* 23:227–248.

Carriquiry, J.D., M.J. Risk, and H.P. Schwarcz. 1988. Timing and temperature record from stable isotopes of the 1982-1983 El Niño warming event in eastern Pacific corals. *Palaios* 3:359–364.

Chansang, H., and N. Phongsuwan. 1993. Health of fringing reefs of Asia through a decade of change: A case history from Phuket Island, Thailand. In *Proceedings of the Colloquium on Global Aspects of Coral Reefs: Health, Hazards and History,* compiler R.N. Ginsburg, 286–292. Rosenstiel School of Marine and Atmospheric Science, University of Miami, Florida.

Chesher, R.H. 1969. Destruction of Pacific corals by the sea star *Acanthaster planci*. *Science* 165:280–283.

Clement, A., C.R. Seager, and M.A. Cane. 1999. Orbital controls on the El Niño/Southern Oscillation and the tropical climate. *Paleoceanography* 14:441–456.

Coffroth, M.A., H.R. Lasker, and J.K. Oliver. 1990. Coral mortality outside of the eastern Pacific during 1982–1983: Relationship to El Niño. In *Global Ecological Consequences of the 1982–83 El Niño-Southern Oscillation,* ed. P.W. Glynn, 141–182. Amsterdam: Elsevier Oceanography Series 52.

Cohen, A.L., G.D. Layne, S.R. Hart, and P.S. Lobel. 2001. Kinetic control of skeletal Sr/Ca in a symbiotic coral: Implications for the paleotemperature proxy. *Paleoceanography* 16:20–26.

Cohen, A.L., K.E. Owens, G.D. Layne, and N. Shimizu. 2002. The effect of algal symbionts on the accuracy of Sr/Ca paleotemperatures from coral. *Science* 296: 331–333.

Coles, S.L., and P.L. Jokiel. 1978. Synergistic effects of temperature, salinity and light on the hermatypic coral *Montipora verrucosa. Mar. Biol.* 49:187–195.

Colgan, M.W. 1990. El Niño and the history of eastern Pacific reef building. In *Global Ecological Consequences of the 1982–83 El Niño-Southern Oscillation,* ed. P.W. Glynn, 183–232. Amsterdam: Elsevier Oceanography Series 52.

Connell, J.H. 1997. Disturbance and recovery of coral assemblages. *Coral Reefs* 16:S101–S113.

Crowley, T.J., 2000. Causes of climate change over the past 1000 years. *Science* 289:270–277.

Crowley, T.J., and G.R. North. 1991. Time series analysis of paleoclimate records. In *Paleoclimatology,* 132–151. New York: Oxford Monographs on Geology and Geophysics, No. 16.

D'Elia, C.F., R.W. Buddemeier, and S.V. Smith. 1991. Workshop on coral bleaching, coral reef ecosystems, and global change: Report of proceedings. Maryland Sea Grant Coll. Publ. No. UM-SG-TS-91-03. College Park: University of Maryland.

de Villiers, S., B.K. Nelson, and A.R. Chivas. 1995. Biological controls on coral Sr/Ca and $\delta^{18}O$ reconstructions of sea surface temperatures. *Science* 269:1247–1249.

DeVries, T.J. 1987. A review of geological evidence for ancient El Niño activity in Peru. *J. Geophys. Res.* 92:14,471–14,479.

Done, T.J. 1987. Simulation of the effects of *Acanthaster planci* on the population structure of massive corals in the genus *Porites*: Evidence of population resilience? *Coral Reefs* 6:75–90.

Dunbar, R.B., G.M. Wellington, M.W. Colgan, and P.W. Glynn. 1994. Eastern Pacific sea surface temperature since 1600 A.D.: The $\delta^{18}O$ record of climate variability in Galápagos corals. *Paleoceanography* 9:291–315.

Eakin, C.M. 1996. Where have all the carbonates gone? A model comparison of calcium carbonate budgets before and after the 1982–1983 El Niño at Uva Island in the eastern Pacific. *Coral Reefs* 15:109–119.

Eakin, C.M. 2001. A tale of two ENSO events: Carbonate budgets and the influence of two warming disturbances and intervening variability, Uva Island, Panamá. *Bull. Mar. Sci.* 69:171–186.

Enfield, D.B. 1992. Historical and prehistorical overview of El Niño/Southern Oscillation. In *El Niño: Historical and Paleoclimatic Aspects of the Southern Oscillation,* eds. H.F. Diaz and V. Markgraf, 95–117. Cambridge: Cambridge University Press.

Enfield, D.B. 2001. Evolution and historical perspective of the 1997-1998 El Niño-Southern Oscillation event. *Bull. Mar. Sci.* 69:7–25.

Enfield, D.B., and M.A. Mestas-Nuñez. 1999. Multiscale variabilities in global sea surface temperatures and their relationships with tropospheric climate patterns. *J. Climate* 12:2719–2733.

Enfield, D.B., and D.A. Mayer. 1997. Tropical Atlantic sea surface temperature variability and its relation to El Niño-Southern Oscillation. *J. Geophys. Res.* 102:929–945.

Fabricius, K.E., J.C. Mieog, P.L. Colin, D. Idip, and M.J.H. van Oppen. 2004. Identity and diversity of coral endosymbionts (zooxanthellae) from three Palauan reefs with contrasting bleaching, temperature and shading histories. *Mol. Ecol.* 13:2445–2458.

Feingold, J.S. 2001. Responses of three coral communities to the 1997–98 El Niño-Southern Oscillation: Galápagos Islands, Ecuador. *Bull. Mar. Sci.* 69:61–77.

Fitt, W.K., B.E. Brown, M.E. Warner, and R.P. Dunne. 2001. Coral bleaching: Interpretation of thermal tolerance limits and thermal thresholds in tropical corals. *Coral Reefs* 20:51–65.

Fong, P., and P.W. Glynn. 1998. A dynamic size-structured population model: Does disturbance control size structure of a population of the massive coral *Gardineroseris planulata* in the eastern Pacific. *Mar. Biol.* 130:663–674.

Gates, R.D., and P.J. Edmunds. 1999. The physiological mechanisms of acclimation in tropical corals. *Am. Zool.* 39:30–43.

Ginsburg, R.N., (compiler). 1993. *Proceedings of the Colloquium on Global Aspects of Coral Reefs: Health, Hazards and History.* Rosenstiel School of Marine and Atmospheric Science, University of Miami.

Gleason, D.F., and G.M. Wellington. 1993. Ultraviolet radiation and coral bleaching. *Nature* 365:836–838.

Gleason, D.F., and G.M. Wellington. 1995. Variation in UVB sensitivity of planula larvae of the coral *Agaricia agaricites* along a depth gradient. *Mar. Biol.* 123:693–703.

Glynn, P.W. 1976. Some physical and biological determinants of coral community structure in the eastern Pacific. *Ecol. Monogr.* 46:431–456.

Glynn, P.W. 1984. Widespread coral mortality and the 1982/83 El Niño warming event. *Environ. Conserv.* 11:133–146.

Glynn, P.W. 1985. El Niño-associated disturbance to coral reefs and post disturbance mortality by *Acanthaster planci. Mar. Ecol. Prog. Ser.* 26:295–300.

Glynn, P.W. 1988. El Niño warming, coral mortality and reef framework destruction by echinoid bioerosion in the eastern Pacific. *Galaxea* 7:129–160.

Glynn, P.W. (ed.). 1990a. *Global Ecological Consequences of the 1982–83 El Niño-Southern Oscillation.* Amsterdam: Elsevier Oceanography Series 52.

Glynn, P.W. 1990b. Coral mortality and disturbances to coral reefs in the tropical eastern Pacific. In *Global Ecological Consequences of the 1982–83 El Niño-Southern Oscillation*, ed. P.W. Glynn, 55–126. Amsterdam: Elsevier Oceanography Series 52.

Glynn, P.W. 1993. Coral reef bleaching: Ecological perspectives. *Coral Reefs* 12:1–17.

Glynn, P.W. 1994. State of coral reefs in the Galápagos Islands: Natural vs anthropogenic impacts. *Mar. Pollut. Bull.* 29:131–140.

Glynn, P.W. 2000. El Niño-Southern Oscillation mass mortalities of reef corals: A model of high temperature marine extinctions? In *Carbonate Platform Systems: Components and Interactions*, eds. E. Insalaco, P.W. Skelton, and T.J. Palmer, 117–133. Geological Society, London, Special Publications.

Glynn, P.W. 2002. Effects of the 1997–98 El Niño-Southern Oscillation on eastern Pacific corals and coral reefs: An overview. *Proc. Ninth Int. Coral Reef Symp., Bali* 2:1169–1174.

Glynn, P.W. 2003. Coral communities and coral reefs of Ecuador. In *Latin American Coral Reefs*, ed. J. Cortés, 449–472. Amsterdam: Elsevier.

Glynn, P.W., and S.B. Colley. 2001. A collection of studies on the effects of the 1997–98 El Niño-Southern Oscillation event on corals and coral reefs in the eastern tropical Pacific. *Bull. Mar. Sci.* 69:1–288.

Glynn, P.W., S.B. Colley, J.H. Ting, J.L. Maté, and H.M. Guzmán. 2000. Reef coral reproduction in the eastern Pacific: Costa Rica, Panamá and Galápagos Islands (Ecuador). IV. Agariciidae, recruitment and recovery of *Pavona varians* and *Pavona* sp. a. *Mar. Biol.* 136:785–805.

Glynn, P.W., J. Cortés, H.M. Guzmán, and R.H. Richmond. 1988. El Niño (1982-83) associated coral mortality and relationship to sea surface temperature deviations in the tropical eastern Pacific. *Proc. Sixth Int. Coral Reef Symp., Australia* 3:237–243.

Glynn, P.W., and L. D'Croz. 1990. Experimental evidence for high temperature stress as the cause of El Niño-coincident coral mortality. *Coral Reefs* 8:181–191.

Glynn, P.W., R. Imai, K. Sakai, Y. Nakano, and K. Yamazato. 1992. Experimental responses of Okinawan (Ryukyu Islands, Japan) reef corals to high sea temperature and UV radiation. *Proc. Seventh Int. Coral Reef Symp., Guam* 1:27–37.

Glynn, P.W., and G.E. Leyte Morales. 1997. Coral reefs of Huatulco, west México: Reef development in upwelling Gulf of Tehuantepec. *Rev. Biol. Trop.* 45:1033–1047.

Glynn, P.W., J.L. Maté, A.C. Baker, and M.O. Calderón. 2001. Coral bleaching and mortality in Panamá and Ecuador during the 1997–1998 El Niño-Southern Oscillation event: Spatial/temporal patterns and comparisons with the 1982–1983 event. *Bull. Mar. Sci.* 69:79–109.

Glynn, P.W., M. Perez, and S.L. Gilchrist. 1985. Lipid decline in stressed corals and their crustacean symbionts. *Biol. Bull.* 168:276–284.

Glynn, P.W., H.V. Prahl, and F. Guhl. 1982. Coral reefs of Gorgona Island, Colombia, with special reference to corallivores and their influence on coral community structure and development. *An. Inst. Invest. Mar. Punta Betín* 12:185–214.

Glynn, P.W., J.E.N. Veron, and G.M. Wellington. 1996. Clipperton Atoll (eastern Pacific): Oceanography, geomorphology, reef-building coral ecology and biogeography. *Coral Reefs* 15:71–99.

Glynn, P.W., and G.M. Wellington. 1983. *Corals and Coral Reefs of the Galápagos Islands*. Berkeley: University of California Press.

Goreau, T.J., and R.L. Hayes. 1994. Coral bleaching and ocean "hot spots". *Ambio* 23:176–180.

Goreau, T.J., and A.H. Macfarlane. 1990. Reduced growth rate of *Montastraea annularis* following the 1987–1988 coral-bleaching event. *Coral Reefs* 8:211–215.

Goreau, T., T. McClanahan, R. Hayes, and A. Strong. 2000. Conservation of coral reefs after the 1998 global bleaching event. *Conserv. Biol.* 14:5–15.

Graham, N.E. 1995. Simulation of recent global temperature trends. *Science* 267:666–671.

Guilderson, T., and D.P. Schrag. 1998. Abrupt shift in subsurface temperatures in the tropical Pacific associated with changes in El Niño. *Science* 281:240–243.

Guzmán, H.M., and J. Cortés. 1992. Cocos Island (Pacific of Costa Rica) coral reefs after the 1982–83 El Niño disturbance. *Rev. Biol. Trop.* 40:309–324.

Guzmán, H.M., and J. Cortés. 2001. Changes in reef community structure after fifteen years of natural disturbance in the eastern Pacific (Costa Rica). *Bull. Mar. Sci.* 69:133–149.

Guzmán, H.M., J. Cortés, R.H. Richmond, and P.W. Glynn. 1987. Efectos del fenómeno de "El Niño Oscilación Sureña" 1982/83 en los arrecifes coralinos de la Isla del Caño, Costa Rica. *Rev. Biol. Trop.* 35:325–333.

Hansen, D.V. 1990. Physical aspects of the El Niño event of 1982–1983. In *Global Ecological Consequences of the 1982–1983 El Niño-Southern Oscillation*, ed. P.W. Glynn, 1–20. Amsterdam: Elsevier Oceanography Series 52.

Harmelin-Vivien, M.L., and P. Laboute. 1986. Catastrophic impact of hurricanes on atoll outer reef slopes in the Tuamotu (French Polynesia). *Coral Reefs* 5:55–62.

Harvell, C.D., K. Kim, J.M. Burkholder, R.R. Colwell, P.R. Epstein, J. Grimes, E.E. Hofmann, E.K. Lipp, A.D.M.E. Osterhaus, R. Overstreet, J.W. Porter, G.W. Smith, and G.R. Vasta. 1999. Emerging marine diseases—Climate links and anthropogenic factors. *Science* 285:1505–1510.

Harvell, C.D., K. Kim, C. Quirolo, J. Weir, and G. Smith. 2001. Coral bleaching and disease contributors to 1998 mass mortality in *Briareum asbestinum* (Octocorallia, Gorgonacea). *Hydrobiologia* 460:97–104.

Hirose, M., and M. Hidaka. 2000. Reduced reproductive success in scleractinian corals that survived the 1998 bleaching in Okinawa. *Galaxea* 2:17–21.

Hoegh-Guldberg, O. 1999. Coral bleaching, climate change and the future of the world's coral reefs. *Mar. Freshwater Res.* 50:839–866.

Hoegh-Guldberg, O., and B. Salvat. 1995. Periodic mass-bleaching and elevated sea temperatures: Bleaching of outer reef slope communities in Moorea, French Polynesia. *Mar. Ecol. Prog. Ser.* 121:181–190.

Hudson, J.H. 1981. Growth rates in *Montastraea annularis*: A record of environmental change in Key Largo Marine Sanctuary, Florida. *Bull. Mar. Sci.* 31:444–459.

Hudson, J.H., E.A. Shinn, R.B. Halley, and B. Lidz. 1976. Sclerochronology—A tool for interpreting past environments. *Geology* 4:361–364.

Hughen, K.A., D.P. Schrag, S.B. Jacobsen, and W. Hantoro. 1999. El Niño during the last interglacial period recorded to a fossil coral from Indonesia. *Geophys. Res. Lett.* 26:3129–3132.

Hughes, T.P., A.H. Baird, D.R. Bellwood, M. Card, S.R. Connolly, C. Folke, R. Grosberg, O. Hoegh-Guldberg, J.B.C. Jackson, J. Kleypas, J.M. Lough, P. Marshall, M. Nyström, S.R. Palumbi, J.M. Pandolfi, B. Rosen, and J. Roughgarden. 2003. Climate change, human impacts, and the resilience of coral reefs. *Science* 301:929–933.

IPCC. 1992. Climate change 1992: The supplementary report to the IPCC scientific assessment. Intergovernmental Panel on Climate Change, eds. J.T. Houghton, B.A. Callander, and S.K. Varney. Cambridge: Cambridge University Press.

IPCC. 1996. Climate change 1995: The science of climate change. Contribution of working group I to the second assessment report of the Intergovernmental Panel on Climate Change, eds. J.T. Houghton, L.G. Meira Filho, B.A. Callander, N. Harris, A. Kattenberg, and K. Maskell. Cambridge: Cambridge University Press.

IPCC. 2001. Climate change 2001: The scientific basis. Contribution of working group I to the third assessment report of the Intergovernmental Panel on Climate Change, eds. J.T. Houghton, Y. Ding, D.J. Griggs, M. Noguer, P.J. van der Linden, X. Dai, K. Maskell, and C.A. Johnson. Cambridge: Cambridge University Press.

Jackson, J.B.C. 1997. Reefs since Columbus. *Coral Reefs* 16:S23–S32.

Jiménez, C., J. Cortés, A. León, and E. Ruíz. 2001. Coral bleaching and mortality associated with the 1997-98 El Niño in an upwelling environment in the eastern Pacific (Gulf of Papagayo, Costa Rica). *Bull. Mar. Sci.* 69:151–169.

Jokiel, P.L., and S.L. Coles. 1990. Response of Hawaiian and other Indo-Pacific reef corals to elevated temperature. *Coral Reefs* 8:155–162.

Jones, R.J., O. Hoegh-Guldberg, W.D. Larkum, and U. Schreiber. 1998. Temperature-inducing bleaching of corals begins with impairment of the carbon dioxide fixation mechanism in zooxanthellae. *Plant Cell Environ.* 21:1219–1230.

Jones, R.J., S. Ward, A.Y. Amri, and O. Hoegh-Guldberg. 2000. Changes in quantum efficiency of Photosystem II of symbiotic dinoflagellatateses after heat stress, and of bleached corals sampled after the 1998 Great Barrier Reef mass bleaching event. *Mar. Freshwater Res.* 51:63–71.

Karl, T.R., W. Knight, and B. Baker. 2000. The record breaking global temperatures of 1997–1998: Evidence for an increase in the rate of global warming? *J. Geophys. Res.* 27:719–722.

Kinzie, R.A., III, M. Takayama, S.R. Santos, and M.A. Coffroth. 2001. The adaptive bleaching hypothesis: Experimental tests of critical assumptions. *Biol. Bull.* 200:51–58.

Kleypas, J.A., R.W. Buddemeier, D. Archer, J.-P. Gattuso, C. Langdon, and B. Opdyke. 1999. Geochemical consequences of increased atmospheric CO_2 on coral reefs. *Science* 284:118–120.

Kleypas, J.A., R.W. Buddemeier, and J.-P. Gattuso. 2001. The future of coral reefs in an age of global change. *Int. J. Earth Sci. (Geol. Rundsch.)* 90:426–437.

Knutson, D.W., R. Buddemeier, and S.V. Smith. 1972. Coral chronometers: Seasonal growth bands in reef corals. *Science* 177:270–272.

Koutavas, A., J. Lynch–Stieglitz, T.M. Marchitto, Jr., and J.P. Sachs. 2002. El Niño-like pattern in ice age tropical Pacific sea surface temperature. *Science* 297:226–229.

Kushmaro, A., Y. Loya, M. Fine, and E. Rosenberg. 1996. Bacterial infection and coral bleaching. *Nature* 380:396.

Laboute, P. 1985. Evaluation of damage done by the cyclones of 1982–1983 to the outer slopes of the Tikehau and Takapoto Atolls (Tuamotu Archipelago). *Proc. Fifth Int. Coral Reef Congr., Tahiti* 3:323–329.

LaJeunesse, T.C., W.K.H. Loh, R. van Woesik, O. Hoegh-Guldberg, G.W. Schmidt, and W.K. Fitt. 2003. Low symbiont diversity in southern Great Barrier Reef corals, relative to those of the Caribbean. *Limnol. Oceanogr.* 48:2046–2054.

Lau, K.-M., and H. Weng. 1999. Interannual, decadal-interdecadal, and global warming signals in sea surface temperature during 1955–1997. *J. Climate* 12:1257–1267.

Levitus, L., J.I. Antonov, P.B. Boyer, and C. Stephens. 2000. Warming of the world ocean. *Science* 287:2225–2229.

Linsley, B.K., G.M. Wellington, and D.P. Schrag. 2000. Decadal sea surface temperature variability in the subtropical South Pacific from 1726 to 1997 A. D. *Science* 290:1145–1148.

Lirman, D., P.W. Glynn, A.C. Baker, and G.E. Leyte Morales. 2001. Combined effects of three sequential storms on the Huatulco coral reef tract, Mexico. *Bull. Mar. Sci.* 69:267–278.

Lough, J.M. 1994. Climate variation and El Niño-Southern Oscillation events on the Great Barrier Reef: 1958 to 1987. *Coral Reefs* 13:181–195.

Lough, J.M. 2000. 1997–98: Unprecedented thermal stress to coral reefs? *Geophys. Res. Lett.* 27:3901–3904.

Ma, T.Y.H. 1933. On the seasonal change of growth in some Palaeozoic corals. *Proc. R. Acad. Tokyo* 9:407–409.

Margalef, R. 1968. *Perspectives in Ecological Theory.* Chicago: University of Chicago Press.

Marshall, P.A., and A.H. Baird. 2000. Bleaching of corals on the Great Barrier Reef: Differential susceptibilities among taxa. *Coral Reefs* 19:155–163.

Mascarelli, P., and L. Bunkley-Williams. 1999. An experimental field evaluation of healing in damaged, unbleached and artificially bleached star coral, *Montastrea annularis. Bull. Mar. Sci.* 65:577–586.

McPhaden, M.J. 1999. Genesis and evolution of the 1997–98 El Niño. *Science* 283:950–954.

McWilliams, J.P., I.M. Côté, J.A. Gill, W.J. Sutherland, and A.R. Watkinson. 2005. Accelerating impacts of temperature-induced coral bleaching in the Caribbean. *Ecology* 86:2055–2060.

Meesters, E.H., and R.P.M. Bak. 1993. Effects of coral bleaching on tissue regeneration potential and colony survival. *Mar. Ecol. Prog. Ser.* 96:189–198.

Meibom, A., M. Stage, J. Wooden, B.R. Constantz, R.B. Dunbar, A. Owen, N. Grumet, C.R. Bacon, and C.P. Chamberlain. 2003. Monthly strontium/calcium oscillations in symbiotic coral aragonite: biological effects limiting the precision of the paleotemperature proxy. *Geophys. Res. Lett.* 30(7, 1418), doi:10.1029/2002GLO16864. 71-1–71-4.

Murphy, R.C. 1926. Oceanic and climatic phenomena along the west coast of South America during 1925. *Geogr. Rev.* 16:26–54.

Nakamura, T., and R. van Woesik. 2001. Water-flow rates and passive diffusion partially explain differential survival of corals during the 1998 bleaching event. *Mar. Ecol. Prog. Ser.* 212:301–304.

Odum, H.T., and E.P. Odum. 1955. Trophic structure and productivity of a windward coral reef community on Eniwetok Atoll. *Ecol. Monogr.* 25:291–320.

Ogden, J., and R. Wicklund. (eds.). 1988. Mass bleaching of coral reefs in the Caribbean: A research strategy. *NOAA's Undersea Research Program, Res. Rep.* 88–2.

Omori, M., H. Fukami, H. Kobinata, and M. Hatta. 2001. Significant drop of fertilization of *Acropora* corals in 1999: An after-effect of heavy coral bleaching? *Limnol. Oceanogr.* 46:704–706.

Ortlieb, L., and J. Machare. 1993. Former El Niño events: Records from western South America. *Global Planet. Change* 7:181–202.

Pandolfi, J.M., R.H. Bradbury, E. Sala, T.P. Hughes, K.A. Bjorndal, R.G. Cooke, D. McArdle, L. McClenachan, M.J.H. Newman, G. Paredes, R.R. Warner, and J.B.C. Jackson. 2003. Global trajectories of the long-term decline of coral reef ecosystems. *Science* 301:955–958.

Philander, S.G.H. 1990. *El Niño, La Niña, and the Southern Oscillation.* New York: Academic Press.

Pittock, A.B. 1999. Coral reefs and environmental change: Adaptation to what? *Am. Zool.* 39:10–29.

Podestá, G.P., and P.W. Glynn. 1997. Sea surface temperature variability in Panamá and Galápagos: Extreme temperatures causing coral bleaching. *J. Geophys. Res.* 102:15,749–15,759.

Podestá, G.P., and P.W. Glynn. 2001. The 1997-98 El Niño event in Panamá and Galápagos: An update of thermal stress indices relative to coral bleaching. *Bull. Mar. Sci.* 69:43–59.

Porter, J.W. (ed.). 2001. *The Ecology and Etiology of Newly Emerging Marine Diseases.* Dordrecht: Kluwer Academic Publishers.

Prahl, H. von. 1983. Blanqueo masivo y muerte de corales en la Isla de Gorgona, Pacífico Colombiano. *Cespedesia* 12:125–129.

Prahl, H. von. 1985. Blanqueo y muerte de corales hermatípicos en el Pacífico Colombiano atribuidos al fenómeno de El Niño 1982–83. *Bol. ERFEN* 12:22–24.

Quinn, W.H., and V.T. Neal. 1995. The historical record of El Niño events. In *Climate Since A.D. 1500*, eds. R.S. Bradley and P.D. Jones, 623–648. New York: Routledge.

Quinn, W.H., V.T. Neal, and S.E. Antunez de Mayolo. 1987. El Niño occurrences over the past four and a half centuries. *J. Geophys. Res.* 92:14449–14461.

Rasmusson, E.M., and T.H. Carpenter. 1982. Variations in tropical sea surface temperature and surface wind fields associated with the Southern Oscillation/El Niño. *Mon. Weather Rev.* 110:354–384.

Reaka-Kudla, M.L., J.S. Feingold, and P.W. Glynn. 1996. Experimental studies of rapid bioerosion of coral reefs in the Galápagos Islands. *Coral Reefs* 15:101–107.

ReefBase. 2002. Online map. In *ReefBase: A global information system on coral reefs*, eds. J. Oliver and M. Noordeloos. World Wide Web electronic publication. 6 August 2002.

Reyes-Bonilla, H. 1993. The 1987 coral bleaching at Cabo Pulmo reef, Gulf of California. *Bull. Mar. Sci.* 52:832–837.

Reyes-Bonilla, H. 2001. Effects of the 1997–1998 El Niño-Southern Oscillation on coral communities of the Gulf of California, Mexico. *Bull. Mar. Sci.* 69:251–266.

Reyes-Bonilla, H., J.D. Carriquiry, G.E. Leyte-Morales, and A.L. Cupul-Magaña. 2002. Effects of the El Niño-Southern Oscillation and the anti-El Niño event (1997–1999) on coral reefs of the western coast of México. *Coral Reefs* 21:368–372.

Reynolds, R.W., and T.M. Smith. 1994. Improved sea surface temperature analysis using optimal interpolation. *J. Climate* 7:929–948.

Riedinger, M.A., M. Steinitz-Kannan, W.M. Last, and M. Brenner. 2002. A ~ 6100 ^{14}C yr record of El Niño activity from the Galápagos Islands. *J. Paleolimnol.* 27:1–7.

Robbart, M.L., P. Peckol, S.P. Scordilis, H.A. Curran, and J. Brown-Saracino. 2004. Population recovery and differential heat shock protein expression for the corals *Agaricia agaricites* and *A. tenuifolia* in Belize. *Mar. Ecol. Prog. Ser.* 283:151–160.

Robinson, G. 1985. Influence of the 1982–83 El Niño on Galápagos marine life. In *El Niño in the Galápagos Islands: The 1982-1983 event*, eds. G. Robinson and E.M. del Pino, 153–190. Charles Darwin Foundation for the Galápagos Islands, Quito, Ecuador.

Rodbell, D.T., G.S. Seltz, D.M. Anderson, M.B. Abbott, D.B. Enfield, and J.H. Newman. 1999. An ~15,000-year record of Niño-driven alluviation in southwest Ecuador. *Science* 283:516–520.

Rowan, R. 2004. Thermal adaptation in reef coral symbionts. *Nature* 430:742.

Rowan, R., N. Knowlton, A. Baker, and J. Jara. 1997. Landscape ecology of algae symbionts creates variation in episodes of coral bleaching. *Nature* 388:265–269.

Sandweiss, D.H., J.B. Richardson, III, E.J. Reitz, H.B. Rollins, and K.A. Maasch. 1996. Geoarchaeological evidence from Perú for a 5000 years B.P. onset of El Niño. *Science* 273:1531–1533.

Schrag, D.P. 1999. Rapid analysis of high-precision Sr/Ca ratios in scleractinian corals and other marine carbonates. *Paleoceanography* 14:97–102.

Schrag, D.P., and B.K. Linsley. 2002. Corals, chemistry and climate. *Science* 296:277–278.

Scott, P.J.B., M.J. Risk, and J.D. Carriquiry. 1988. El Niño, bioerosion and the survival of east Pacific reefs. *Proc. Sixth Int. Coral Reef Symp., Australia* 2:517–520.

Shein, K.A. (ed.). 2006. State of the climate in 2005. *Bull. Amer. Meteorol. Soc. (Spec. Suppl.)* 87:S1–S102.

Smith, S.V., and R.W. Buddemeier. 1992. Global change and coral reef ecosystems. *Ann. Rev. Ecol. Syst.* 23:89–118.

Stott, L., C. Poulsen, S. Lund, and R. Thunell. 2002. Super ENSO and global climate oscillations at millennial time scales. *Science* 297:222–226.

Stott, P.A., S.F.B. Tett, G.S. Jones, M.A. Allen, J.F.B. Mitchell, and G.J. Jenkins. 2000. External control of 20th century temperature by natural and anthropogenic forcings. *Science* 290:2133–2137.

Strong, A., C.B. Barrientos, C. Duda, and J. Sapper. 1997. Improved satellite techniques for monitoring coral reef bleaching. *Proc. Eighth Int. Coral Reef Symp., Panamá* 2:1495–1498.

Szmant, A.M., and N.J. Gassman. 1990. The effects of prolonged "bleaching" on the tissue biomass and reproduction of the reef coral *Montastraea annularis*. *Coral Reefs* 8:217–224.

Trenberth, K.E. 1997. The definition of El Niño. *Bull. Am. Meteorol. Soc.* 78:2771–2777.

Trenberth, K.E., and T.J. Hoar. 1996. The 1990-1995 El Niño-Southern Oscillation event: Longest on record. *Geophys. Res. Lett.* 23:57–60.

Tudhope, A.W., C.P. Chilcott, M.T. McCulloch, E.R. Cook, J. Chappell, R.M. Ellam, D.W. Lea, J.M. Lough, and G.B. Shimmield. 2001. Variability in the El Niño-Southern Oscillation through a glacial-interglacial cycle. *Science* 291:1511–1517.

Vargas-Ángel, B., F.A. Zapata, H. Hernández, and J.M. Jiménez. 2001. Coral and coral reef responses to the 1997-98 El Niño event on the Pacific coast of Colombia. *Bull. Mar. Sci.* 69:111–132.

Walker, G.T., and E.W. Bliss. 1930. World weather IV: Some applications to seasonal fore-shadowing. *Mem. R. Meteorol. Soc.* 3:82–95.

Warner, M.E., W.K. Fitt, and G.W. Schmidt. 1996. The effects of elevated temperature on the photosynthetic efficiency of zooxanthellae *in hospite* from four species of reef coral: A novel approach. *Plant Cell Environ.* 19:291–299.

Warner, M.E., W.K. Fitt, and G.W. Schmidt. 1999. Damage to photosystem II in symbiotic dinoflagellates: A determinant of coral bleaching. *Proc. Natl. Acad. Sci. USA* 96:8007–8012.

Wellington, G.M., and W.K. Fitt. 2003. Influence of UV radiation on the survival of lar-vae from broadcast-reef corals. *Mar. Biol.* 143:1185–1192.

Wellington, G.M., and P.W. Glynn. 1983. Environmental influences on skeletal banding in eastern Pacific (Panamá) corals. *Coral Reefs* 1:215–222.

Wellington, G.M., P.W. Glynn, A.E. Strong, S.A. Navarrete, E. Wieters, and D. Hubbard. 2001. Crisis on coral reefs linked to climate change. *EOS* 82:1,5.

Wells, J.W. 1957. Coral reefs. In *Treatise on Marine Ecology and Paleoecology*, ed. J.W. Hedgpeth. Geol. Soc. Amer. Mem. 67:609–631.

West, J.M., and R.V. Salm. 2003. Resistance and resilience to coral bleaching, implications for coral reef conservation and management. *Conserv. Biol.* 17:956–967.

Wilkinson, C.R. (ed.). 1990. *Acanthaster planci. Coral Reefs* (special issue) 9:93–172.

Wilkinson, C.R. (ed.). 2000. *Status of Coral Reefs of the World: 2000.* Cape Ferguson, Queensland: Australian Institute of Marine Science.

Wilkinson, C.R. (ed.). 2002. *Status of Coral Reefs of the World: 2002.* Cape Ferguson, Queensland: Australian Institute of Marine Science.

Wilkinson, C.R., and I.G. Macintyre. (eds.). 1992. The *Acanthaster* debate. *Coral Reefs* (special issue)11:51–122.

Williams, E.H., Jr., and L. Bunkley-Williams. 1988. Bleaching of Caribbean coral reef symbionts in 1987–1988. *Proc. Sixth Int. Coral Reef Symp., Australia* 3:313–318.

Williams, E.H., Jr., and L. Bunkley-Williams. 1990. The world-wide coral reef bleaching cycle and related sources of coral mortality. *Atoll Res. Bull.* 335:1–71.

Wyrtki, K. 1973. Teleconnections in the equatorial Pacific Ocean. *Science* 180:66–68.

12. Constraints on Predicting Coral Reef Response to Climate Change

Joan A. Kleypas

12.1 Introduction and Caveats

Over the last two decades, a primary concern within coral reef science has been the fate of coral reefs in an environment significantly altered by both direct anthropogenic impacts at the regional scale and indirect impacts associated with changes in Earth's atmospheric chemistry. Regional impacts, such as destructive fishing, deforestation, and nutrient loading, can greatly degrade reefs over very short time periods, but in many cases it should be possible to engineer solutions to these problems through mitigative practices such as the formation of marine parks, establishment of fishing regulations, runoff control, etc. The global impacts of greenhouse gas emissions on coral reefs are much more difficult to understand and mitigate. However, unless there are drastic changes in the fossil fuel-based world economy and politics, or engineering breakthroughs that either reduce our dependence on fossil fuels or sequester fossil fuel emissions, atmospheric CO_2 concentration will increase to double the preindustrial concentration by the middle of this century, and other greenhouse gases (CH_4, N_2O, H_2O) will also increase (Houghton et al. 2001). A great challenge within coral reef science is to assess how coral reefs will respond to warmer, more acidic oceanic conditions. This chapter addresses two major global threats facing coral reefs today—increased water temperatures and changing seawater chemistry—by systematically examining current predictions of (1) environmental change, (2) the response of coral reefs to that environmental change, and (3) the likely long-term implications of the response.

Making a reasonable prediction of what will happen to reefs in the future requires knowledge of what future environmental conditions will be like and the rate that future conditions will change, and the response of reef-building organisms to such changes. Unfortunately, neither of these is well constrained, and predictions must be based on analogues, experiments which simulate the environmental change, or modeling efforts. Usually, predictions of future environmental conditions are based on mathematical models, and predictions of organism and community response to these conditions are based on analogues. Although some information about an organism's response to environmental change can also be derived from knowledge of its biochemistry and genetic makeup, this chapter focuses on information obtained from the geologic record of reefs and from climate modeling.

The best way to predict how an ecosystem will respond to environmental change is to look in both space and time for areas or periods with environmental conditions similar to those projected for the future. For example, as a first approximation and based on present-day distributions of ecosystems, ecologists assume that global warming will cause species to migrate poleward in parallel with shifts in ideal temperature regimes, and ecosystems, as a result, to be dismantled to some degree, although the exact nature of that shift is complicated by other environmental and ecological changes (e.g., precipitation, ocean currents, biotic interactions, substrate availability). However, predicting how and in what configuration coral reefs and other tropical ecosystems will migrate in response to global warming is complicated because while the distributions of these ecosystems reflect their lower thermal limits, they do not necessarily reflect their upper thermal limits (because they occur throughout the tropics). If one assumes that coral reef development at higher latitudes is limited by minimum water temperature (and this assumption may not be entirely correct; see Section 12.4), then coral reef ecosystems are likely to expand into higher latitudes. But will reefs also be eliminated from regions that are already near their upper thermal limits? Certainly the geographically extensive mass coral bleaching episodes of the last two decades indicate that reefs are limited by rapid elevation of the normal maximum monthly temperature, but there is little evidence that maximum temperature limits the present-day distribution of coral reefs (Fig. 12.1).

When environmental conditions exceed those that shaped the current ecological map of Earth, one must look to other time periods when similar environmental conditions existed. This requires paleontological records of not only past ecosystems but also the environmental conditions that shaped them. This approach is limited by how well those records are known, which almost always degrades with progressively older time periods. Comparisons of ancient communities with those of today carry additional assumptions related to limitations in paleontological and paleoenvironmental data: (1) decreased temporal resolution with increased age; (2) diagenetic alterations to proxy material; and (3) evolutionary differences in organisms, communities, and ecosystems.

Human-induced changes in atmospheric chemistry are forcing Earth's climate and ocean chemistry toward conditions that have not occurred for hundreds of thousands, and probably millions, of years. In addition, the current suite of reef

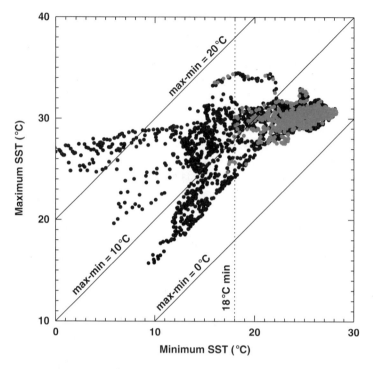

Figure 12.1. Distribution of annual minimum versus annual maximum water temperatures for coral reef regions (gray circles), compared with the distribution for tropical continental shelf waters (black circles). The 18 °C minimum is indicated by a dotted line. No maximum is shown because coral reefs occur in even the warmest waters of the tropics, and hence their maximum temperature tolerance cannot be discerned from this distribution. The diagonal lines designate where annual temperature variation (max–min) of 0 °C, 10 °C, and 20 °C fall on the graph. Most reefs occur where the annual temperature range is less than 10 °C. SST data were extracted from weekly SST data set of Reynolds et al. (2002).

builders has evolved considerably over that time, so that using paleoanalogues to gain insights into how coral reefs will adapt to future environmental change is further limited to observations at the scale of a reef, rather than the scale of a reefbuilder. In fact, most of today's important reef-building coral families appeared sometime in the Eocene, and acroporids were not dominant on coral reefs until the Pleistocene (ca. 2 Ma; Wood 2001). Thus, reefs of the early Tertiary (Paleocene and Eocene, ca. 65–35 Ma), when atmospheric CO_2 levels were probably at least 500 ppmV (parts per million volume; this is roughly equivalent to the partial pressure of CO_2 [pCO_2] in the atmosphere expressed in µatm), may provide important clues in terms of certain physical reef characteristics such as calcification rates, distribution patterns, etc., but they are probably less useful as analogues in terms of ecological response. Another way to think of this is to examine the variety of present-day coral reefs: to what extent can we compare Brazilian reefs to "mainstream" Caribbean or Indo-Pacific reefs, which support a different coral fauna?

12.2 Atmospheric CO_2 Concentration

12.2.1 Predictions of Future CO_2 Concentration

According to the Intergovernmental Panel on Climate Change (IPCC) Special Report on Emission Scenarios (SRES; Nakicenovic and Swart 2000), atmospheric CO_2 concentrations are predicted to reach between about 555 and 825 ppmV within this century. This wide range in predictions reflects the uncertainty in various inputs used to derive future emissions, such as population growth, development of energy technology, and economic growth. The lower end of the range essentially represents a doubling of the preindustrial concentration and the upper limit a tripling. The uncertainty implied by this range alone is much greater than the known glacial–interglacial fluctuations of CO_2 of about 100 ppmV (Petit et al. 1999).

12.2.2 Records of Past CO_2 Concentration

The best records of past atmospheric concentrations are from ice cores. CO_2 concentrations, sampled directly from air trapped within the ice, can be measured quite precisely, although some caution is warranted because it takes several years for the air to be permanently sealed off from the atmosphere ("diffusional smoothing"), and chemical reactions within the ice can alter the chemistry of the trapped air. The longest record is from the Vostok ice core, which extends back nearly a half-million years (Petit et al. 1999; Fig. 12.2A). This record indicates that over several major glaciations, atmospheric CO_2 concentrations remained between 180 and 300 ppmV. These values are corroborated by leaf stomatal index data for the last interglacial (Eemian; 115–130 ka), which indicate more variable but similarly low atmospheric concentrations (Rundgren and Bennike 2002).

CO$_2$ data extending back millions of years are scarce. The leaf stomatal index from fossil trees (*Gingko* and *Metasequoia*, both of which have modern counterparts) dating from nearly 60 to 50 Ma, indicates that atmospheric concentrations in the early Tertiary remained between 300 and 450 ppmV, only slightly higher than those of recent interglacials (Royer et al. 2001). However, data from the marine environment (calcium isotopes [De La Rocha and DePaolo 2000], boron isotopes [Pearson and Palmer 2000; Pearson et al. 2001], alkenones [Pagani, Arthur, and Freeman 1999]) all indicate that atmospheric CO_2 concentrations were probably much higher in the early Tertiary and decreased to below 300 ppmV by the Early Miocene (24 Ma; see Fig. 12B). In addition, the GEOCARB model, which hindcasts atmospheric CO_2 levels over the entire Phanerozoic by combining geological, geochemical, biological, and climatological data, suggests early Tertiary CO_2 levels were up to five times those of preindustrial levels (Fig. 12C; Berner 1994, 1997; Berner and Kothavala 2001). Recent isotopic and temperature proxy data from foraminifera (Kennett and Stott 1991; Zachos et al. 2003) indicate that the Paleocene/Eocene Thermal Maximum (an abrupt increase in sea-surface temperatures [SSTs] about 55 Ma) coincides with a sudden increase in greenhouse gas concentration in the atmosphere, as evidenced by negative carbon-isotope excursion of >2.5%, and a deep-sea horizon of carbonate dissolution (Zachos et al. 2005).

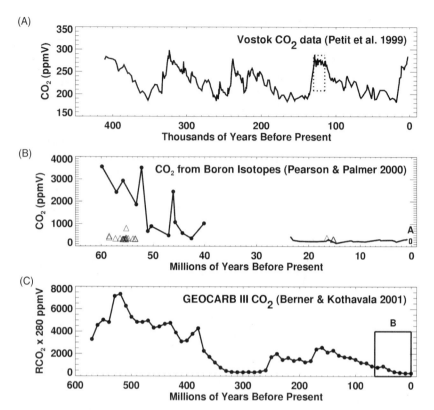

Figure 12.2. Historical atmospheric CO_2 concentrations determined from: (**A**) the Vostok ice core record (Petit et al. 1999); rectangle indicates range of Leaf Stomatal Index (LSI) values from early Tertiary samples (Royer et al. 2001); (**B**) boron isotope data from foraminifera (Pearson and Palmer 2000), with LSI data from late Pleistocene samples shown as triangles (Rundgren and Bennike 2002), and range of (a) indicated by rectangle; (**C**) results from the GEOCARB III model (Berner and Kothavala 2001), and range of (b) indicated by rectangle.

12.3 Temperature and Coral Bleaching

Large-scale mass bleaching of corals and other reef organisms is a phenomenon with no known precedent prior to 1982 (Glynn 1993). Many factors have been implicated in these mass-bleaching events (e.g., temperature, UV-B, nutrients, hydrodynamics; see summaries by Hoegh-Guldberg 1999 and Glynn 2000), but water temperature appears to be the primary factor, with solar radiation playing a key role as well (see review by Fitt et al. 2001). Patterns of coral bleaching are consistent with the hypothesis that photosynthesis in corals at elevated temperatures can cause photochemical damage. Bleaching often begins on coral surfaces that receive the most light; and some corals appear to have photo-protective adaptations to slow photosynthesis during light-induced bleaching (Brown 1997).

12.3.1 Predictions of Future Temperature Change

Surface temperatures on the Earth are related to the atmosphere's capacity to act as a greenhouse. Certain gases are transparent to incoming short-wave radiation from the sun but hinder the escape of reradiated long-wave radiation derived from heating of the Earth's surface. Without this natural greenhouse effect, global temperatures would be substantially cooler. Predictions of surface temperature due to the greenhouse effect go hand-in-hand with those of atmospheric CO_2 and other greenhouse gas concentrations. Virtually all SST predictions for the near future are based on projections of atmospheric composition, with changes in CO_2 concentration exerting the greatest increase in radiative forcing (CO_2 exerts about 60% of the current increase in radiative forcing relative to conditions in 1750; CH_4, 20%; halocarbons, 14%; and N_2O, 6%; Houghton et al. 2001).

Atmospheric paleotemperatures are often inferred from paleo-CO_2 records as well (although $\delta^{18}O$ measurements from ice cores provide independent estimates of atmospheric temperature [e.g., Petit et al. 1999]). The CO_2–temperature correlation probably held true for most of the Phanerozoic (the past 560 Myr), but the relationship appears to have broken down for certain brief periods in the early Cenozoic (Royer et al. 2001) and in the Paleozoic (Veizer, Godderis, and Francois 2000; Crowley and Berner 2001). Other factors (e.g., tectonic events) are thought to have been the main influence on global climate during these times (Zachos et al. 2001). For the late Cenozoic, however, warmer climates did prevail during periods of higher atmospheric CO_2.

Records/Analogues of Ocean Temperature and El Niño–Southern Oscillation

Temperature

For periods prior to the widespread instrumental weather observations (late 19th century) various proxy temperature records are available for atmosphere and ocean. The most common sources of information are: tree-ring width and density, isotopic and geochemical tracers in coral skeletons and foraminifera, and information from cores of marine and lake sediments. The best records of seawater temperature on reefs are from the corals themselves, as recorded by both $\delta^{18}O$ values and Sr/Ca ratios in the annual bands of the skeleton (e.g., McConnaughey 1989; Cole and Fairbanks 1990; Beck et al. 1992). Both techniques can provide high-resolution records of seawater temperatures, and both have been used on living corals as well as fossil corals as old as 130 kyr (Tudhope et al. 2001).

Coral records have provided useful reconstructions of past ocean temperature variability, but like all proxy climate records they are imperfect. Sources of error include problems of separating mutliple environmental effects (e.g., $\delta^{18}O$ can record a mixture of both salinity and temperature), reliance on single coral samples, and variations in signal preservation through time (Bard 2001). Secondary precipitation of inorganic cements can contaminate both the $\delta^{18}O$ and Sr/Ca signals (Müller, Gagan, and McCulloch 2001; Ribaud-Laurenti et al. 2001), even in skeletons of living corals (Enmar et al. 2000); and zooxanthellae can impose a strong signature on the Sr/Ca signal (Cohen et al. 2002). Despite these potential shortcomings and mismatches with other paleotemperature proxies (Crowley 2000), coral paleo-SST records have proven invaluable at capturing climatic histories such as the variability of the El Niño–Southern Oscillation (ENSO) over the past few hundred years (Dunbar 2000).

One of the main problems with using corals as records of SST is that temperatures are not recorded if a coral discontinues skeletal growth, such as during severe stress or with mortality. The Galápagos record, for example, terminated abruptly during the 1982/1983 bleaching and mass mortality (Shen et al. 1992). The elimination of high-temperature excursions from coral records would certainly bias reconstructions of past temperature variations.

ENSO

The El Niño–Southern Oscillation is the best-understood source of short-term climate variability that significantly affects tropical SSTs and hence coral reefs (see Riegl, Chapter 10; Wellington and Glynn, Chapter 11). During the warm phases of ENSO (El Niño events) large parts of the tropical oceans are unusually warm and have recently been associated with mass coral bleaching events. The cold phases of ENSO (La Niña events) are generally associated with unusually cool SSTs although warmer than normal waters do occur in some areas (e.g., coral reefs lying in the South Pacific Convergence Zone). The response of ENSO to increased greenhouse gases is unclear at present. The El Niño events of 1982 to 1983 and 1997 to 1998 were the strongest since 1950 and probably during the instrumental record period. This does not necessarily indicate, however, that El Niño events are increasing in intensity due to global warming, and the limited data available for years prior to 1950 suggest that similarly intense events occurred in the late 1800s (Enfield 2001). Tudhope et al. (2001) presented coral records spanning the last 130 kyr which show that (1) although ENSO events persisted throughout the glacial/interglacial periods, their intensity waned during colder periods, and (2) current ENSOs are more intense than at any other time in the coral records. Some researchers use these results as evidence that frequency and/or intensity of the ENSO will increase under increased greenhouse gas forcing (but see discussion of modeling ENSO events in the next section).

Modeled Predictions of SST and ENSO

Model Advantages and Limitations

The response of Earth's climate to increased atmospheric greenhouse gas concentrations is complex, and forecasting future changes in surface temperature, precipitation, ocean circulation, and other climate characteristics necessarily rely on mathematical models. These numerical models are primarily based on physics, but many components are parameterized (i.e., the physical processes are formulated as model constants or functional relations) based on empirical observations (e.g., freshwater runoff). A model's accuracy is often judged by how well it simulates conditions of both the present and the recent past, periods for which we have accurate records. Modeled increases in global ocean temperature over the

20th century, for example, show increases consistent with recent analysis of observations (e.g., Levitus et al. 2000).

Given computing constraints, all climate models are limited by both spatial and temporal resolution, as well as by gaps in scientific understanding. For example, one of the most important hurdles in climate modeling today is parameterization of cloud processes and feedbacks. Many aspects of cloud formation and cloud feedbacks to radiative forcing (e.g., albedo) remain poorly understood, so that current cloud parameterizations could be introducing substantial error in climate predictions. Another condition of the Earth system that has been difficult to model, but which could have profound impacts on climate, is the thermohaline circulation. The oceanic "conveyor belt" (sensu Broecker 1987) is driven by the formation of dense (colder and/or saltier) water in polar regions. Some models predict a slowing or virtual shutting down of the thermohaline circulation in response to future climate change (Rahmstorf 1995; Stocker and Schmittner 1997), whereas others predict little change. This large range of predictions arises from rather small differences in temperature and freshwater fluxes of the various models (see reviews by Gent 2001 and Clark et al. 2002).

Nonetheless, global climate models contribute significantly to our understanding of various feedbacks in the climate system, and several large efforts are underway that combine (or "couple") models of the major components of the Earth system—atmosphere, ocean, land, ice—so that they interact simultaneously, allowing feedbacks to occur by passing forcing fields back and forth across the interfaces (atmosphere–ocean, atmosphere–land, ocean–ice, etc.). Earth system models are thus better than analogues at understanding the various *mechanisms* of climate change. They also allow quantification, determination of the limits of change, regional assessments, global coverage, and separation of the effects of covarying parameters.

Modeled SSTs

The Third IPCC Report (Houghton et al. 2001) based its forecasts of future climate on a suite of Earth system models, most of which allowed feedbacks between the major components: atmosphere, ocean, land, and ice. All of the models predict an increase in surface temperature in response to increased greenhouse gas concentrations. Although there is near-consensus that increased greenhouse gas concentrations will cause or are already causing global warming, there is not a consensus on how much the temperature will rise. The range of predicted global temperature increase among models included in the Third IPCC Report is large: 1.4 to 5.8 °C for the period 1990 to 2100 (Houghton et al. 2001). These differences arise from both the differences in the climate models and the range of scenarios of future greenhouse gas concentrations. These same models indicate that warming of tropical SSTs (1 to 3 °C, Houghton et al. 2001) will be less than the global average.

Almost all coupled models indicate greater warming at high latitudes than in the tropics (Fig. 12.3). Despite recent evidence that SSTs during the middle Cretaceous were as high as 33 °C (Wilson, Norris, and Cooper 2002), observations and model results tend to support the "thermostat hypothesis." This hypothesis states that various atmospheric and oceanic feedbacks prevent SSTs in the open

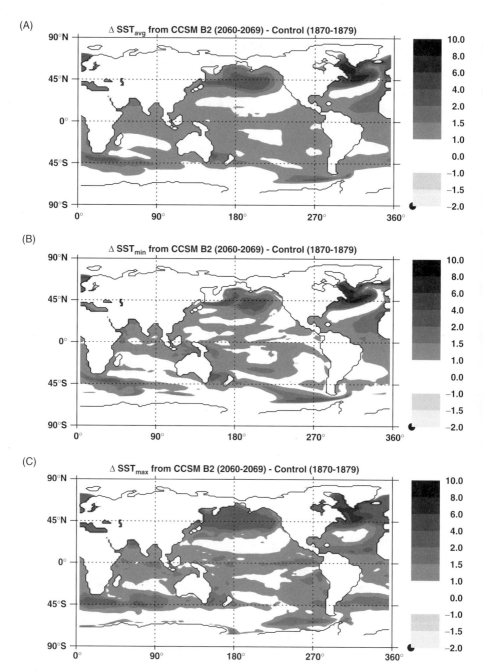

Figure 12.3. Changes in sea surface temperature predicted by the Community Climate System Model version 1.0 (CCSM 1.0; Boville and Gent 1998) of the National Center for Atmospheric Research for the 10-year period 2060 to 2069. Each frame shows the difference between the model run with IPCC's SRES B2 forcings, when atmospheric CO_2 = 517 ppmV, and the CONTROL run, in which atmospheric CO_2 was held constant at 280 ppmV. (**A**) Annual average SST; (**B**) annual minimum SST; and (**C**) annual maximum SST.

ocean from exceeding 31 to 32 °C. Three mechanisms for an ocean thermostat have been proposed: (1) evaporative cooling or evaporation–wind–SST feedback (Newell 1979); (2) the cloud–SST feedback or cloud shortwave radiative forcing (Ramanathan and Collins 1991); and (3) ocean dynamics and heat transport (Li, Hogan, and Chan 2000; Loschnigg and Webster 2000). Maximum SSTs today are around 31 to 32 °C in the open ocean, and about 33 to 34 °C in enclosed seas such as the Red Sea. The thermostat may play an important role in how coral reefs fare within a greenhouse earth, because it is likely that the thermostat temperatures vary from region to region, as well as across spatial scales, and consequently reefs that exist in waters that are already near the temperature set by the thermostat may not experience as much warming as reefs in cooler waters.

Modeled ENSO Events

Most global models are not explicitly designed to predict future changes in ENSO, although many predict relatively greater warming of SSTs in the eastern compared with the western tropical Pacific as well as increases in tropical Pacific precipitation (Houghton et al. 2001). Both of these changes are consistent with increased ENSO development or "more El Niño-like conditions." The few coupled ocean–atmosphere models that have been used to examine ENSO sensitivity to atmospheric greenhouse gas concentrations have produced mixed results (e.g., Tett 1995; Knutson, Manabe, and Gu 1997; Timmerman et al. 1999; Otto-Bliesner and Brady 2001). Timmerman et al. (1999) predicted that El Niño conditions will be more frequent and La Niñas more intense with further increases in atmospheric CO_2. In contrast, the ENSO response modeled by E. Brady and B. Otto-Bliesner (unpublished data) was much less sensitive to CO_2 change. Changes in ENSO under $2 \times CO_2$ conditions were insignificant, and under $6 \times CO_2$, ENSO intensity actually decreased. Such results reflect the fact that ENSO frequency and intensity are not merely a function of temperature change, but also reflect the influence of other climatic responses such as monsoons (Otto-Bliesner et al. 2003).

12.3.2 Coral Reef Responses to Rising Temperatures

Regardless of whether there is consensus that the present warming trend is due to increased atmospheric CO_2 forcing, nearly all coral reef scientists agree that mass coral reef bleachings have increased in frequency during the last two decades and are unprecedented within this century and probably for several preceding centuries (Hoegh-Guldberg 1999; Aronson et al. 2000). Recent mass coral bleaching events are clearly associated with anomalously warm SSTs. Given the acceleration in global warming over the same time period there is a strong suggestion that global climate change due to the greenhouse effect is leading to an increased frequency of coral reef bleaching (Lough 2000). These mass bleachings have been shown to be exacerbated by other factors such as subaerial exposure during unusually low tide (Glynn 1984), increased penetration of visible and UV light (Gleason and Wellington 1993), and decreased water circulation (Nakamura and van Woesik 2001). Temperature and the nonanthropogenic exacerbating factors are all associated with climatic features typical of the El Niño phase of ENSO, and the three most notable recent episodes of coral bleaching occurred during ENSO events: 1982 to 1983, 1987, 1997 to 1998. If mass coral bleaching events continue to increase in frequency, major changes in coral community structure are inevitable (Hughes et al. 2003), and some researchers predict a collapse of coral reef ecosystems over the next few decades (e.g., Hoegh-Guldberg 1999). How reef-building organisms will fare under continued high-temperature stress depends mainly on their ability to adapt to those changes as discussed below.

Records/Analogues of Coral Bleaching

To establish a history of past episodes of bleaching events caused by warm water temperatures, records of both SST and coral bleaching are needed. As mentioned earlier, the best records of past reef-water temperatures are obtained from the corals themselves. Both stable isotopes and Sr/Ca ratios measured chronologically along cores from massive corals can reveal detailed information about past SSTs experienced by those corals. One classic example of this is the record obtained from massive *Pavona* spp. in the Galápagos (Dunbar et al. 1994). The oxygen isotope record revealed that for the time period represented by this record (years 1586–1953) SSTs had remained below that which caused the near-total die-off of Galápagos corals during the 1982 to 1983 bleaching event. Note that this approach is not yet suitable for branching and foliose corals.

Coral bleaching records are much more difficult to obtain. In human recorded history, reports of isolated bleaching events can be found in the literature, but there are no historical records of regional-scale mass bleaching events prior to 1982 (Glynn 1993). Glynn (2000) outlined a variety of markers in the geological record that could be used as indicators of past mass bleaching events, including isotopic and trace-metal markers in coral cores indicative of ENSO events, alterations in skeletal banding, protuberant growths on massive corals (signaling where surviving patches of colonies continued to grow; cf. DeVantier and Done, Chapter 4), and indications of accelerated bioerosion in reef sediments. To date, these have not been used to derive past records of coral bleaching or mass mortality (see review in Wellington and Glynn, Chapter 11).

A combined ENSO-bleaching signal has been derived from coral cores as an excursion in the normal covariance of carbon and oxygen isotopic signals (Carriquiry, Risk, and Schwarcz 1994). Here, the $\delta^{18}O$ signal indicates warmer than normal conditions whereas the $\delta^{13}C$ signal indicated bleaching. This interpretation relies on environmental records of both temperature and salinity, however, and extrapolation to historical times is therefore difficult. Another obvious drawback of a skeletal "coral bleaching signal" is that, like temperature, such a signal is not likely to be preserved when the bleaching leads to coral mortality. There is also evidence that some bleached corals fail to secrete a growth band (Halley and Hudson, Chapter 6). In either case, coral skeletons may underestimate the past frequency and intensity of bleaching.

How corals and coral reefs fared under greenhouse conditions versus icehouse conditions of the last 30 million years is a question not readily answerable with the paleontological record. In general, corals and other aragonite producers appear to dominate reef-building during cool rather than warm periods of Earth history (Sandberg 1983). The reasons for their dominance may not be strictly related to lower temperature but to some other environmental factor such as the seawater Mg:Ca ratio (Stanley and Hardie 1998). Scleractinian corals were dominant reef-builders in the Jurassic, but rudist bivalves largely displaced them sometime in the Early to Middle Cretaceous. Rosen and Turnsek (1989) also noted that the highest coral diversity in the Cretaceous was shifted poleward to 35° to 45°N. Why were corals rare in low-latitude regions during the Cretaceous, and indeed during many other Phanerozoic periods (Kiessling 2001)? There are many hypotheses for the coral–rudist shift that are based on both environmental and ecological arguments, but there is no strong evidence to favor one hypothesis over another (Gili, Masse, and Skelton 1995; Johnson and Kauffman 2001). Glynn (2000) also raised the interesting hypothesis that some past extinction events could have been caused by mass bleaching. Unfortunately, recent records of coral bleaching are scant and paleontological records of coral bleaching are essentially nonexistent, so the question of whether massive coral bleachings occurred during warm periods in the past cannot be answered yet. It is even more difficult to assess how quickly corals and/or their algal symbionts might adapt to higher SSTs.

Models of Coral Adaptation to Higher Temperature

Assuming that corals cannot adapt to increasing temperature, future global warming is likely to lead to more frequent mass coral bleaching events. Even without global warming, if ENSO frequency and/or intensity increase, then the probability of mass coral bleachings will also increase, at least in certain geographic areas. The current large-scale bleaching response to increased SSTs indicates that many reef-building corals (either the coral animals, the symbiotic algae, or both) are not adapting adequately to the SST increase. There is, however, evidence that reef-building corals and other symbiotic organisms can adapt to increasing temperatures, either through acclimation (short-term adaptation to environmental change at the individual level), acclimatization (medium-term adaptation), or evolution (long-term adaptation through natural selection; Coles 2001).

The distribution of corals in time and space illustrates this fact. First, scleractinian corals survived the high temperatures of the Cretaceous, although their distribution was shifted to higher latitudes (Kiessling 2001). Second, corals can exist and indeed thrive under extreme temperature regimes. Enclosed regions such as the Red Sea and Persian Gulf support species that tolerate much greater temperature ranges than do their conspecifics in the Indo-Pacific (Coles and Fadlallah 1991). These two examples probably illustrate adaptation over long periods of time, such as adaptation via natural selection.

The "adaptive bleaching hypothesis" (Buddemeier and Fautin 1993; Fautin and Buddemeier 2004) is an adaptive mechanism that acts over shorter time scales. Given that several species of zooxanthellae (and indeed several clades within species) inhabit corals as symbionts, this hypothesis is based on the possibility of a variety of coral–zooxanthellae combinations, each with a different environmental tolerance for the surrounding environmental conditions. Bleaching could thus be considered an adaptive mechanism if the expelled zooxanthellae are replaced by another species or type of zooxanthellae better suited to the new environment. Baker (2001) suggested the additional hypothesis that because bleaching expels all of the less-tolerant algae, competition with incoming algae is thus eliminated. Manipulation experiments have shown that some corals and anemones that bleached subsequently acquired new algal symbionts better suited to their new environment (Baker 2001; Kinzie et al. 2001). Adaptive bleaching is more diffi-cult to detect under natural conditions, mainly because of the influence of other environmental factors.

There is evidence that corals in some areas are adapting to the recent warming trend. Bleaching on some eastern Pacific reefs was much worse during the El Niño event of 1982 to 1983 than that in 1997 to 1998, even though temperature excursions during the two events were similar, offering evidence for adaptation and perhaps support for the adaptive bleaching hypothesis (Guzmán and Cortés 2001; Podestá and Glynn 2001). Glynn et al. (2001) also found that *Pocillopora* spp. at Gorgona Island in the eastern Pacific fared better in 1997 to 1998 than in 1982 to 1983, and suggested that this reflected evolutionary selection for more heat-irradiance-tolerant individuals that survived the 1982 to 1983 bleaching event. Finally, Brown et al. (2002) demonstrated adaptation in corals exposed to combinations of elevated SST and UV radiation, but without any associated changes in the zooxanthellae.

If one accepts that corals can adapt to increased SST, one must then determine how much and how rapidly they can adapt. Ware (1997) tested how different rates of acclimation might affect the frequency of coral bleaching in the future. He first assumed that corals maintain upper temperature tolerances close to the upper thermal limits to which they are normally accustomed. This is probably a valid assumption given that coral bleaching seems to occur when temperatures exceed the average climatological maximum by about 1 to 2 °C, rather than some absolute value (Coles and Brown 2003). Ware then assumed that the "acclimation temperature" of a coral at a particular location was based on the average annual maximum temperature over some past interval of time. Corals with relatively short acclimation periods (e.g., <25 years, Fig. 12.4) should experience fewer and milder bleaching episodes than those that require longer acclimation periods. Given a long enough temperature versus bleaching record, one could theoretically derive the acclimation period and then use this information to determine the prob-ability that coral bleaching will occur in the future (Ware 1997).

Determining the acclimation period is complicated because (1) there is probably some absolute maximum temperature beyond which corals can physiologically adapt, which is likely to vary across species, and (2) there is probably significant variation in adaptive potential between species and the rate at which they can adapt. Above its absolute maximum, the future of a species will depend on its ability to adapt genetically, through evolutionary processes, and this is also likely to vary across species as a function of gene flow (Hughes et al. 2003). The impact

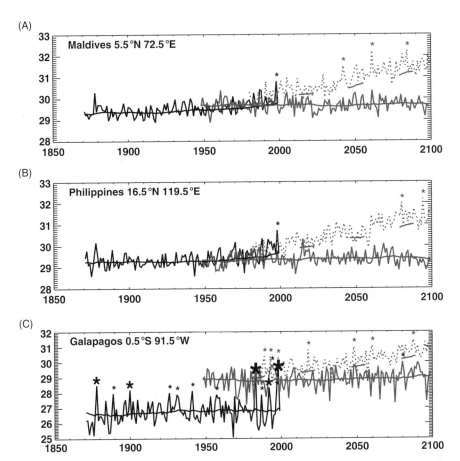

Figure 12.4. Comparison of observed and modeled annual maximum SSTs for three reef regions: (**A**) Philippines, (**B**) Bahamas, and (**C**) Galápagos. Black solid lines for the period 1970–2000 represent annual maxima derived from the HADISST 1×1° lat/lon reconstruction of observed SSTs (Rayner et al. 2003). Gray solid lines show annual SST maxima for years 1948–2000 of the CCSM 1.0 CONTROL integration (atmospheric CO_2 concentration held constant at 280 ppmV). Gray dotted lines show annual SST maxima for years 1980–2100 of the CCSM 1.0 SRES B2 integration, a moderate global warming scenario. The 25-year temperature of acclimation (sensu Ware 1997) is also shown for each SST series. Possible bleaching thresholds are indicated by stars where maximum temperature exceeds the acclimation temperature by 1.0 °C (small star), 1.5 °C (medium star), and 2.0 °C (large star). Even where absolute maximum temperatures are not duplicated by the model (e.g., for the Galápagos), a comparison between the HADISST and CONTROL data shows that the model does a good job of capturing interannual variability of maximum SSTs, which is more limiting to reefs than average maximum temperature.

of temperature *variability* on a coral's ability to adapt is also difficult to assess; a coral may be much better at adapting to increasing temperatures in regions where the annual maxima do not vary much from year to year.

12.4 Seawater Chemistry

12.4.1 Background

The oceans currently take up about about a third of the fossil-fuel-related increases in atmospheric CO_2 (Sabine et al. 2004). The increase in seawater-pCO_2, which is greatest in the well-mixed upper ocean, has measurable effects on seawater pH and the abundance of carbonate ions that are used by organisms that secrete calcium carbonate ($CaCO_3$).

The atmospherically driven acidification of the oceans and the potential effects on $CaCO_3$ precipitation were first raised as concerns several decades ago. Early fears that the normally "$CaCO_3$-saturated state" of the surface ocean would be driven to undersaturation, with grave consequences for marine calcifiers (Fairhall 1973; Zimen and Altenhein 1973), were alleviated by calculations showing that a 10-fold increase in atmospheric CO_2 would be required to achieve $CaCO_3$ undersaturation in surface waters (Whitfield 1974; Skirrow and Whitfield 1975). However, Smith and Buddemeier (1992) pointed out that future changes in the *degree* of $CaCO_3$ saturation state could affect calcification rates on reefs, and many experiments have since illustrated that calcification rates in both corals and coral reef communities are indeed proportional to $CaCO_3$ saturation state (e.g., Gattuso et al. 1998; Langdon et al. 2000; Leclercq, Gattuso, and Jaubert 2000, 2002; Langdon et al. 2003; Marubini, Ferrier-Pages, and Cuif 2003). The acidification of the oceans is therefore an important consequence of increased atmospheric CO_2, because it is likely to affect calcification of corals and algae, and the reef structures that they build.

Seawater pH is largely buffered by the equilibrium chemistry of the carbonate system (Fig. 12.5). Dissolved inorganic carbon (DIC; also called total CO_2 or TCO_2), is partitioned into several different ionic states in seawater, and the relative concentrations of these states constantly adjust to maintain charge balance between total cation and anion concentrations. Dissolving CO_2 in seawater increases TCO_2, but does not alter alkalinity (total charge of cations such as Na^+, Mg^{2+}, Ca^{2+}, etc.). In response to the increased TCO_2, the carbonate system shifts to ensure charge balance by converting some of the CO_3^{2-} (carbonate) ions to HCO_3^- (bicarbonate). The CO_3^{2-} concentration thus decreases as CO_2 is dissolved in seawater. $CaCO_3$ saturation state, a product of the concentrations of both Ca^{2+} and CO_3^{2-} in seawater, therefore, also decreases.

Factors which affect the CO_2 concentration in seawater also affect $CaCO_3$ saturation state. In the surface ocean, the first-order control on CO_2 concentration in seawater is the concentration of CO_2 in the overlying atmosphere, while second-order controls include temperature and biological activity. At regional

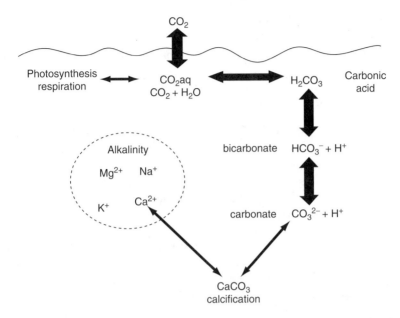

Figure 12.5. The carbonate system in seawater. As CO_2 is driven into the ocean, it quickly forms carbonic acid, which is a weak acid. Most of this rapidly dissociates to either bicarbonate (HCO_3^-) or carbonate (CO_3^{2-}). Total CO_2 (TCO_2) is the sum of HCO_3^-, CO_3^{2-}, and dissolved CO_2 in seawater, while total alkalinity (T_{Alk}) is the excess positive charge in seawater. The proportion of HCO_3^- to CO_3^{2-} adjusts to balance this positive charge. As a first approximation, the carbonate ion concentration can be estimated as $T_{Alk}-TCO_2$. Note that by adding CO_2, TCO_2 increases and the carbonate ion concentration decreases. From Kleypas and Langdon (2003).

scales, warmer waters hold less CO_2 than colder waters, which explains why tropical waters have higher saturation states than temperate waters. At local scales, photosynthesis and respiration consume and release CO_2 in the water column, respectively. $CaCO_3$ precipitation also affects CO_2 concentration because it reduces alkalinity which causes the carbonate equilibrium to shift toward higher CO_2 concentration. $CaCO_3$ dissolution has the opposite effect (Fig. 12.5).

Compared with the uncertainty of how SST will respond to increasing atmospheric CO_2, the response of ocean chemistry to future CO_2 increases is more predictable, at least over the next 100 years or so. Through processes of air–sea gas exchange, surface ocean seawater chemistry responds relatively quickly to changes in gas concentrations in the atmosphere, reaching equilibrium within about a year. The thermodynamic behavior of the carbonate system in seawater is well characterized, and measured changes in seawater chemistry from large field programs such as the World Ocean Circulation Experiment (WOCE) and the Joint Global Ocean Flux Survey (JGOFS) are consistent with predicted changes due to increases in atmospheric CO_2 (Sabine et al. 1999, 2002, 2004). Carbon isotopic measurements from oceanic waters that trace the $\delta^{13}C$ signature from CO_2 emissions derived from fossil fuel burning also support these findings (Quay et al. 2003). Predictions of future carbonate ion concentration in the surface ocean depend mainly on predictions of future atmospheric CO_2 concentrations, and secondarily on predictions of future SSTs (Figs. 12.6–12.9).

The pCO_2 in the surface ocean currently varies between about 80 µatm lower and 120 µatm higher than that of the atmosphere. Regions where surface ocean pCO_2 is lower than atmospheric pCO_2 occur where CO_2 is exported to the ocean interior, either through downwelling or through net biological uptake of CO_2 from the water column. Conversely, surface ocean pCO_2 is higher than that of the atmosphere where upwelling brings deep CO_2-rich water to the surface (the deep sea is isolated from air–sea gas exchange and becomes progressively enriched in CO_2 from the continued decay of organic matter). Upwelling in the eastern equatorial Pacific delivers waters to the surface, with pCO_2 more than 100 µatm higher than that of the atmosphere (Sakamoto et al. 1998). These patterns are reflected in the overall distribution of DIC and alkalinity in the Pacific (Fig. 12.7).

A study by Gattuso et al. (1998) was one of the first to directly test the effects of $CaCO_3$ saturation state on coral calcification, by altering the Ca^{2+} concentration in seawater. Recent experiments on calcification versus saturation state have instead altered CO_3^{2-} through a variety of manipulations of the carbonate system (see review by Langdon 2003). Langdon (2003) determined that calcification rate in the Biosphere 2 coral reef mesocosm responded mainly to CO_3^{2-} concentration rather than to pH or some other carbonate species. However, there is no convenient way to determine CO_3^{2-} concentration directly from seawater and there is no proxy for determining its concentration in the past oceans.

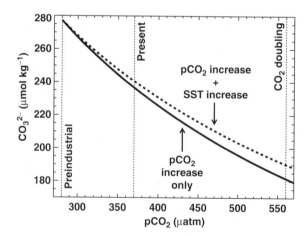

Figure 12.6. Changes in carbonate ion concentration in response to pCO_2 forcing, illustrating the effect of increasing temperature.

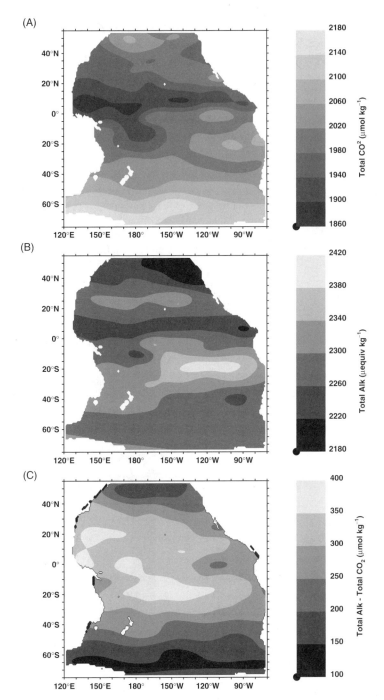

Figure 12.7. Distribution of major carbonate system parameters in the Pacific Ocean: (**A**) TCO_2, (**B**) T_{Alk}, and (**C**) T_{Alk}–TCO_2 as a rough approximation of carbonate ion concentration (see Fig. 12.5 for explanation). Data for panels (**A**) and (**B**) were obtained from a synthesis of the Global Ocean CO_2 Survey conducted over the last two decades (Sabine et al. 1999, 2002).

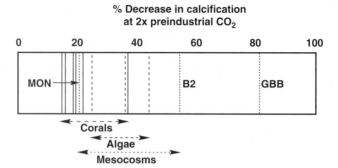

Figure 12.8. Experimentally derived estimates of calcification response of corals (solid lines), calcareous algae (dashed lines), and whole "communities" in mesocosm and field experiments (dotted lines) to changes in seawater chemistry. Response is standardized to calcification at 2× preindustrial atmospheric CO_2 concentrations, as percent decrease relative to that at preindustrial levels. MON = Monaco mesocosm (Leclercq, Gattuso, and Jaubert 2000); B2 = Biosphere 2 (Langdon 2003); GBB = Great Bahama Bank (Broecker and Takahashi 1966). Data are derived from a summary by Langdon (2003) and from Marubini, Ferrier-Pages, and Cuif (2003).

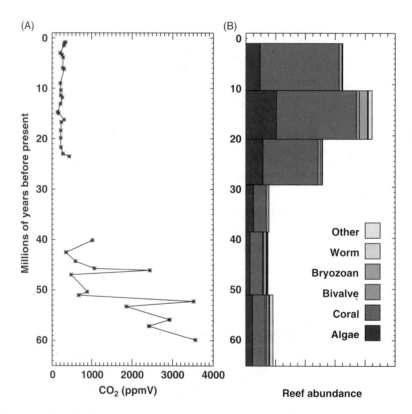

Figure 12.9. (**A**) The Pearson and Palmer (2000) record of atmospheric CO_2 for the Tertiary, and (**B**) a compilation of relative reef abundance (cumulative number of reefs in a given period), and the organisms that built them, for the same time period (derived from Kiessling, Flügel, and Golonka 1999; note that absolute numbers of reefs were not provided in their original figure, but bars range from about 30 to 175 reefs). Reef abundance appears to have increased around 30 Ma.

12.4.2 Past versus Future Seawater Chemistry Change

Records/Analogues

The carbonate chemistry of a given parcel of seawater can be determined if any two of the following carbonate parameters are known: pH, TCO_2, $TAIK$, and pCO_2, as well as temperature, pressure, and salinity of the water. It is now generally accepted by the scientific community that atmospheric pCO_2 was much higher during the Cretaceous and has declined throughout the Tertiary. However, one cannot simply use estimates of atmospheric CO_2 concentration to infer ocean carbonate chemistry during the Tertiary as $TAIK$, for example, may have also been higher. Records of past ocean chemistry are few, and they rely on proxies of pH and other ion concentrations. A proxy for one of the parameters alone is technically not sufficient to extrapolate the other carbonate parameters, and in such cases a second parameter must be assumed.

Over glacial time scales, there is evidence that ocean pH and CO_3^{2-} concentration were higher during glacial periods (Sanyal et al. 1995; Broecker and Clark 2002), but there is also evidence that ocean pH was relatively stable over the glacial–interglacial cycles (Anderson and Archer 2002). The dichotomy of these results, even over the relatively short time scale of millennia, illustrates the difficulty in reconstructing past ocean chemistry. Despite these problems, Pearson and Palmer (2000) used boron isotopes of deep-sea foraminifera to determine the paleo-pH of the ocean for periods extending back 49 million years. They assumed that ocean alkalinity over that period changed little, and thus used their data to infer atmospheric concentrations over the same period (Fig. 12.2). Broecker and Peng (1982) originally suggested that ocean alkalinity at the last glacial maximum was slightly higher than it is today, but there is no direct proxy for alkalinity, and the assumption of constant alkalinity over time needs to be verified.

Models

Predicting changes in seawater chemistry in the surface ocean over the short term (e.g., 100–200 years) is fairly straightforward. These predictions are complicated, however, by the response of ocean biology to increased pCO_2. Calcification of some of the most important $CaCO_3$ secreters in the oceans (e.g., coccolithophores and foraminifera, as well as reef-building organisms) decreases as pCO_2 increases, while the production of organic carbon could increase, at least in coccolithophores (Riebesell et al. 2000). This not only alters predictions of changes in surface seawater chemistry, but it can also affect our ability to predict changes in the ratio of organic carbon versus inorganic carbon delivered to the deep sea (the "rain ratio"). Changes in the rain ratio in turn affect oceanic uptake of atmospheric CO_2 (Archer et al. 2000a; Sarmiento et al. 2002). Accurately predicting surface changes in seawater chemistry over the next one to two centuries therefore depends primarily on how well we predict how atmospheric CO_2 will change over that time period, and secondarily on how well we predict the biological responses and feedbacks of the ocean.

Attempts to model Pleistocene fluctuations in atmospheric pCO_2 have not adequately captured the mechanisms for the glacial–interglacial differences (e.g., the 80 ppmV increase in pCO_2 since the last glacial maximum). There are many hypotheses to explain these fluctuations, and most concern changes in either the organic carbon pump (storage or burial of organic carbon in the deep sea) or the alkalinity pump (storage or burial of $CaCO_2$ in the deep sea). Even sophisticated models that combine ocean physics, biogeochemistry, and sediment chemistry have been unable to capture the magnitude of glacial–interglacial fluctuations in atmospheric CO_2 concentration (Archer et al. 2000b).

Over longer time periods, additional factors that affect seawater chemistry come into play. For example, deep-sea circulation can enhance CO_2 neutralization through dissolution of deeper-water carbonate; and continental weathering both removes CO_2 from the atmosphere and releases cations that ultimately contribute to ocean alkalinity. However, as demonstrated by results from the Hamburg Model of the Ocean Carbon Cycle (HAMOCC) coupled with a carbonate sediment diagenesis model, these processes would require thousands of years to bring the carbonate system back to preindustrial conditions (Archer, Kheshgi, and Maier-Reimer 1997).

12.4.3 Coral Reef Responses to Seawater Chemistry Change

Natural Chemical Variation on Reefs

Two early field studies tracked the change in seawater carbonate chemistry through time within semienclosed regions: Broecker and Takahashi (1966), on the Bahama Banks; Smith and Pesret (1974), in Fanning Island Lagoon. Both studies found that calcification rates decreased as the calcium carbonate saturation state of the reef waters decreased.

Carbonate chemistry in reef waters is known to fluctuate, sometimes dramatically, in response to both physical forcing and biological processes (Table 12.1). Bates, Samuels, and Merlivat (2001) sampled fCO_2 hourly for nearly one month on a Bermuda coral reef flat and found that fCO_2 fluctuated greatly (340–470 µatm over the sampling period, with a maximum range of about 60 µatm on any one day). These findings are consistent with measurements of Gattuso et al. (1996) on a reef flat in Moorea (240–420 µatm) and others (Suzuki and Kawahata 1999; Kayanne et al. 2003), where CO_2 rarely exceeded 500 µatm.

Fluctuations in carbonate chemistry of reef waters reflect not only calcification, but also dissolution (Chisholm and Barnes 1998; Halley and Yates 2000). Dissolution of carbonate minerals buffers the carbonate system in seawater because Ca^{2+} and CO_3^{2-} ions are released back into the water column. As saturation state declines, high-magnesium calcite (HMC) is the first mineral to dissolve. As atmospheric CO_2 increases in a system with little exchange with open-ocean seawater, such as in an aquarium or lagoon, buffering due to HMC dissolution cannot bring the system back to preindustrial values, but it can maintain the

Table 12.1. Variation of pCO_2 in coral reef environments. Note that pCO_2 is the partial pressure of CO_2, and fCO_2 is the fugacity of CO_2; the two values are very similar under natural surface seawater conditions

Study	Location	pCO_2 µatm	Comments
Bates, Samuels, and Merlivat (2001)	Bermuda reef flat	340–470	Maximum range of hourly measurements taken over 24 d
Gattuso et al. (1996)	Moorea lagoon	240–420	Maximum range of hourly measurements taken over 3 d
Suzuki and Kawahata (1999)	Palau barrier reef	366–414	Difference between lagoon and open ocean values
	Majuro Atoll	345–370	Difference between lagoon and open ocean values
	South Male Atoll	362–368	Difference between lagoon and open ocean values
Kayanne et al. (2003)	Shiraho Reef (Ryukyu Is.)	200–600	Range of nearly continuous measurements taken over 3 d

seawater saturation state at levels near that of the HMC saturation state. Such a steady state was observed in the Biosphere 2 mesocosm when the CO_3^{2-} concentration approached about 125 µmol kg^{-1}, but this corresponds to more than a doubling of pCO_2, when coral calcification rates would be at least 10% to 20% lower than they were under preindustrial conditions (C. Langdon, personal communication). Numerical modeling designed to examine changes in shallow-water carbonate dissolution and its effect on seawater chemistry found that although dissolution of metastable carbonates was likely, it was insufficient to buffer alkalinity or carbonate saturation state of shallow marine environments (Andersson, Mackenzie, and Ver 2003).

Experiments

Information on possible coral reef responses to changes in seawater chemistry is primarily based on laboratory experiments in which the carbonate chemistry of the seawater is manipulated in various ways (Fig. 12.8; see summary in Langdon 2003). Most of the experiments were performed on single organisms or in aquarium-size mesocosms, and conducted over hours to days. Manipulations of the Biosphere 2 coral reef mesocosm, however, were conducted over weeks to years. All of the experiments, regardless of the size, duration, or type of manipulation, found a reduction in calcification in response to decreased carbonate ion concentration.

Seawater manipulations are difficult in the field, but Schneider and Erez (1999) were able to demonstrate that the calcification response of *Acropora* in the Red Sea was similar to that in laboratory experiments. Halley and Yates (2000) conducted novel field experiments in Hawaii in which they placed large incubation chambers over reef sections and tracked changes in seawater chemistry for more than a day.

They found that as pCO_2 increased in the chambers, calcification rates decreased, but dissolution rates also increased, particularly of HMC. They estimated that on their particular reef the dissolution rate will equal the calcification rate once atmospheric CO_2 concentrations reach double preindustrial levels.

The calcification response of corals and algae thus appears to be affected by changes in the carbonate chemistry of seawater, despite biochemical evidence that corals, at least, exert significant control on the internal chemistry of their tissues (see review by Gattuso, Allemand, and Frankignoulle 1999). The photosynthetic response of zooxanthellae to increased CO_2 concentration is less clear, but it appears that coral zooxanthellae mainly utilize bicarbonate as a substrate for photosynthesis (Al-Moghrabi et al. 1996; Goiran et al. 1996). Zooxanthellar photosynthesis does not increase with increasing pCO_2, and does not appear to be stimulated by the CO_2 released by calcification (Gattuso et al. 2000). However, zooxanthellar photosynthesis does respond to increases in HCO_3^- (Marubini and Thake, 1999; note that HCO_3^- increases as pCO_2 increases). Since zooxanthellar photosynthesis is thought to enhance coral calcification, increased pCO_2 could simultaneously affect calcification in two ways: (1) by decreasing the CO_3^{2-} concentration and (2) by increasing the HCO_3^- concentration. The calcification experiments discussed above (as summarized by Langdon 2003) measured the net effect of both processes, but future experiments should be designed to measure the contribution of each separately.

Paleorecords/Analogues

Geologic Evidence

The dominant form of precipitated $CaCO_3$ has changed several times in the geological past. Such mineralogical shifts are interpreted as markers for major changes in seawater chemistry and/or biota. Two main hypotheses to explain these shifts involve the influence of seawater chemistry on the precipitation of one $CaCO_3$ mineralogy over another. Sandberg (1983) observed that the mineralogy of nonskeletal carbonates has "oscillated" in the Phanerozoic in concert with changes in atmospheric CO_2 and/or sea-level change; in essence, aragonite precipitation occurred when pCO_2 was low ("icehouse" conditions), but did not occur when pCO_2 was high ("greenhouse" conditions). A more recent hypothesis ties the historical calcite:aragonite oscillations to shifts in the magnesium to calcium ratio (Mg:Ca) in seawater (Stanley and Hardie 1998). High Mg^{2+} concentration tends to inhibit calcite precipitation (Berner 1975).

During the Cretaceous and for some time after the Cretaceous–Tertiary (K-T) event and the extinction of rudists, environmental conditions probably favored calcite precipitation. However, reef-building taxa that survived the K-T extinction, scleractinian corals and calcareous algae, secrete skeletons of aragonite or HMC. Indeed, reef building after the K-T event was apparently subdued for some 20 Myr until the Miocene (Kiessling, Flügel, and Golonka 1999; Fig. 12.9), although there are certainly other possible reasons for this delay in reef development, such as the decrease in reef-building taxa in the mass extinction event (Perrin 2002).

Given that atmospheric CO_2 concentrations prior to the Miocene probably remained higher than today, and given that Mg:Ca was probably lower (Wilson and Opdyke 1996), ocean chemistry of the near future cannot be adequately compared to any Tertiary time period (B.N. Opdyke, personal communication). Ocean chemistry of the near future will be extraordinary, mainly because the rapid rate of increase in atmospheric CO_2 will drive the system out of equilibrium. Given time, processes such as weathering will nudge the system back toward steady-state conditions, but in the short term (the next several thousand years) ocean chemistry is likely to be quite different from that of the last 65 Myr.

Recent Field Evidence

Based on experimental data of others (Section 12.4.1), Kleypas et al. (1999) estimated that the average calcification rate on coral reefs might have declined by 6% to 14% as atmospheric CO_2 concentration increased from 280 ppmV to the present-day value of 370 ppmV. However, two studies have shown that calcification rates derived from massive *Porites* in the Great Barrier Reef (GBR) and in Moorea have increased rather than decreased over the latter half of the 20th century (Lough and Barnes 1997, 2000; Bessat and Buigues 2001). Both studies demonstrate a strong linear relationship between temperature and calcification rate, and because temperature and saturation state are strongly correlated in the surface ocean, these studies also offer insight into the relative effects of temperature and seawater chemistry on coral calcification. Since warmer water holds less CO_2, aragonite saturation state increases with increasing temperature (Fig. 12.10A). The Lough and Barnes (2000) analysis of hundreds of *Porites* spp. cores showed that for every 1 °C increase in SST, the calcification rate increased by about 3.5% (Fig. 12.10B), and indeed found a nearly 5-fold increase in calcification rate between *Porites* growing in 23 °C (0.6 g $CaCO_3$ cm^{-2} yr^{-1}) and those growing in 29 °C (2.4 g $CaCO_3$ cm^{-2} yr^{-1}). From a thermodynamic standpoint, at present-day concentrations of atmospheric CO_2, a 1 °C increase in temperature causes about a 2.5% increase in carbonate ion concentration, and by inference from coral calcification-versus-carbonate ion experiments, an increase in calcification. However, the increase in calcification expected from a temperature-induced increase in saturation state can explain only a fraction of the calcification-versus-temperature regressions from the cores (Fig. 12.10B). Simply put, the response of *Porites* calcification to temperature is both biological and geochemical, and the biological response is obviously greater. McNeil, Matear, and Barnes (2004) modeled future coral reef calcification by combining the calcification-versus-SST relationship of Lough and Barnes (2000) and the experimental calcification-versus-saturation state relationship of Langdon et al. (2000). They concluded that increased SSTs in the future will outweigh the effects of decreased saturation state, so that future coral reef calcification will actually increase over the coming century. However, they assumed that coral calcification will increase indefinitely with temperature increase. In reality, calcification in corals peaks near the average summertime temperature, then decreases as temperatures exceed that point (Jokiel and Coles, 1977; Marshall and Clode 2004). Thus, even with the assumption that coral bleaching

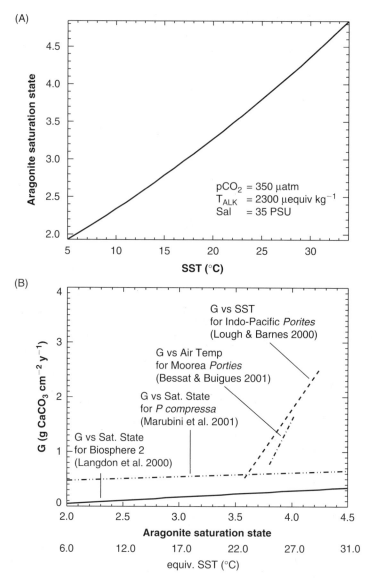

Figure 12.10. (A) Aragonite saturation state as a function of SST when other major variables are held constant (pCO_2, total alkalinity, and salinity). Changes in SST explain much of the latitudinal gradient in aragonite saturation state. (B) Calcification rate (G) versus aragonite saturation state as inferred from four studies. Note that calcification rate was determined differently in each study, and the results are shown here to illustrate trends and not exact relationships. The solid line represents the G-versus-saturation state relationship determined for long-term (1.9 months to 2.3 years) experiments in the Biosphere 2 mesocosm (Langdon et al. 2000). The dotted-dashed curve represents the G-versus-saturation state relationship for laboratory experiments on *Porites compressa*. The two remaining G-versus-saturation state relationships were inferred from G-versus-temperature relationships

will not occur, while temperature-induced calcification rates may increase in the short term, they are likely to decrease after that (Kleypas et al. 2005). Of course, this suggestion assumes that only temperature and seawater chemistry affect calcification, but in reality other factors should also be considered (e.g., light, water quality, nutrients, food supply). A strong recommendation arising from these results is that future calcification experiments should include both temperature and saturation state as variables. One such study has been performed on *Stylophora pistillata*, with confusing results: at normal temperature, calcification rate did not change with increasing saturation state, but at elevated temperatures, calcification rate decreased by 50% (Reynaud et al. 2003).

Current-day coral reef development is limited at higher latitudes. Coral communities exist at latitudes beyond 30°, but they lose their capacity to build reefs. The calcium carbonate budgets of these coral reef communities must be affected by lower $CaCO_3$ production due to decreased temperature, light, or saturation state; increased $CaCO_3$ removal from increased erosion and off-reef transport, or dissolution; or both lower production and increased removal. Identifying why these communities fail to accumulate calcium carbonate is essential to making predictions of how coral reefs will respond to future changes in seawater chemistry.

12.4.4 Implications for Future Reef Development

With a lowered calcification rate, calcifying organisms would extend their skeletons more slowly and/or make less dense skeletons. Because calcification rates in live specimens are usually measured as either weight changes in individual organisms or via changes in seawater chemistry, how reduced calcification rate is expressed in coral skeletons has rarely been determined and is likely to vary with species. Lough and Barnes (2000) found that slower calcification rates in massive *Porites* were correlated with a decrease in extension rate rather than a decrease in density (density was in fact negatively correlated with calcification rate), while Carricart-Ganivet (2004) found that *Montastraea annularis* maintained its skeletal extension rate despite a change in calcification rate. A reduced extension rate would reduce an individual's ability to compete for space on a reef, whereas reduced density ("osteoporosis") would mean less resistance to breakage, and greater susceptibility to both physical breakdown and bioerosion.

The role of secondary cementation and its response to changes in seawater chemistry could be an important factor in determining both the density of coral skeletons and overall reef calcification rates. Secondary cementation here refers to the inorganic precipitation of aragonite and HMC that occurs within skeletons

---

Figure 12.10. (*continued*) derived from coral cores, by applying the aragonite saturation state-versus-SSTrelationship shown in (**A**) (note that the temperatures equivalent to the aragonite saturation state on the x axis are shown). The Lough and Barnes (2000) G-versus-SST regression was derived from several hundred *Porites* spp. cores from the Great Barrier Reef and Indo-Pacific. The Bessat and Buigues (2001) G-versus-air temperature regression was derived from a single *Porites lutea* colony from French Polynesia.

and reef structures after the original $CaCO_3$ deposition by organisms. Marine cements are prevalent on most reefs. They contribute to reef building by increasing reef resistance to erosion and by simply adding $CaCO_3$ to the structure. Buddemeier and Oberdorfer (1997) estimated that secondary cementation on Davies Reef, a middle-shelf reef on the GBR, contributes about 100 g $CaCO_3$ m^{-3} yr^{-1}. This is much slower than biological precipitation, but over time, secondary cementation in corals and within reefs can thus contribute significantly to the total reefal $CaCO_3$ (Enmar et al. 2000). In general, secondary cementation within reefs increases with the degree of hydrographic exposure (Marshall 1985), and this is reflected both within individual reefs and across the shelf within reef provinces. Because inorganic precipitation of $CaCO_3$ is primarily controlled by thermodynamic rather than biological factors, a reasonable hypothesis is that secondary cementation will decrease in response to increased pCO_2. However, whether future increases in pCO_2 will result in decreased reef cementation, and whether decreased cementation affects reef growth overall, remain essentially unknown.

Reef building itself is the net sum of carbonate deposition, primarily via $CaCO_3$ precipitation, less its removal through erosion, off-reef sediment transport, and dissolution. A reduction in $CaCO_3$ precipitation by whatever means—organism mortality, lowered calcification rates, lowered cementation rates, increased dissolution rates—ultimately lowers the net carbonate deposition on a reef, and thus its reef-building potential (see review by Kleypas, Buddemeier, and Gattuso 2001). The structure of slower-growing reefs would, therefore, be at a greater risk from a reduction in calcification rate than faster-growing reefs.

Regional Effects

Carbonate accumulation on some reefs is naturally low, due to either a low $CaCO_3$ production rate or a high $CaCO_3$ removal rate. High-latitude reefs are often considered marginal in terms of their reef-building capacity, and their slow reef-building state is thought to be a consequence of cooler temperatures, shallower light penetration, or lower saturation state (Kleypas et al. 1999). Although ocean warming will likely allow the migration of corals poleward by a few degrees latitude, future changes in seawater chemistry may drive carbonate budgets in these areas to levels below that required for reef building (Guinotte, Buddemeier, and Kleypas 2003).

Some reefs that are considered marginal may experience less change in surface ocean chemistry than others. Upwelling regions, for example, tend to be dominated by the chemistry of the upwelled waters. Upwelled waters of high pCO_2, for example, strongly affect the Galápagos reefs (other factors such as low temperature and high nutrients are also important). The pCO_2 of these waters can exceed 500 µatm (120 µatm higher than current atmospheric pCO_2). The Galápagos reefs, and other reefs which are dominated by upwelling, may actually be less affected by changes in seawater chemistry in the future. In fact, the Galápagos and other eastern Pacific reefs may provide present-day analogues for studying reef response to increased pCO_2. These reefs tend to be slow-growing and are poorly cemented (Cortés 1997; Glynn and Macintyre 1977). Following the 1982 to 1983 bleaching and mass mortality event in the Galápagos, the reef structure was nearly destroyed by bioerosion (Glynn 1988; Eakin 1996, 2001). This leads one to question whether the extremely rapid loss of $CaCO_3$ on Galápagos reefs was influenced by their poor cementation or increased dissolution under the naturally higher pCO_2 conditions.

12.5 Sea-Level Rise

Sea-level rise as a consequence of global warming is not considered a major threat to coral reef development. This is because the projected rise of 0.1 to 0.8 m by the end of the 21st century (Houghton et al. 2001) is two orders of magnitude less than the 120-m rise since the last glacial maximum. Reefs are not, therefore, considered to be directly threatened by sea-level rise, at least in terms of "drowning" due to decreased light-dependent calcification rates in deeper water. However, indirect effects of sea-level rise could have an impact on some reefs, in cases in which nutrients and sediments released from newly flooded coastlines will lead to degradation of water quality (Neumann and Macintyre 1985).

12.6 Summary and Conclusions

Future increases in greenhouse gas emissions will create a suite of environmental conditions that lie outside those experienced by modern humans. The current rapid rate of increase in atmospheric CO_2 concentration is effectively a shock to Earth's climate system and to the carbonate system in seawater, because the rates at which natural feedbacks can return these systems to equilibrium are far slower than the rate of atmospheric CO_2 increase. This chapter considers possible future coral reef environments by systematically looking at future projections of atmospheric CO_2, sea-surface temperature, and seawater chemisty, and by examining whether analogues for these projections occur, either within present-day environments or in the past.

Predicting coral reef responses to future climate change and changing seawater chemistry is hampered by uncertainty at multiple levels. Uncertainty begins with projections of future CO_2 emissions (Table 12.2). The range of projected greenhouse-gas emissions varies greatly because of the difficulty of predicting future population growth, economic growth, political stability, and technological advances. Additional uncertainty arises from climate modeling efforts; modeled increases in average global surface temperature, as reported in the Third IPCC Report (Houghton et al. 2001), range between 1.4 and 5.8 °C. This range reflects variations in different global climate models as well as future greenhouse-gas emission scenarios. Finally, predicting coral reef responses to warming SSTs is fraught with uncertainty as to how coral reef organisms and ecosystems will respond to these changes as well as other possible changes in regional climates (e.g., changing frequencies and intensities of ENSO events; tropical storms, rainfall and river runoff, etc.). Predicting changes in seawater chemistry is considerably easier than predicting temperature changes but there remains considerable

Joan A. Kleypas

Table 12.2. Summary of preindustrial versus projected values for several climate-change-related variables that can affect coral reefs. Analogues from both present-day and the past environments are listed

Variable	Preindustrial values	Projected values	Present-day analogues	Past analogues	Notes
Atm. CO_2	280 ppmV	555–825 ppmV	None	Early Tertiary, Cretaceous	Records most reliable for past 400 kyr. Older records rely on chemical and biological proxies.
Max. SST	30–31 °C (higher in enclosed seas)	31–33 °C? (higher in enclosed seas)	Red Sea / Persian Gulf (34 °C)	Early Tertiary, Cretaceous	Role of ocean "thermostat" in mediating maximum SST is uncertain. Higher latitudes will experience greater change.
$[CO_3^{2-}]$	200–350 μmol kg^{-1}	150–250 μmol kg^{-1}	Extratropics	Early Tertiary?	Knowing atmospheric CO_2 concentration alone is insufficient for determining carbonate equilibria in seawater; also need pH, TCO_2, or alkalinity. Effects of other seawater chemistry parameters, e.g. Mg/Ca, probably play a role in precipitation of various carbonate mineralogies.
Sea-level rise	Stable	< 1 m	Reefs in subsiding regions	Holocene, Pleistocene glacial–interglacial sea-level rise	Sea level per se is not considered a threat to reefs. Sea-level-related increases in coastal erosion and nutrient inputs considered a threat to some reefs.

uncertainty as to how calcifying organisms will respond to seawater-chemistry changes, particularly in combination with other environmental changes.

Present-day environments can offer some clues about how reefs will survive in the future. Corals in semienclosed seas already tolerate higher-than-normal SSTs, as do corals in tidal pools. The reasons for their higher temperature tolerance—whether it is through acclimation, acclimatization, or adaptation—could provide clues to the time scale of coral adaptation to future temperature increases. Similarly, non-reef-building coral communities that already exist under high pCO_2 conditions (e.g., high-latitude reefs and the Galápagos) may provide clues about the importance of carbonate saturation state to reef building.

The most recent time period of Earth history that had atmospheric CO_2 concentrations similar to those of future projections appears to be the early Tertiary, sometime prior to the Miocene and the onset of major glaciations. Can reef scientists look to the early Tertiary as a guide to what future reefs will look like? Reefs did exist during this period of higher SSTs and higher atmospheric CO_2 concentrations. However, high atmospheric CO_2 alone does not mean that the carbonate-ion concentration of the early Tertiary oceans was lower. Higher alkalinities, for example, would have counteracted the effects of higher pCO_2. Because of this incomplete knowledge of the carbonate system of past oceans, the question of whether coral reefs ever existed under lower carbonate-ion concentrations cannot yet be answered. Another difficulty with using this time period as an analogue for future reef environments has to do with reef ecology. Reefs prior to the Miocene were likely quite different from those of today; the acroporids, for example, were not yet major reef components.

Warming SSTs and changes in ENSO frequency and/or intensity are probably the most significant projected climate changes that will affect coral reefs. It is almost certain that average tropical SSTs will continue to increase in the future by an average of 1 to 3 °C. Even with an SST thermostat of 30 °C, many parts of the tropical ocean have maximum SSTs significantly below this threshold and could experience continued warming. Increases in SST maxima experienced by some corals will exceed their capacity to adapt, and mass coral bleaching is likely to occur with increasing frequency on those reefs. Whether ENSO frequency or intensity will increase in response to global warming is still uncertain. Attempts to quantify the potential effects of future SST increases on coral bleaching rely not only on broad assumptions about future changes in SST extremes and variability, but also on assumptions about the abilities of corals to adapt to increases in SST. Predictions, therefore, range from almost no bleaching mortality, through changed coral reef community structure (favoring more thermally resistant species), to the total collapse of reef ecosystems. Current-day observations drive predictions of future coral reef response to bleaching. Records of past coral bleaching have proven elusive, either because bleaching was rare in the past or because it is still too difficult to detect within the coral skeletal record. It is clear, however, that the occurrence and magnitude of mass coral-bleaching events have increased in recent decades with only a relatively modest SST warming in comparison to that projected by the end of this century.

Possible effects of changes in seawater chemistry on reef communities and on reef building are largely based on geochemical arguments. Essentially all of the experiments that have manipulated seawater chemistry and measured effects on corals, calcifying algae, and coral reef mesocosms indicate that calcification approaches a linear response to decreased carbonate saturation state. This suggests a strong geochemical control over calcification. In contrast, calcification rates derived from cores of *Porites* heads illustrate a stronger response to temperature changes than to declining saturation state, at least within two separate regions. Regardless of the biological factors that control $CaCO_3$ precipitation, it is very likely that dissolution will increase on reefs as atmospheric CO_2 concentrations increase. At the very least, $CaCO_3$ budgets will decrease on coral reefs overall (particularly if coral cover on reefs declines) and reef building will decrease.

Reef scientists are not likely to lower the uncertainty of predictions about the future Earth climate system, but future research *can* lower the uncertainty about reef responses to changes in climate and atmospheric chemistry. Some of the best analogues for reef responses to changes in climate and atmospheric chemistry can be found on present-day Earth, by looking at reefs living in high-temperature semi-enclosed seas and in regions with naturally low carbonate saturation states. Although the closest paleo-analogue for our future climate appears to be the early Tertiary, the current uncertainties associated with estimates of atmospheric CO_2, tropical SSTs, seawater chemistry, and reef ecology for that time period are too poorly constrained to provide an adequate analogue for predicting the future of coral reefs within the next decade or so. The urgent need to understand reef response to climate change, and to design effective mitigative and managerial strategies, will need to rely more heavily on present-day observations and experimental manipulations.

Acknowledgments. This chapter was greatly improved by comments from Janice Lough, Rich Aronson, and Bob Buddemeier. The following persons are thanked for providing advance copies of manuscripts and unpublished results: B. Riegl, and G. Wellington and P. Glynn (their own chapters), E. Brady and B.E. Bliesner (unpublished model results), and B. N. Opdyke (unpublished manuscript on Tertiary seawater chemistry). Gokhan Danabasoglu is also appreciated for his advice on interpreting coupled climate model results.

References

Al-Moghrabi, S., C. Goiran, D. Allemand, N. Speziale, and J. Jaubert. 1996. Inorganic carbon uptake for photosynthesis by the symbiotic coral/dinoflagellate association. II. Mechanisms for carbonate uptake. *J. Exp. Mar. Biol. Ecol.* 199:227–248.

Anderson, D.M., and D. Archer. 2002. Glacial-interglacial stability of ocean pH inferred from foraminifer dissolution rates. *Nature* 416:70–73.

Andersson, A.J., F.T. Mackenzie, and L.M. Ver. 2003. Solution of shallow-water carbonates: An insignificant buffer against rising atmospheric CO_2. *Geology* 31:513–516.

Archer, D., G. Eshel, A. Winguth, and W. Broecker. 2000a. Atmospheric CO_2 sensitivity to the biological pump in the ocean. *Global Biogeochem. Cycles* 14:1219–1230.

Archer, D., H. Kheshgi, and E. Maier-Reimer. 1997. Multiple timescales for neutralization of fossil fuel CO_2. *Geophys. Res. Lett.* 24:405–408.

Archer, D., A. Winguth, D. Lea, and N. Mahowald. 2000b. What caused the glacial / inter-glacial pCO$_2$ cycles? *Rev. Geophys.* 38:159–189.

Aronson, R.B., W.F. Precht, I.G. Macintyre, and T.J.T. Murdoch. 2000. Coral bleach-out in Belize. *Nature* 405:36.

Baker, A.C. 2001. Reef corals bleach to survive change. *Nature* 411:765–766.

Bard, E. 2001. Comparison of alkenone estimates with other paleotemperature proxies. *Geochem. Geophys. Geosyst.* 2: art. no. 2000GC000050.

Bates, N.R., L. Samuels, and L. Merlivat. 2001. Biogeochemical and physical factors influencing seawater *f*CO$_2$ and air-sea CO$_2$ exchange on the Bermuda coral reef. *Limnol. Oceanogr.* 46:833–846.

Beck, J.W., R.L. Edwards, E. Ito, F.W. Taylor, J. Recy, F. Rougerie, P. Joannot, and C. Henin. 1992. Sea-surface temperature from coral skeletal strontium/calcium ratios. *Science* 257:644–647.

Berner, R.A. 1975. The role of magnesium in the crystal growth of calcite and aragonite from seawater. *Geochim. Cosmochim. Acta* 39:489–504.

Berner, R.A. 1994. GEOCARB II: A revised model of atmospheric CO$_2$ over Phanerozoic time. *Am. J. Sci.* 294:56–91.

Berner, R.A. 1997. The rise of plants and their effect on weathering and atmospheric CO$_2$. *Science* 276:544–546.

Berner, R.A., and Z. Kothavala. 2001. GEOCARB III: A revised model of atmospheric CO$_2$ over Phanerozoic time. *Am. J. Sci.* 301:182–204.

Bessat, F., and A.D. Buigues. 2001. Two centuries of variation in coral growth in a massive *Porites* colony from Moorea (French Polynesia): A response of ocean-atmosphere variability from south central Pacific. *Palaeogeogr. Palaeoclimatol. Palaeoecol.* 175:381–392.

Boville, B.A., and P.R. Gent. 1998. The NCAR climate system model, version one. *J. Climate* 11:1115–1130.

Broecker, W.S. 1987. The biggest chill. *Nat. Hist.* 96:74–82.

Broecker, W.S., and E. Clark. 2002. Carbonate ion concentration in glacial-age deep waters of the Caribbean Sea. *Geochem. Geophys. Geosyst.* 3:U1–U14.

Broecker, W.S., and T.-H. Peng. 1982. *Tracers in the Sea.* Lamont-Doherty Geological Observatory, Columbia University, Palisades, NY.

Broecker, W.S., and T. Takahashi. 1966. Calcium carbonate precipitation on the Bahama Banks. *J. Geophys. Res.* 71:1575–1602.

Brown, B.E. 1997. Coral bleaching: Causes and consequences. *Coral Reefs* 16: S129–S138.

Brown, B.E., R.P. Dunne, M.S. Goodson, and A.E. Douglas. 2002. Experience shapes the susceptibility of a reef coral to bleaching. *Coral Reefs* 21:119–126.

Buddemeier, R.W., and D.G. Fautin. 1993. Coral bleaching as an adaptive mechanism — A testable hypothesis. *BioScience* 43:320–326.

Buddemeier, R.W., and J.A. Oberdorfer. 1997. Hydrogeology of Enewetak Atoll. In *Geology and Hydrogeology of Carbonate Islands,* eds. H.L. Vacher and T.M. Quinn, 671–693. Amsterdam: Elsevier.

Carricart-Ganivet, J.P. 2004. Sea surface temperature and the growth of the West Atlantic reef-building coral *Montastraea annularis. J. Exp. Mar. Biol. Ecol.* 302:249–260.

Carriquiry, J.D., M.J. Risk, and H.P. Schwarcz. 1994. Stable isotope geochemistry of corals from Costa Rica as proxy indicator of the El Niño/Southern Oscillation (ENSO). *Geochim. Cosmochim. Acta* 58:335–351.

Chisholm, J.R.M., and D.J. Barnes. 1998. Anomalies in coral reef community metabolism and their potential importance in the reef CO$_2$ source-sink debate. *Proc. Natl. Acad. Sci. USA*: 95:6566–6569.

Clark, P.U., N.G. Pisias, T.F. Stocker, and A.J. Weaver. 2002. The role of thermohaline circulation in abrupt climate change. *Nature* 415:863–869.

Cohen, A.L., K.E. Owens, G.D. Layne, and N. Shimizu. 2002. The effect of algal symbionts on the accuracy of Sr/Ca paleotemperatures from coral. *Science* 296:331–333.

Cole, J.E., and R.G. Fairbanks. 1990. The Southern Oscillation recorded in the oxygen isotopes of corals from Tarawa Atoll. *Paleoceanography* 5:669–683.

Coles, S.L. 2001. Coral bleaching: What do we know and what can we do? In: *Proceedings of the Workshop on Mitigating Coral Bleaching Impact through MPA Design*, May 29–31, 2001, Honolulu, HI, eds. R. Salm and S.L. Coles, 25–35.

Coles, S.L., and B.E. Brown. 2003. Coral bleaching—Capacity for acclimatization and adaptive selection. *Adv. Mar. Biol.* 46:183–223.

Coles, S.L., and Y.H. Fadlallah. 1991. Reef coral survival and mortality at low temperatures in the Arabian Gulf: New species-specific lower temperature limits. *Coral Reefs* 9:231–237.

Cortés, J. 1997. Biology and geology of eastern Pacific coral reefs. *Coral Reefs* 16:S39–S46.

Crowley, T.J. 2000. CLIMAP SSTs re-revisited. *Climate Dynamics* 16:241–255.

Crowley, T.J., and R.A. Berner. 2001. CO_2 and climate change. *Science* 292:870–872.

De La Rocha, C.L., and D.J. DePaolo. 2000. Isotopic evidence for variations in the marine calcium cycle over the Cenozoic. *Science* 289:1176–1178.

Dunbar, R.B. 2000. El Niño—Clues from corals. *Nature* 407:956–959.

Dunbar, R.B., G.M. Wellington, M.W. Colgan, and P.W. Glynn. 1994. Eastern Pacific sea surface temperature since 1600 AD: The $\delta^{18}O$ record of climate variability in Galápagos corals. *Paleoceanography* 9:291–315.

Eakin, C.M. 1996. Where have all the carbonates gone? A model comparison of calcium carbonate budgets before and after the 1982–1983 El Niño. *Coral Reefs* 15:109–119.

Eakin, C.M. 2001. A tale of two ENSO events: Carbonate budgets and the influence of two warming disturbances and intervening variability, Uva Island, Panama. *Bull. Mar. Sci.* 69:171–186.

Enfield, D.B. 2001. Evolution and historical perspective of the 1997–1998 El Niño-Southern Oscillation event. *Bull. Mar. Sci.* 69:7–25.

Enmar, R., M. Stein, M. Bar-Matthews, E. Sass, A. Katz, and B. Lazar. 2000. Diagenesis in live corals from the Gulf of Aqaba. I. The effect on paleo-oceanography tracers. *Geochim. Cosmochim. Acta* 64:3123–3132.

Fairhall, A.W. 1973. Accumulation of fossil CO_2 in the atmosphere and the sea. *Nature* 245:20–23.

Fautin, D.G., and R.W. Buddemeier. 2004. Adaptive bleaching: A general phenomenon. *Hydrobiologia* 530:459–467.

Fitt, W.K., B.E. Brown, M.E. Warner, and R.P. Dunne. 2001. Coral bleaching: Interpretation of thermal tolerance limits and thermal thresholds in tropical corals. *Coral Reefs* 20:51–65.

Gattuso, J.-P., D. Allemand, and M. Frankignoulle. 1999. Photosynthesis and calcification at cellular, organismal and community levels in coral reefs: A review on interactions and control by carbonate chemistry. *Am. Zool.* 39:160–183.

Gattuso, J.-P., M. Frankignoulle, I. Bourge, S. Romaine, and R.W. Buddemeier. 1998. Effect of calcium carbonate saturation of seawater on coral calcification. *Global Planet. Change* 18:37–46.

Gattuso, J.-P., M. Pichon, B. Delesalle, C. Canon, and M. Frankignoulle. 1996. Carbon fluxes in coral reefs. I. Lagrangian measurement of community metabolism and resulting air-sea CO_2 disequilibrium. *Mar. Ecol. Prog. Ser.* 145:109–121.

Gattuso, J.-P., S. Reynaud-Vaganay, P. Furla, S. Romaine-Lioud, J. Jaubert, I. Bourge, and M. Frankignoulle. 2000. Calcification does not stimulate photosynthesis in the zooxanthellate scleractinian coral *Stylophora pistillata*. *Limnol. Oceanogr.* 45:246–250.

Gent, P.R. 2001. Will the North Atlantic Ocean thermohaline circulation weaken during the 21st century? *Geophys. Res. Lett.* 28:1023–1026.

Gili, E., J.-P. Masse, and P.W. Skelton. 1995. Rudists as gregarious sediment-dwellers, not reef-builders, on Cretaceous carbonate platforms. *Palaeogeogr. Palaeoclimatol. Palaeoecol.* 118: 245–267.

Gleason, D.F., and G.M. Wellington. 1993. Ultraviolet radiation and coral bleaching. *Nature* 365:836–838.

Glynn, P.W. 1984. Widespread coral mortality and the 1982/83 El Niño warming event. *Environ. Conserv.* 11:133–146.

Glynn, P.W. 1988. El Niño warming, coral mortality and reef framework destruction by echinoid bioerosion in the eastern Pacific. *Galaxea* 7:129–160.

Glynn, P.W. 1993. Coral reef bleaching: Ecological perspectives. *Coral Reefs* 12:1–7.

Glynn, P.W. 2000. El Niño-Southern Oscillation mass mortalities of reef corals: A model of high temperature marine extinctions? In *Carbonate Platform Systems: Components and Interactions*, eds. E. Insalaco, P.W. Skelton, and T.J. Palmer, 117–133. Geological Society of London, Special Publication 178.

Glynn, P.W., and I.G. Macintyre. 1977. Growth rate and age of coral reefs on the Pacific coast of Panama. *Proc. Third Int. Coral Reef Symp., Miami* 2:251–259.

Glynn, P.W, J.L. Maté, A.C. Baker, and M.O. Calderón. 2001. Coral bleaching and mortality in Panama and Ecuador during the 1997–1998 El Niño-Southern Oscillation event: Spatial/temporal patterns and comparisons with the 1982-1983 event. *Bull. Mar. Sci.* 69:79–109.

Goiran, C., S. Al-Moghrabi, D. Allemand, and J. Jaubert. 1996. Inorganic carbon uptake for photosynthesis by the symbiotic coral/dinoflagellate association. I. Photosynthesis performances of symbionts and dependence on sea water bicarbonate. *J. Exp. Mar. Biol. Ecol.* 199:207–225.

Guinotte, J., R.W. Buddemeier, and J.A. Kleypas. 2003. Future coral reef habitat marginality: Temporal and spatial effects of climate change in the Pacific basin. *Coral Reefs* 22:551–558.

Guzmán, H.M., and J. Cortés. 2001. Changes in reef community structure after fifteen years of natural disturbances in the eastern Pacific (Costa Rica). *Bull. Mar. Sci.* 69:133–149.

Halley, R.B., and K.K. Yates. 2000. Will reef sediments buffer corals from increased global CO_2? *Proc. Ninth Int. Coral Reef Symp., Bali,* Abstract.

Hoegh-Guldberg, O. 1999. Climate change, coral bleaching and the future of the world's coral reefs. *Mar. Freshwater Res.* 50:839–866.

Houghton, J.T., Y. Ding, D.J. Griggs, M. Noguer, P.J. van der Linden, and D. Xiaosu. (eds.). 2001. *IPCC Third Assessment Report: Climate Change 2001: The Scientific Basis.* Cambridge University Press. [Also see: *Summary for Policymakers and Technical Summary.*]

Hughes, T.P., A.H. Baird, D.R. Bellwood, M. Card, S.R. Connolly, C. Folke, R. Grosberg, O. Hoegh-Guldberg, J.B.C. Jackson, J. Kleypas, J.M. Lough, P. Marshall, M. Nyström,

S.R. Palumbi, J.M. Pandolfi, B. Rosen, and J. Roughgarden. 2003. Climate change, human impacts, and the resilience of coral reefs. *Science* 301:929–933.

Johnson, C.C., and E.G. Kauffman. 2001. Cretaceous evolution of reef ecosystems; A regional synthesis of the Caribbean tropics. In *The History and Sedimentology of Ancient Reef Ecosystems*, ed. G.D. Stanley, Jr., 311–349. New York: Kluwer Academic/Plenum Publishers.

Jokiel, P.L., and S.L. Coles. 1977. Effects of temperature on the mortality and growth of Hawaiian reef corals. *Mar. Biol.* 43:201–208.

Kayanne, H., S. Kudo, H. Hata, H. Yamano, K. Nozaki, K. Kato, A. Negishi, H. Saito, F. Akimoto, and H. Kimoto. 2003. Integrated monitoring system for coral reef water pCO_2, carbonate system and physical parameters. *Proc. Ninth Int. Coral Reef Symp., Bali* 2:1079–1084.

Kennett, J.P., and L.D. Stott. 1991. Abrupt deep-sea warming, palaeoceanographic changes and benthic extinctions at the end of the Palaeocene. *Nature* 353:225–229.

Kiessling, W. 2001. Paleoclimatic significance of Phanerozoic reefs. *Geology* 29:751–754.

Kiessling, W., E. Flügel, and J. Golonka. 1999. Paleoreef maps: Evaluation of a comprehensive database on Phanerozoic reefs. *AAPG Bull.* 83:1552–1587.

Kinzie, R.A., III, M. Takayama, S.R. Santos, and M.A. Coffroth. 2001. The adaptive bleaching hypothesis: Experimental tests of critical assumptions. *Biol. Bull.* 200:51–58.

Kleypas, J.A., R.W. Buddemeier, D. Archer, J.-P. Gattuso, D. Langdon, and B.N. Opdyke. 1999. Geochemical consequences of increased atmospheric CO_2 on coral reefs. *Science* 284:118–120.

Kleypas, J.A., R.W. Buddemeier, C.M. Eakin, J.-P. Gattuso, J. Guinotte, O. Hoegh-Guldberg, R. Iglesias-Prieto, P.L. Jokiel, C. Langdon, W. Skirving, and A.E. Strong. 2005. Comment on "Coral reef calcification and climate change: The effect of ocean warming." *Geophys. Res. Lett.* 32:L08601, doi:10.1029/2004GL022329.

Kleypas, J.A., R.W. Buddemeier, and J.-P. Gattuso. 2001. The future of coral reefs in an age of global change. *Int. J. Earth Sci.* 90:426–437.

Kleypas, J.A., and C. Langdon. 2003. Overview of CO_2-induced changes in seawater chemistry. *Proc. Ninth Int. Coral Reef Symp., Bali* 2:1085–1090.

Kleypas, J.A., J.W. McManns, and L.A.B. Meñez. 1999. Environmental limits to Coral reef development: Where do we draw the line? *Am. Zool.* 39:146-159.

Knutson, T.R., S. Manabe, and D. Gu. 1997. Simulated ENSO in a global coupled ocean-atmosphere model: Multidecadal amplitude modulation and CO_2 sensitivity. *J Climate* 10:138–161.

Langdon, C. 2003. Review of experimental evidence for effects of CO_2 on calcification of reef builders. *Proc. Ninth Int. Coral Reef Symp., Bali* 2:1091–1098.

Langdon, C., W.S. Broecker, D.E. Hammond, E. Glenn, K. Fitzsimmons, S.G. Nelson, T.-S. Peng, I. Hajdas, and G. Bonani. 2003. Effect of elevated CO_2 on the community metabolism of an experimental coral reef. *Global Biogeochem. Cycles* 17:1011, doi: 10.1029/2002GB001941.

Langdon, C., T. Takahashi, C. Sweeney, D. Chipman, J. Goddard, F. Marubini, H. Aceves, H. Barnett, and M.J. Atkinson. 2000. Effect of calcium carbonate saturation state on the calcification rate of an experimental coral reef. *Global Biogeochem. Cycles* 14:639–654.

Leclercq, N., J.-P. Gattuso, and J. Jaubert. 2000. CO_2 partial pressure controls the calcification rate of a coral community. *Global Change Biol.* 6:329–334.

Leclercq, N., J.-P. Gattuso, and J. Jaubert. 2002. Primary production, respiration, and calcification of a coral reef mesocosm under increased CO_2 partial pressure. *Limnol. Oceanogr* 47:558–564.

Levitus, S., J.I. Antonov, T.P. Boyer, and C. Stephens. 2000. Warming of the world ocean. *Science* 287:2225–2229.

Li, T., T.F. Hogan, and C.-P. Chan. 2000. Dynamic and thermodynamic regulation of ocean warming. *J. Atmos. Sci.* 57:3353–3365.

Loschnigg, J., and P.J. Webster. 2000. A coupled ocean-atmosphere system of SST modulation for the Indian Ocean. *J. Climate* 13:3342–3360.

Lough, J.M. 2000. 1997–98: Unprecedented thermal stress to coral reefs? *Geophy. Res. Lett.* 27:3901–3904.

Lough, J.M., and D.J. Barnes. 1997. Several centuries of variation in skeletal extension, density and calcification in massive *Porites* colonies from the Great Barrier Reef: A proxy for seawater temperature and a background of variability against which to identify unnatural change. *J. Exp. Mar. Biol. Ecol.* 211:29–67.

Lough, J.M., and D.J. Barnes. 2000. Environmental controls on growth of the massive coral *Porites*. *J. Exp. Mar. Biol. Ecol.* 245:225–243.

Marshall, A.T., and P. Clode. 2004. Calcification rate and the effect of temperature in a zooxanthellate and an azooxanthellate scleractinian reef coral. *Coral Reefs* 23:218–224.

Marshall, J.F. 1985. Cross-shelf and facies related variations in submarine cementation in the Central Great Barrier Reef. *Proc. Fifth Int. Coral Reef Congr. Tahiti* 3:221–226.

Marubini, F., H. Barnett, C. Langdon, and M.J. Atkinson. 2001. Interaction of light and carbonate ion on calcification of the hermatypic coral *Porites compressa*. *Mar. Ecol. Prog. Ser.* 220:153–162.

Marubini, F., C. Ferrier-Pagès, and J.P. Cuif. 2003. Suppression of skeletal growth in scleractinian corals by decreasing ambient carbonate-ion concentration: A cross-family comparison. *Proc. R. Soc. London Ser. B* 270:179–184.

Marubini, F., and B. Thake. 1999. Bicarbonate addition promotes coral growth. *Limnol. Oceanogr.* 44:716–720.

McConnaughey, T.A. 1989. $\delta^{13}C$ and $\delta^{18}O$ isotopic disequilibrium in biological carbonates. I. Patterns. *Geochim. Cosmochim. Acta* 53:151–162.

McNeil, B.I., R.J. Matear, and D.J. Barnes. 2004. Coral reef calcification and climate change: The effect of ocean warming. *Geophys. Res. Lett.* 31, L22309, doi:10.1029/2004GL021541.

Müller, A., M.K. Gagan, and M.T. McCulloch. 2001. Early marine diagenesis in corals and geochemical consequences for paleoceanographic reconstructions. *Geophys. Res. Lett.* 28:4471–4474.

Nakamura, T., and R. van Woesik. 2001. Water-flow rates and passive diffusion partially explain differential survival of corals during the 1998 bleaching event. *Mar. Ecol. Prog. Ser.* 212:301–304.

Nakicenovic, N., and R. Swart. (eds.). 2000. *Emission Scenarios 2000*. Special Report of the IPCC. Cambridge University Press.

Neumann, A.C., and I. Macintyre. 1985. Reef response to sea level rise: Keep-up, catch-up or give-up. *Proc. Fifth Int. Coral Reef Congr., Tahiti* 3:105–110.

Newell, R.E. 1979. Climate and the ocean. *Am. Sci.* 67:405–416.

Otto-Bliesner, B.L., and E.C. Brady. 2001. Tropical Pacific variability in the NCAR Climate System Model. *J. Climate* 14:3587–3607.

Otto-Bliesner, B.L., E.C. Brady, S.I. Shin, Z.Y. Liu, and C. Shields. 2003. Modeling El Niño and its tropical teleconnections during the last glacial-interglacial cycle. *Geophys. Res. Lett.* 30: art. no. 2198.

Pagani, M., M.A. Arthur, and K.H. Freeman. 1999. Miocene evolution of atmospheric carbon dioxide. *Paleoceanography* 14:273–292.

Pearson, P.N., P.W. Ditchfield, J. Singano, K.G. Harcourt-Brown, C.J. Nicholas, R.K. Olsson, N.J. Shackleton, and M.A. Hall. 2001. Warm tropical sea surface temperatures in the Late Cretaceous and Eocene epochs. *Nature* 413:481–487.

Pearson, P.N., and M.R. Palmer. 2000. Atmospheric carbon dioxide concentrations over the past 60 million years. *Nature* 406:695–699.

Perrin, C. 2002. Tertiary: The emergence of Modern reef ecosystems. In *Phanerozoic Reef Patterns*, eds. W. Kiessling, E. Flügel, and J. Golonka. SEPM Spec. Publ. 72.

Petit, J.R., J. Jouzel, D. Raynaud, N.I. Barkov, J.-M. Barnola, I. Basile, M. Bender, J. Chappellaz, M. Davis, G. Delaygue, M. Delmotte, V.M. Kotlyakov, M. Legrand, V.Y. Lipenkov, C. Lorius, L. Pepin, C. Ritz, E. Saltzman, and M. Stievenard. 1999. Climate and atmospheric history of the past 420,000 years from the Vostok ice core, Antarctica. *Nature* 399:429–436.

Podestá, G.P., and P.W. Glynn. 2001. The 1997-98 El Niño event in Panama and Galápagos: An update of thermal stress indices relative to coral bleaching. *Bull. Mar. Sci.* 69:43–59.

Quay, P., R. Sonnerup, T. Westby, J. Stutsman, and A. McNichol. 2003. Changes in the $^{13}C/^{12}C$ of dissolved inorganic carbon in the ocean as a tracer of anthropogenic CO_2 uptake. *Global Biogeochem. Cycles* 17(1), 1004, doi:10.1029/2001GB001817.

Rahmstorf, S. 1995. Bifurcations of the Atlantic thermohaline circulation in response to changes in the hydrological cycle. *Nature* 378:135–149.

Ramanathan, V., and W.D. Collins. 1991. Thermodynamic regulation of ocean warming by cirrus clouds deduced from observations of the 1987 El Niño. *Nature* 351:27–32.

Rayner, N.A., D.E. Parker, E.B. Horton, C.K. Folland, L.V. Alexander, D.P. Rowell, E.C. Kent, and A. Kaplan. 2003. Global analyses of sea surface temperature, sea ice and night marine air temperature since the late nineteenth century. *J. Geophys. Res.* 108: art. no. 4407.

Reynaud, S., N. Leclercq, S. Romaine-Lioud, C. Ferrier-Pages, J. Jaubert, and J.-P. Gattuso. 2003. Interacting effects of CO_2 partial pressure and temperature on photosynthesis and calcification in a scleractinian coral. *Global Change Biol.* 9:1660–1668.

Reynolds, R.W., N.A. Rayner, T.M. Smith, D.C. Stokes, and W. Wang. 2002. An improved in situ and satellite SST analysis for climate. *J. Climate* 15:1609–1625.

Ribaud-Laurenti, A., B. Hamelin, L. Montaggioni, and D. Cardinal. 2001. Diagenesis and its impact on Sr/Ca ratio in Holocene *Acropora* corals. *Int. J. Earth Sci.* 90:438–451.

Riebesell, U., I. Zondervan, B. Rost, P.D. Tortell, R.E. Zeebe, and F.M.M. Morel. 2000. Reduced calcification of marine plankton in response to increased atmospheric CO_2. *Nature* 407:364–368.

Rosen, B.R., and D. Turnsek. 1989. Extinction patterns and biogeography of scleractinian corals across the Cretaceous/Tertiary boundary. *Mem. Assoc. Australas. Paleontol.* 8:355–370.

Royer, D.L., S.L. Wing, D.J. Beerling, D.W. Jolley, P.L. Koch, L.J. Hickey, and R.A. Berner. 2001. Paleobotanical evidence for near present-day levels of atmospheric CO_2 during part of the Tertiary. *Science* 292:2310–2313.

Rundgren, M., and O. Bennike. 2002. Century-scale changes of atmospheric CO_2 during the last interglacial. *Geology* 30:187–189.

Sabine, C.L., R.A. Feely, N. Gruber, R.M. Key, K. Lee, J.L. Bullister, R. Wanninkhof, C.S. Wong, D.W.R. Wallace, B. Tilbrook, F.J. Millero, T.-H. Peng, A. Kozyr, T. Ono, and A.F. Rios. 2004. The oceanic sink for anthropogenic CO_2. *Science* 305:367–371.

Sabine, C.L., R.A. Feely, R.M. Key, J.L. Bullister, F.J. Millero, K. Lee, T.-H. Peng, B. Tilbrook, T. Ono, and C.S. Wong. 2002. Distribution of anthropogenic CO_2 in the Pacific Ocean. *Global Biogeochem. Cycles* 16:1083, doi:10.1029/2001GB001639.

Sabine, C.L., R.M. Key, K.M. Johnson, F.J. Millero, A. Poisson, J.L. Sarmiento, D.W.R. Wallace, and C.D. Winn. 1999. Anthropogenic CO_2 inventory of the Indian Ocean. *Mar. Chem.* 13:179–198.

Sakamoto, C.M., F.J. Millero, W. Yao, G.E. Friederich, and F.P. Chavez. 1998. Surface seawater distributions of inorganic carbon and nutrients around the Galápagos Islands: Results from the PlumEx experiment using automated chemical mapping. *Deep-Sea Res. II* 45:1055–1071.

Sandberg, P.A. 1983. An oscillating trend in Phanerozoic non-skeletal carbonate mineralogy. *Nature* 305:19–22.

Sanyal, A., N.G. Hemming, G.N. Hanson, and W.S. Broecker. 1995. Evidence for a higher pH in the glacial ocean from boron isotopes in foraminifera. *Nature* 373:234–236.

Sarmiento, J.L., J. Dunne, A. Gnanadesikan, R.M. Key, K. Matsumoto, and R. Slater. 2002. A new estimate of the $CaCO_3$ to organic carbon export ratio. *Global Biogeochem. Cycles* 16:10.1029/2002/GB001919.

Schneider, K., and J. Erez. 1999. Effects of carbonate chemistry on coral calcification, and symbiotic algae photosynthesis and isotopic fractionation. Abstract published in supplement to *Eos, Trans. Am. Geophys. Union,* vol. 80, no. 49.

Shen, G.T., J.E. Cole, D.W. Lea, L.J. Linn, T.A. McConnaughey, and R.G. Fairbanks. 1992. Surface ocean variability at Galápagos from 1936–1982: Calibration of geochemical tracers in corals. *Paleoceanography* 7:563–588.

Skirrow, G., and M. Whitfield. 1975. The effect of increases in the atmospheric carbon dioxide content on the carbonate ion concentration of surface ocean water at 25 °C. *Limnol. Oceanogr.* 20:103–108.

Smith, S.V., and R.W. Buddemeier. 1992. Global change and coral reef ecosystems. *Ann. Rev. Ecol. Syst.* 23:89–118.

Smith, S.V., and F. Pesret. 1974. Processes of carbon dioxide flux in the Fanning Island Lagoon. *Pac. Sci.* 28:225–245.

Stanley, S.M., and L.A. Hardie. 1998. Secular oscillations in the carbonate mineralogy of reef-building and sediment-producing organisms driven by tectonically forced shifts in seawater chemistry. *Palaeogeogr. Palaeoclimatol. Palaeoecol.* 144:3–19.

Stocker, T.F., and A. Schmittner. 1997. Influence of CO_2 emission rates on the stability of the thermohaline circulation. *Nature* 388:862–865.

Suzuki, A., and H. Kawahata. 1999. Partial pressure of carbon dioxide in coral reef lagoon waters: Comparative study of atolls and barrier reefs in the Indo-Pacific Oceans. *J. Oceanogr.* 55:731–745.

Tett, S. 1995. Simulation of El Niño-Southern Oscillation variability in a global AOGCM and its response to CO_2 increase. *J. Climate* 8:1473–1502.

Timmermann, A., J. Oberhuber, A. Bacher, M. Esch, M. Latif, and E. Roeckner. 1999. Increased El Niño frequency in a climate model forced by future greenhouse warming. *Nature* 398:694–697.

Tudhope, A.W., C.P. Chilcott, M.T. McCulloch, E.R. Cook, J. Chappell, R.M. Ellam, D.W. Lea, J.M. Lough, and G.B. Shimmield. 2001. Variability in the El Niño-Southern Oscillation through a glacial-interglacial cycle. *Science* 291:1511–1517.

Veizer, J., Y. Godderis, and L.M. Francois. 2000. Evidence for decoupling of atmospheric CO_2 and global climate during the Phanerozoic eon. *Nature* 408:698–701.

Ware, J.R. 1997. The effect of global warming on coral reefs: Acclimate or die. *Proc. Eighth Int. Coral Reef Symp., Panama* 1:527–532.

Whitfield, M. 1974. Accumulation of fossil CO_2 in the atmosphere and in the sea. *Nature* 247:523–525.

Wilson, P.A., R.D. Norris, and M.J. Cooper. 2002. Testing the Cretaceous greenhouse hypothesis using glassy foraminiferal calcite from the core of the Turonian tropics on Demerara Rise. *Geology* 30:607–610.

Wilson, P.A., and B.N. Opdyke. 1996. Equatorial sea surface temperatures for the Maastrichtian revealed through remarkable preservation of metastable carbonate. *Geology* 24:555–558.

Wood, R. 2001. *Reef Evolution*. Oxford: Oxford University Press.

Zachos, J., M. Pagani, L. Sloan, E. Thomas, and K. Billups. 2001. Trends, rhythms, and aberrations in global climate 65 Ma to present. *Science* 292:686–693.

Zachos, J.C., U. Röhl, S.A. Schellenerg, A. Sluijs, D.A. Hodell, D.C. Kelly, E. Thomas, M. Nicolo, I. Raffi, L.J. Lourens, D. Kroon, and H. McCarren. 2005. Rapid acidification of the ocean during the Paleocene-Eocene Thermal Maximum. *Science* 308:1611–1615.

Zachos, J.C., M.W. Wara, S. Bohaty, M.L. Delaney, M.R. Petrizzo, A. Brill, T.J. Bralower, and I. Premoli-Silva. 2003. A transient rise in tropical sea surface temperature during the Paleocene-Eocene Thermal Maximum. *Science* 302:1551–1554.

Zimen, K.E., and F.K. Altenhein. 1973. The future burden of industrial CO_2 on the atmosphere and the oceans. *Z. Naturforsch.* 28a:1747–1753.

Index

Ecological Studies

Volumes published since 2001

Printed in the United States of America